# The Montana Mathematics Enthusiast Monographs in Mathematics Education

## Monograph 12

*Crossroads in the History of Mathematics
and Mathematics Education*

## THE MONTANA MATHEMATICS ENTHUSIAST
## MONOGRAPH SERIES IN MATHEMATICS EDUCATION

*Series Editor*
Bharath Sriraman, The University of Montana

*International Perspectives on Social Justice in Mathematics Education*
   Edited by Bharath Sriraman

*Mathematics Education and the Legacy of Zoltan Paul Dienes*
   Edited by Bharath Sriraman

*Beliefs and Mathematics: Festschrift in Honor of Guenter Toerner's 60th Birthday*
   Edited by Bharath Sriraman

*Creativity, Giftedness, and Talent Development in Mathematics*
   Edited by Bharath Sriraman

*Interdisciplinary Educational Research in Mathematics and its Connections*
*to the Arts and Sciences*
   Edited by Bharath Sriraman, Claus Michelsen, Astrid Beckmann,
   Viktor Freiman

*Critical Issues in Mathematics Education*
   Edited by Paul Ernest, Brian Greer, Bharath Sriraman

*Interdisciplinarity, Creativity, and Learning: Mathematics with Literature,*
*Paradoxes, History, Technology, and Modeling*
   Edited by Bharath Sriraman, Viktor Freiman, Nicole Lirette-Pitre

*Relatively and Philosophically E*ª*rnest: Festschrift in Honor of Paul Ernest's*
*65th Birthday*
   Edited by Bharath Sriraman, Simon Goodchild

*The Role of Mathematics Discourse in Producing Leaders of Discourse*
   By Libby Knott

*Interdisciplinarity for the 21st Century: Proceedings of the 3rd International Symposium on Mathematics*
*and its Connections to the Arts and Sciences, Moncton 2009*
   Edited by Bharath Sriraman

# The Montana Mathematics Enthusiast Monographs in Mathematics Education

## Monograph 12

*Crossroads in the History of Mathematics and Mathematics Education*

**Edited by**
**Bharath Sriraman**
*The University of Montana, USA*

INFORMATION AGE PUBLISHING, INC.
Charlotte, NC • www.infoagepub.com

Copyright © 2012 Information Age Publishing Inc. & The Montana Council of Teachers of Mathematics

ISBN: 978-1-61735-704-6   Paperback
978-1-61735-705-3   Hardcover
978-1-61735-706-0   ebook

Permission to photocopy, microform, and distribute print or electronic copies may be obtained from:
    Bharath Sriraman, Ph.D.
    Editor, *The Montana Mathematics Enthusiast*
    The University of Montana
    Missoula, MT 59812
    Email: sriramanb@mso.umt.edu
    (406) 243-6714

Printed in the United States of America

# CONTENTS

# PREFACE

**Bharath Sriraman**
*The University of Montana*

This is an uneven book—uneven in many respects. The fonts, the layouts, the spacing, and the content are a potpourri for several good reasons that are not elucidated here. However, the substance is interesting enough that these superficial distractions will hopefully not impede the reader's attention on the several focal points of the book.

The history of mathematics has often been relegated to the same category as the popularization of mathematics. Historians of mathematics have nevertheless pursued their interests and published their research in scholarly outlets such as books and journals, which have drawn the attention of math history aficionados and pedagogues. The latter consists of mathematicians and teachers of mathematics drawn to the history of the subject for seeking clarity, for stimulating interest among students by alluding to quirky historical tidbits, and most recently mathematics educators who seek a different approach to basic questions plaguing the field of mathematics education on the nature of mathematical thinking, of teaching, and of learning.

The book has been divided into three sections depending on the interest of the reader, and has been put together to be accessible to undergraduate mathematics students as well as the lay person. Some of the chapters are reprints of articles previously published in the (Montana) Mathematics Enthusiast. The opening section which is also the largest section is on topics in the history and didactics of mathematics, has a pedagogical slant to it and material useful for the teaching of Calculus. The second section on the history and didactics of geometry and number is more esoteric and accessible to those with a rudimentary understanding of Euclid's Elements, and some number theory. The last section contains writings of mathematics educators trying to discern and justify the role of history in mathematics education.

It is foolish to attempt causal links to pedagogy based on subsuming a small number of studies purporting to address historical topic "x," with the didactical effect "y." In fact it is foolish to even attempt correlations of say "clusters of historical topics" in an area of mathematics with say increased/decreased "interest levels among students in related mathematics content." The purpose of this book is not to propose any panaceas for teaching or learning mathematics, nor prescribe a particular pedagogical approach to a topic in Calculus or Geometry or Number theory. The book is eclectic and its only purpose is to stimulate the reader to look into the history of mathematics as a valuable and untapped resource for furthering their understanding of a mathematical topic. Last but not least, one can always adopt a *Whiggish* stance and use history as classroom pastiche—an outcome not unlikely in the age diminished attention spans in the lacuna of social media.

# Topics in History and Didactics of Calculus and Analysis

# A note on the institutionalization of mathematical knowledge or, "What was and is the Fundamental Theorem of Calculus, really"?

Eva Jablonka and Anna Klisinska
Luleå University of Technology, Sweden

## Background

The question posed in the title arose from our interest in knowledge recontextualisation (e.g. Bernstein, 1990; 1996) or didactical transposition (Chevallard 1985; 1991) in tertiary education. Both Bernstein and Chevallard are concerned with the selection and transformation of elements of this practice for the purpose of teaching. Scholarly knowledge is first transformed into knowledge to be taught, then into the actual knowledge taught, and eventually into learnt knowledge (see e.g. Bosch & Gascon, 2006, for an overview of these four levels of bodies of knowledge). The didactic transposition implies an objectification and public exposure of the knowledge so that social control of the learners by developing systems for testing is made possible (Chevallard, 1991, pp. 61-62). Bernstein theorizes knowledge as a form of discourse with a focus on its social base and the differential valorization and distribution of its different forms, and on how these are maintained and changed. In the course of recontextualisation, the store of valorized knowledge is transformed into a pedagogic discourse, which in turn is transformed into evaluative criteria for what has to be attained by the learner. In both theories the production of knowledge on the one hand, and the recontextualisation or transposition of knowledge on the other hand, are associated with different fields of activity. The first sequence of transformations of knowledge from the field of its production is taking place in the *noosphere* (Chevallard), a non-structured set of experts, educators, politicians, curriculum, developers, recommendations to teachers, textbooks etc., or in the *recontextualising field* (Bernstein), comprising the official field of specialized departments, state agencies and local educational authorities, and the pedagogic unofficial field, consisting of professional educators (e.g. at departments of education) as well as specialized educational media. Especially when the teaching at universities is concerned, there are actors who operate in both fields.

Consequently, we were interested in how the process of institutionalization of knowledge in a specific field of knowledge production is affected in cases where its members act as its recontextualising agents, and how the relationship between producers and transposers of knowledge changed in the course of history. We were also interested in the role of pedagogic discourse for the social construction of mathematical knowledge. In order to say that a field of study is institutionalized, it must be possible to identify works clearly as belonging or as not belonging to it. Institutionalization can be achieved through specialization of notation and methods, a methodology for relating the propositional statements to each other (e.g. by proofs), and reference to the same intellectual roots. All these mark the area as distinct from others. In addition, the use of common names for the field of study and for important theorems, especially for its "fundamental" theorem, can be taken as a sign for its institutionalization, as well as the emergence of textbooks addressed to a wider audience. As to the people working in an institutionalized area, one can expect a form of affiliation to the group. Such an affiliation is of course hard to trace in a historical context where most researchers worked in a diversity of areas, much of the knowledge was exchanged in personal correspondence, through privately circulated manuscripts and private meetings. Groups are mostly constituted geographically in the form of school building.

We have chosen calculus as an example to investigate, because calculus courses have an outstanding position in undergraduate teaching within a diversity of academic programs. Steen (1988, p. xi) even claims that "calculus is a dominating presence in a number of vitally important educational and social systems". In particular, we investigate the emergence and institutionalization of the Fundamental Theorem of Calculus (FTC), as it has an important role in the characterization of calculus as a delineated body of "scholarly knowledge".

As a complement to the historical study, we present outcomes of interviews about the FTC with mathematicians who work in different areas. We were interested in whether their affiliation with a specific sub-area of mathematics makes a difference in their interpretation of the relevance and meaning of this theorem. Further, some of the interviewees have worked as teachers in undergraduate calculus courses, while others did not. As Bernstein (1990, pp. 196-198) notes:

> The recontextualising field brings together discourses from fields which are usually strongly classified, but rarely brings together the agents. On the whole, although there are exceptions, those who produce the original discourse, the effectors of the discourse to be recontextualised, are not agents of its recontextualisation. It is important to study those cases where the producers or effectors of the discourse are also its recontextualisers.

The outcomes of the historical account and the mathematicians' views are discussed from the perspective of didactic transposition and knowledge recontextualisation. We also discuss some uses that have been made of historical accounts in research in mathematics education. The historical investigation and the interview study draw largely on the work of Klisinska (2009).

## The historical investigation

The historical investigation does not aim at providing new views or insights into the history of calculus, but aims at portraying the emergence of the calculus, and the FTC as its "fundamental" theorem, with a focus on its institutionalization. We include some level of original detail, which is necessary to avoid a presentation of the development as following and "evolution with some meanders". For the most part, we rely on classical works on the history of calculus. However, for tracing the FTC in the first textbooks on calculus these sources turned out to be less useful, as the focus of such historical overviews is the advancement of knowledge and not its institutionalization and recontextualisation.

In order to trace the institutionalization of the FTC, our historical investigation includes a study of the emergence of names for basic concepts and theorems as well as for the whole sub-area. For the study of the propositions related to the FTC and of the names used for basic concepts in calculus we looked into all well-known early textbooks and a mathematical handbook. As "textbooks" we took publications that are written by researchers in the field intended for an audience with less specialized knowledge in mathematics. The use of a shared name (in some variation) for the fundamental theorem was taken as a criterion for the selection of more recent textbooks, which comprise examples of English, German, Swedish and Polish textbooks.

## Interviews with mathematicians

The eleven mathematicians who participated in the interview study have diverse backgrounds in education and are working in various areas of mathematics. They also differ in their teaching experiences. All participants have volunteered to participate. They have been informed about the purpose of the study, and in particular that the reporting and storage of the data is anonymous. The set of questions were given to the participants in printed form at the beginning of the interviews. They were invited to read all of the questions before answering, if they liked, or responding directly after reading each question.

The interviews took place in each mathematician's office at their mathematics departments and were tape-recorded. The interviews were carried out in English. They have been transcribed and analyzed through an open coding in search for similarities and differences in answers to each question.

## The FTC in the context of the development of calculus

The present classification into differential calculus, integral calculus, theory of infinite series, and theory of differential equations stems from a relatively well-developed conception of calculus. In its beginning, the classification was linked to types of problems to be dealt with, such as mechanical problems, especially those dealing with accelerated movement and with locating centres of gravity, which were linked to developments in physics and of production techniques in early capitalism. A more mathematical classification included the study of curves, areas and solids and the problem of finding tangent lines. There was quite an excessive body of work that dealt with variables and limits in attempts to produce manageable forms for handling a variety of problems.

For the design of the infinitesimal calculus, knowledge from different areas was necessary. This includes knowledge of the Antique geometrical continuous limit processes based on geometrical interpretations, of the methods employed by Kepler and Cavalieri, in addition to familiarity with the algebraic methods of Vieta, Descartes and Fermat, amongst others. Also, some more direct forms of proof were needed, as opposed to a preference of indirect proof in Greek philosophy. At the end of the 17th century a relatively delineated body of calculus methods was established. Later, in the 18th century, work on the concept of function started, a beginning theory of differential equations and variation calculation emerged. Also, disputes about the infinitely small began. Many of the methods associated with the field of study referred to as *the calculus* since the 17th century were known to work long before justification of their functioning by given standards of proof was at issue, and many of its theorems were established long before all results necessary for building the calculus on solid conceptual foundations were integrated into a body of scholarly knowledge.

### *Predecessors*

Much of the literature on the history of mathematics portrays calculus as one of the most important achievements of European mathematics that begins with the contributions and compilations of work produced in Classical Antiquity, in particular of Antiphon, Eudoxus, Euclid and Archimedes (for a challenge of this focus on European roots, see for example Joseph, 1991; for Arabian contributions to the development of calculus see Juschkewitsch & Rosenfeld, 1963).

In the 16th century Latin translations became available of Greek mathematics, as for example the translations of Euclid's Elements and Proclus' Commentary on Euclid, an edition of the of *Conics* of Apollonius, and a Latin edition of works of Archimedes. The invention and skilful use of the famous method of exhaustion is attributed to Eudoxus of Cnidus (410/408-355/347 BEC[1]), a student at Plato's Academy in Athens (Boyer, 1959). Archimedes of Syracuse (287?-212 BEC) applied this method to plane curves (e.g. parabolas) and solid figures (e.g. spheres and cones). His demonstration that the volume of a cone and of a pyramid is 1/3 that of a cylinder and a prism of the same height and base, respectively, features a prominent example. In proving a number of specific results by exhaustion, he did not show or indicate how he found them in the first place (Katz, 1998). Direct influences from reading Archimedes can be seen in the works of Kepler, Galilei, Torricelli and Cavalieri (Wußnig, p. 152). Oresme, Galilei and Kepler took up some of the typical problems that had been posed (Becker, 1975). Japanese mathematicians (most prominently Seki Takakazu, a contemporary of Leibniz and Newton) also elaborated the method of exhaustion.

---

[1] We use the abbreviations BCE (Before the Common Era) and CE (Common Era).

Various types of calculus problems studied in the 17$^{th}$ century were inherited from Greek mathematics. These include problems of quadratures and cubatures (finding areas and volumes), questions concerning centers of gravity, problems of finding tangent lines and problems about extreme values. Also, ideas about producing figures and solids through "flowing" elements were taken up ("forma fluens", "fluxus formae"; cf. Becker, 1975, p.144). The method of exhaustion through using inscribed and circumscribed rectangles together with an estimation of the decrease of the error when increasing the number of stripes, can indeed be seen as one basic idea of integral calculus. Whiteside (1964) even claims that the technique is equivalent to the definite Cauchy-Riemann integral on a convex set of points in the plane.

Generalization and algorithmization of the methods of calculus rely on a range of prerequisites that had yet to be developed. These include the invention of variables and formulas for symbolic representation of mathematical propositions with a minimum amount of everyday language, analytical geometry as well as a theory of functions and their differentiation and integration that allow to see commonalities in different phenomena (such as velocity, centre of mass/ gravity, slope of a curve, slope of its tangent etc.).

### *Prerequisites*

In the Late Middle Ages some ideas related to a beginning theory of functions developed, which were mostly related to graphical representations. Nicole Oresme (ca. 1323-1382) presented some general rules for representing the quantity of a given quality (e.g. velocity or heat). His methods for representing functional dependencies graphically entered courses at late-scholastic universities under the name *Latitudines Formarum* (latitude of forms) (Kaiser and Nöbauer, 1984, p. 31). He probably for the first time interpreted the area under a velocity-time curve as representing the covered distance. Further, he formulated the proposition that the distance traversed by a body starting from rest and moving with uniform acceleration, is the same as the distance the body would cover if it were to move for the same interval of time with a uniform velocity of one-half of the final velocity. Oresme also worked with infinite series - another concept essential in the development of calculus. For example, he considered a body moving with uniform velocity for half of a period of time, with double of this velocity for the next quarter of the time, three times this velocity for the next eighth and so on. By comparison of areas corresponding to the distances, he found that the total distance would be four times that covered in the first half of the time (Boyer, 1959).

Geometrical limit processes were discussed in the context of the problem of falling bodies, in Italy and also in the Netherlands. Some important work, for example, is that of Isaac Beeckman (1588 – 1637), which remained unpublished but was only disseminated through personal contacts. Beeckman gave a graphical derivation of the time-squared law for falling bodies (Wußing, p. 155). It is reminiscent of Oresme's theory of latitude of forms.

Another prerequisite for the development of the calculus was the generalization of numbers, that is, the introduction of symbols for the quantities involved in algebraic relations. In the 13$^{th}$ century, Jordanus Nemorarius had already used letters as symbols for quantities. But quantities assumed to be known were not distinguished from those to be found, that is, parameters were not distinguished from variables (Boyer, 1968). The French mathematician François Viéte (1540-1603) started to use consonants to represent known quantities and vowels for the unknown (Edwards, 1979). This symbolism was essential to the progress of analytic geometry and the calculus in the following centuries because it made possible to use the concepts of variability and functionality. The convention to use letters near the beginning of the alphabet to represent known quantities, while letters near the end to represent unknown quantities, was introduced later by Descartes in *La Gèometrie* (Edwards, 1979).

Methods with (flowing) indivisibles and with infinitively small entities featured prominently amongst the ways by which various types of calculus problems were studied in the 17th century._Kepler, in his version of infinitesimal geometry, used the concept of infinitely small entities, such as in the famous example of deriving the area of a circle through similar triangles. But Kepler also talked about areas that flow into solids and the areas are the indivisibles of the solids. Indivisibles also had a role in scholastic philosophy, especially in their Aristotelain view that "infinitum actu non datur" (there is no actual infinity).

Most prominently, Cavalieri Bonaventura Cavalieri's (a student of Galilei) elaborated the concept of indivisibles, presented in his *Geometria indivisibilibus continuorum nova quadam ratione promota* (1635, new edition in 1653). He developed a method for representing continuous "wholes" (lines, areas and volumes) by indivisibles of a lower dimension than the continuous thing that is made by them (Edwards, 1979). Cavalieri thought of an area as being made up of components which were intersecting lines parallel to a tangent; the "flowing" of one characteristic line (the *regula*) produces the area (Becker, 1975). He stated that if two solids have the same height then their volumes will be proportional to the areas of their bases ("the theorem of Cavalieri"). With his method, in modern terms, he worked out the integration of $f(x) = x^n$.

Gilles de Roberval (1602-1675) considered problems of the same type as Cavalieri. He conceptualised a curve as a path of a moving point. A tangent line at a given point then showed the direction of the motion at that point. Roberval looked at the area between a curve and a line as being made up of an infinite number of infinitely narrow rectangular strips (Edwards, 1979).

Evangelista Torricelli (1608-1647, the last student of Galilei) was acquainted with both methods, infinitely small entities as well as indivisibles. Notably, he managed the cubature of a rotational hyperboloid. In modern terms this would mean an indefinite integral, which appeared as a somewhat paradox thing, that is, an infinite body with a finite volume (Wußnig, 2008, p. 161).

Pierre de Fermat (1601-1665) devoted a considerable part of his work to finding tangents. He first developed a method for extreme values and then expanded this method for finding tangent lines (the method of *l'adégalité*). How far one could come with this method was a matter of disputes, with Descartes being one of the critics (Kaiser & Nöbauer, pp185-186). Further, Fermat invented a general and systematic method for the quadrature of higher parabolas and hyperbolas (Boyer, 1959).

Blaise Pascal (1632-1662) worked with infinitesimally small rectangles and triangles instead of simple indivisibles and derived some results equivalent to trigonometric integrals. This idea was taken up by Leibniz (Baron, 1987; Becker, 1975). When Leibniz read Pascal's 1659 *Traité des sinus des quarts du cercle* he discovered a "light the author himself has not seen" (Becker, p. 158). Leibniz found that similar proportions and area equivalences would apply to all curves and not only to circles. The name characteristic triangle (*triangulum characteristicum*) is due to Leibniz (Wußing, p. 168).

In the 30ies and 40ies of the 17th century, some methods of indivisibles were arithmetisezed. Cavalieri's method was not appropriate for surfaces and arc lengths. Fermat, for example, did a quadrature of general parabolas, and Wallis (known for his famous product formula for pi) calculated the quadrature for power functions by an inductive approach, influenced by Bacon. Wallis also introduced the symbol for infinity in the form of a horizontal "8". After Wallis, on the British Islands Barrow became important, who can be attributed a version of the FTC (see next section).

At the time in which many of the methods described above were established, there was a Latin translation available of the works of a Greek physician who is affiliated with the tradition of "skepticism". This is the complete Latin edition of the work of Sextus Empiricus (ca. 160-210 CE) with Gentian Hervet as translator, published in 1569. In their work on Newton's Trinity Notebook, McGuire and Tammy (1983) state that Newton has been reading this work, which was quite popular in these times. Empiricus writes:

Nonnulli autem dicunt ex uno puncto constare corpus: ab hoc enim puncto *fluente* effici lineam: ab linea autem *fluente effici* superficiem: eam autem *motam ad profundidatem*, gignere corpus quod triplex habet spatium ac dimensionem. (Empiricus 1718, p. 679, our emphasis)

[Some, on the other hand, say the solid is established through a single point, because this point creates a line *through flowing*. This line in turn creates a surface *through flowing*. This in turn generates a three-dimensional solid *through being moved towards the depth*.][2]

As the idea also appears in the works of John Neper (1550-1617) and Isaac Barrow (1630-1677), one can suspect that they all might have been inspired by reading Sextus Empiricus. The idea is almost as old as Eudoxos' (4th century BCE) idea of exhaustion.

## The "FTC" and the calculus

Isaac Barrow's (1630-1677) is generally credited to be the first who conceptualised differentiation and integration as inverse operations. He reduced inverse-tangent problems to quadratures, but not the reverse. The propositions related to the FTC are formulated and proved in his *Lectiones Opticae et Geometricae* (1669, Lectio X, Prop. 11, pp. 30-32) (Edwards, 1979). The book contains methods for finding areas bounded by curves and for drawing tangent. The geometrical disguise makes the text hard to understand:

Let *ZGE* (Fig.1) be any curve of which the axis is *VD* and let there be perpendicular ordinates to the axis (*VZ, PG, DE*) continually increasing from the initial ordinate *VZ*; also let *VIF* be a line such that, if any straight line *EDF* is drawn perpendicular to *VD*, cutting the curves in the points *E, F*, and *VD* in *D*, the rectangle contained by *DF* and a given length *R* is equal to the intercepted space *VDEZ*; also let *DE:DF = R:DT*, and join (*T* and *F*). Then *TF* will touch the curve *VIF*. (Struik, 1986, p. 255)

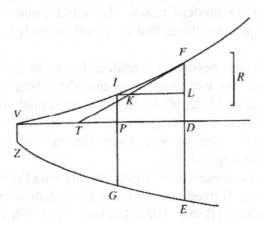

**Figure 1**: Reconstruction of Barrow's approach (Struik, 1986, p.225)

In modern symbolism $y = f(x)$ stands for the curve ZGE and $z = g(x)$ for the curve VIF. Then Barrow's theorem reads that TF is the tangent at F. One can understand $\frac{DF}{DT}$ as the slope of the second curve at the point F, which is equivalent to the formulation that $y:z = R:z\frac{dx}{dz}$ and $y = \frac{d}{dx}\int_a^x ydx$ if $Rz = \int_a^x ydx$ .

Barrow probably did not attribute a fundamental nature to the two theorems presented in his text. The lack of an analytic representation of the operations restricted the effective use of the inverse relationship (Jahnke, 2003). The use of the classical definition of the tangent may have caused a rather static interpretation that did not suggest handling problems about rates of change.

Similar ideas are attributed to James Gregory (1638-1675), a mathematician of Scottish origin who worked for a period in Italy. Prag (1939) classifies Gregory's *Geometriae pars*

---

[2] If not stated otherwise, all translations in this chapter are our own translations.

*universalis* (1668) as an early, if not the first, attempt to write a systematic textbook on what became to be called calculus. The work includes a proof that the method of tangents was inverse to the method of quadratures.

At that time, the common techniques still resembled a fragmented collection and not systematically integrated, which would allow algorithmic treatment of different types of problems, such as the study of motion, determining the speed and acceleration of a moving body or finding the speed and distance travelled by the body at any instant of time if the acceleration is unknown. Some techniques remained unpublished and hence were less influential. Quadrature had become recognized as the geometric model for all integration processes, while differential processes were connected to tangent methods and problems of finding extreme values.

Isaac Newton and Gottfied Wilhelm Leibniz have been accorded a central role in the invention of the calculus. Newton was the first to apply calculus to more general problems and Leibniz developed much of the notation that is still used in calculus today. Leibniz started with integration and Newton with differentiation. But the calculus developed by Newton and Leibniz had not the form students see today: the main object of their study was not functions but curves. Both Newton and Leibniz realized that a whole variety of problems follow from two basic problems and that these two problems were the inverse of each other, that is, the propositional content of what became known as the FTC.

When Newton and Leibniz first published their results there emerged, as it is well known, a controversy over who deserved the credit. Newton derived his results first, but Leibniz published earlier. At his death Newton left around 5000 pages of unpublished mathematical manuscripts that first appeared in the *Cambridge edition of The Mathematical Papers of Isaac Newton* edited by D. T. Whiteside in 1964.

The controversy on the priority of the invention of the calculus eventually caused an isolation of British mathematics from further developments on the continent. British mathematicians did rather explain and apply Newton's work than attempt to develop it, while the Leibnizian calculus continued to develop. After some time it could not anymore be easily translated into the Newtonian style, terminology and notation. Leibniz used the name *Calculus differentialis et calculus integrali* (Kaiser & Nöbauer, 1984, p. 47), while Newton called his calculus *The science of fluxions* (Boyer, 1959).

Newton wrote a tract on fluxions in October 1666, not published at the time, but seen by many mathematicians. The work that had a major influence on the direction calculus was to take (Edwards, 1979).

Soe if $y^e$ curve bee $x^3 = byy$. Then is $\mathfrak{X} = 3x^3$. $\mathfrak{X} = -2byy$. $\mathfrak{X} = 6x^3$. $\mathfrak{X} = -2[b]yy$. $\mathfrak{X} = 0$. And therefore $ck^{(95)} = 3y + \dfrac{4xx}{3y}$, which hath no least nor the curve any least crookednesse.[96]

*Prob 5$^t$. To find $y^e$ nature of $y^e$ crooked line whose area is expressed by any given equation.*

That is, $y^e$ nature of $y^e$ area being given to find $y^e$ nature of $y^e$ crooked line whose area it is.

*Resol.* If $y^e$ relation of $ab = x$, & $\triangle abc = y$ bee given & $y^e$ relation of $ab = x$, & $bc = q$ bee required ($bc$ being ordinately applyed at right angles to $ab$). Make $de \| ab \perp ad \| be = 1$. & $y^n$ is $\square abed = x$. Now supposing $y^e$ line $cbe$ by parallel motion from $ad$ to describe $y^e$ two superficies $ae = x$, & $abc = y$; The velocity $w^{th}$ $w^{ch}$ they increase will bee, as $be$ to $bc$: $y^t$ is, $y^e$ motion by $w^{ch}$ $x$ increaseth being $be = p = 1$, $y^e$ motion by $w^{ch}$ $y$ increaseth will

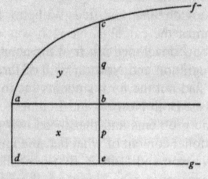

bee $bc = q$. which therefore may bee found by prop: 7$^{th}$. viz: $\dfrac{-\mathfrak{X}y}{\mathfrak{X}x} = q = bc$.[97]

---

(95) Or $\dfrac{[(3x^3)^2 y^2 + (-2by^2)^2 x^2] \cdot -2by^2}{-(3x^3)^2 (-2y^2 b) y - (-2by^2)^2 (6x^3) y}$.

(96) Newton entered this in cancellation of 'which is least when $x = -\dfrac{4b}{27}$'. Where

$$ck = 3y + 4x^2/3y = 3y + \tfrac{4}{3}b^{\frac{2}{3}}y^{\frac{1}{3}}, \quad \text{then} \quad d/dx(ck) = 3 + 4b/9x$$

which increases uniformly as $x$ decreases and is zero for $x = -4b/27$. However Newton wishes here to consider only real points on the curve and so restricts $x$ to the interval $[0, \infty]$. Hence this 'minimal' value for curvature defined by $x = -4b/27$ is not admissible in his scheme. More directly, we may calculate the radius of curvature to be

$$cm = \dfrac{(9x^2 + 4bx)^{\frac{3}{2}}}{6bx}, \quad \text{so that} \quad \dfrac{d(cm)}{dx} = \dfrac{\sqrt{[9x^2 + 4bx]}}{2bx^2}(9x^2 + 6bx)$$

and $cm$ has therefore an extreme value for $x$ in $[0, \infty]$ at $x = \infty$ (which, in fact, defines the inflexion point at infinity on the semicubic parabola $by^2 = x^3$). We may also show that $cm$ increases with $x$ in the region of $x = 0$, so that the curvature at a real point takes on an apparent maximum at $x = 0$, $y = 0$. (Compare the difficulties which Newton had in the winter of 1664/5 in considering the extreme values of curvature in the case of a conic. See 4, §2.5 above.)

(97) Since, where $\mathfrak{X} \equiv f(x, y) = 0$, we have $\mathfrak{X} = xf_x$, $\mathfrak{X} = yf_y$ and $f_x + f_y \dfrac{dy}{dx} = 0$,

therefore $q = -\dfrac{\mathfrak{X}y}{\mathfrak{X}x} = \dfrac{dy}{dx}$. Hence the area '$\triangle abc$' $= \int q \cdot dx = y$, a statement of the

fundamental theorem of the calculus that $\int \left(\dfrac{dy}{dx}\right) d$ ...

**Figure 2**: Newton's fundamental theorem (Whiteside, 1964, Vol. I, p. 427)

Newton's first manuscript notes date from 1665, in which he used "pricked" letters, which he had started to use consequently in late 1691. In 1710 William Jones made a transcript of the 1671 work on fluxions (*Methodus fluxionum et Serierum Infinitarum*) and inserted the dot notation, where the fluxions of the fluents $x,y,z$ are $\dot{x}, \dot{y}, \dot{z}$ respectively. This transcript was subsequently copied in all published editions. Newton acknowledges that he had been influenced by Wallis' work. Studying Wallis's *Arithmetica Infinitorum* (1655) he had also discovered the binomial series. Newton stated that he was in possession of his fluxionary calculus in 1665-1666, but the first notice of his calculus was given in 1669 in *De Analysi Per Aequationes Numero Terminorum Infinitas* (Boyer, 1959; Scott, 1960). In this work he did not use the fluxionary idea but worked with the idea of an indefinitely small rectangle or "moment" of area to find the quadratures of numerous curves. Newton began this short summary of his discoveries with three rules (Jahnke, 2003, p.76):

Rule 1: If $y = ax^{\frac{m}{n}}$, then the area under y is $\dfrac{an}{(n+m)} x^{\frac{m}{n}+1}$

Rule 2: If $y$ is given by the sum of more terms (also infinite number of terms) $y = y_1 + y_2 + ...$, then the area under y is given by the sum of the areas of the corresponding terms.

Rule 3: In order to calculate the area under a curve $f(x,y) = 0$ one must expand $y$ as a sum of terms of the form $ax^{\frac{m}{n}}$ and apply Rule 1 and Rule 2.

Later in this work he gave "a general" procedure for finding the relation between the quadrature of a curve and its ordinate in which it is possible to see that Newton recognized the inverse relationship of integration and differentiation. The procedure is the following (Whiteside, 1964, Vol. II, pp.242-245):

Let area $ABD = z$, $BD = y$, $AB = x$, $B\beta = o$, $BK = v$ chosen in such way that area$BD\delta\beta$ = area$BKH\beta = ov$.

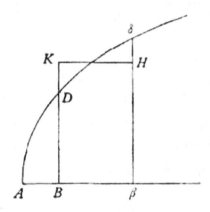

**Figure 3**: Reconstruction of Newton's approach (Whiteside, 1964, Vol. II, p. 242)

And then follows an example with the curve $z = \frac{2}{3} x^{\frac{3}{2}}$. If one substitutes $x + o$ for $x$ and $z + ov$ for $z$ one gets:

$$(z+ov)^2 = \frac{4}{9}(x+o)^3$$

$$z^2 + 2zov + o^2v^2 = \frac{4}{9}(x^3 + 3x^2o + 3ox^2 + o^3).$$

Removing terms without $o$ (which are equal) and dividing both sides by $o$, results in

$$2zv + ov^2 = \frac{4}{9}(3x^2 + 3xo + o^2).$$

Now, Newton takes $B\beta$ as "infinitely small", so $v = y$ and terms with $o$ vanish:

$$2zy = \frac{4}{3} x^2.$$

Using the first expression for z results in $y = x^{\frac{1}{2}}$.

After the example Newton used the same method to show that if $\frac{n}{m+n}ax^{\frac{m+n}{n}} = z$ then

$ax^{\frac{m}{n}} = z$, and he finished by saying "And similarly in other cases".

The essential element in the procedure is the substitution of the "small" increments $o$ and $ov$ for $x$ and $z$ in the equation. Newton made use of this method also in the determination of maxima and minima, tangents and curvature. Only later he reformulated these algorithms in terms of fluents and fluxions. So if one considers the point $D$ as moving along the curve then the corresponding ordinate $y$, abscissa x, quadrature $z$ (or any other variable quantity connected with the curve) would increase or decrease, that is, "flow". Newton called such flowing quantities fluents and their rate of change with respect to time fluxion. Newton often made the additional assumption that one of the variables, say $x$, moves uniformly so $\dot{x} = 1$. It is possible to do so because he was not interested in the values of the fluxions but in their ratio $\frac{\dot{y}}{\dot{x}}$, which is equal to the slope of the tangent. He explains that the ratio of the fluxions $\frac{\dot{y}}{\dot{x}}$ is equal to the "prime" (when $x$ and $y$ come into existence) or "ultimate" ratio (when they cease to exist) of augments or decrements of $y$ and $x$ (Struik, 1986). From these concepts emerges one of the foundational questions in the debate and critique of the calculus: Do prime or ultimate ratios exist?

For Newton, time was the universal independent variable. Variables were conceptualised as moving quantities and the focus was on their velocities. He regarded $o$ as a small time interval and used $p$ for the velocity of the variable $x$ (later with the dot), so the change in $x$ over the time interval $o$ was $op$. In the opening chapter of the *Methodus fluxionum et sericum infinitarum* (1671) Newton states two basic problems (Jahnke, 2003):

1. Given the length of the space continuously (that is, at every time) to find the speed of motion at any time proposed [modern explanation: given an expression for distance in terms of time $s = f(t)$, compute the velocity $v(t) = \frac{ds}{dt} f'(t)$].

2. Given the speed of motion continuously, to find the length of the space described at any time proposed [modern explanation: given an expression for velocity in terms of time $v(t) = \frac{ds}{dt} = \Phi(t)$, compute the distance $s = \int_{o}^{t} \Phi(x)dx$].

Newton did not use a single notation for the area under a curve. Usually he used phrases as "the area of" and sometimes a symbol of a rectangle with the term inside it (Jahnke, 2003). He also used a small vertical bar above $x$ for notating the integral of $x$ (Cajori, 1923).

Altogether, Newton's method of fluxions was based on change:

- a fluent as a changing quantity,
- the fluxion of the fluent: its rate of change,
- the moment of a fluent: the infinitely small amount of change experienced by a fluent in an infinitely small time interval.

Leibniz's conception of the foundations was different. While Newton considered variables changing with time, Leibniz thought of variables $x$, $y$ as ranging over sequences of infinitely close values. He introduced $dx$ and $dy$ as differences between successive values of these sequences. The following basic ideas can be said to underlie Leibniz's invention of the calculus (Baron, 1987):

- Leibniz's interest in symbolism and notation (in connection with his idea of a general symbolic language),
- the insight that summing of sequences and taking their differences are inverse operations and so determining quadratures and tangents as inverse operations,
- the use of the characteristic triangle in deriving general transformations of quadratures.

Leibniz's idea about integrals, derivatives and calculus were derived from close analogies with finite sums and differences. He combined these ideas in a series of studies on the analytic treatment of infinitesimal problems.

His notation appeared to be more clear than Newton's (Boyer, 1959). He is responsible for introducing the integral and differential sign $dx$. The symbols $dx, dy, \frac{dx}{dy}$ were introduced in a manuscript from November 11, 1675 (Cajori, 1923). Before introducing the integral symbol, Leibniz wrote *omn.* in front of the term to be integrated. The manuscript from 29th of October contains the following result, which he comments to be "a very fine theorem, and one that is not at all obvious" (Edwards, 1979, p.252):

$$\frac{\overline{omn.\, l}^2}{2} = omn.\, \frac{\overline{l}}{\overline{omn.\, l}\, a}$$

Overbars are used in place of parentheses, $a$ is a constant. In modern notation this would be:

$$\frac{1}{2}\left(\int dy\right)^2 = \int \left(\int dy\right) dy.$$

The integral symbol was first used in this unpublished manuscript. Later in 1675, he proposed the use of the integral symbol in a letter to Oldenburg (secretary of the Royal Society). The first appearance of the integral symbol in print was in an article in the *Acta Eruditorum* from 1686 but it did not look exactly as the one used today. The symbol missed the lower part and was similar to the letter "f" (Cajori, 1923). The modern definite integral symbol was first introduced in 1822 by Jean Baptiste Fourier in his work *The Analytical Theory of Heat*. He also extended the integral symbol with the lower and upper interval notation (Cajori, 1923, p. 35).

Leibniz and Newton are considered as the inventors of the calculus, but they did not invent the same calculus. The differences between Leibniz's and Newton's versions of the calculus can be summarised as follows (Baron, 1987):

- The conception of variable: Newton considered variables as changing in time (flowing quantities), Leibniz considered them as ranging over sequences of infinitely close values.
- The fundamental concepts: Newton's fluxion is a kinematic concept that draws on the velocity or rate of change; Leibniz's differential is the difference of two successive values in the sequence that could not be finite and had to be infinitely small; both are not directly related to problems about quadratures and tangents.
- While Leibniz worked with infinitely small quantities, Newton's fluxion was a finite velocity.
- The conception of the integral and the fundamental theorem: for Newton integration meant finding fluent quantities for given fluxions and so the fundamental theorem is implied in the definition of integration; Leibniz conceived integration as summation and so the inverse relation between differentiation and integration is not implied.
- Notation: Leibniz used separate symbols $d$, $\int$ which are more easy to use in complicated formulae; Newton worked with dots for fluxions but did not have a special symbol for integration and his notation requires that all variables are considered as functions of time.

## Arithmetization of the calculus

One conception for describing and explaining calculus is by infinitesimals, which are objects that can be treated like infinitely small numbers, which on a number line have zero distance from zero. Any multiple of an infinitesimal is still infinitely small. Then calculus is a collection of techniques for manipulating infinitesimals. In the course of exactification, in the 19th century infinitesimals were replaced by limits. Calculus becomes a collection of techniques for manipulating certain limits. The development of the limit concept was crucial for the institutionalization of the calculus and linked to the unresolved discussion on the ontological status of the infinitesimals, initiated by Berkeley's critic in 1734 (see e.g. Katz, 1998, pp. 628-632). Some details of this historical development are therefore worth mentioning.

An early description of a limit notion similar to present use is Jean Baptiste d'Alembert's in 1754 in the *French Encyclopedia*, though not very influential. He calculated the slope of the tangent as the limit of the slope of the secants exactly as it is done today, but without using a formal definition of limit (Struik, 1986, p. 343).

Also with minor influence was Simon l'Huilier, who introduced the notation "lim." for limit in his *Exposition élémentaire des principes des calculs supérieures* from 1786 and used the notion of limit to define the derivative: $Lim.\frac{\Delta y}{\Delta x}$ (see Cajori, 1993, Vol. II, p. 257). More influential was Joseph Louis Lagrange, even if his first outline of the differential calculus from 1772 (*Sur une nouvelle espèce de calcul relatif á la differentiation et á l'intégration des quantités variables*) was not used, though much praised. However, his book from 1797 (*Théorie des functions analytiques, contenant les principes du calcul ….*), where he further developed the derivatives (*fonctions dérivées*), was the most important attempt at the time to form a basis for the calculus (Boyer, 1959, p. 260; Katz, 1998, p. 633). Then, through the work of Sylvestre François Lacroix, building on Lagrange, limits became familiar by his shorter version from 1802 (*Traité élémentaire*) of his earlier text from 1797 (*Traité du calcul différentiel et du calcul integral*). This shorter version was the most famous and ambitious textbook of the time, translated into many languages (Boyer, 1959, p. 264-265). Both Lagrange and Lacroix expressed serious concern for the role of notations used in the calculus (Cajori, 1993, Vol. II, p. 213-215).

Another popular text using limits was Lazare Carnot's *Oeuvres mathématiques du Citoyen Carnot* from 1797, though more focused on practical use than on foundation issues, and not building on the function concept as Augustin Louis Cauchy did later. The work of Bernhard Bolzano (1817) was not much recognized in his time, and it is unclear if his non-geometric definition of continuity by limits and of the derivative as a limit of a quotient, similar to l'Huilier and almost identical to Cauchy, did influence Cauchy (Katz, 1998, p. 770). An important idea was his use of $\frac{dy}{dx}$ as a symbol for a single function and not as a quotient of infinitesimals.

With Cauchy (1821; 1823) the non-geometrical arithmetic definition of limit and the derivative, for which he used the notation $f'(x)$ for the derivative of $y$ with respect to $x$, the development moves beyond the ontological interpretation of the infinitesimals, to be completed with the present ε-δ-definitions of Karl Theodor Weierstrass in 1854, where also the notation $\underset{n=\infty}{Lim.p_n}$ was introduced.

The arrow notation for limits was not introduced until 1905 by John Gaston Leathem in his *Volume and surface integrals used in physics*, and then promoted by Godfrey Harold Hardy in 1908 (see Cajori, 1993, Vol. II, p. 257).

For most of the 18th century, mathematicians were mainly interested in the results and the methods of the calculus, especially in its applications in physics, in particular mechanics, and astronomy. The foundations of the basic concepts of the calculus became a major issue for d'Alembert and Lagrange. In the 19th century a major task for mathematicians like Cauchy, Bolzano and Weierstrass was the rigorisation of the foundations. Some contradictions that derived from calculating with infinite series drew attention to the concept of convergence.

However, standardisation for the purpose of teaching was another motivation for rigorisation. Universities became the centres of mathematical training and research, which led to the development of pure mathematics as an independent field (Jahnke, 2003).

In the same period the ideas of calculus were generalized to Euclidean space and to the complex plane. Lebesgue generalized the notion of the integral so that any function has an integral, while Schwarz extended differentiation in the same way (Boyer, 1959).

Cauchy is regarded as the founder of the differential calculus in its modern version (Boyer, 1959). The critical mathematics of the 19th century aimed at developing foundations that comprised the objects to work with without relying on informal understanding and sensory perception (Becker, 1975).

The major ideas of modern calculus – derivative, continuity, integral, convergence/divergence of sequences and series, became defined in terms of limits. Limit is to be seen as the fundamental concept of calculus, which distinguishes it from other branches of mathematics.

In 1821 Cauchy was searching for a rigorous development of calculus to be presented to his engineering students at the École Polytechnique in Paris. He started his calculus course with a definition of the limit. In his writings Cauchy used limits as the basis for definitions of continuity and convergence, the derivative and the integral:

> When the values successively attributed to a particular variable approach indefinitely a fixed value so as to differ from it by as little as one wishes, this latter value is called the limit of the others (Cauchy, 1821, series 2, vol. 3, p.19).

Cauchy's work transformed the calculus into a specialized mathematical discipline, integrated by definitions and proofs. Equipped with the definition for the integral of any continuous function he gives a proof of the FTC.

Riemann, in 1854 in a paper on trigonometric series, generalized Cauchy's work to include the integrals of bounded functions. He defines the oscillation of the function in an interval as the difference between the greatest and least value of the function on that interval and with this approach he could integrate functions with an infinite number of discontinuities (Kline, 1972). However, the modern notion of Riemann sums was not actually completed by Riemann himself but by Gaston Darboux, who defined the upper and lower Riemann sums and showed that a function is integrable only when upper and lower Riemann sums approach the same value as the maximum subinterval approaches zero in any partition over the interval (Kline, 1972).

Even after Riemann generalized Cauchy's concept of the integral, mathematicians discovered functions that could not be integrated, for example the Dirichlet-function that takes the value 1 on the rational numbers, and 0 on the irrationals. Riemann integration could not handle such functions because of the large number of discontinuities in any interval one may choose for integration.

## *The FTC in 20th century mathematics*

For some mathematicians who were trying to systematize the calculus in the 18th century, the approach of seeing the integral as antiderivative was preferable, because it is algebraic and formal with no references to geometric ideas. When seeing differentiation and integration as operations, the inverse relationship between them is no longer a problem. Such a shift from a geometric to an algebraic approach is for example found in the works of Euler and Johann Bernoulli.

While the approach of conceptualising the integral as antiderivative leads to formalization, conceptualisation as area goes in the direction of generalization to wider applications. The notion of area of geometric regions is generalized to the notion of measure as a function defined on an abstract space. This approach led to the definition of the Lebesgue integral and measure theory. The Lebesgue integral is considered a quite difficult mathematical concept but also a very important tool that can be used in a number of various applications in modern mathematics. As Norbert Wiener formulated it (1956):

> The Lebesgue integral is not easy conception for the layman to grasp (…) It is easy enough to measure the length of an interval along a line or the area inside a circle or other smooth, closed curve. Yet, when one tries to measure sets of points which are scattered over an infinity of segments or curve-

bound areas, or sets of points so irregularly distributed that even this complicated description is not adequate for them, the very simplest notions of area and volume demand high-grade thinking for their definition. The Lebesgue integral is a tool for measuring such complex phenomena. The measurement of highly irregular regions is indispensable to the theories of probability and statistics; and these two closely related theories seemed to me, even in those remote days before the war, to be on the point of taking over large areas of physics (ibid., pp. 22-23).

The development of the function concept in the 19$^{th}$ century with the expansion of arbitrary functions into Fourier series led to Riemann's definition of the integral in which he used a new class of functions: integrable functions. But still Riemann's integration and differentiation were found not to be completely reversible because the process of differentiation of a function $f$ leads to a bounded derivative $f'$ which is not necessarily Riemann integrable (Jahnke 2003, p. 271). This problem led to another conceptualisation. Henry Lebesgue introduced his theory of integration in 1902 (in his doctoral thesis *Intégrale, longueur, aire*) and expended it in 1904 in his book *Leçons sur l'integration et la recherché des functions primitives*.

Presently, the standard integral in advanced mathematics is the the Lebesgue integral which generalizes the Riemann integral in the sense that any function that is Riemann-integrable is also Lebesgue-integrable and integrates to the same value, but the class of integrable functions is much larger; there are functions whose improper Riemann integral exists but which are not Lebesgue-intagrable. Lebesgue started from the problem of measurability of sets, Lebesgue's integral uses a generalization of length, which is called the measure of a set. A function is measurable if the inverse image of any open interval is a measurable set. Lebesgue's theory inverts the roles played by inputs and function values in defining the integral by looking on the range of a function instead for its domain. In his definition he was interested in some properties for the integral (Intégrale, longueur, aire, p.253):

- Riemann's definition should be a special case.
- The definition can be used in the cases of one and of several variables.
- It should guarantee the solution in the fundamental problem of calculus.

Lebesgue proved that a bounded function on a closed interval is Riemann-integrable if and only if its set of discontinuities has measure zero (almost everywhere) but every bounded measurable function on a closed interval is Lebesgue-integrable. Lebesgue's integral can be extended to the case of several variables with the proper measure on R, but even with the Lebesgue-integral it cannot be proved that $\int_a^b f' = f(b) - f(a)$ without some additional assumptions for $f$. In the introduction of his thesis he wrote (p. 203):

It is known that there are derivatives which are not integrable, if one accepts Riemann's definition of the integral; the kind of integration as defined by Riemann does not allow in all cases to solve the fundamental problem of calculus: find a function with a given derivative. It thus seems to be natural to search for a definition of the integral which makes integration the inverse operation of differentiation in as large a range as possible.

Lebesgue's Fundamental Theorem of Calculus:

The integral of $f$ over a set E is written as $\int_E f(x)dm$. In one dimension it is equivalent to $\int_\alpha^\beta f(t)dt$ when $E = (\alpha,\beta)$. A non-negative function $f$ is called integrable if it is Lebesgue measurable and $\int_E f(x)dm < \infty$. Then, if the function $f:[a,b] \rightarrow R$ is Lebesgue integrable and $F(x) = \int_a^x f(t)dt$ is its indefinite integral then for almost every $x \in [a,b]$, $F'(x)$ exists and is equal to $f(x)$.

That does not characterize indefinite integrals. If a function $G$ has a derivative almost everywhere and its derivative is an integrable function $f$, this does not imply that $G$ differs from the indefinite integral of $f$ by a constant. This is satisfied for another type of function: absolutely continuous functions.

The following theorems state that differentiation and Lebesgue integration are inverse operations on very large classes of functions:

*Theorem*: Let $f:[a,b] \to$ R be integrable.

The indefinite integral $F$ of $f$ is absolutely continuous. For almost every $x$, $F'(x)$ exists and is equal $f(x)$. If $G$ is an absolutely continuous function and $G'(x) = f(x)$ for almost every $x$, then $G$ differs from $F$ by a constant.

In the next theorem he proved that integration and differentiation are reversible if only the derivative is bounded.

*Theorem*: If a function $f$ on $[a,b]$ has a bounded derivative $f$, then $f'$ is Lebesgue integrable and $\int_a^b f'(x)dx = f(b) - f(a)$.

By the 19$^{th}$ century mathematics had moved far beyond of being simply a tool for science and it became possible to define something as abstract as the Lebesgue measure and integration.

## *Name giving and textbooks*

The short outline of the history of the FTC above did not focus on the issue of terminology and textbooks. The following section concentrates on the emergence of shared names and on the appearance of the FTC in textbooks.

Before the official publication of Leibniz's work, there existed several textbooks in calculus. Naturally, these cannot be expected to contain what we now call the FTC, but are traces of the first didactic transposition of the new knowledge about tangents and areas.

The first printed textbook in differential calculus appeared in Paris in 1696 *Analyse des infiniment petits pour l'intelligence des lignes courbes*. This book was edited anonymously but was written by Marquis de l'Hospital with the help of Johann Bernoulli (Kaiser & Nöbauer, 1984, p. 49). Maligranda (2004) states that some parts (about 30 pages) of it were indeed written in 1691-1692 by Bernoulli as the material for the private lectures given by him to de l'Hospital. It was only in the second editions (which exist from 1715 and from 1716) that l'Hospital was named as the author. The book consists of ten chapters about rules of differential calculus, applications of differentials for determinations of tangents, maximum and minimum problems and other problems and also contains a formula known later as l'Hospital's Rule, but is missing any formulation of the fundamental theorem of calculus. The preface of the second edition contains historical comments as well as an overview of the contents of the work:

> The type of analysis we shall describe in this work presupposes an acquaintance with ordinary analysis, but is very different from it. Ordinary analysis deals only with finite quantities whereas we shall be concerned with infinite ones. We shall compare infinitely small differences with finite quantities; we shall consider the ratios of these differences and deduce those of the finite quantities, which, by comparison with the infinitely small quantities are like so many infinities. We could never say that our analysis takes us beyond infinity because we shall consider not only these infinitely small differences but also the ratios of the differences of these differences, and those of the third differences and the fourth differences and so on, without encountering any obstacle to our progress. So we shall not only deal with infinity but with an infinity of infinity or an infinity of infinities [De forte qu'elle n embrasse pas seulement l'infini ; mais l'infini de l'infini ou une infinité d'infinis]." (l'Hospital, 1716, p. iii-iv, translation from The MacTutor History of Mathematics Archive, created by John J O'Connor and Edmund F Robertson).

The new methods here are introduced as calculations with ratios of infinitely small differences. In terms of modern notions, the textbook is about differential calculus. The introduction continues:

> All this is only the first part of M Leibniz's work on calculus, which consists of working down from integral quantities to consider the infinitely small differences between them and comparing these infinitely small differences with each other, whatever their type: this part is called Differential Calculus. The other part of M Leibniz's work is called the Integral Calculus, and consists of working up from these infinitely small quantities to the quantities of totals of which they are the differences: that is, it consists of finding their sums. I had intended to describe this also. But M Leibniz wrote to me to say that he himself was engaged upon describing the integral calculus in a treatise he calls De Scientia infiniti, and I did not wish to deprive the public of such a work, ... (l'Hospital, 1716, p. xii, translation from The MacTutor History of Mathematics Archive, created by John J O'Connor and Edmund F Robertson).

However, the second part of the planned textbook about the integral calculus never appeared, probably because Leibniz died in 1716. From the introduction (above) it becomes clear that the name *Integral Calculus* [*Calcul integral*] was already in use. Thus, by having a specific name it had gained an official status as a connected part of knowledge to which one could easily refer. However, it is not clear whether the name initially referred to Leibniz's method. With reference to Boyer (1939), Domingues (2008, p. 140) points out that the name *Integral* was used by the Bernoullis (in a printed work of Jakob), not for denoting Leibniz's definition as the sum of infinitesimally narrow rectangles, but as indicating the definition of the integral as antiderivative. According to Gerhardt (1860) the name *Calculus integralis* was officially sanctioned by Leibniz, and eventually superseded the *Calculus summatorius*. In the *Elementa universae: commentationem de methodo mathematica...* (Wolff, 1732), a mathematics handbook, there is a section on *Calculo integrali seu summatorio*. But this section still does only refer to the integral as antiderivative (see Fig. 4).

The first English translation of l'Hospital's book by John Colson appeared in London in 1736 as *The method of fluxions and infinite series with its applications to the geometry of curve-lines*. This re-naming of the title *Analyse des infiniment petits pour l'intelligence des lignes courbes* obviously reflects a cultural identity that links the works in calculus to Newton. A much later English translation from 1930 curiously has the title *Method of fluxions both direct and inverse: the former being a translation from the celebrated Marquis De L'Hospital's Analyse des infinements petits and the latter supply'd by the translator*. This Newtonian version of the title was also supplied by the translator.

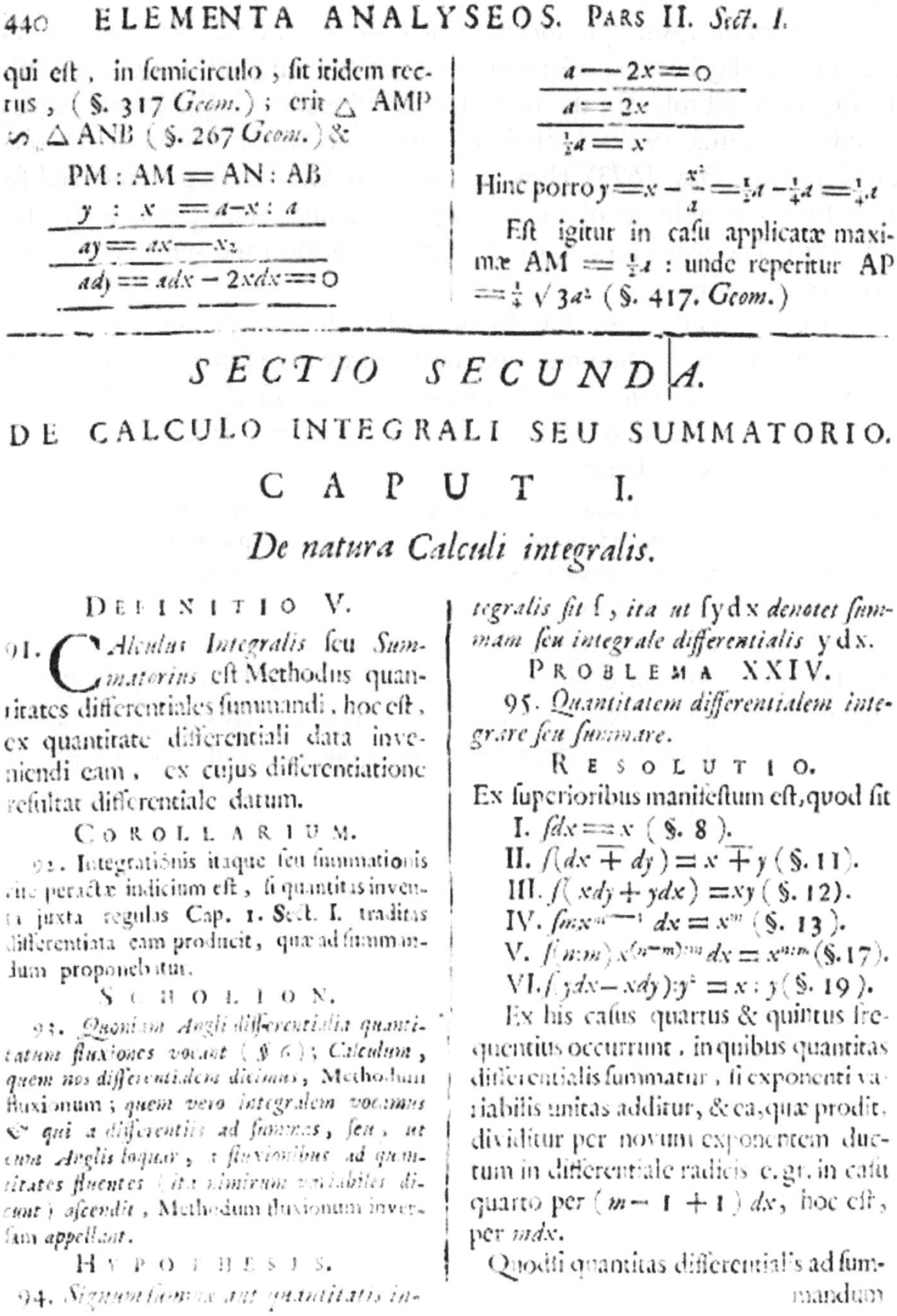

440  ELEMENTA ANALYSEOS. Pars II. *Sect. I.*

qui est, in femicirculo, fit itidem rec-
tus, (§. 317 *Geom.*); erit $\triangle$ AMP
$\backsim$ $\triangle$ ANB (§. 267 *Geom.*) &

$$PM : AM = AN : AB$$
$$y : x = a-x : a$$
$$ay = ax - x^2$$
$$ady = adx - 2xdx = 0$$

$$a - 2x = 0$$
$$a = 2x$$
$$\tfrac{1}{2}a = x$$

Hinc porro $y = x - \dfrac{x^2}{a} = \tfrac{1}{2}a - \tfrac{1}{4}a = \tfrac{1}{4}a$

Est igitur in cafu applicatæ maxi-
mæ AM $= \tfrac{1}{4}a$ : unde reperitur AP
$= \tfrac{1}{4} \sqrt{3a^2}$ (§. 417. *Geom.*)

---

## SECTIO SECUNDA.

### DE CALCULO INTEGRALI SEU SUMMATORIO.

## CAPUT I.

### De natura Calculi integralis.

DEFINITIO V.

91. *Calculus Integralis feu Sum-matorius* est Methodus quantitates differentiales fummandi, hoc est, ex quantitate differentiali data inveniendi eam, ex cujus differentiatione refultat differentiale datum.

COROLLARIUM.

92. Integrationis itaque feu fummationis rite peractæ indicium est, fi quantitas inventa juxta regulas Cap. I. Sect. I. traditas differentiata eam producit, quæ ad fummandum proponebatur.

SCHOLION.

93. *Quoniam Angli differentialia quantitatum fluxiones vocant* (§ 6); *Calculum, quem nos differentialem dicimus, Methodum fluxionum; quem vero integralem vocamus & qui a differentiis ad fummas, feu, ut cum Anglis loquar, a fluxionibus ad quantitates fluentes (ita nimirum variabiles dicunt) afcendit, Methodum fluxionum inverfam appellant.*

HYPOTHESIS.

94. *Signum fummæ aut quantitatis in-* tegralis fit $\int$, ita ut $\int y\,dx$ denotet fummam feu integrale differentialis $y\,dx$.

PROBLEMA XXIV.

95. *Quantitatem differentialem integrare feu fummare.*

RESOLUTIO.

Ex fuperioribus manifeftum est, quod fit

I. $\int dx = x$ (§. 8).

II. $\int (dx \mp dy) = x \mp y$ (§. 11).

III. $\int (x\,dy + y\,dx) = xy$ (§. 12).

IV. $\int m x^{m-1}\, dx = x^m$ (§. 13).

V. $\int (n:m) x^{(n-m):m}\, dx = x^{n:m}$ (§.17).

VI. $\int (y\,dx - x\,dy):y^2 = x : y$ (§. 19).

Ex his cafus quartus & quintus frequentius occurrunt, in quibus quantitas differentialis fummatur, fi exponenti variabilis unitas additur, & ea, quæ prodit, dividitur per novum exponentem ductum in differentiale radicis e. gr. in cafu quarto per $(m - 1 + 1)\, dx$, hoc est, per $m\,dx$.

Quodfi quantitas differentialis ad fum-mandum

---

**Figure 4**: Page 440 of Wolff's Handbook

Leibniz's work from 1684 was printed in Acta Eruditorum with the title *Nova methodus pro maximis et minimis, itemque tangentibus quae nec fractas nec irrationals quantitates moratur, et singulare pro illis calculi genus*. It cannot be called a textbook but it contains a definition of the differential together with the rules for computing differentials of powers, products and quotients. Consequently the year 1684 is by some officially accepted as the beginning of the differential calculus. Newton's 1671 book *Methodus fluxionum et sericum infinitarum* was first published only nine years after his death in an English translation by

John Colson as *The Method of Fluxions and Infinite Series with its Applications to the Geometry of Curve-lines* (1736).

Euler's *Introductio in analysin infinitorum* from 1748 is the first textbook in calculus that can easily be read even today because of the modern notation and terminology and of the function concept playing a central role; for the first time functions instead of curves were the principal objects of study (Edwards, 1979). Leibniz had used the term function (*Methodus tangentium inversa, seu de functionibus*, 1673). However, the term was not in general used for the regular relationship between an ordinate of a curve depending and its abscissa, as this is described as a relation (*relatio*). The word function is by Leibniz used in the sense as of a task that some part in a mechanism fulfils.

Cauchy's Course of analysis of the École Polytechnique (*Cours d'analyse de l'École Polytechnique*, 1821) and the Summary of the lectures given at the École Polytechnique on the infinitesimal calculus (*Résumé des leçons données a l'École Polytechnique sur le Calcul Infinitesimal*, 1823) were the first textbooks in which calculus appeared in a general character as an integrated body of knowledge:

> In the integral calculus, it has appeared necessary to me to demonstrate generally the existence of the integrals or primitive functions before making known their diverse properties. In order to attain this objective, it was found necessary to establish at the outset the notion of integrals taken between given limits or definite integrals (Cauchy in Oeuvres (2), IV, pp. ii-iii).

Cauchy is also viewed as being responsible for changing the attitude towards the value of dealing with the foundations of calculus. In his books he stressed the definitions of the basic concepts and included many examples of a new style of reasoning. He demonstrated the necessity for rigour not only in defining basic concepts but also in proving theorems (Grabiner, 1981).

# VINGT-SIXIÈME LEÇON.

## INTÉGRALES INDÉFINIES.

Si, dans l'intégrale définie $\int_{x_0}^{X} f(x)\,dx$, on fait varier l'une des deux limites, par exemple la quantité X, l'intégrale variera elle-même avec cette quantité; et, si l'on remplace la limite X devenue variable par $x$, on obtiendra pour résultat une nouvelle fonction de $x$, qui sera ce qu'on appelle une intégrale prise à partir de l'*origine* $x = x_0$. Soit

$$(1) \qquad \mathfrak{F}(x) = \int_{x_0}^{x} f(x)\,dx$$

cette fonction nouvelle. On tirera de la formule (19) (vingt-deuxième Leçon)

$$(2) \qquad \mathfrak{F}(x) = (x - x_0)\, f[x_0 + \theta(x - x_0)], \qquad \mathfrak{F}(x_0) = 0,$$

$\theta$ étant un nombre inférieur à l'unité, et de la formule (7) (vingt-troisième Leçon)

$$\int_{x_0}^{x+\alpha} f(x)\,dx - \int_{x_0}^{x} f(x)\,dx = \int_{x}^{x+\alpha} f(x)\,dx = \alpha\, f(x + \theta\alpha)$$

ou

$$(3) \qquad \mathfrak{F}(x + \alpha) - \mathfrak{F}(x) = \alpha\, f(x + \theta\alpha).$$

Il suit des équations (2) et (3) que, si la fonction $f(x)$ est finie et continue dans le voisinage d'une valeur particulière attribuée à la variable $x$, la nouvelle fonction $\mathfrak{F}(x)$ sera non seulement finie, mais encore continue dans le voisinage de cette valeur, puisqu'à un accrois-

**Figure 5**: Cauchy in Oeuvres (2), IV, p. 151

With two operations (differentiation and integration) defined independently of each other, Cauchy established the inverse relationship between them without relying on an informal concept of area. In his *Twenty sixth lesson* (ibid., pp.151-155, see fig. 4, 5), Cauchy presents the FTC.

He showed that if *f(x)* is a continuous function, the function defined as the definite integral $F(x) = \int_{x_o}^{x} f(x)\,dx$ has as its derivative the function *f(x)*. This was perhaps the first rigorous demonstration of the proposition known as the fundamental theorem of calculus (Boyer, 1959, pp.279-280).

**152 RÉSUMÉ DES LEÇONS SUR LE CALCUL INFINITÉSIMAL.**

sement infiniment petit de $x$ correspondra un accroissement infiniment petit de $\mathcal{F}(x)$. Donc, si la fonction $f(x)$ reste finie et continue depuis $x = x_0$ jusqu'à $x = X$, il en sera de même de la fonction $\mathcal{F}(x)$. Ajoutons que, si l'on divise par $\alpha$ les deux membres de la formule (3), on en conclura, en passant aux limites,

$$(4) \qquad \qquad \mathcal{F}'(x) = f(x).$$

Donc l'intégrale (1), considérée comme fonction de $x$, a pour dérivée la fonction $f(x)$ renfermée sous le signe $\int$ dans cette intégrale. On prouverait de la même manière que l'intégrale

$$\int_x^X f(x)\,dx = -\int_X^x f(x)\,dx,$$

considérée comme fonction de $x$, a pour dérivée $-f(x)$. On aura donc

$$(5) \qquad \frac{d}{dx}\int_{x_0}^x f(x)\,dx = f(x) \qquad \text{et} \qquad \frac{d}{dx}\int_x^X f(x)\,dx = -f(x).$$

Si aux diverses formules qui précèdent on réunit l'équation (6) de la septième Leçon, il deviendra facile de résoudre les questions suivantes.

PROBLÈME I. — *On demande une fonction $\varpi(x)$ dont la dérivée $\varpi'(x)$ soit constamment nulle. En d'autres termes, on propose de résoudre l'équation*

$$(6) \qquad \qquad \varpi'(x) = 0.$$

*Solution.* — Si l'on veut que la fonction $\varpi(x)$ reste finie et continue depuis $x = -\infty$ jusqu'à $x = +\infty$, alors, en désignant par $x_0$ une valeur particulière de la variable $x$, on tirera de la formule (6) (septième Leçon)

$$\varpi(x) - \varpi(x_0) = (x - x_0)\,\varpi'[\,x_0 + \theta(x - x_0)\,] = 0$$

et, par suite,

$$(7) \qquad \qquad \varpi\, x = \varpi(x_0).$$

**Figure 6**: Cauchy in Oeuvres (2), IV, p. 152

## *Emergence of a shared name for different versions of the FTC*

In the early textbooks on calculus the propositions similar to the theorem presently called the fundamental theorem are not named. The origin may well be from the French tradition, from the French word *fondamentale* for something basic.

The volumes of the series *Course d'analyse mathématiques* (volume 1 printed in 1902) by the French mathematician Eduard Goursat started to be translated into English already in 1904 (*A course in mathematical analysis*). It was widely spread and was based on his university lectures (Osgood, 1903) and consequently can be considered a textbook. In volume II of this book the phrase **the fundamental formula of the integral calculus** (p. 63) refers to volume I, where the expression **fundamental theorem** is used for the fact that

every continuous function *f(x)* is the derivative of some other function" (p. 140) and "the fundamental formula becomes $\int_a^b f(x)dx = F(b) - F(a)$" (p. 155). Later, in volume II, this name of the theorem is also used for complex analysis: "the fundamental formula of the integral calculus can be extended to the case of complex variables: $\int_{z_1}^{z_2} f(z)dz = F(z_2) - F(z_1)$ (p. 72).

In the same vein, Charles Jean de la Vallée Poussin in *Cours d'analyse infinitésimale* from 1921, uses the name **relation fondamentale pour le calcul des integrals définies** (p. 211) and writes

$\int_a^b f(x)dx = F(b) - F(a)$ C'est la formule fondamentale pour le calcul des integrals définies.

Possibly, also Ernest Hobson was influenced by the well known book by Goursat in *The theory of functions of a real variable & the theory of Fourier series*, published in 1907. In this book a whole chapter has the title **The fundamental theorem of the integral calculus**. Later in the same book there is one chapter with the title **The fundamental theorem of the integral calculus for the Lebesgue integral** and another chapter called **The fundamental theorem of the integral calculus for the Denjoy integral**.

In the textbook *An introduction to the summation of differences of a function* by Benjamin Feland Groat, printed in 1902, the expression **the fundamental theorem of the integral calculus** is used, as well as more shortly **fundamental theorem**:

To find the limits of sums of the form $\sum \phi(x)\Delta x$, it was necessary to have an identity of the form:

$\phi(x)\Delta x = \psi(x) - \psi(x + \Delta x) + F(x,\Delta x)\Delta x^2$. The fundamental theorem of the integral calculus puts into mathematical language a rule for finding the limit of any sum, of the kind considered, provided an identity of the right form can be found; and the rules an formulae of the integral calculus afford a method for the discovery of the essential form of the identity when it exists.

33. Fundamental theorem.     $\int_a^b f'(x)dx = f(b) - f(a)$. Or, more explicitly,

$\underset{\Delta x=0}{\text{Lt.}} \sum_a^b \psi'(x)\Delta x = \psi(b) - \psi(a)$, where $\psi'(x)$ is any function of $x$ and $\psi(x)$ any function whose differential coefficient with regard to $x$ is $\psi'(x)$ (pp. 40-41).

In addition to this quote, the fundamental theorem is also expressed without employing mathematical formulae. This indicates the effort of the author to address an audience that is not fluent to read specialized technical language, that is, an attempt of a didactic transposition. In his textbook, *A course of pure mathematics* from 1908, In a paragraph called *areas of plane curves* (derivatives and integral) the FTC is proved but not named. Then, in a paragraph on *Definite Integrals. Areas of curves* (in the chapter *Additional theorems in the differential and integral calculus*) it is referred to as **the fundamental theorem of the integral calculus** with reference to the proof in the passage where it is not named (p. 293):

(10) The Fundamental Theorem of the Integral Calculus.

The function $F(x) = \int_a^x f(t)dt$ has a derivative equal to *f(x)*.

This has been proved already in § 145, but it is convenient to restate the result here as a formal theorem. It follows as a corollary, as was pointed out in § 157, that *F(x)* is a continuous function of *x*.

In this case, the name seems to be linked to the statement rather than to the statement along with its proof. The listing of "additional" theorems without proofs (but with names), of which the proofs are given earlier in the book, points to a distinct didactical rationale for this way of structuring the topic.

A reference to the name that seemed to be common as well as a formulation of the FTC is found in *Differential and Integral Calculus* (1931) by Richard Courant (pp.113-114):

> The question about the group of all primitive functions is answered by the following theorem, sometimes referred to as the fundamental theorem of the differential and integral calculus:
>
> The difference of two primitives $F_1(x)$ and $F_2(x)$ of the same function $f(x)$ is always a constant:
>
> $F_1(x) - F_2(x) = c.$
>
> Thus, from any one primitive function $F(x)$ we can obtain all the others in the form
>
> $F(x) + c.$
>
> by suitable choice of the constant $c$. Conversely, from every value of the constant $c$ the expression $F_1(x) = F(x) + c$ represents a primitive function of $f(x)$.

From this theorem Courant derives the formula $\int_a^b f(u)du = F(b) - F(a)$, also stated as the **important rule**:

> If $F(x)$ is any primitive of the function $f(x)$ whatsoever, the definite integral of $f(x)$ between the limits $a$ and $b$ is equal to the difference $F(b) - F(a)$. (p. 117)

Paley and Wiener (1934) refer several times to **the fundamental theorem of the calculus** in a contribution to *Fourier Transforms in the Complex Domain*. That this name has become institutionalized is evident from the classical book *What is mathematics?* where Courant and Robbins (1941) use the chapter title "The fundamental theorem of the calculus", and write:

> There is no separate differential calculus and integral calculus, but only one calculus. It was the great achievement of Leibniz and Newton to have first clearly recognized and exploited this *fundamental theorem of the calculus* (p. 436, our emphasis).

The reason for skipping the modification "the integral" is evident from the quotation above. This stresses the role of the FTC and its proof as a means of integrating two related but still distinct sub-areas of *the calculus*. In later textbooks, the definite article has been left out, which can be taken as a reference to a general field and not to a particular method.

In commercial textbook production, different traditions of formulating and proving the FTC seem to have developed for different markets. In the U.S. a division of the theorem into two parts is common. An early example is to be found in Morrey (1962), where the **first form** of the FTC refers to the formula for computing the definite integral by the primitive function and the **second form** to the derivation of the integral. In modern U.S. calculus textbooks it seems to be standard to refer to this second form as **the first part of the FTC** and the computational formula as **the second part of the FTC** (e.g. Adams, 2006). In other countries this division is less common.

In many places it is also common that researching mathematicians who teach undergraduate calculus courses produce their own "textbooks" in the form of lecture notes or local publications and do not choose to use commercial and common texts. Some examples of these texts (selected as a convenience sample), show that these are less standardised in their approach and display unusual versions of the FTC.

The early Swedish textbook by Björling (1877) has an outline of the definite integral similar to that of Cauchy. Björling gives the name **Grundsats** [*ground theorem*] to the computational formula $\int_a^b f(x)dx = F(b) - F(a)$, i.e. the 'second part' of the FTC. In another early Swedish textbook, which is one example of how notes from a lecture series have been compiled to a printed book, the FTC is not named though described by the words (Malmquist, 1923, p. 266):

> Differentiation och integration äro fullständigt motsatta processer, vadan följande likheter gälla:
>
> $d\int f(x)dx = f(x)dx$ , $\int df(x) = f(x) + C$

[Differentiation and integration are completely opposite processes, from which the following equalities hold:]

The same author is later involved in a calculus textbook (Malmquist, Stenström, & Danielsson, 1951), where a theorem named **Integralkalkylens fundamentalsats** [*fundamental theorem of the integral calculus*] refers to the following proposition:

> If the derivative of a function equals zero on an interval, then the function is constant on that interval.

The formula $\int_a^b f(x)dx = F(b) - F(a)$, usually included in the FTC, is contained in the same section of the book, but not named. In more modern Swedish textbooks the name **integralkalkylens huvudsats** [*main theorem of the integral calculus*] is commonly used for the computational formula $\int_a^b f(x)dx = F(b) - F(a)$, with the assumption that $f$ is continuous (e.g. Hyltén-Cavallius & Sandgren, 1968; Domar, Haliste, Wallin, & Wik, 1969; Ullemar, 1972; Hellström, Molander, & Tengstrand, 1991). It can be noted that in an earlier edition of Hyltén-Cavallius and Sandgren (1968), i.e. Hyltén-Cavallius and Sandgren (1956), the name "integralkalkylens huvudsats" did not appear.

When the name **analysens huvudsats** is used in Persson and Böiers (1990) and in Forsling and Neymark (2004), it refers only to the 'first part' of the FTC, i.e. the proposition that $\frac{d}{dx}\int_a^x f(t)dt = f(x)$ for any continuous function $f$, though the name indicates the key role of the theorem for the coherence of the calculus (in Swedish 'analys') more than does the name "integralkalkylens huvudsats". However, in this textbook the second part is also named, using the term **insättningsformeln** [*the insert formula*].

A completely different approach is used in Eriksson, Larsson and Wahde (1975), where the definite integral from $a$ to $b$ of a continuous function $f$ is *defined* as $F(b) - F(a)$, where $F$ is any primitive function to $f$. It is then proved, from a previous definition of area measure, that this formula computes the area under the curve $f$. No name is given to this proposition.

In a common German textbook (von Mangoldt & Knopp, 1932; 13th edition from 1967) only the computational part is called the **Hauptsatz der Differential- und Integralrechnung** [*main theorem of the differential and integral calculus*], whereas in a textbook from the German Democratic Republic (Belkner & Bremer, 1984) this computational part is called the **Umkehrung des Hauptsatz der Differential- und Integralrechnung** [*converse of the main theorem of the differential and integral calculus*].

In a Polish textbook from 1929, *The differential and integral calculus*, written by Stefan Banach there is a chapter (§9) about definite integrals and primitive functions, in which a formulation of what can be recognised as the FTC is presented without naming it; it is divided into three parts. It is also interesting to look on the proof Banach proposes for these theorems. The first part is formulated as the following theorem:

> If a function *f(x)* is integrable on *(a,b)*, $a \le \alpha \le b$ and $a \le x \le b$, then the derivative of the function $\int_\alpha^x f(t)dt$ exists and it is equal to a function under the integral in every point at which the function $f$ is continuous.

> *Proof.* Let *f(x)* be a continuous function at $x_0$ in an interval *(a,b)*. Taking an arbitrary $\varepsilon > 0$ we can find $\eta > 0$ such that every point $x$ in *(a,b)* is satisfying the inequality

$$|x - x_0| \le \eta; \qquad (1)$$

> it follows that

$$|f(x) - f(x_0)| \le \varepsilon,$$

> that is

$$f(x_0) - \varepsilon \le f(x) \le f(x_0) + \varepsilon. \qquad (2)$$

> Putting $F(x) = \int_\alpha^x f(t)dt$ we see that

$$\frac{F(x)-F(x_0)}{x-x_0} = \frac{1}{x-x_0}\int_{x_0}^{x} f(t)dt,$$

thus if $x$ satisfies inequality (1), then by inequality (2) and the theorem from §6 we get:

$$f(x_0) - \varepsilon \le \frac{F(x)-F(x_0)}{x-x_0} \le f(x_0)+\varepsilon$$

Since $\varepsilon$ was arbitrary, it follows that

$$\lim_{x\to\infty} \frac{F(x)-F(x_0)}{x-x_0} = f(x_0),$$

i.e. $F'(x_0)$ exists $F'(x_0) = f(x_0)$.

From this theorem there follows directly another one:

A continuous function $f(x)$ on an interval $(a,b)$ has primitive function on this interval. The primitive function is $F(x)\int_a^x f(t)dt+c$.

And the third theorem is formulated and proved as follows:

If $F(x)$ is a primitive function of $f(x)$, which is a continuous function on the interval $(a,b)$, then $\int_\alpha^x f(t)dt = F(x)-F(\alpha)$, $a \le \alpha \le b$, $a \le x \le b$.

*Proof.* Because both $F(x)$ and $\int_a^x f(t)dt$ are primitive functions of $f(x)$ they differ only by a constant. Hence

$$\int_a^x f(t)dt = F(x)+c.$$

Taking $x = \alpha$ we obtain $0 = F(\alpha)+c$ There it follows that $c = -F(\alpha)$ Hence

$$\int_\alpha^x f(t)dt = F(x)-F(\alpha).$$

Remark: If we put $x=b$ and $\alpha = a$ in the last equality, then we obtain

$$\int_a^b f(t)dt = F(b)-F(a).$$

The last formula allows to compute a definite integral if we know a primitive function, that is, the indefinite integral.

## The FTC in the eyes of mathematicians: a case study

Eleven mathematicians were interviewed. They worked in different areas of mathematics, that is in Computational Group Theory, Dynamical Systems, Ergodic Theory, Functional Analysis, Harmonic Analysis, Homological Algebra and Category Theory, Logic, Mathematical Physics, Number Theory, Partial Differential Equations, Philosophy of Mathematics, and Variational Analysis. Some of the interviewees also were involved in teaching calculus courses. Thir educational background varied in terms of the countries in which they had studied at undergraduate or graduate level.

In the following, we only report those parts of the outcomes of the interview study that are relevant for our question posed in the title of the chapter. The interviews were quite extensive. Altogether, there were ten questions discussed. The following account comprises only three of these questions.

---

*Question 1*: What is the "Fundamental Theorem of Calculus", as you understand it? Do you know several versions of it? In what version have you first heard about it as a student?

---

In responding to this question, a couple of the interviewees did not explicitly separate their answers to the first and the last sub-question, that is, the meaning of the FTC for them was related to how they encountered it as a student. Six of the mathematicians used the word 'inverse' in some form, for example in saying that differentiation and integration are 'inverse processes', 'inverse operations', or simply 'inverses'. It is in most of the statements unclear to what type of inversion they exactly referred. The named inverse relationship could be interpreted in a trivial sense, seeing the integral only as the antiderivative. In only one interview it was pointed out that, in order to make the FTC interesting, the integral should be defined as a limit of a sum and not simply as an antiderivative. Two persons mentioned the aspect of finding the area. For one, the FTC consists in providing a way for finding the area under a curve, while another more generally talks about the FTC as a connection between area and function.

Many of the interviewees used rather informal formulations of the theorem, including hedges, and did not specify assumptions. Their statements include phrases such as "*sort of inverse processes*", "*you have like inverse functions basically*", "*So in that sense you have the two inverse operations in their two orders*", "*a connection between differentiation and integration so that they are roughly speaking inverses to each other*". However, those mathematicians who had reported to have experience in teaching calculus courses, generally provided more formal formulations of the FTC, often like it is stated in common textbooks, as composed of two parts and valid for continuous functions.

Three persons stated that there exist other versions of the FTC, for multivariable functions or with weaker assumptions. One person (with a background in philosophy of mathematics) approached the issue from an epistemological perspective. He connected the FTC with the value of an ambiguity in having two different ways to look at the same phenomenon, also with reference to history. For him the FTC represents "One of those deep theorems that you keep learning more about as you have more familiarity with mathematics".

In the next question, the interviewees were presented with diverse formulations of propositions related to the FTC, including historical ones, and asked for their preferences. The stateent (1) this can be characterised as folklore, (2) is a literal rendering of the first one by using symbols for integration and derivation, (3) represents what in textbooks often is named "the first part" of the FTC, without any assumptions (4), (5) and (6) represent historical formulations from Leibniz and Newton, (7) and (8) are common textbook formulations of the "two parts", (9) represents a different approach with a "right" and "left" primitive function (Kuratowski, 1977), (10) is a multidimensional formulation in form of the divergence theorem and (11) is from the same source (Thomas and Finney, 1996), but at the end of the chapter as an attempt to present an informal formulation of the same theorem.

---

*Question 2:* Which, if any, of the following statements would you consider as the closest to the FTC as you understand it?

**1.** Differentiation and integration are inverse operations.

**2.** $d \int y dx = y$.

**3.** If $F(x) = \int_a^x f(t)dt$, then $F'(x) = f(x)$.

**4.** The rate of change of the area under a curve, as a function of the abscissa of the curve, is the same as the function which describes the curve. So, to calculate the area under a given curve, it is enough to find the antiderivative of the function which describes the curve.

**5.** To find the antiderivative of a function, it is enough to find the area under the curve as a function of the abscissa.

**6.** Area under a curve is to the abscissa as the ordinate of the curve is to the abscissa.

**7.** If $f$ is continuous on [a,b] and if $F$ is an antiderivative of $f$ on [a,b] then
$\int_a^b f(x)dx = F(b) - F(a)$.

---

**8.** Let $f$ be a continuous function on an interval I, and let $a$ be a point in I. If $F$ is defined by $F(x) = \int_a^x f(t)dt$, then $F'(x) = f(x)$ at each point x in the interval I .

**9.** Let $f$ be a continuous function on a closed interval $a \le x \le b$. Then by "the definite integral of $f(x)$ from $a$ to $b$" denoted $\int_a^b f(x)dx$, we understand the number $F(b) - F(a)$, where $F$ is any primitive function of $f$ in the interval $(a, b)$, and $F'_+(a) = f(a)$ and $F'_-(b) = f(b)$.

**10.** The flux of a vector field $\mathbf{F} = M\mathbf{i} + N\mathbf{j} + P\mathbf{k}$ across a closed oriented surface S in the direction of the surface's outward unit normal field $\mathbf{n}$ equals the integral of $\nabla \cdot F$ over the region D enclosed by the surface
$$\iint_S F \cdot n\, do = \iiint_D \nabla \cdot F\, dV .$$

**11.** The integral of a differential operator acting on a field over a region equals the sum of the field components appropriate to the operator over the boundary of the region.

---

In response to this prompt, most of the interviewed mathematicians commented on all or several of these formulations. By most of the mathematicians, all formulations were accepted, apart from in some cases (1), (9), (10), (11), and (5), and one person did not accept (2) and (6). Two of the mathematicians explicitly accepted all formulations.

The most preferred versions were (3), (7) and (8). The reasons given for these choices were more or less elaborated, due to the open format of the question. Mathematicians with experience in teaching calculus usually choose (7) and (8), or sometimes (3) and they stated that the FTC is usually formulated in two parts. However, three talked only about the first part and one person only about the second part.

One mathematician who selected (3) as the closest to his interpretation of the FTC, gave the following comments:

> I think of it as basically number three…with a picture so to speak [ ] where I look at a graph and f of x is the area so far under the graph and we look at how quickly that area is changing. I like to think about things geometrically. So I mean that has a nice geometric feeling for me.

The phrase "a nice geometric feeling for me" indicates a very personalised view.

One interviewee chose without any hesitation (7) and (8), on the grounds that he used them in his teaching. Another person, who selected these two, also refers to students in his choice:

> Number seven is very often the way you see it but really it's as I said before, it's part of it only…This is the way the student would study first.

Another interviewee who chose (7) mentioned his own time as a student:

> When I first studied calculus the way in which I was taught and the way in which I understood it was [number] seven.

Similar choices, that is (3), (8) and (7), (9), respectively, are made by another person, who however is more precise in her comments:

> So there are two parts. One part is three and eight, and the other part is seven and nine, and again the continuity [assumptions in seven, eight and nine] is not necessary if you use Riemann integrable. The others of course are just applications in various contexts.

When (3), (7) and (8) were chosen, then mostly also (1) was accepted, but two persons explicitly opposed to (1) and they said:

> To say that they [integration and derivation] are inverse operations, I think, misses the point.

> The first one not, because if you take not continuous functions, this is completely not good.

Still another interviewee accepted the correctness of the statement (1), though not as a proposition but as a definition.

The reasons for not accepting the formulations (9), (10) and (11) varied. One person does not like (9) because it is a bit "*pedantic*", another says that "*number nine misses the whole point as far as I am concerned*" while one simply states that "*nine is one* that I do not like". Some comments on (10) and (11) are rather vague, such as "*This is an abstract version or something*", or "*a little bit too much generalized*".

The formulations stemming from Newton (4) and Leibniz (5 and 6) were not accepted as formal versions of the FTC. They were not classified as wrong, but accepted as "*useful in terms of explaining or motivating*", "*how to show where it comes from*". One person uttered some doubts about (5):

> Finding the area is more or less the other way around. First you find the antiderivative and then you use it to find the area rather than the other way.

However, one of the mathematicians ranked (4) and (5) as outdated and commented, "*This is some old language. I don't like it*" and one criticized (5), "*because you have the sign of the function, positive and negative. When the function is negative you have problems.*"

Still, in another prompt, the interviewees were asked what they think about two proofs of the FTC. The reconstruction of Leibniz's formulation and proof of FTC (question 6) was taken from Laubenbacher and Pengelley (1999, pp. 133-134), but we added additional diagrammes. The other text (question 7) shows Newton's reasoning as described by Fauvel and Gray (1987, p. 384). However, the reference to Leibniz was not made explicit.

---

*Question 6*: What do you think about the following reasoning?

We show that the general problem of finding areas under curves can be reduced to the finding of a line that has a given law of tangency.

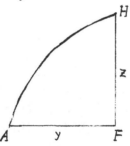

Suppose we want to find the area under the curve AH, described by the ordinate FH = z. Let AF = y be the abscissa of the curve.

Suppose the line g is such that the tangent TC at point C of g satisfies

(1) $\dfrac{TB}{BC} = \dfrac{FH}{a}$ , where a is a constant.

This means that the law of tangency of g is the same as the ordinate z, which describe the curve AH.

Denote FC = x.

---

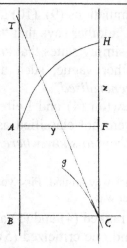

We will show that the sum of all lines z (i.e. the area under the curve AH) is equal to ax, and hence it is determined by the curve g.

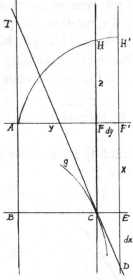

We do it by way of "motion": we extend the curve AH to H'.
By similarity of triangles TBC and CED, ED:EC = TB:BC.
By (1), TB:BC = FH:a.
And EC = FF'.
Now, FF' = dy and ED = dx are increments of y and x, resp.
Then a dx = z dy.
Therefore the sum of all rectangles a by dx is equal to the sum of all rectangles x by dy.
$$\int adx = \int zdy$$
Since the sum of all rectangles a by dx is equal to ax, we get
$$\int zdy = ax$$
i.e. once we know a curve whose law of tangency is equal to z, we know the area under the curve with ordinate z.

Question 7: What do you think about the following reasoning, attributed to Newton? How does it differ from the present day reasoning?

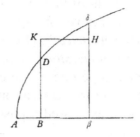

Let any curve $Ad\delta$ have base $AB = x$, perpendicular ordinate $BD\ 0\ y$ and area $ABD = z$. Take $B\beta = o$, $BK = \delta$ and rectangle $B\beta HK(o\delta)$ equal to the space $B\beta\delta D$.

It is, therefore, $A\beta = x +o$ and $A\delta\beta = z +o\delta$. With these premises, from any arbitrarily assumed relationship between x and z I seek y in the way you see following.

Take at will $\frac{2}{3}x^{\frac{3}{2}} = z$ or $\frac{4}{9}x^3 = z^2$. Then, when $x +o(A\beta)$ is substituted for x and $z +o\delta(A\delta\beta)$ for z, there arises (by the nature of the curve)

$$\frac{4}{9}\left(x^3 + 3x^2o + 3xo^2 + o^3\right) = z^2 + 2zo\delta + o^2\delta^2.$$

On taking away equal quantities ($\frac{4}{9}x^3$ and $z^2$) and dividing the rest by $o$, there remains

$\frac{4}{9}\left(3x^2 + 3xo + o^2\right) = 2z\delta + o\delta^2$. If we now suppose $B\beta$ to be infinitely small, that is, $o$ to be zero, $\delta$ and $y$ will be equal and terms multiplied by $o$ will vanish and there will consequently

remain $\frac{4}{9} \times 3x^2 = 2z\delta$ or $\frac{2}{3}x^2(= zy) = \frac{2}{3}x^{\frac{3}{2}}y$, that is, $x^{\frac{1}{2}}\left(= x^2/x^{\frac{3}{2}}\right) = y$. Conversely therefore

if $x^{\frac{1}{2}} = y$, then will $\frac{2}{3}x^{\frac{3}{2}} = z$.

Or in general if $\left[n/(m+n)\right]ax^{(m+n)/n} = z$, that is, by setting $na/(m+n) = c$ and $m+n = p$, if $cx^{p/n} = z$ or $c^n x^p = z^n$, then when $x +o$ is substituted for $x$ and $z +o\delta$ (or, what is its equivalent, $z +oy$) for $z$ there arises $c^n\left(x^p + pox^{p-1}...\right) = z^n + noyz^{n-1}...$, omitting the other terms, to be precise, which would ultimately vanish.

Now, on taking away the equal terms $c^n x^p$ and $z^n$ and dividing the rest by $o$, there remains $c^n px^{p-1} = nyz^{n-1}\left(= nyz^n/z\right) = nyc^n x^p/cx^{p/n}$. That is, on dividing by $c^n x^p$, there will be $px^{-1} = ny/cx^{p/n}$ or $pcx^{(p-n)/n} = y$; in other words, by restoring $na/(m+n)$ for $o$ and for $p$, that is, $m$ for $p= m$ and $na$ for $pc$, there will come $ax^{m/n} = y$. Conversely therefore if $ax^{m/n} = y$, then will $\left[n/(m+n)\right]ax^{(m+n)/n} = z$. As was to be proved.

Here in passing may be noted a method by which as many curves as you please whose areas are known may be found: namely, by assuming any equation at will for the relationship between the area z and from it in consequence seeking the ordinate y. So if you should suppose $\sqrt{\left[a^2 + x^2\right]} = z$, by computation you will find $x/\sqrt{\left[a^2 + x^2\right]} = y$. And similarly in other cases.

In their comments, the interviewees often compared the two versions. Leibniz's proof evoked both confusion and appreciation. Some knew that the proof was from Leibniz. It seems that the proof is far remote from what counts as mathematical representation and reasoning today: *"I am not familiar with it"*, or, after some time spent reading it, *"Well, very strange"*. One person displayed a quite strong reaction: *"Wow, this is crazy. This one is nothing for me. I don't understand."* The most common point of view was that the reasoning is very geometric and that it was difficult to understand it: "*I never thought of it geometrically, in this way*", and "*This is complicated, but it's kind of geometrical*". One mathematician did not want to read the proof with the comment that "*I never understood Leibniz before*". Clearly, the geometrical representation used by Leibniz seemed to be one of the causes for difficulties to understand the argument. Newton's is "*probably easier to read*", as one person expressed it.

However, some of the interviewees also expressed a positive view on the kind of reasoning used in the proof:

> It would take me a little bit longer to absorb it but it looks like a good argument. [ ] It's a nice geometric way to prove it.

> First of all the proof is clever.

One person explained that usually opinions about mathematical work of others are based upon one's own experience and because the proof of Leibniz is very far from what they have seen before it is difficult to accept it.

Comparing the texts by Leibniz and Newton initiated comments on what constitutes a proof. Everybody agreed that both texts are quite removed from today's reasoning but that the by Newton is much easier to understand. One person classified this proof as belonging to physics: "*That's how it's done in physics books all the time*". Another person found Newton's way of working more modern than most of the others: "*It's similar to the modern day proof*". That a geometric way of proving is seen to be more remote, is expressed in the following statement about Leibniz proof: "*That is a very unusual technique which never went anywhere*".

Even though most of the interviewees prefer Newton's way of reasoning, most of them find that it does not count as a proof according to today's standards. One person finds that the "*intuition is better in Newton's*". Another appreciated the strategy of taking one example at the beginning and then generalizing it for arbitrary cases. For a few, it would be easier to accept Leibniz's version as a proof.

One interviewee expanded on the differences between what people think and what they write down, and that what constitutes a proof is not always clear. One has to choose what to write down in a proof and what to leave implicit. He states that the difference between a mathematician and a student in this respect is that a mathematician can fill in all the details but a student usually cannot.

## Discussion

### *Institutionalization and recontextualisation*

Our case study of the emergence of the delineated sub-area that came to be called 'calculus', together with its "fundamental" theorem, shows that the institutionalization of this mathematical sub-area along with its concomitant knowledge claims evolves in a dynamic relationship between the field of production and pedagogic recontextualisation of knowledge. In the early history of the calculus an independent recontextualising field did not exist. The first printed textbook on differential calculus appeared in Paris in 1696, by de l'Hospital with the help of Johann Bernoulli. From the introduction of the 1716 edition it becomes clear that the name Integral Calculus (Calcul integral) was already in use but what was signified by this name changed considerably.

The acceptance of a shared name can be taken as an expression of attributing value and affiliation with a group. This is, for example, witnessed by the discussion between Bernoulli and Leibniz about the preference of "calculus integralis" over "calculus summatorius". The development from rather long names denoting two sub-areas (calculus differentialis et calculus integralis) into one (the calculus, or more general without the definite article) mirrors the development of the area. Our study of textbooks shows that the FTC long after its invention, in its basic form at the elementary undergraduate level has been and is still being given different or no names (also within the same country), different formulations, and different proofs. The overall development reflects a move from a name referring to integration only, to a name emphasising the role of the FTC as a fundamental link between differentiation and integration, reflecting its role in defining calculus as a distinct sub-area of mathematics.

The *École Polytechnique* in Paris was established to increase the number of engineers needed to maintain the new French Republic and the school was kept after the counterrevolution to serve the military. Mathematical knowledge was considered important for these purposes (see Fourcy, 1828). Cauchy's *Cours d'analyse* from 1821 and *Résumé* from 1823, written for the *École Polytechnique*, were to meet the demands arising from this expansion of higher education. The intention of re-organizing and re-describing a set of related outcomes of research in different sub-areas for the purpose of presenting it in a coherent way contributed to the institutionalization of the calculus. The borderline between what constitutes a "textbook" and a publication addressed to an audience of other researchers with less specialized knowledge is not easy to draw. Reference to Cauchy's scholarly work is commonly made by drawing on his textbooks. If one interprets a textbook as an attempt to provide access to a knowledge code promoted in the field of mathematical knowledge production, then there was no different code for the initiation into the fields in which the knowledge was supposed to be applied. This intention is explicitly stated by himself in the introduction. He intends to present a text that is useful for both "Professeurs et aux Élèves des Colléges royaux" (1921, p. ii). But Cauchy's approach at the *École Polytechnique* was criticized as too theoretical and in an evaluation report the replacement of the limit concept with infinitely large and infinitely small quantities was suggested. These were thought to be more suitable to the practical needs of engineers (Barany, 2011).

The study of the FTC reveals that institutionalization rather emerges out of the field of knowledge recontextualisation than out of research. The meaning and the relevance of the FTC and of its proof is dependent on the mathematical area in which it is embedded. Only with the commercial mass production of textbooks on calculus a standardized version of the FTC emerged, while in more personalised lecture notes and early textbooks an almost bewildering variety of versions appeared. Further, the interviews show that only when referring to teaching, there was agreement about what constitutes the FTC, really. The personal views of the mathematicians about the essence of the theorem and about standards for its proof, differed considerably.

Distinctiveness of a field can be achieved through use of a particular notational or representational system, shared basic notions, specific methods and objects of study, a methodology for relating the propositional statements to each other, and by referring to common intellectual roots. The history of the calculus and of the FTC shows that there are indeed developments of all of these parts, in parallel and sometimes overlapping with developments in other areas, until the sub-area of calculus became established. That the institutionalization of an area is also linked to the intellectual roots can be seen in much of the popular history writings about calculus with Newton and Leibniz as the central figures. Modern expansions of integral calculus are even linked to the name of the creators, that is, to Riemann and Lebesgue. The institutionalization of the calculus is also linked to its achieved standards of "rigour" associated with (i) well defined basic concepts, (ii) a functional specialised language and notational system, and (iii) deductive forms of argument. A sign of a lower degree of rigour would, for example, be the use of natural language for basic concepts (such as Newton writing about "the area of"), wide use of inductive argument and reference to geometrical intuition, for example through referring to popular forms of diagrams. It is usually Cauchy who is identified as a central actor in this respect, often enthusiastically, as for example by Grabiner (1981), who states, "Augustin-Louis Cauchy neither began nor completed the rigorization of analysis. But more than any other mathematician, he was responsible for the first great revolution in mathematical rigor since the time of the ancient Greeks." (p. 166). By using limits as the basis for definitions, Cauchy's work established indeed new knowledge criteria by integrating definitions and proofs with applicable methods. In the 19th century the formal ε-δ definition of limit by Weierstrass, the definition of the Riemann-integral and a set theoretic definition of function were added in accordance with the development of the criteria for legitimate knowledge towards more "rigour" in relation to its foundations.

The historical study also points to the emergence of an independent recontextualization field for higher education out of the field of production. There is a profound difference in historical and in the present context of higher education. While in the historical context the criteria for the two discourses (the unmediated and the mediated) match or are developed in dynamic relationship between producers and transmitters, in the present context the mathematicians suggest a change of criteria as soon as pedagogic discourse is at issue. Some parts of the interview study with researching mathematicians that we have reported elsewhere even show that the mathematicians suggest different recontextualization principles for different groups of students and thus construct future insiders (students of pure mathematics) and outsiders (engineering students) (Bergsten, Jablonka & Klisinska, 2010).

In a review of Chevallard's (1985) exposition of his theory, Freudenthal (1986) questions the very idea of a didactic transposition, among other things because of the fuzziness of the notion of *savoir* (knowledge), and in particular of the *savoir savant* (translated to scholarly knowledge by Chevallard). Referring to the example given by Chevallard in the text (1985, further elaborated with M-A. Johsua in the 1991 edition), Freudenthal asks whether scholarly knowledge refers to the good mathematics that some mathematicians with a great reputation – veritable scholars – enunciate to a bewildered public and that now has to be transposed to the level of understanding of the youth (Freudenthal, 1986, p. 325). He asks where the scholarly mathematical knowledge starts and at which point one cannot be sure anymore to have a transposition. As our study shows, this question clearly cannot be answered only with reference to the knowledge structure, but only in relation to the social base of its production and distribution.

## A short remark on historical studies and "epistemological obstacles" in mathematics education

Historical accounts of mathematical concepts, theories and methods within the field of knowledge production and communication have informed knowledge recontextualisation by professional mathematics educators in many ways, for example, in the search for some "essence" of a mathematical concept in order to design situations, where the learners can grasp this essence when working with a (a set of) problem(s) (Wagenschein, 1970; Freudenthal, 1973; Brousseau, 1997). That a reconstruction of the historical development is seen informative for decisions on the selection, sequencing and criteria for topics and problems, relies on some version of an assumed parallelism between historical and individual knowledge development. In its most general form, this assumption has become known as the "genetic principle", which emerged in response to a critique of an axiomatic-deductive style of teaching mathematics (Mosvold, 2003). In essence, the principle demands that the teaching of a topic should mirror the principles that have determined its historical development. The question remains, through what type of historical analysis such principles could be revealed.

One question in relation to a historic-genetic principle for learning, is how to interpret those instances, which from a modern point of view, can be seen as pitfalls and problems in the historical development that operated as hindrances for progression towards more "advanced" and "better" knowledge. This question draws attention to the notion of *epistemological obstacle*, which is attributed to Gaston Bachelard (1938), who provided a critique of empiricism through pointing out that scientific knowledge develops through solving theoretical problems, and that the history of science is marked by epistemological ruptures that evolve from overcoming epistemological obstacles. In his view, advancement is not an achievement of individuals, but solutions are worked out within given epistemological constraints of the collective unconscious. Further, the evolution of scientific knowledge is not seen as a continuous process but involves rejection of previous forms of knowledge that become obsolete. In drawing on Bachelard, the notion of epistemological obstacle was introduced as a concept in mathematics education by Brousseau (1983; 1997, pp. 85-114). The learners' misunderstandings are seen as functional, as they result from constraints in the sphere of mathematical knowledge within which they are able to conceive an optimal solution to a given problem.

It was not the purpose of our historical study to identify epistemological obstacles or qualitative "jumps" in the history of calculus in order to develop rationales for teaching the FTC, because we do not have access to a theoretical foundation for such claims. However, out of some naïve reading of the history, one can indeed phrase it as a history of overcoming obstacles, if not epistemological ones. This interpretation is based on a rough distinction between geometrical and arithmetic-algebraic procedures.

The physical or geometrical interpretation of the integral as the area of a region (in the form of a concrete numerical value suggesting a static image) and the derivative as a slope (as changes of magnitude suggesting a dynamic view) refer to different phenomena and thus there is no relation between the concomitant mathematical operations. One could interpret the historical development such as that seeing the integral as measuring an area provided an obstacle, a "geometrical obstacle". As a means of overcoming this obstacle the FTC can be seen as meaningful and as a significant result in the development of mathematics.

On the other hand, if the integral were from the outset conceptualized as antiderivative, then any version of the FTC would appear as trivial, but the "geometrical obstacle" could be avoided. For the mathematicians who attempted to reorganize and systematize analysis in the 18$^{th}$ century and were concerned with a more rigorous foundation of calculus, an algebraic formal approach without reference to geometric interpretations, but with seeing differentiation and integration as "operations", was indeed suitable. Such a change from a geometric to an algebraic approach is reflected in the works Johann Bernoulli and Euler, when the methods of calculus were already proven to work for classes of different problems and it was seen as important to deal with the formal, logical grounds for the topic. While the second approach leads to specialization through formalization, the first allows further generalizations through expanding the possible range of applications.

As to the teaching of calculus, a start with antidifferentiation (antiderivative; primitive functions) before defining the definite integral as a limit of a Riemann sum, would limit the justification function of the proof of the first part of the FTC. This effect is also supported by a "notational obstacle", when the standard integral symbol is used to denote primitive functions (without any reference to the integral as a limit of a Riemann sum). The propositional content of the FTC is inherited in the symbol used (now for the definite integral). However, the choice of starting with a problem to develop a technique for calculating for example an area, or computing distance by way of a non constant velocity over a specific time, may evoke a need for justification of how the sum (leading to an integral) can be evaluated by primitive functions, if the notion of primitive function and techniques for finding them has not yet been taught.

The extent to which epistemological obstacles, assumed one can identify these through historical studies, should be deliberately made visible or rather avoided in teaching is a disputed issue. Sierpinska (1994) does not see the obstacles as necessary steps in individual knowledge development, but as obstacles that could make modifications of their theories about mathematical concepts difficult. A parallelism is not to be seen in the supposed recapitulation of phylogenesis in ontogenesis, but rather in a commonality of mechanisms of

36

these developments. In drawing on Skarga (1989), Sierpinska (1994) argues that these include "the preservation, in linguistic tradition and the metaphorical use of words, of the past senses" (ibid., p. 122). Artigue (1992) draws attention to the problem that what might have been identified by researchers as an epistemological obstacle, is often closely related to the choices and characteristics of the educational system. The obstacle is then of a didactical and not of an epistemological nature (Artigue 1992, p. 110).

Radford, Boero and Vasco (2003) point out that interpreting the history of mathematics for the purpose of understanding individual learners' difficulties relies on implicit assumptions about the epistemology of mathematical knowledge. Damerow (2007) shows that establishing a connection between historical developments and individuals' cognition depends on how much cultural relativism is attributed to the development of mathematical concepts. He develops some principles of a historical epistemology of logico-mathematical thought. Fried (2008) also points to a fundamental problem with writing a history of mathematics, as with other historical studies:

> More generally, even though it is interested in understanding how the present has come to be as it is, the practice of history studiously avoids measuring the past according to modern conceptions of what mathematics is and modern standards of what is mathematically significant. Failure to do this leads one to what is known in historiography as "Whiggism," after the tendency of certain British historians to see history as marching ever towards the liberal values and aspirations of the Whig party (ibid., p. 4).

Given the high level of specialization of present-day mathematics, accompanied by increasing differentiation and segmentation, it is indeed not easy to see towards which shared values and accepted forms mathematical knowledge is marching.

## Appendix: List of textbooks (chronological order)

Gregory (Gregorius), J. (1668). *Geometriae pars universalis: inserviens quantitatum curvarum transmutationi & mensurae*. Patavii: Typis heredum Pauli Frambotti

l'Hospital, M. de (1716, 2nd ed., 1696) *Analyse des infiniment petits pour l'intelligence des lignes courbes*. Paris: imprimerie royale.

Wolff (Wolfius), C. (1732). *Elementa universae. Qui commentationem de methodo mathematica, arithmeticam, geometriam, trigonometrian planam, & analysim, tam finitorum quam infinitorum complexitur*. Genevae: Bousquet & Socios.

Euler, L. (1748). *Introductio in analysin infinitorum*. Lausanne: Marcum Michaelem Bousquet & Socios.

Cauchy, A-L. (1821). *Cours d'Analyse de l'École Royale Polytechnique; 1re Partie*. Analyse Algebrique. Paris: Debure.

Cauchy, A-L. (1823). *Résumé des Leçons données a l'École Royale Polytechnique, sur le Calcul Infinitésimal*. Paris: Debure.

Björling, C.F.E. (1877). *Lärobok i integral-kalkyl*. Upsala: Lundequistska bokhandeln.

Groat, B.F (1902). *An introduction to the summation of differences of a function*. Minneapoliis: H.W. Wilson.

Goursat, E. (1902-13). *Cours d'Analyse mathématique, 3 vol*. Paris: Gauthier-Villars.

Goursat, G. (1904). *Vol. II, part 1 Derivatives and differentials. Definite integrals. Expansion in series. Applications to geometry* (translated by Earle Raymond Hedrick and Otto Dunkel). Boston: Ginn and Company.

Hobson, E. (1907). *The theory of functions of a real variable and the theory of Fourier series*. Cambridge: University Press.

Hardy G. H. (1908). *A course of pure mathematics*. Cambridge: University Press.

Vallée Poussin, Charles Jean de la (1921). *Cours d'analyse infinitésimale*. Louvain, Paris: A. Uystpruyst-Dieudonné, Gauthier-Villars.

Malmquist, J. (1923). *Föreläsningar i matematik (redigerad av T. Sällfors, civ.ing., fil.kand.)*. Stockholm: Teknologernas handelsförenings förlag.

Banach, S. (1929). *Rachunek różniczkowy i całkowy*. Lwów: Zakład Narodowy im. Ossolińskich.

Mangoldt, H. v., & Knopp, K. (1932; 13th edition from 1967). *Einfuhrung in die Höhere Mathematik. Fur Studierende und zum Selbststudium*. Stuttgart: S. Hirzel Verlag.

Courant, R. (1934, 1936). *Differential and integral calculus. Translated by E. J. McShane, vol. 1*, 1934; vol. 2, 1936. Nordemann.

Malmquist, J., Stenström, V., & Danielsson, S. (1951/1959). *Matematisk analys, del I, differential- och integralkalkyl*. Stockholm: Natur och Kultur.

Hyltén-Cavallius, C., & Sandgren, L. (1956). *Matematisk analys*. Lund: Lunds studentkårs intressebyrå.

Morrey Jr., C. B. (1962). *University calculus with analytic geometry*. Reading, MA: Addison-Wesley.

Hyltén-Cavallius, C., & Sandgren, L. (1968). *Matematisk analys, 1968/Band 2*. Lund: Studentlitteratur.

Domar, T., Haliste, K., Wallin, H., & Wik, I. (1969). *Analys I, band 2*. Lund: Gleerups.

Ullemar, L. (1972). Funktioner av en variabel. Lund: Studentlitteratur.

Eriksson, F., Larsson, E., & Wahde, G. (1975). *Grundläggande matematisk analys för tekniska högskolor. Sjätte upplagan*. Göteborg: University of Gothenburg.

Björling, C.F.E. (1877). *Lärobok i integral-kalkyl*. Upsala: Lundequistska bokhandeln.

Belkner, H., & Brehmer, S. (1984). *Riemannsche Integrale*. Berlin: VEB Deutscher Verlag der Wissenschaften.

Persson, A., & Böiers, L-C. (1990). *Analys i en variabel*. Lund: Studentlitteratur.

Hellström, L., Morander, S., & Tengstrand, A. (1991). Envariabelanalys. Lund: Studentlitteratur.

Forsling, G., & Neymark, M. (2004). *Matematisk analys: en variabel*. Stockholm: Liber.

Adams, R. A. (2006). Calculus: a complete course. 6th edition. Toronto, Ont.: Pearson/Addison Wesley.

## References:

Artigue, M. (1992). Functions from an algebraic and graphic point of view: Cognitive difficulties and teaching practices. In E. Dubinsky & G. Harel (Eds.), *The concept of function: Aspects of epistemology and pedagogy* (pp. 109–132). Washington, DC: M.A.A.

Bachelard, G. (1938). *La formation de l'esprit scientifique. Contribution à une psychanalyse de la connaissance objective*. Vrin: Paris.

Barany, M. (2011). God, king, and geometry: revisiting the introduction to Cauchy's Cours d'analyse. *Historia Mathematica, 38* (3), 368-388.

Baron, M. E. (1987). *The origins of the infinitesimal calculus*. New York: Dover.

Becker, O. (1975). *Grundlagen der Mathematik in geschichtlicher Entwicklung*. Franfurt am Main: Suhrkamp.

Bergsten, C., Jablonka, E., & Klisinska, A. (2010). Reproduction and distribution of mathematical knowledge in higher education: constructing insiders and outsiders. In U. Gellert, E. Jablonka, & C. Morgan (Eds.), *Mathematics education and society. Proceedings of MES6, Berlin 20-25 March 2010* (pp. 130-140). Freie Universität, Berlin.

Bernstein, B. (1990). The structuring of pedagogic discourse. Class, codes and control, Vol. IV. London: Routledge.

Bernstein, B. (1996). *Pedagogy, symbolic control and Identity. Theory, research, critique*. London: Taylor & Francis.

Bosch, M., & Gascon, J. (2006). 25 years of the didactic transposition. *ICMI Bulletin, No. 58* (pp. 51-65), June 2006.

Boyer, C.B. (1939). *The concepts of the calculus – a critical and historical discussion of the derivative and the integral*. New York: Columbia University Press.

Boyer, C. B. (1959). *The history of the calculus and its conceptual development*. New York: Dover Publications, Inc.

Boyer, C. B. (1968). *A history of mathematics*. New York: John Wiley & Sons.

Brousseau, G. (1983). Les obstacles épistémologiques et les problèmes en mathématiques. *Recherches en Didactique des Mathématiques, 4*(2), pp.165-198.

Brousseau, G. (1997). *Theory of didactical situations in mathematics*. Dordrecht: Kluwer Academic Publishers.

Burton, L. (2004). *Mathematicians as enquirers. Learning about learning mathematics*. Boston: Kluwer Academic Publishers.

Cajori, F. (1923). The history of notations of the calculus. *The Annals of Mathematics, 25*(1), 1-46.

Chevallard, Y. (1985). *La transposition didactique du savoir savant au savoir enseigné*. Grenoble: La Pensée Sauvage.

Chevallard, Y. (1991*). La transposition didactique du savoir savant au savoir enseigné. 2nd éd*. Grenoble: La Pensée Sauvage.

Courant, R. & Robbins, H. (1941). *What is mathematics? : an elementary approach to ideas and methods*. London : Oxford University Press.

Damerow, P. (2007). The material culture of calculation. A theoretical frameowrk for a historical epistemology of the concept of number. In U. Gellert & E. Jablonka (Eds.), *Mathematization and de-mathematization: social, philosophical and educational ramifications* (pp. 19-56). Rotterdam: Sense Publishers.

Domingues, J.C, (2008). *Lacroix and the calculus*. Basel: Birkhäuser.

Edwards, Jr. C. H. (1979). *The historical development of the calculus*. New York: Springer-Verlag.

Fourcy, A. (1828). *Historie de l'École Polytéchnique*. A Paris, Chez l'Auteur, A l'Ecole Polytéchnique.

Fried, M. (2008). *History of mathematics and the future of mathematics education*. Paper presented at the Symposium on the Occasion of the 100th Anniversary of ICMI (WG5), Rome, 5-8 march 2008.

Freudenthal, H. (1973). *Mathematik als Pädagogische Aufgabe, I, II*. Stuttgart: Klett.

Freudentahl, H. (1986). Book review. Yves Chevallard, La Transposition Didactique du Savoir Savant au Savoir Enseigné, Editions La Pensée Sauvage, Grenoble 1985. *Educational Studies in Mathematics, 17*, 323 -327.

Gerhardt, C.I. (Ed.) (1860). *Briefwechsel zwischen Leibniz und Christian Wolf. Aus den Handschriften der Königlichen Bibliothek zu Hannover*. Halle: H.W. Schmidt.

Grabiner, J. V. (1981). *The origins of Cauchy's rigorous calculus*. Cambridge, Mass: The MIT Press.

Jahnke, H. N. (Ed.). (2003). *A history of analysis*. Providence, RI: American Mathematical Society.

Joseph, G. (1991). *Crest of the Peacock: Non-European Roots of Mathematics*. London: I.B. Taurus.

Juschkewitsch, A., & & Rosenfeld, B. (1963). *Die Mathematik der Länder des Ostens im Mittelalter*. Berlin: Deutscher Verlag der Wissenschaften.

Kaiser, H., & Nöbauer, W. (1984). *Geschichte der Mathematik für den Schulunterricht*. Wien: Hölder-Pichler-Tempsky.

Katz, V. J. (1998). *A history of mathematics. An introduction (2nd edition)*. Reading, Mass.: Longman.

Kline, M. (1972). *Mathematical thought from ancient to modern times*. Oxford: Oxford University Press.

Klisinska, A. (2009). *The fundamental theorem of calculus. A case study on the didactic transposition of proof*. Doctoral thesis. Luleå: Luleå University of Technology.

Laubenbacher, R., & Pengelley, D. (1999). *Mathematical Expeditions: Chronicles by the Explorers* (Undergraduate Texts in Mathematics). New York: Springer.

Lebesgue, H. (1902). *Intégrale, longueur, aire*. Doctoral thesis. Published in Annali di Matematica, Milano.

Lebesgue, H. (1904). *Leçons sur l'integration et la recherché des functions primitives*. Paris: Gauthier-Villars.

Maligranda, L. (2004). Guillaume Francois Antoine de l'Hospital (1661-1704) -- in tercentenary of death (in Polish). In *Famous mathematical works and anniversaries*, XVIII All Polish Conference on History of Mathematics, Bialystok-Suprasl, Poland, May 31-June 4, 2004, (pp.81-123).

McGuire, J.E., & Tamny, M. (Eds.) (1983). *Certain philosophical questions: Newton's Trinity Notebook*. Cambridge: Cambridge University Press.

Mosvold, R. (2003). Genesis principles in mathematics education. In O. Bekken & R. Mosvold (Eds.), *Study the masters* (pp. 85–96). Göteborg: Nationel Centrum för Matematikutbildning.

Newton, I. (1736). *The Method of Fluxions and Infinite Series with its Applications to the Geometry of Curve-lines translated from the author's Latin original not yet made publick*. (translator John Colson). London: Henry Woodfall.

Osgood, W. F. (1903). *A modern French Calculus*. Bulletin of the American Mathematical Society, 9(10), 547-555.

Paley, R.E., & Wiener, N. (1934). *Fourier Transforms in the Complex Domain*. New York: Amer. Math. Soc.

Prag, A. (1993). On James Gregory's Geometriae pars universalis. In *The James Gregory Tercentenary Memorial Volume* (pp. 487-509). London.

Radford, L., Boero, P., & Vasco, C. (2003) Epistemological assumptions framing interpretations of students' understanding of mathematics. In J. Fauvel & J. van Maanen (Eds.), *History in mathematics education. The ICMI study.* (pp. 162-167). Dordrecht: Kluwer.

Sextus Empiricus (1718). *Adversus Mathematicos X. In Sexti Empirici opera Graece et Latine*. Leipzig: Jo. Albertus Fabricius.

Sierpinska, A. (1994). *Understanding in mathematics*. Washington, DC: Falmer Press.

Skarga, B. (1989). *Granice historycznosci*. Warsaw: Panstwowy Instytut Wydawniczy.

Struik, D. J. (Ed.) (1986). *A source book in mathematics, 1200-1800*. Princeton, NJ: Princton University Press.

Thomaidis, Y., & Tzanakis, C. (2007). The notion of historical "parallelism" revisited: historical evolution and students' conception of the order relation on the number line. *Educational Studies in Mathematics, 66*(2), 165–183.

Wagenschein, M. (1970). *Ursprüngliches Verstehen und exaktes Denken, I, II*. Stuttgart: Klett.

Whiteside, D. (1964). *The mathematical works of Isaac Newton, Vol. I- II*. New York: Johnson Reprint Cooperation.

Wiener, N. (1964). *I am a mathematician*. Cambridge, Mass.: MIT Press.

Wußing, H. (2008). *Vorlesungen zur Geschichte der Mathematik*. Frankfurt am Main: Wissenschaftlicher Verlag Harri Deutsch.

# Transitioning students to calculus: Using history as a guide
## Nicolas Haverhals & Matt Roscoe

### Precalculus: A Misnomer?

Most American high schools and universities offer a course entitled "Precalculus". As the name implies, these courses are meant to prepare students for the subsequent study of differential and integral calculus. A typical catalog description obtained at a large, public university in the western United States summarizes the course content:

> Functions of one real variable are introduced in general and then applied to the usual elementary functions, namely polynomial and rational functions, exponential and logarithmic functions, trigonometric functions, and miscellaneous others. Inverse functions, polar coordinates and trigonometric identities are included. (The University of Montana, 2011)

As the description makes clear, the study of precalculus is primarily concerned with the study of the various families of elementary mathematical functions. And while a firm notion of function is surely a prerequisite to the study of calculus, it is the assertion of these authors that the study "precalculus" has significantly migrated away from the historical account of the discovery of the subject. In this paper we advocate for a historical perspective towards precalculus as a means of bridging the calculus divide that is described below.

Perhaps most central in the study of calculus is the limit definition of the derivative. Stated formally, the derivative of a function $f$ at a point $a$ is the limit of the difference quotient, if the limit exists, as $h$ goes to zero:

$$f'(a) = \lim_{h \to 0} \frac{f(a+h) - f(a)}{h}.$$

An understanding of the limit of the difference quotient is central to the development of the rules of differentiation: the difference rule, the sum rule, the product rule, the power rule, and the chain rule to name a few. The limit of the difference quotient is also employed as justification for the derivative rules associated with polynomial functions, trigonometric functions, exponential functions, and logarithmic functions. Finally, the limit of the difference quotient is central to the development of the fundamental theorem of calculus which reveals the inverse relationship between derivative and integral.

Given the centrality of the limit of the difference quotient in the study of differential and integral calculus, an important question that mathematics educators must address is how we might better prepare students for deep understanding of this object, especially in courses carrying the title "Precalculus". A survey of popular textbooks reveals that the object does not go unaddressed in precalculus course materials. For example, in Bittenger et al.'s (2009) *Precalculus: Graphs and Models*, we find the following problems included in a chapter on rational functions (p. 45):

**Synthesis**

*Simplify.*

73. $\dfrac{(x+h)^2 - x^2}{h}$

74. $\dfrac{\frac{1}{x+h} - \frac{1}{x}}{h}$

75. $\dfrac{(x+h)^3 - x^3}{h}$

76. $\dfrac{\frac{1}{(x+h)^2} - \frac{1}{x^2}}{h}$

**Figure 1**

In Hornsby's (2003) *Precalculus with Limits*, we find the difference quotient addressed directly in the context of chapter 2.5 (p. 151):

**EXAMPLE 4**   *Finding the Difference Quotient*

Let $f(x) = 2x^2 - 3x$. Find the difference quotient and simplify the expression.

**Solution**   To find $f(x+h)$, replace $x$ in $f(x)$ with $x+h$ to get
$$f(x+h) = 2(x+h)^2 - 3(x+h).$$

Then,

$$\frac{f(x+h) - f(x)}{h} = \frac{2(x+h)^2 - 3(x+h) - (2x^2 - 3x)}{h}$$

$$= \frac{2(x^2 + 2xh + h^2) - 3x - 3h - 2x^2 + 3x}{h} \quad \text{Square } x+h; \text{ distributive property.}$$

$$= \frac{2x^2 + 4xh + 2h^2 - 3x - 3h - 2x^2 + 3x}{h}$$

$$= \frac{4xh + 2h^2 - 3h}{h} \quad \text{Combine terms.}$$

$$= \frac{h(4x + 2h - 3)}{h} \quad \text{Factor out } h.$$

$$= 4x + 2h - 3. \quad \text{Divide.}$$

**Figure 2**

Hornsby's presentation is accompanied by a sidebar which describes the importance of the difference quotient in the study of calculus. Perhaps most exceptional is Stewart's (2004) *Precalculus: Mathematics for Calculus*, where we find the following presentation (p. 175):

## AVERAGE RATE OF CHANGE

The **average rate of change** of the function $y = f(x)$ between $x = a$ and $x = b$ is

$$\text{average rate of change} = \frac{\text{change in } y}{\text{change in } x} = \frac{f(b) - f(a)}{b - a}$$

The average rate of change is the slope of the **secant line** between $x = a$ and $x = b$ on the graph of $f$, that is, the line that passes through $(a, f(a))$ and $(b, f(b))$.

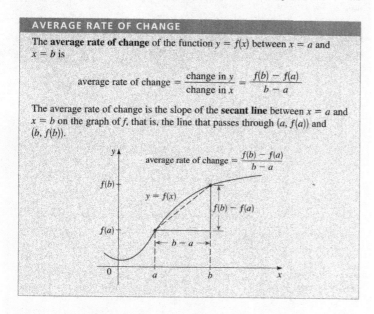

**EXAMPLE 1 ▪ Calculating the Average Rate of Change**

For the function $f(x) = x^2 + 4$, find the average rate of change of the function between the following points:

(a) $x = 2$ and $x = 6$
(b) $x = 5$ and $x = 10$
(c) $x = a$ and $x = a + h$   $(h \neq 0)$

**Figure 3**

Note that Stewart introduces the form of the difference quotient in part (c) of example 1. Further, the form of the difference quotient is presented in the context of the "average rate of change" of a function which is certainly a developmental means of introducing the object. Stewart presents the solution to this problem along with a short explanation of its significance (p. 176):

(b) Average rate of change $= \dfrac{f(10) - f(5)}{10 - 5}$      Definition

$\qquad\qquad\qquad\qquad = \dfrac{(10^2 + 4) - (5^2 + 4)}{10 - 5}$      Use $f(x) = x^2 + 4$

$\qquad\qquad\qquad\qquad = \dfrac{104 - 29}{5} = \dfrac{75}{5}$      Simplify

$\qquad\qquad\qquad\qquad = 15$

(c) Average rate of change $= \dfrac{f(a + h) - f(a)}{(a + h) - a}$

$\qquad\qquad\qquad\qquad = \dfrac{[(a + h)^2 + 4] - (a^2 + 4)}{h}$      Use $f(x) = x^2 + 4$

$\qquad\qquad\qquad\qquad = \dfrac{[a^2 + 2ah + h^2 + 4] - (a^2 + 4)}{h}$      Expand

$\qquad\qquad\qquad\qquad = \dfrac{2ah + h^2}{h}$      Simplify numerator

$\qquad\qquad\qquad\qquad = \dfrac{(2a + h)h}{h}$      Factor

$\qquad\qquad\qquad\qquad = 2a + h$      Cancel $h$ ∎

The average rate of change calculated in Example 1(c) is known as a *difference quotient*. In calculus we use difference quotients to calculate instantaneous rates of change. (See Section 12.2.)

**Figure 4**

While Stewart's nod to the importance of this object in the study of precalculus is certainly commendable, we, the authors, assert that such presentation leaves the average student with a less-than-complete understanding of the purpose for the inclusion of this object in the course of study. Further, in the case of Stewart, the study of the difference quotient does not arise again until ten chapters after its initial introduction, perhaps erroneously indicating the non-centrality of the study of this object to the student.

Our point is just this: if we are to prepare students for the study of calculus then we would do well to focus not only on elementary functions but also on *elementary notions of limit*. Historical notions of the limit found in the work of Descartes, Hudde, Fermat and Barrow should be central in a course that seeks to prepare students for the rigorous study of calculus as these accounts accomplish two important tasks. First, they motivate a need for a strong understanding of mathematical functions, currently central in the study of precalculus. Second, they serve as a bridge for the creation of the new, and certainly not trivial, idea of the notion of "limit" in the student.

Motivation for the inclusion of historical accounts in the study of precalculus is supported by the literature. Moreno and Waldegg (1991) point out that an historico-critical analysis of concepts can illuminate the learning process. In their study of student concepts of infinity, they point out that history provides the mathematics educator with "general view of the evolution of a concept" leading to historical-epistemic stages of understanding which can be employed to analyze student understanding. Just as Moreno and Waldegg (1991) found that, "…current school instruction does not help the evolution of student response schemes towards the construction of formal Cantorian definitions…(p. 212)" of infinity, we assert that current school instruction does not help the evolution of student understanding of the formal notion of the limit of the difference quotient in the study of precalculus.

Similarly, Ely (2007), in his study of student conceptions of foundational calculus concepts of function, limit, continuity and the number line, found that clusters of commonly co-occurring obstacles to understanding parallel historical epistemological dispositions. Ely comments in his conclusion, "This study suggests that a more deliberate and historically informed approach would help students establish consistent and powerful conceptions. (p. 85)"

To this end, we present here three historical methods for finding the slope of the tangent to a function at a point: Descarte's (and Hudde's) intersection method, Fermat's subtangent method, and Barrow's characteristic triangle. The presentation of each is similar. A short biographical account of each figure is provided. This is followed by a theoretical presentation of each method. Finally, a polynomial example of each method is displayed. We follow this historical presentation of precalculus methods of tangency with examples of student work obtained from an undergraduate history of mathematics course. Finally, we present several excerpts from student reflections upon the question, "What insight is gained from the study of historical methods of tangency such as those of Descartes, Hudde, Fermat, and Barrow?"

## Descartes and Hudde

The first few methods of performing calculus tasks without the direct use of limits that will be examined were developed by René Descartes (1596 – 1650). Descartes was born in France, but spent much of his adult life in Holland. Among his associates were Marin Mersenne and Christiaan Huygens. His perhaps most famous mathematical work was the development of analytic geometry. It is here we find our first example of a derivative without the use explicit of limits. Suppose one wants to find the line tangent to a given curve at a given point. The first of Descartes' methods which will be examined here can be described as follows:

Imagine you have a curve that is intercepted by a circle. Under most circumstances, the circle would intercept the curve in two points. (See Figure 5.)

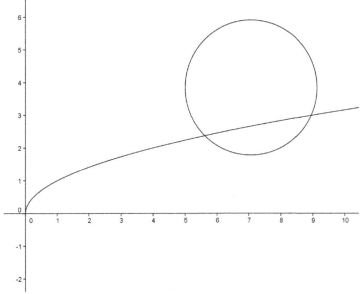

**Figure 5**

However, the circle can be transformed such that its center falls on the *x*-axis and intercepts the curve at exactly one point. (See Figure 6.). Note that as this happens, the two points of intersection get closer together until they merge and the circle is tangent to the curve.

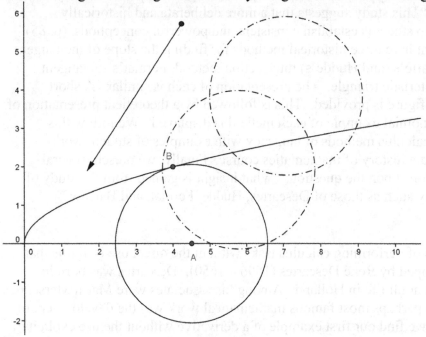

**Figure 6**

When the circle is in this position, the line perpendicular to the AB (that passes through point B) is the line tangent to the curve at B. (Note that the choice of the *x*-axis is not necessary, but will make later calculations easier).

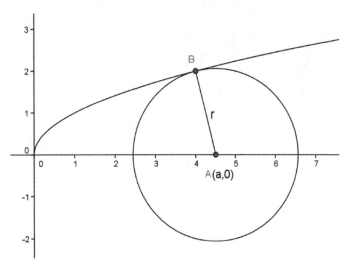

**Figure 7**

In order to accomplish this analysis, consider the curve in question $f(x)$ and the circle $g(x)$. Here, $g(x)$ is taken to be the appropriate half (upper or lower) of the circle. Then, points of intersection will be those which satisfy the equation $f(x) - g(x) = 0$. In the case of the functions to which this method will be applied, this difference results in a polynomial. Thus the polynomial in general will have two real roots. However, the particular point we will be looking for will be a double root.

Having introduced the general theory of this method, an example is in order. Consider the curve $y = \sqrt{x}$ (see Figure 7). Suppose that we would like to find the line tangent to this curve at the point (4,2). To do this, we will construct the appropriate circle with center ($a$,0) and radius $r$. From this, we can see that the equations of the given curves can be expressed as

$$y^2 = x \text{ and } y^2 + (x-a)^2 - r^2 = 0.$$

By substitution, we arrive at

$$x + (x-a)^2 - r^2 = x^2 + x(1-2a) + a^2 - r^2 = 0.$$

We would like to find values for $a$ and $r$ such that this polynomial has a double root at $x = 4$. Algebraically speaking, find $a$ and $r$ such that

$$x^2 + x(1-2a) + a^2 - r^2 = (x-4)^2 = x^2 - 8x + 16.$$

By comparing coefficients, we see that

$$1 - 2a = -8,$$

or

$$a = 4.5.$$

Thus, the slope of the line connecting the center of the circle and the point on the given curve, respectively (4.5,0) and (4,2), is

$$m = \frac{2-0}{4-4.5} = -4.$$

Hence, the line tangent to the curve at (4,2) is $y = \frac{1}{4}(x-4) + 2$, which agrees with the result obtained in differential calculus.

This method works for polynomials as well. Let us consider the line tangent to $y = x^2 + 4$ at (1,5). Again, if we label the center of our circle ($a$,0) and its radius $r$, then we arrive at the equation

$$(x^2 + 4)^2 + (x-a)^2 - r^2 = 0.$$

Upon expanding the left side, we get the fourth degree polynomial

$$x^4 + 0x^3 + 9x^2 + 2ax + 16 + a^2 - r^2 = 0.$$

Again, we wish to find the $a$ value that will give a double root at $x = 1$. However, knowledge of the double root does not give all the coefficients as before. So instead of setting this polynomial equal to $(x-1)^2$ (a quadratic), it must set equal to $(x-1)^2(x^2 + Bx + C)$ (a quartic polynomial). After expanding, the equation becomes

$$x^4 + 0x^3 + 9x^2 - 2ax + 16 + a^2 - r^2 = x^4 + (B-2)x^3 + (C-2B+1)x^2 + (B-2C)x + C.$$

From this, a system of equations is generated which allows for the solving for $a$. For example, we can see that

$$B - 2 = 0,$$

or

$$B = 2.$$

Since

$$9 = C - 2B + 1,$$

we have

$$C = 12.$$

Finally,

$$-2a = B - 2C$$

implies that

$$a = 11.$$

Then, the slope between (11,0) and (1,5) is -0.5, leaving us with $y = 2x + 3$ as our tangent line.

Typical motivation for a simplified method is generally the execution of an overly strenuous task the hard way. For a more laborious example, the reader is referred to an article by Suzuki (2005, p. 341 – 342). Although the preceding example was not overly complicated, it is hoped that the reader will believe that an increase of degree in the curve will increase the tediousness of the calculation. Luckily, Descartes devised a more efficient method. In the improved version, Descartes intelligently considers a line that intersects the curve at two places normally but once at the desired point. In other words, if $f(x)$ is the curve in question and one is looking for the tangent line at $(x_0, f(x_0))$, one would find the value for $m$ such that $x = x_0$ is a double root for the equation $f(x) - (m(x - x_0) + f(x_0)) = 0$.

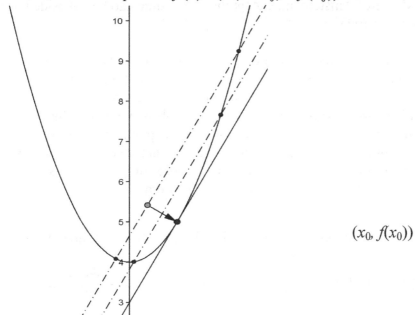

$$(x_0, f(x_0))$$

**Figure 8**

Let us apply this simplified approach to the previous example of $y = x^2 + 4$ at (1,5). We are to consider the equation $x^2 + 4 - (m(x - 1) + 5) = 0$. This leads to

$$x^2 - mx + m - 1 = (x - 1)^2, \text{ or } x^2 - mx + m - 1 = x^2 - 2x + 1.$$

Again, a comparison of coefficients leads rather quickly to $m = 2$. Thus the tangent line is $y = 2x + 3$ as before. Even with this improved method, the reader would likely not be hard-pressed to construct a function for which this method would be unwieldy. For this reason, a more efficient method for finding double roots would be helpful. To this end, we will next examine the work of Dutch mathematician Jan Hudde (1628 – 1704).

Hudde studied law at the University of Leiden, in the Netherlands. There, he met and was tutored by mathematics teacher Frans van Schooten. Van Schooten had earlier met Descartes there and was a promoter of Cartesian geometry. After his time spent on mathematics, between 1654 and 1663, Hudde dedicated his life to public service. In 1672, he was appointed to the Amsterdam city council (composed of four Burgomasters). From then until his death, he would leave and be reappointed to the post 21 times. This was due to a law which stated that one could serve as Burgomaster only two out of every three years. In the off years, Hudde worked in other civic positions. His long stay in government was certainly due to his early success. When Louis XIV invaded the Netherlands in 1672, Hudde gained national acclaim for his help in directing their national defenses. However, the Netherlands gain was mathematics' loss. When presented with the brachistochrone problem by Johann Bernoulii, Gottfried Leibniz responded:

If Huygens lived and was healthy, the man would rest, except to solve your problem. Now there is no one to expect a quick solution from, except for the Marquis de l'Hôpital, your brother [Jacob Bernoulli], and Newton, and to this list we might add Hudde, the Mayor of Amsterdam, except that some time ago he put aside these pursuits. (as cited in Suzuki, 2005, p. 344)

The work of Hudde that will be of interest to this chapter was given in the form of two letters to van Schooten. Van Schooten was working on his second edition of translation Descartes' *La Géométrie* from French into Latin. This edition featured, beyond translation, extensions and explanations of Descartes' work. The edition was published in two volumes, and Hudde's contributions to the current topic appeared in the first volume, published in 1659.

As mentioned above, the ability to find roots of multiplicity greater than 1 (in particular, double roots) is crucial to the application of Descartes' method in more complex settings. To this end Hudde presents the following theorem, presented here in modern notation:

*Theorem:*

Let $f(x) = \sum_{k=0}^{n} a_k x^k$ be a polynomial with root $x = r$ of multiplicity greater than 1, and let

$\{b_k\}_{k=0}^{n}$ be any arithmetic progression. Then, $x = r$ will also be a root of $g(x) = \sum_{k=0}^{n} b_k a_k x^k$ .

The new polynomials formed (stated as $g(x)$ in the theorem) will be referred to as Hudde polynomials. Note that for the same $f$, different Hudde polynomials may be formed by different arithmetic progressions (for those unfamiliar with this term, an arithmetic progression is a finite sequence in which successive terms differ by a constant). The proof for this theorem is taken from Suzuki (2005). It assumes $f$ is of degree five, but hopefully the reader will find its generalization clear.

*Proof:*

Let $\{b_k\}_{k=0}^{5}$ be an arithmetic progression and let $f(x)$ be the product of the third degree polynomial $x^3 + px^2 + qx + r$ and the second degree polynomial $h(x) = (x-d)^2 = x^2 - 2dx + d^2$, so that $d$ is the double root. Then $f$ can be written as

$$f(x) = (x^2 - 2dx + d^2)x^3$$
$$+ (x^2 - 2dx + d^2)px^2$$
$$+ (x^2 - 2dx + d^2)qx$$
$$+ (x^2 - 2dx + d^2)r$$

The natural temptation, at least for these authors, is to expand each of these terms and group those of like degree. For the purpose of this proof, this is disadvantageous. Consider the first line of the above equation. When $f$ is used to construct a Hudde polynomial, the first three terms of $\{b_k\}_{k=0}^{15}$ ($b_0$, $b_1$, and $b_2$) are multiplied, in order, by the terms of $h$ ($x^2$, $-2dx$ and $d^2$ respectively). In the second line, the terms in parenthesis are multiplied by $b_1$, $b_2$, and $b_3$, respectively. In general, note that the terms in each individual factor of $h$ correspond to terms of descending degree in $f$. Thus, these terms (when grouped with respect to the factor from which they came) will be multiplied by successive terms of $\{b_k\}_{k=0}^{15}$ (call them $b_j$, $b_j + t$ and $b_j + 2t$). Hence, the individual factors, the $h$'s, become

$$b_j x^2 - 2d(b_j + t)x + (b_j + 2t)d^2.$$

Note that this last expression is identically 0 when $x = d$. Thus all coefficients of the Hudde polynomial will be zero when $x = d$, and $d$ will be root of it as well. □

An example: consider $f(x) = x^3 - 5x^2 + 8x - 4$. Now, quadratic polynomials are easier to deal with than are cubic, so multiply the coefficients of $f(x)$ by the progression $\{0, 1, 2, 3\}$. This yields the Hudde polynomial

$$f_0(x) = -5x^2 + 16x - 12.$$

Application of the quadratic formula yields two roots of $f_0(x)$: 2 and $\frac{6}{5}$. Now, $\frac{6}{5}$ is not a root of $f$, but 2 is. In fact, $f$ factors as $(x-2)^2(x-1)$.

Recall the example of finding (via Descartes' first method) the tangent line of $y = x^2 + 4$ at (1,5). This led to the need to find the value of $a$ that makes 1 a double root of

$$x^4 + 0x^3 + 9x^2 - 2ax + 16 + a^2 - r^2 = 0.$$

This becomes quicker with a handy choice of arithmetic progression. Consider the progression $\{-4, -3, -2, -1, 0\}$. (Note that the progression was chosen to remove $r$.) This yields the Hudde polynomial:

$$-4x^4 - 18x + 2ax = 0.$$

With $x = 1$, we arrive at:

$$-4 - 18 + 2a = 0,$$

requiring $a = 11$ as before. The more laborious example referenced from Suzuki's article is dispatched with similar ease using this approach.

If one applies this method with Descartes' second (linear) method for finding tangents, complicated problems become much more approachable. Consider the function

$$f(x) = x^5 - 5x^2 - 7x + 5$$

and find the line tangent to $f$ at (1,-6). Recall that in Descartes second method, one would subtract from $f(x)$ the expression $m(x-1) - 6$ and set this equal to a fifth degree polynomial with a double root at 1:

$$x^5 - 5x^2 + x(-7-m) + 11 + m = (x-1)^2(x^3 + Bx^2 + Cx + D).$$

Next, one would have to expand the right-hand side completely to get a system of equations in order to solve for $m$. Things are significantly simplified, however, if the polynomial on the left-hand side of the previous equation is modified with the progression {4, 3, 2, 1, 0, -1}. (The choice of this progression is fairly arbitrary.) Note that the lack of fourth and third degree terms in the polynomial does not preclude the need for $n + 1$ terms in the progression. Having done this, one need only solve the equation

$$4x^5 - 5x^2 - 11 - m = 0$$

with $x = 1$. This gives $m = -12$ which matches the result obtained using the derivative rules from modern calculus: $5(-1)^4 - 10(-1) - 7 = -12$.

Hudde's first letter offers more work on polynomials than is relevant here. Also included was the use of the Euclidean algorithm (and a clever variation) to find the greatest common divisor of two polynomials. Hudde's second letter, however, does offer more that will be of interest here. In it, Hudde offers a method for finding extreme values of functions (another topic generally reserved for calculus). Like Descartes' methods outline above, the method is a blending of algebra and geometry. Suppose you have a local (or global) maximum (or minimum) for a polynomial equation. Then, imagine the graph of this polynomial is being intersected by the horizontal line $y = M$. (See Figure 9.) Then, the $x$-values of the points of intersection will be roots of the polynomial $f(x) - M$. As $M$ increases or decreases to the extreme of $f$ we arrive, as before, at a double root of $f(x) - M = 0$. This appears to require knowledge of the extreme value. However, this is one of the advantage's of Hudde's Theorem. If one chooses the arithmetic progression in such a way as to multiply the constant term of $f(x) - M$ by zero, the value of $M$ becomes irrelevant.

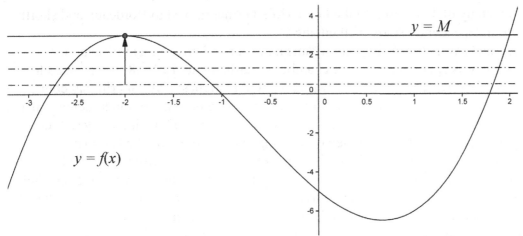

**Figure 9**

For example, let $f(x) = x^3 + 2x^2 - 4x - 5$. What are the local maximum and minimum values of $f(x)$? First we need to set $f$ equal to $M$. Then, use the progression {3, 2, 1, 0} to convert

$$f(x) - M = x^3 + 2x^2 - 4x - 5 - M$$

into the following Hudde polynomial:

$$3x^3 + 4x^2 - 4x = x(3x^2 + 4x - 4).$$

Notice that with this choice of progression (where the terms differ by -1 and the final term is 0), one will always end up with the Hudde polynomial $x \cdot f'(x)$, where $f'(x)$ is the modern derivative of $f(x)$. The zeros of this polynomial are $x = 0$, $x = -2$, and $x = \frac{2}{3}$. As with modern calculus techniques, not all these critical values correspond to maxima or minima. They do, however, encompass the local maximum and minimum. In this case, a local maximum occurs at (-2,3) and a local minimum occurs at $\left(\frac{2}{3}, \frac{95}{27}\right)$. If one was interested in a particular interval, one would need to check the relevant end points, again similar to the technique associated with the modern the method.

Before moving on, note that the idea of a secant line moving and transforming itself into a tangent line is a common way to motivate the idea of a limit in calculus courses. This idea is present in Descartes' methods. A crucial difference exists, however, in that students are allowed to think of this limiting process in purely geometric terms. This is advantageous because once the reasoning shifts to algebraic manipulations, the work to be done is on solid ground and students are in a mathematical realm within which they are comfortable.

**Fermat's Subtangent Method**

Pierre de Fermat was born August 17, 1601 in Beaumont-de-Lomagne, France. His father was a wealthy merchant and provided Fermat with substantial educational support as a child. Fermat's early education took place in a nearby Franciscan monastery. He later went on to study at the University of Toulouse. In the late 1620's Fermat moved to Bordeaux and shortly thereafter began to independently study mathematics.

After studying law in Orléans Fermat became a lawyer and government official in Toulouse. In 1631 he was appointed to lower chamber of parliament. In 1638 he advanced to the higher chamber of parliament. In 1651 he was appointed to highest level of criminal court. In spite of Fermat's successful career as a lawyer and judge, there is some reason to believe that Fermat was distracted from his chosen career. In a letter written in 1653, Colbert notes, "Fermat, a man of great erudition, has contact with men of learning everywhere…but he is rather preoccupied, he does not report cases well and is confused. (O'Connor & Robinson, 2011)"

The source of Fermat's distraction, of course, was his near fanatical devotion to the study of mathematics. Fermat's work in the subject traversed a great diversity of topics: optics, the motion of falling bodies, maxima-minima problems, methods of tangency, number theory, and probability. Though Fermat's contributions are now widely recognized (i.e. Fermat's Principle, Fermat's Last Theorem, Fermat's Little Theorem) he did not enjoy such notoriety during his lifetime. While some historians attribute this phenomena to Fermat's timid and isolated nature or his use of antiquated mathematical symbols there is evidence that Fermat's characterization of *La Dioptrique* as "groping about in shadows" initiated a life-long controversy with its well-known and well-reputed author: Rene Descartes. This dark cloud is said to have followed Fermat for the duration of his life and may have suppressed his status as a scientist and mathematician among his contemporaries. Fermat died January 12, 1665 in Castres, France.

Fermat's method for finding the slope of the tangent to a curve at a point is presented in *Methodus ad Disquirendam Macimam et Minimam* (Method of Finding Maxima and Minima) which was written in 1637. Suppose we desire to find the slope of the tangent to the curve $f$ at the point $P$.

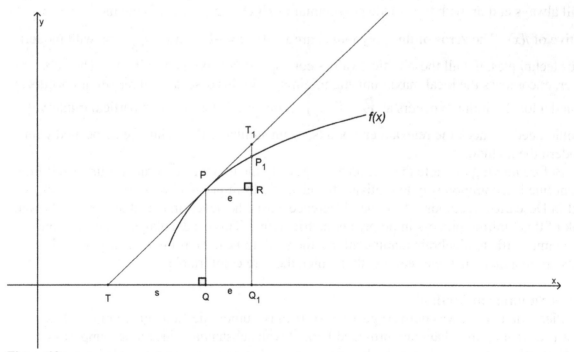

**Figure 10**

Then, the problem can be resolved if we can find the length of the subtangent, $TQ$, of the line of tangency (for $PQ$ is provided by evaluating the function). To find the length of the subtangent, Fermat adds an increment $e$ to $TQ$ to establish $Q_1$. Constructing the segment $Q_1T_1$ perpendicular to the horizontal, Fermat notes that triangle $TPQ$ and triangle $PT_1R$ are similar, which provides for the equality

$$\frac{TQ}{PQ} = \frac{e}{T_1R}.$$

Here, Fermat argues that, since $e$ is small, $T_1R$ is almost the same as $P_1R$ and since

$$P_1R = P_1Q_1 - RQ_1 = P_1Q_1 - PQ,$$

Therefore

$$\frac{TQ}{PQ} = \frac{e}{P_1Q_1 - PQ}.$$

If we solve the proportion for the desired subtangent $TQ$ we arrive at

$$TQ = \frac{PQ \cdot e}{P_1Q_1 - PQ},$$

which can be written

$$TQ = \frac{PQ}{(P_1Q_1 - PQ)/e}.$$

Modern functional notation identifies

$$PQ = f(x),$$

and

$$P_1Q_1 = f(x+e),$$

thus making the appropriate modern substitutions we arrive at

$$TQ = \frac{f(x)}{(f(x+e)-f(x))/e},$$

or just

$$s = \frac{f(x)}{(f(x+e)-f(x))/e}.$$

Fermat then solves for $s$ by setting the value of $e$ to zero (Edwards, 1979; Kline, 1972).

As always, an example of Fermat's technique helps to illuminate the procedure. Let us employ Fermat's method to find the subtangent, $s$, to the curve $y = x^2 + 2x - 5$. We have

$$s = \frac{f(x)}{(f(x+e)-f(x))/e}$$

$$= \frac{x^2 + 2x - 5}{((x+e)^2 + 2(x+e) - 5 - x^2 - 2x + 5)/e}$$

$$= \frac{x^2 + 2x - 5}{(x^2 + 2ex + e^2 + 2x + 2e - 5 - x^2 - 2x + 5)/e}$$

$$= \frac{x^2 + 2x - 5}{(2ex + e^2 + 2e)/e}$$

$$= \frac{x^2 + 2x - 5}{(2x + e + 2)}$$

Here, Fermat would set the value of $e$ to zero to arrive at

$$s = \frac{x^2 + 2x - 5}{2x + 2}.$$

To determine the slope of the tangent line we substitute the expression for the subtangent into the ratio

$$\frac{PQ}{TQ} = \frac{f(x)}{s}$$

$$= \frac{x^2 + 2x - 5}{\left(\dfrac{x^2 + 2x - 5}{2x + 2}\right)}$$

$$= 2x + 2$$

Readers will recognize this expression as the derivative of the function chosen for the example. Fermat's method, at least in this case, arrives at the accepted expression which provides the slope of the tangent line to the function.

Several interesting points arise in the method of subtangent that Fermat presents. The first involves the potential impasse presented by divisibility in the denominator of the expression for the subtangent:

$$s = \frac{f(x)}{(f(x+e) - f(x))/e}.$$

That is, for what functions does $e$ divide $f(x+e) - f(x)$? For, if $e$ does not divide the difference, then, Fermat's method fails to arrive at a definitive conclusion and instead results in division by zero. It can be shown that for polynomial functions, the principle object of investigation during Fermat's time, the division can always be performed. Limitations or difficulties associated with this divisibility on which the method relies certainly points towards the need for a more formal understanding of the substitution of zero for the value of $e$, that is, a formalization of the notion of a limit which can be applied in more diverse settings involving functions other than polynomials. In other words, this question naturally motivates a need for the study of limits as a more general approach to the problem of tangency: certainly a nice result from a pedagogical standpoint.

Also interesting in Fermat's method is the emergence of the difference quotient in the denominator of the expression for subtangent. An interesting investigation might demonstrate the equivalence of Fermat's method with the modern limit definition of the derivative for polynomial functions. Such an investigation surely would provide the student with insight into the development of the definition as well as motivate its extension into other functional settings such as trigonometric functions, exponential functions and logarithmic functions.

## Barrow's Characteristic Triangle

Sir Isaac Barrow (1630-1677) was born in London in 1630, the son of a linen draper. Isaac's father, Thomas, planned for him to become a scholar at a very early age. Thomas sent Isaac to good schools and paid for extra tutoring to insure that his aspirations would be met. Isaac, however, had a reputation as a headstrong and pugnacious child and did not respond well to his father's plans for him, changing schools more than once.  He arrived at Felsted school, which had a reputation for strict discipline, in 1640.  He responded positively to the new environment and made great progress.

In 1643 he was admitted to Cambridge and studied a curriculum that included Greek, Latin, Hebrew, French, Spanish, Italian, literature, history, geography and theology.  He studied some mathematics including geometry, optics and arithmetic.  He graduated in 1649.  In the same year he gave a speech which criticized the lack of mathematics and science that was included in the classical education presented in England's best schools.  Three years later, in 1652, Barrow graduated with his Master's of Arts and continued as a fellow at Cambridge studying theology and Greek.  His study of theology led him to geometry and the study of Euclid's Elements.  In 1655 Barrow published a simplified version of the Elements which remained in print as a standard textbook for the next 50 years.  In the same year Barrow received a travel award from the university to conduct studies abroad.  Barrow spent the next three years in Paris, Florence and Constantinople studying with local scholars in mathematics and theology.  Upon return, Barrow was appointed as Professor of Greek at Cambridge.  The position did not pay well, though, so Barrow supplemented his income by taking a position at Gresham College as a professor of Geometry.  In 1663, the now famous Lucasian Professor of Mathematics position was created at Cambridge and Barrow was the first to occupy the chair.

The years that followed his appointment as professor of mathematics represent those in which Barrow worked most productively in mathematics.  In 1664 Barrow began the delivery of six of his most well-known lectures covering topics as diverse as optics, divisibility, congruence, equality, time, space, measurement, proportion and ratio.  Most notably, Barrow lectures were designed to inspire interest in mathematics and draw other scholars to its study while pointing the field in new and interesting directions.  Barrow was one of the first to classify the study of mathematics in different branches.  Geometry, it seems, was Barrow's most revered of the areas of mathematics, commenting that, "It is the basic mathematical science, for it includes arithmetic, and mathematical numbers are simply the signs of geometrical magnitude (O'Connor & Robinson, 2011)."

Barrow was not an active publisher but many of his lectures were published by other mathematicians.  *Lectiones Opticae* and *Lectiones Geometricae* were each published in 1670.  *Lectiones Mathematicae* was published in 1683.  In 1669 Barrow resigned from the Lucasian Chair.  His vacancy made room for Sir Isaac Newton.  Barrow died in 1677 after having contracted malignant fever.  He is buried in Westminster Abbey.

Both Descartes and Fermat had developed methods for finding tangents at a point on a curve previous to Barrow's work on the subject.  In *Lectiones Geometricae* Barrow presents a method for finding the tangent of a curve at a point but relegates the technique to the end of the lecture commenting that, "I do so on the advice of a friend, and all the more willingly, because it seems to be more profitable and general than those which I have discussed (Edwards, 1979, p. 172)."  The friend that Barrow mentions in the prologue to his technique is Sir Isaac Newton who undoubtedly was influenced by Barrow's presentation.

Barrow's method for finding the tangent to a curve requires that the curve can be implicitly defined according to $f(x, y) = 0$.  Suppose that we desire to find the tangent to the curve at some point $A = (x, y)$.  Barrow asks us to consider an "indefinitely small arc" from the point A to some nearby point B on the curve (see figure 11).

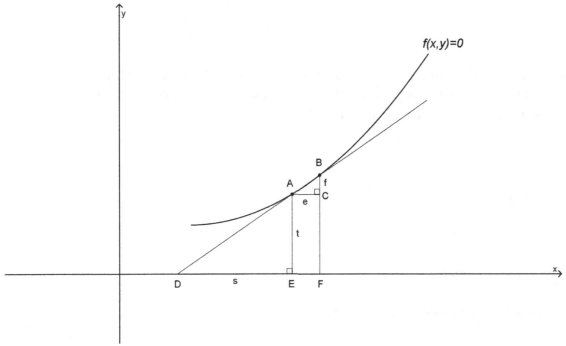

**Figure 11**

From the above drawing we can infer

$A = (x, y)$

$B = (x + e, y + f)$

Barrow makes the claim that if AB is an "indefinitely small arc" we have

$$f(x + e, y + f) - f(x, y) = 0$$

This relationship yields an algebraic relationship in four variables: *x, y, e,* and *f.* In this equation, Barrow deletes, "all terms containing a power" of *e* or *f* or "products of these for these terms have no value" (Edwards, 1979). Finally, noting the similarity of triangle *ABC* and *DAE* Barrow solves the equation

$$f(x + e, y + f) - f(x, y) = 0$$

for the quantity

$$\frac{e}{f}$$

which, by similiarity of triangles, is equal to

$$\frac{e}{f} = \frac{s}{t}$$

which is taken as the slope of the tangent line at A  (Edwards, 1979; Kline, 1972).

An example of this technique is helpful.  Let us examine the cubic equation $y = x^3 + 2x + 8$.  We first define the relation implicitly as

$$f(x, y) = y - x^3 - 2x - 8 = 0.$$

Using this implicit relation we solve

$$f(x+e, y+f) - f(x, y) = 0,$$

which in this case is written

$$(y+f) - (x+e)^3 - 2(x+e) - 8 - y + x^3 + 2x + 8 = 0.$$

This expression, after expansion and distribution, is equivalent to

$$y + f - x^3 - 3ex^2 - 3e^2x - e^3 - 2x - 2e - 8 - y + x^3 + 2x + 8 = 0$$

deleting all powers and products of a and e and combining like terms we arrive at

$$f - 3ex^2 - 2e = 0,$$

which solves to

$$\frac{f}{e} = 3x^2 + 2.$$

Thus, according to Barrow's characteristic triangle method, the slope of the tangent line of the cubic

$$y = x^3 + 2x + 8,$$

at a point

$$A = (x, y),$$

is given by

$3x^2 + 2$.

Students of modern calculus will recognize the result as the derivative of the original cubic function, thus, Barrow's technique confirms the modern technique for finding the slope of the tangent to a curve at a point.

It is worth noting that Barrow's technique does offer at least one unique advantage over the modern technique for finding the slope of the tangent line to a curve at a point. The fact that Barrow's characteristic triangle method requires functions to be defined implicitly at the onset negates the requirement of a "special" technique for implicit relations. This is not the case in the modern presentation of the derivative of a function. Here, techniques of differentiation are largely restricted to *explicit* functions which requires new theory and accompanying techniques for implictly defined functions.

On the other hand, Barrow's method does fall short of the modern method in several areas. One might question why powers and products of the infinitesimals $e$ and $f$ are deleted because they "have no value" but the values of $e$ and $f$ *themselves* are not deleted in the analysis. This is all the more disconcerting considering that Barrow is considering an "infinitetesimal arc" on the curve, one where $e$ and $f$ are, by definition, very small in value. This conversation, we submit, is exactly the kind of conversation that one wants to have in a class given the name *Precalculus* for it identifies the need for a more formal definition of the notion of the limit.

### Student responses

As mentioned earlier, the authors presented the procedures above to a senior-level History of Mathematics class as a means of piloting this sort of instruction. Because the authors are advocating the teaching of the historical methods listed above, it is worth noting the reactions students had to the techniques. To this end, examples of students' work and reflective comments are provided below. The examples and commentary came from papers written by the students as part of their grade for the course.

Of course, the student work provided sheds little light on how prepared typical precalculus students would be to handle the concepts involved. However, authors believe that there are no ideas beyond the reach of such students. The mathematics included in the student work provided ought to be familiar to any student who has advanced to a course such as precalculus. To be sure, the student-authors of the some of the work below performed operations beyond what would be expected of students who had recently completed an intermediate algebra or even a precalculus course. However, it is the authors' belief that these students would not find basic, carefully chosen examples out of their reach. With that said, the reader is encouraged to examine the work below and reach his or her own conclusion.

### Examples of student work
*Descartes' circle method*

62

# Descartes Method

We choose to find the tangent line of the function

$$f(x) = x^3 - 5x^2 - 2x + 24$$

at the value of $x = 5$. Descartes Method finds the intersection point of a circle and the function and also a value of $a$ which as shown in the following graph.

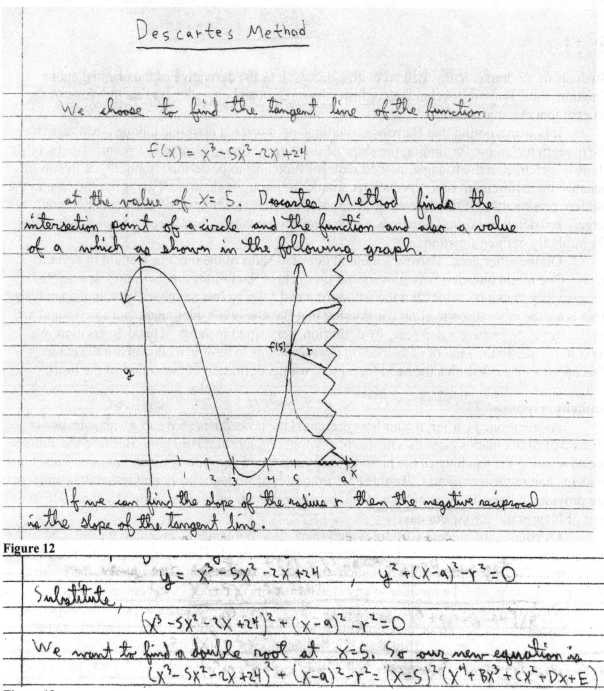

If we can find the slope of the radius $r$ then the negative reciprocal is the slope of the tangent line.

**Figure 12**

$$y = x^3 - 5x^2 - 2x + 24 \quad , \quad y^2 + (x-a)^2 - r^2 = 0$$

Substitute,

$$(x^3 - 5x^2 - 2x + 24)^2 + (x-a)^2 - r^2 = 0$$

We want to find a double root at $x = 5$. So our new equation is

$$(x^3 - 5x^2 - 2x + 24)^2 + (x-a)^2 - r^2 = (x-5)^2(x^4 + Bx^3 + Cx^2 + Dx + E)$$

**Figure 13**

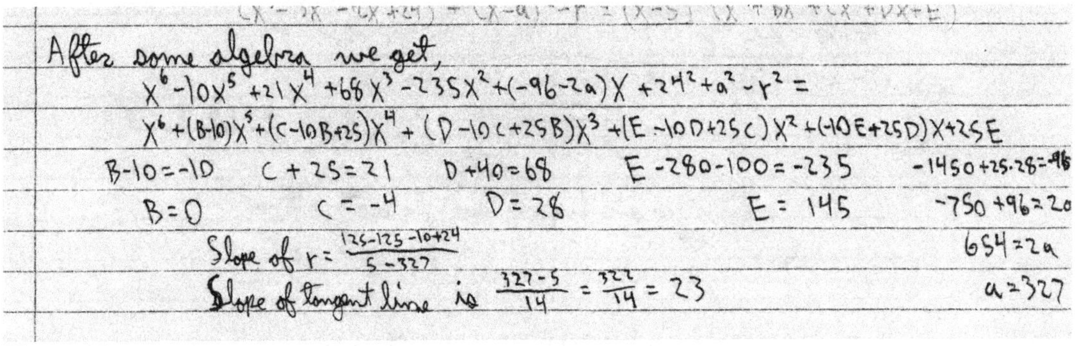

After some algebra, we get,

$$x^6 - 10x^5 + 21x^4 + 68x^3 - 235x^2 + (-96 - 2a)x + 24^2 + a^2 - r^2 =$$
$$x^6 + (B-10)x^5 + (C - 10B + 25)x^4 + (D - 10C + 25B)x^3 + (E - 10D + 25C)x^2 + (-10E + 25D)x + 25E$$

| $B - 10 = -10$ | $C + 25 = 21$ | $D + 40 = 68$ | $E - 280 - 100 = -235$ | $-1450 + 25 \cdot 28 = -96$ |
|---|---|---|---|---|
| $B = 0$ | $C = -4$ | $D = 28$ | $E = 145$ | $-750 + 96 = 2a$ |

Slope of $r = \dfrac{125 - 125 - 10 + 24}{5 - 327}$

$654 = 2a$

Slope of tangent line is $\dfrac{327 - 5}{14} = \dfrac{322}{14} = 23$

$a = 327$

**Figure 14**

Note that this matches what one would find if one were to evaluate the derivative of $f$ at $x = 5$: If

$$f(x) = x^3 - 5x^2 - 2x + 24,$$

then

$$f'(5) = 3(5)^2 - 10(5) - 2 = 75 - 50 - 2 = 23.$$

*Descartes' Line Method*

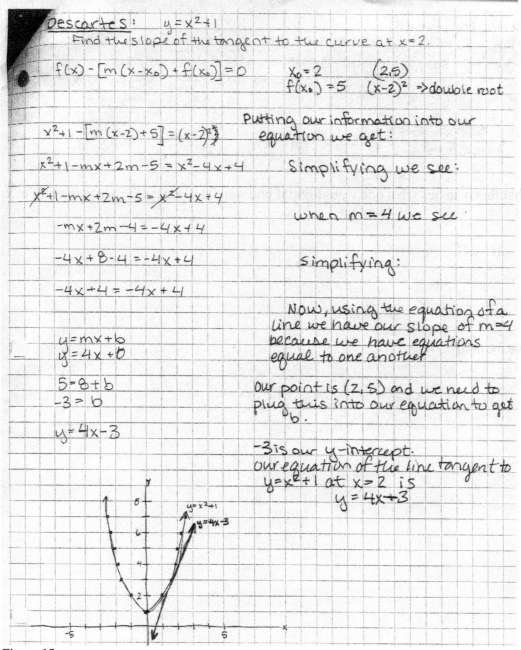

Descartes:  $y = x^2 + 1$
Find the slope of the tangent to the curve at $x = 2$.

$f(x) - [m(x - x_0) + f(x_0)] = 0$        $x_0 = 2$        $(2,5)$
                                           $f(x_0) = 5$    $(x-2)^2 \Rightarrow$ double root

Putting our information into our equation we get:

$x^2 + 1 - [m(x-2) + 5] = (x-2)^2$

Simplifying we see:

$x^2 + 1 - mx + 2m - 5 = x^2 - 4x + 4$

$x^2 + 1 - mx + 2m - 5 = x^2 - 4x + 4$

$-mx + 2m - 4 = -4x + 4$

when $m = 4$ we see

$-4x + 8 - 4 = -4x + 4$

Simplifying:

$-4x + 4 = -4x + 4$

Now, using the equation of a line we have our slope of $m = 4$ because we have equations equal to one another

$y = mx + b$
$y = 4x + b$

$5 = 8 + b$
$-3 = b$

our point is $(2,5)$ and we need to plug this into our equation to get $b$.

$y = 4x - 3$

$-3$ is our y-intercept.
our equation of the line tangent to $y = x^2 + 1$ at $x = 2$ is
$y = 4x + 3$

**Figure 15**

Again, modern calculus techniques allow for a quick check of the result: If

$f(x) = x^2 + 1$,

then

$f'(2) = 2(2) = 4$.

## Fermat's Method

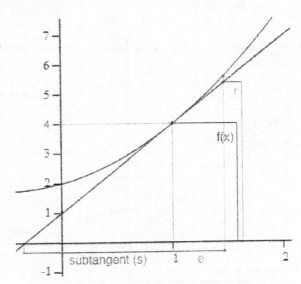

### Fermat's Method of Tangency

Polynomial: $f(x) = x^2 + x + 2$
Want to find the line tangent to $f(x)$ at $x = 1$.
We can solve to see that $f(1) = 4$.

Let s be the sub-tangent as depicted.
Let r be the point on the tangent line at $x + e$.

Fermat's method says that: $\frac{s}{s+e} = \frac{f(x)}{r}$

This is a result of similar triangles formed by the x-axis and the tangent line.

For small values of $e$, $r \approx f(x + e)$.

We substitute for $r$ and get $\frac{s}{s+e} = \frac{f(x)}{f(x+e)}$.

Let's find the sub-tangent for f(x) above:

$$s = \frac{f(x)}{[f(x+e) - f(x)]/e} = \frac{x^2 + x + 2}{[((x+e)^2 + (x+e) + 2) - (x^2 + x + 2)]/e}$$
$$= \frac{x^2 + x + 2}{[x^2 + 2ex + e^2 + x + e + 2 - x^2 - x - 2]/e} = \frac{x^2 + x + 2}{[2ex + e^2 + e]/e} = \frac{x^2 + x + 2}{2x + e + 1}$$

Now, we let $e$ be sufficiently small to be 0 and substitute $x = 1$:

$$s = \frac{1^2 + 1 + 2}{2(1) + 0 + 1} = \frac{4}{3}$$

Now, to find the slope of the tangent line at $x = 1$ and we calculate $\frac{f(1)}{s} = \frac{4}{4/3} = 3$. Thus, the slope is 3.

Using the power rule, let's confirm this is true.
$f(x) = x^2 + x + 2$.         $f'(x) = 2x + 1$.
$f'(1) = 2(1) + 1 = 3$.

Thus, the calculation of the slope of the tangent line to $f(x)$ at $x = 1$ produced by Fermat's method is in fact valid and accurate.

**Figure 16**

## Barrows' Method

66

## Barrow's Characteristic Triangle Method

Barrow worked with the implicit form of functions, $f(x,y) = 0$. To understand Barrow's Method we will consider an arbitrary curve (——) whose tangent line we wish to find at some $x$.

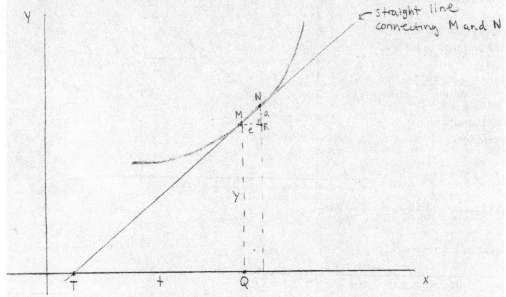

To find the slope at M, we consider the "infinitely small" (Barrow's words) arc MN on $f(x,y) = 0$.

For $a$ and $e$ "infinitely small"

$$M(x,y) \approx N(x+e, y+a)$$

thus

$$f(x,y) \approx f(x+e, y+a) = 0 \quad \dots \text{(Equation 1)}$$

In his work he makes a few assumptions or statements

i) he deletes any powers of $e$, $a > 1$, or products of the two since they are essentially 0. (Emplies a limit on $f(x,y)$)

ii) he notes $\triangle TQM \sim \triangle MRN$

iii) because of ii) $\quad slope = \dfrac{y}{t} = \dfrac{a}{e}$

Therefore if we can solve Equation 1 for $\frac{a}{e}$, we have the slope.

**Figure 17**

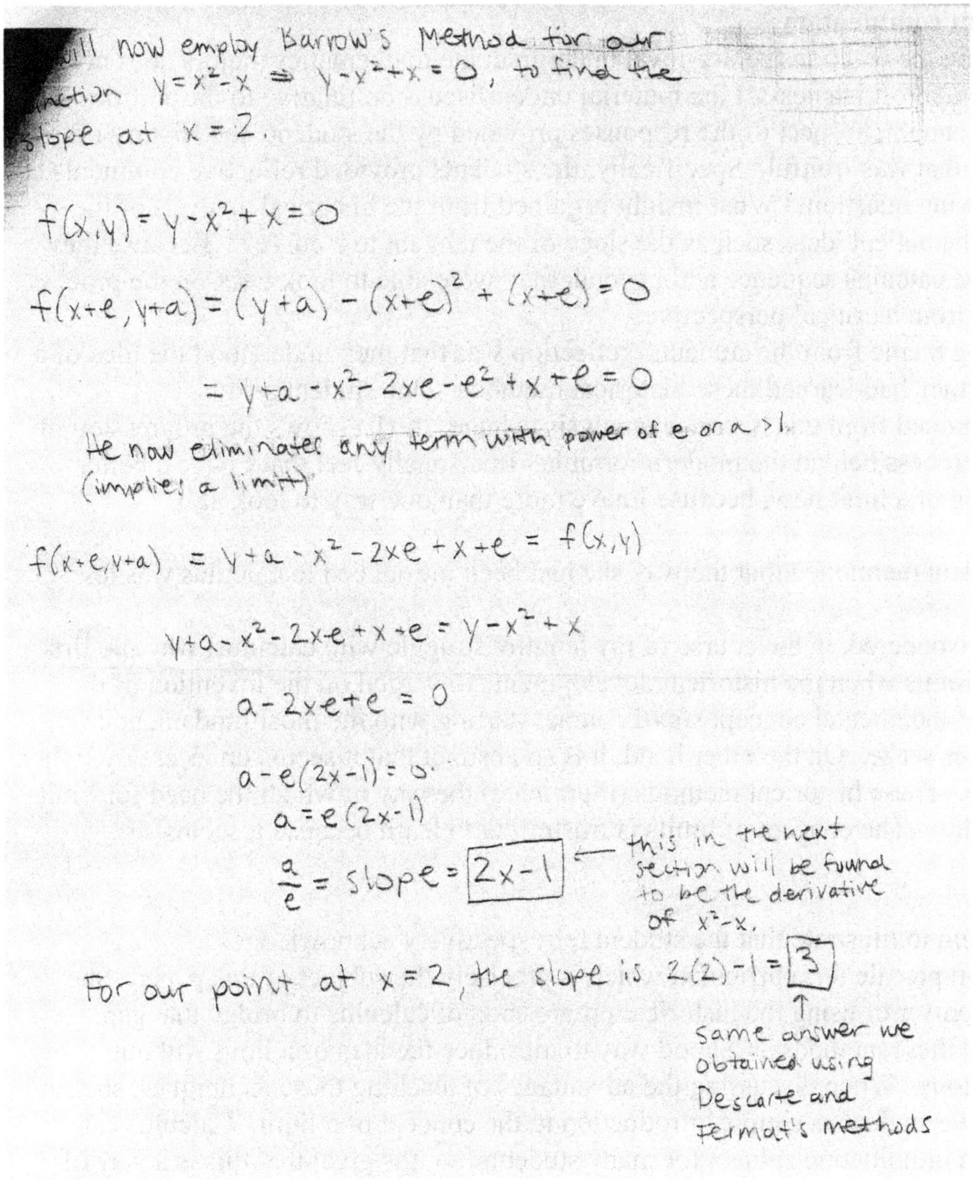

will now employ Barrow's Method for our

nction $y = x^2 - x \Rightarrow y - x^2 + x = 0$, to find the

slope at $x = 2$.

$$f(x,y) = y - x^2 + x = 0$$

$$f(x+e, y+a) = (y+a) - (x+e)^2 + (x+e) = 0$$

$$= y + a - x^2 - 2xe - e^2 + x + e = 0$$

He now eliminates any term with power of $e$ or $a > 1$.
(implies a limit)

$$f(x+e, y+a) = y + a - x^2 - 2xe + x + e = f(x,y)$$

$$y + a - x^2 - 2xe + x + e = y - x^2 + x$$

$$a - 2xe + e = 0$$

$$a - e(2x - 1) = 0$$
$$a = e(2x-1)$$
$$\frac{a}{e} = \text{slope} = \boxed{2x - 1} \longleftarrow \text{this in the next section will be found to be the derivative of } x^2 - x.$$

For our point at $x = 2$, the slope is $2(2) - 1 = \boxed{3}$
$\uparrow$
same answer we obtained using Descarte and Fermat's methods

**Figure 18**
As before, this result can be checked using the product and sum rule from calculus. If

$y = x^2 - x$,

then

$y' = 2x - 1$,

which equals 3 when $x = 2$.

**Examples of student commentary**

While gauging the reactions senior-level undergraduate mathematics majors does not speak directly to the appropriateness of the material under discussion relative to the abilities of precalculus students, another aspect of the responses provided by the students the History of Mathematics course that was fruitful. Specifically, the students provided reflective comments by answering the following question: "What insight is gained from the historical analysis of the development of mathematical ideas such as the slope of the tangent to a curve?" Because they had all completed the calculus sequence and beyond, they were able to look back on the process of learning calculus from a critical perspective.

One recurring theme from the students' reflection was that they understood the idea of a limit better because they had learned these historical methods. One student said:

The insight gained from this historical analysis is huge. ... (It) shows the progression of the thought process behind the modern formula. I personally feel that I have a better understanding of a limit now, because I have more than one way to look at it.

Another student mentioned that the way she had been introduced to calculus was less than ideal.

I have often wondered, in the course of my lengthy struggle with calculus, why the first subject was limits when the historical development converged on the invention of the limit. It is a fundamental concept so, of course, starting with the most fundamental concept makes sense. On the other hand, it is so abstract that it seems unnecessary (when first learned). These historical methods (illuminate) the way in which the need for limits presents itself. ...The concept of limits is frustrating to learn because it seems so arbitrary.

These comments seem to illustrate that the student retrospectively acknowledges a developmental gap in precalculus curriculua which is precisely the subject of this paper. She also recognizes the power of using the historical progression of calculus to bridge that gap. Another student saw these methods as a good way to introduce the idea of a limit without resorting to abstractions. When discussing the advantages of teaching these techniques, she said:

...it could give students a simple introduction to the concept of a limit. Calculus can already be an intimidating subject for many students, so it is great that this is a way of sneaking in an introduction of the idea of a limit without using scary calculus words like "limit." To see the origins of this concept of a limit could enhance understanding because the students are introduced to a concrete example of finding a limit.

While the vast majority of student reaction was positive, it was not entirely so. One student, for example, felt that we ought to focus on taking advantage of the knowledge passed on to us.

The insight gained by seeing these different methods can be argued as valuable, but I am going to take the alternate route. I am very strongly against recreating the wheel. ... I think instead of focusing on the path math developments took to become what they are today, we should focus on how we can best apply the great tools that others have developed.

The point he makes here is valid and echoes Newton's own famous quote, "If I have been able to see further, it was only because I stood on the shoulders of giants." It is worth noting, however, that there is nothing stopping mathematics educators from having it both ways. Students can be introduced to calculus using these historical methods as a guide and still be taught how to apply what this student so rightly called "the great tools" that came later.

It is worth noting here that not all of the students' comments directly addressed the topics at hand. Some students justified the use of these methods by noting a few of the advantages typically associated with using the history of mathematics in mathematics instruction. One student, a future mathematics teacher, said, "Another benefit is the engaging of the students. The history of the people that worked with these methods is intriguing." This idea seemed very much guided by her own experience with the material because she went onto say, "I enjoyed learning these methods. I was truly intrigued by each method." To be sure, putting a human face on the things when students learn something new in a mathematics classroom can hardly be viewed as a negative. Anecdotally speaking, the authors can cite examples in which students have mentioned that getting some background and context for the mathematics they are learning goes a long way towards making it a more interesting subject.

The same student also saw the teaching of these methods as a way engage students in the discovery process directly, even outside the calculus or precalculus setting.

> I think this could be an excellent opportunity for student exploration with (The Geometer's) Sketchpad or GeoGebra or simply pencil and paper. Students could probably derive these formulas themselves, seeing the similar triangles in Fermat's and Barrow's methods. These methods give students an opportunity to access their repertoires and use what they know about geometry, slope and manipulating algebra. Because these methods are basically geometry and algebra, students in geometry and algebra classes could be performing (these) methods. It is exciting to think that finding the slope of a tangent to a curve is not just reserved for precalculus and calculus classes.

As she mentions, because the methods described in this chapter do not use technically complicated ideas they are accessible to students before they reach calculus. Not only that, but it provides an opportunity for students to be guided through the discovery of the techniques. If this were to happen, one could argue that the idea of limit would be much more intuitive for students once they got to calculus where it is formally developed.

**Conclusion**

For many students who make the leap from the algebra and geometry they see in high school to college-level calculus, that jump is a difficult one. There is a fairly wide conceptual gap between many of the topics taught in precalculus courses and a first semester calculus course. Unfortunately, bridging this gap receives less attention than it deserves in the standard precalculus curriculum.

We have attempted to make the case that this disconnect can be bridged by studying the historical development of calculus for instructional aids for the modern precalculus curriculum. Our investigation to this end led us to the methods developed by Descartes, Fermat and Barrow for finding the line tangent to a given curve at a given point. These methods certainly contain the seeds for the modern notion of a mathematical limit, which is a foundational concept in the study of calculus.

For a number of students, following this natural progression of the idea of a limit led to a greater of understanding of the concept itself, allowed for reflection on how the subject of calculus is presented to new students, created an opportunity for hands-on learning and helped humanize a subject that is all too often seen as rigid and lifeless.

## Reference List

Bittinger, M. L., Beecher J. A., Ellenbogen, D. J. & Penna, J. A. (2009). Precalculus: Graphs and models (4th Ed.), Boston, MA: Addison Wesley.

Curtin, D. J. & Hudde, J. (2005). Jan Hudde and the quotient rule before Newton and Leibniz. The College Mathematics Journal, 36(4), 262-272.

Edwards, C. H. Jr. (1979). The historical development of the calculus, New York, NY: Springer-Verlag.

Ely, R.. Student obstacles and historical obstacles to foundational concepts of calculus. Ph.D. dissertation, The University of Wisconsin - Madison, United States -- Wisconsin. Retrieved June 27, 2011, from Dissertations & Theses: Full Text.(Publication No. AAT 3278849).

Eves, H. (1983). An introduction to the history of mathematics (5th Ed.). New York, NY: Saunders College Publishing.

Hornsby, J., Lial, M. L. & Rockswold, G. K. (2003). A graphical approach to precalculus with limits (3rd Ed.). Boston, MA: Addison Wesley.

Kline, M. (1972). Mathematical thought from ancient to modern times. New York, NY: Oxford University Press.

Merzbach, U. C. & Boyer, C. B. (2011). A history of mathematics. Hoboken, NJ: John Wiley and Sons.

Moreno-Armella, L., & Waldegg, G. (1991). The Conceptual Evolution of Actual Mathematical Infinity. Educational Studies in Mathematics, 22(3), 211–231.

O'Connor, J. J. & Robertson, E. F. (2008). Johann van Waveren Hudde. Retrieved from http://www.gap-system.org/~history/Biographies/Hudde.html .

O'Connor, J. J. & Robertson, E. F. (2008). René Descartes. Retrieved from http://www-history.mcs.st-andrews.ac.uk/Biographies/Descartes.html.

O'Connor, J. J., & Robertson E. F. (2011). Isaac Barrow. Retrieved from: http://www-history.mcs.st-and.ac.uk/Biographies/Barrow.html.

O'Connor, J. J., & Robertson E. F. (2011). Pierre de Fermat. Retrieved from: http://www-history.mcs.st-and.ac.uk/Biographies/Fermat.html.

Stewart, J., Redlin, L., & Watson, S. (2002). Precalculus: Mathematics for calculus (4th Ed.). Pacific Grove, CA: Brooks Cole.

Suzuki, J. (2005). Lost Calculus (1637-1670): Tangency and optimization without limits. Mathematics Magazine, 78(5), 339-353.

The University of Montana. (2011). 2010-2011 Course catalog: Department of mathematical sciences. Retrieved from: http://www.umt.edu/catalog/cat/cas/math.html.

# THE TENSION BETWEEN INTUITIVE INFINITESIMALS AND FORMAL MATHEMATICAL ANALYSIS

## Mikhail Katz & David Tall

Abstract: we discuss the repercussions of the development of infinitesimal calculus into modern analysis, beginning with viewpoints expressed in the nineteenth and twentieth centuries and relating them to the natural cognitive development of mathematical thinking and imaginative visual interpretations of axiomatic proof.

Key words and phrases: infinitesimal, hyperreal, intuitive and formal.

## 1. Klein's reflections on "mystical schemes" in the calculus

Infinitesimal calculus is a dead metaphor. In countless courses of instruction around the globe, students register for courses in "infinitesimal calculus" only to find themselves being trained to perform epsilontic multiple-quantifier logical stunts, or else being told briefly about "the rigorous approach" to limits, promptly followed by instructions not to worry about it.

Anticipating the problem as early as 1908, Felix Klein reflected upon the success of a calculus textbook dealing in "mystical schemes", namely

> the textbook by Lübsen [...] which appeared first in 1855 and which had for a long time an extraordinary influence among a large part of the public [...] Lübsen defined the differential quotient first by means of the limit notion; but along side of this he placed [...] what he considered to be the *true infinitesimal calculus*—a mystical scheme of operating with infinitely small quantities [...] And then follows an English quotation: "An infinitesimal is the spirit of a departed quantity" [6, p. 216-217].

In his visionary way, Klein adds:

> The reason why such reflections could so long hold their place [alongside] the mathematically rigorous method of limits, must be sought probably in the widely felt need of penetrating beyond the abstract logical formulation of the method of limits to the intrinsic nature of continuous magnitudes, and of forming more definite images of them than were supplied by emphasis solely upon the psychological moment which determined the concept of limit [6, p. 217].

## 2. Interesting infinitesimals lead to contradictions

In the closing months of World War II, the teenage Peter Roquette's calculus teacher at Königsberg was an old lady trained in the old school, the regular teacher having been drafted into action. Roquette reminisces in the following terms:

> I still remember the sight of her standing in front of the blackboard w[h]ere she had drawn a wonderfully smooth parabola, inserting a secant and telling us that $\Delta y/\Delta x$ is its slope, until finally she convinced us that the slope of the tangent is $dy/dx$ where $dx$ is infinitesimally small and $dy$ accordingly [14, p. 186].

Roquette recalls his youthful reaction:

> This, I admit, impressed me deeply. Until then our school Math had consisted largely of Euclidean geometry, with so many problems of constructing triangles from some given data. This was o.k. but in the long run that stuff did not strike me as more than boring exercises. But now, with those infinitesimals, Math seemed to have more interesting things in stock than I had met so far [14, p. 186].

But then at the university a few years later,

> we were told to my disappointment that my Math teacher had not been up to date after all. We were warned to beware of infinitesimals since they do not exist, and in any case they lead to contradictions. Instead, although one writes $dy/dx$ [...], this does not really mean a quotient of two entities, but it should be interpreted as a symbolic notation only, namely the limit of the quotient $\Delta y/\Delta x$. I survived this disappointment too [14, p. 186-187].

Then, some decades later, the old lady turned out not to have been so far off the mark:

> when I learned about Robinson's infinitesimals [12], my early school day experiences came to my mind again and I wondered whether that lady teacher had not been so wrong after all. The discussion with Abraham Robinson kindled my interest and I wished to know more about it. Some time later there arose the opportunity to invite him to visit us in Germany where he gave lectures on his ideas, first in Tübingen and later in Heidelberg, after I had moved there [14, p. 187].

The results of the ensuing collaboration were reported in [13] and [15].

Roquette mentions an infinitesimal calculus textbook published as late as 1912, the year of the last edition of L. Kiepert [5]. He speculates [14, p. 192] that his old lady teacher may have been trained using Kiepert's textbook.

## 3. Courant and infinitesimals "devoid of meaning"

Kiepert and other infinitesimal textbooks seem to have been edged out of the market by Courant's textbook [3]. Courant set the tone for the attitude prevailing at the time, when he described infinitesimals as "devoid of any clear meaning" and "naive befogging" [3, p. 81], as well as "incompatible with the clarity of ideas demanded in mathematics", "entirely meaningless", "fog which hung round the foundations", and a "hazy idea" [3, p. 101], while acknowledging Leibniz's masterly use of them:

> In the early days of the differential calculus even Leibnitz[1] himself was capable of combining these vague mystical ideas with a thoroughly clear understanding of the limiting process. It is true that this fog which hung round the foundations of the new science did not prevent Leibnitz or his great successors from finding the right path [3, p. 101].

How is it that they were in a position to find the right path? The Russian mathematician and historian Medvedev asks the million dollar question:

> If infinitely small and infinitely large magnitudes are regarded as inconsistent notions, how could they serve as a basis for the construction of so [magnificent] an edifice of one of the most important mathematical disciplines? [9, 10].

## 4. Vygodskiĭ: from biped back to quadruped?

In a 1931 letter [8] to the mathematician Vygodskiĭ, Luzin presents a hilarious account of the reception of Vygodskiĭ's infinitesimal calculus textbook in Soviet Russia. Vygodskiĭ dared to exploit *actual* infinitesimals. Luzin describes the reactions that ensued, in the following terms:

> I heard talk in Moscow about the restoration of the phlogiston[2] theory in science and charges of decadence [8, p. 68].

A modern reader may need to be reminded that in Stalinist Russia, a charge of bourgeois decadence was not to be trifled with, and could lead to a lengthy term in Siberian bestiaria or worse. The defenders of ideological (and decidedly secular) purity did not stop at invocations of phlogiston:

> In Leningrad [...] I heard talk to the effect that while Darwin [traced] the path of man's evolution from quadruped to biped, efforts are underway in mathematics to reverse this course [8, p. 68].

In a show of solidarity with Vygodskiĭ, Luzin proceeds to endorse a viewpoint strikingly similar to Cauchy's 1821 text [1] (which was apparently unavailable to Luzin):

---

[1] In English speaking countries, the German name 'Leibniz' is often transliterated to 'Leibnitz' to represent the sound rather than the original spelling.

[2] Phlogiston was once thought to be a fire-like element contained within combustible bodies, and released during combustion, but became incommensurable 250 years ago.

Unlike my colleagues, I think that an attempt to reconsider the idea of an infinitesimal as a variable finite quantity is fully scientific, and that the proposal to replace variable infinitesimals by fixed ones, far from having purely pedagogical significance, has in its favor something immeasurably deeper, and that this idea is growing roots in modern analysis [8, p. 68].

Luzin notes that

the idea of the actually infinitely small has certain deep roots in the mind [8, p. 68].[3]

In a possible allusion to despotic pre-revolutionary Russia of his student years, Luzin notes:

The theory of limits entered my mind mechanically and crudely, not in a refined way but rather in a forced, police-like manner [8, p. 70].

The stark choice between Weierstrassian limits and infinitesimals came in Luzin's sophomore year:

When the professors announced that $dy/dx$ is the limit of a ratio, I thought: "What a bore! Strange and incomprehensible. No! They won't fool me: it's simply the ratio of infinitesimals, nothing else." [8, p.70.]

Luzin's appropriately sophomoric attempt to construct a simpler version of Weierstrass's nowhere differentiable curve, by means of a diagonal "saw" (with numerous "steps" climbing along the diagonal of a square, see Figure 1) with infinitely many infinitesimal teeth, was patiently rebuffed by Professor Boleslav Kornelievich Mlodzeevskiĭ[4] (whom we later refer to as M.) on the grounds that "the actually infinite does not exist".

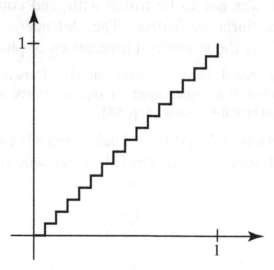

*Figure 1. Luzin's saw*

---

[3] Lakoff & Núñez [7] would certainly agree.
[4] Mlodzeevskiĭ, who brought the ideas of Hilbert and Klein from Göttingen to Moscow and was the first professor in Moscow to lecture on set theory and the theory of functions.

Unfazed, Luzin pursued M., after M.'s lecture on Cantor's cardinalities and $\aleph_0$. Luzin thought that

> These are complete contradictions: in analysis they say that every number is finite and modestly pass over in silence points at infinity on straight lines. In geometry, on the contrary, they keep on talking about points at infinity and deduce marvelous things.[5] A week ago, Boleslav Kornelievich cut me short by explaining that the actually infinite does not exist'. And now he does it himself! [8, p. 72.]

## 5. Luzin and his infinitesimal saw

Encouraged by his Cantorian insight, Luzin confronted M., this time armed with a diagonal saw with what he claimed were countably many teeth. M., patiently, countered with the claim that Luzin's saw is merely

> verbal but not real... it is not genuine. [8, p. 73.]

Luzin countered by asking whether

> the Weierstrass curve, is it genuine or logical?[6] [8, p. 73.]

At this point, M., beginning to lose patience, proceeded to contain Luzin's diagonal saw in a highly eccentric ellipse with tiny minor half-axis $\varepsilon$, and pointed out that as $\varepsilon$ becomes small, the ellipse shrinks down to the diagonal. No room for teeth! Not ready to give up, Luzin responded:

> This is indeed so if $\varepsilon$ is finite, but if $\varepsilon$ is infinitely small... [8, p. 74.]

M.'s "storm of indignation" fell far short of what would one day become the post-revolutionary phlogiston/biped rhetoric:

> I am talking to you for half an hour about limits and not about your actually infinitely small which don't exist in reality. I prove this in my course. Attend it— although for the time being I don't advise you to do so—and you will be convinced of this... [8, p. 74.]

Still unconvinced, Luzin launched into a long soliloquy about filling a cone with gypsum (plaster used for casts), about mathematical idealisation of chemical processes, and how, after removing the cast from the cone,

> we find out that we have not a cone but rather a solid, actually infinitely small... [8, p. 73.]

This was to be Luzin's final comment in that particular conversation. M.'s last suggestion, before stalking away, was that Luzin

> should bring him a jar of that kind of gypsum. [8, p. 76.]

---

[5] Possibly an allusion to projective geometry.
[6] An allusion to Weierstrass's example of a continuous but nowhere diferentiable function.

The remarkable conclusion of this exchange occurred some years later, when student Luzin attended a meeting of the mathematical society on Pfaff equations and sat at the back, unobserved by the professors, with M. sitting further forward. As the speaker adroitly manipulated numerous quantities of the form $dx$, $\delta f$, etc. on the blackboard, M., unaware of Luzin's presence, remarked to his neighbor:

> I have always thought that the symbols for exact differentials are special symbols. Look at how he works with them! In his hands they are simply constant numbers: he adds, subtracts, multiplies, substitutes, and transforms them. One can completely forget their origin and operate with them as if they were constant infinitely smalls. [8, p. 77.]

He continued, saying:

> it is not at all a hopeless attempt, in the spirit of Hilbert,[7] to axiomatically... [8, p. 77.]

but stopped when the speaker was disturbed by the conversation. At this point, Luzin recalled in his letter to Vygodskiĭ that, "A storm erupted in my mind":

> So that's what it is! They teach us, kids, one thing, and they, the grown ups, talk differently to one another. This means that, in fact, to judge by their conversations, things are not so absolutely determined. [8, p. 78.]

With hindsight we know they are not. Luzin continues:

> I looked at them with blazing eyes. I don't know what happened, maybe my stool squeaked … M. suddenly turned around, saw my blazing stare, leaned towards [his neighbor] and said something to him in a low voice. The latter replied in an equally low voice and they [both] fell silent. [8, p. 78.]

As a professional mathematician, Luzin fully understood the need for formalism, but contrasted this with the need for understanding:

> I look at the burning question of the foundations of infinitesimal analysis without sorrow, anger, or irritation. What Weierstrass-Cantor did was very good. That's the way it had to be done. But whether this corresponds to what is in the depths of our consciousness is a very different question. I cannot but see a stark contradiction between the intuitively clear fundamental formulas of the integral calculus and the incomparably artificial and complex work of their "justification" and their "proofs". [8, p. 80.]

We will return to the dialogue between Luzin and Professor M. in Section 10.

## 6. Human thought processes and infinitesimals

Luzin clearly identifies the schism between infinitesimals that seem to make intuitive sense, on the one hand, and the formal definition of limit that gives a

---

[7] Hilbert exploited non-Archimedean extension of the reals at about the turn of the century in proving the independence of his axioms of geometry; M. was apparently aware of such a development.

sound basis for mathematical analysis, on the other. Once the real numbers have been formally constructed as a complete ordered field, it can be proved that there is no room for infinitesimals in the real number system, so their use was widely condemned. And yet ideas of arbitrarily small quantities continue to be useful in thinking about the calculus because they arise from the natural way in which the brain thinks about variables that become arbitrarily small.

Mathematical thinking takes place in the human brain where signals take a few milliseconds to pass between neurons to build up a mental conception. Depending on the connections made it takes around a fortieth of a second to see an object and to recognize it. This process continues in time, and we are able to connect together our perceptions and actions as they change dynamically. It happens naturally when drawing a graph with a continuous stroke of a pencil, or looking along a graph to see its changing slope ([19], [4]).[8] The natural concept of continuity emerges as a dynamic sense of movement over an interval of time and space and certainly not as a formal definition of a limit at a point. In the same way, when we consider a potentially infinite sequence of values

$$x_1, x_2, ..., x_n, ...$$

it is natural for us to imagine not just the distinct numerical values, but to think of the $n$th term $x_n$ as a dynamically changing entity. Empirical evidence shows how both learners and expert mathematicians imagine such a variable entity to be 'arbitrarily small' (see Cornu [2]). Infinitesimal concepts therefore arise naturally in human thought, causing a conflict between the natural thought processes of learners and the formal modes of proof of mathematical analysis.

While mathematicians may learn to share their formal approach and use it with great success, the transition from intuitive mathematics full of imaginative ideas to formal mathematics based on formal definitions and logical step-by-step deduction presents significant difficulties for many learners (see Pinto & Tall [11], Weber [22]).

## 7. A new synthesis of intuition and formalism

Foundational disputes among mathematicians are frequently formulated in purely mathematical terms. However, mathematicians involved in such disputes are not always effective in addressing the transition from intuition to rigour that may be so difficult for learners. Yet the formal approach to mathematics formulated by Hilbert does not ask what the structures are, only

---

[8] This analysis by Tall and Katz, developed in [19], is built on Donald's notion of three levels of consciousness in [4].

what their properties are and what can be deduced from these properties. From this viewpoint, what matters is not what infinitesimals are, but how they behave. An infinitesimal $\varepsilon$ is a (non-zero) element of an ordered field $K$ where $-r < \varepsilon < r$ for all positive rational numbers $r$. If $K$ is an ordered extension field of the real field $\mathbb{R}$, then an infinitesimal will satisfy $-r < \varepsilon < r$ for all positive real numbers $r$.

Infinitesimals cannot fit into the real number system itself, for if $\varepsilon$ is a quantity where $0 < \varepsilon < r$ for all positive real numbers $r$, then $\varepsilon$ cannot be real, for then $r = \frac{1}{2}\varepsilon$ is also real and smaller than $\varepsilon$. However, this does not rule out the possibility that an infinitesimal may be an element of an ordered field $K$ which is an ordered extension of $\mathbb{R}$.

In this case, any element $k$ in an ordered extension field $K$ of $\mathbb{R}$ is either infinite (meaning $k > r$ for all $r \in \mathbb{R}$, or else $k < r$ for all $r \in \mathbb{R}$), or it is finite (meaning it lies between two real numbers $a < k < b$). It is straightforward to prove that a finite element $k$ is precisely of the form $c + \varepsilon$ where c is real and $\varepsilon$ is either zero or infinitesimal.[9]

For any finite element $k$, its 'standard part' is by definition the unique $c \in \mathbb{R}$, written $c = \text{st}(k)$, such that $k = c + \varepsilon$ where $\varepsilon$ is infinitesimal or zero.

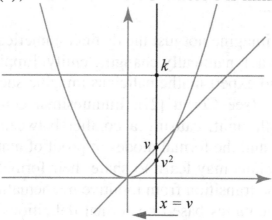

*Figure 2. A helpful sliding vertical line*

As an example, consider the field $\mathbb{R}(x)$ of rational functions in an indeterminate $x$. This field is an extension of the real numbers. The field can be ordered, by defining a rational function $f$ to be 'positive' if it is positive in some open interval $(0, a)$ for some positive real number $a$. A rational function, in this sense, is either zero, 'positive', or else $-f$ is 'positive'.

---

[9] We provide a brief proof. Let $S = \{x \in \mathbb{R} \mid x < k\}$, then $S$ is nonempty since it contains $a$, and is bounded above by $b \in \mathbb{R}$, so $S$ has a least upper bound $c$, and then the difference $\varepsilon = k - c$ can be shown to be infinitesimal.

The ordered field has a visual representation as graphs in the plane. For any positive real number $k$, the rational functions $y = k$, $y = x$, $y = x^2$ are ordered in the relation $0 < x^2 < x < k$.

By drawing the vertical line $x = v$, the three rational functions meet the line in three points $k$, $v$, $v^2$, where the point $k$ is constant as $v$ varies, but $v$ and $v^2$ are variable, see Figure 2.

A further representation can be obtained by imagining the field $\mathbb{R}(\varepsilon)$ as points on a number line. Clearly infinitesimals are not visible to the naked eye. However, the map $m : \mathbb{R}(\varepsilon) \to \mathbb{R}(\varepsilon)$ given by

$$m(x) = \frac{x - c}{\varepsilon}$$

maps $c$ to $0$ and $c + \varepsilon$ to $1$, thus separating out the images of $c$ and $c + \varepsilon$. Following this map by taking the standard part of the image (whenever the image is finite), we obtain $\mathrm{st}(m(x)) \in \mathbb{R}$ (Figure 3).

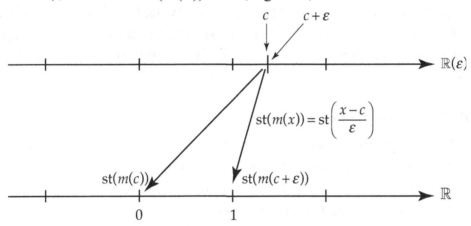

*Figure 3. Visualizing points that differ by an infinitesimal: resolving infinite closeness*

This gives a map to the real numbers which distinguishes the images of the points $c$ and $c + \varepsilon$.

If we imagine such a line as an enhancement of the vertical $y$-axis, then this allows the axis to be imagined as a vertical line with all the elements of $\mathbb{R}(x)$ placed upon it, where now $(0, c + \varepsilon)$ is a fixed point at an infinitesimal distance $\varepsilon$ from the real number $(0, c)$. The point $(0, c + \varepsilon)$ is indistinguishable from $(0, c)$ to the human eye, but it can be distinguished by magnifying the line using the map $m : \mathbb{R}(\varepsilon) \to \mathbb{R}(\varepsilon)$ on the second coordinate and taking the standard part to see a real picture. In this way, we can now imagine the vertical $y$-axis to be a line with fixed infinitesimals that, in a thought experiment, represent the 'final' position of the variable points on the vertical line $x = v$.

This gives four isomorphic representations of an ordered field consisting of rational functions in a single element:

(a) The symbolic system $\mathbb{R}(x)$ of rational functions in an indeterminate $x$;

(b) The graphical system $\mathbb{R}(x)$ of graphs of rational functions in a variable $x$;

(c) The system $\mathbb{R}(v)$ of points on a line where some are constants (the real numbers) and other quantities are variable points determined by where a rational function in $x$ meets the line $x = v$. These include infinitesimals such as $v$ and $v^2$, and infinite elements such as $1/v$.

(d) The elements in the ordered field $\mathbb{R}(\varepsilon)$ where $\varepsilon$ is an infinitesimal, which may be represented as fixed points on an extended number line that may be revealed by an appropriate magnification.

In formal terms, all these systems have isomorphic structures that represent the same underlying axiomatic structure: rational functions in a single indeterminate (or variable) with real coefficients. As representations they have very different meanings, for instance, (c) has 'variable' infinitesimals and (d) has fixed infinitesimals, but as formal structures they are isomorphic. To favour one over another is a matter of choice rather than a matter of the underlying formal structure.

## 8. Extending functions beyond the real numbers

To be able to perform the process of differentiation using infinitesimal increments, one needs to be able to extend a given real function $f$ from its definition at a real value $x$ to a nearby value $x + \varepsilon$ where $\varepsilon$ is an infinitesimal. One can then attempt to define the derivative as

$$ f'(x) = \text{st}\left( \frac{f(x+\varepsilon) - f(x)}{\varepsilon} \right) $$

For example, if $f(x) = x^2$, then $f'(x) = \text{st}(2x + \varepsilon) = 2x$.

If $f$ is a rational function, it is easy to substitute $x + \varepsilon$ for $x$ in the extension field $\mathbb{R}(\varepsilon)$ and compute the value of the function using algebra. However, more general functions such as $\sin x$ or $e^x$ cannot be extended in this field. Meanwhile, both extensions are possible in the field of series $\mathbb{R}((\varepsilon))$ in an infinitesimal $\varepsilon$, which allow a finite number of terms in $1/\varepsilon$, of the form

$$ a_{-N}\varepsilon^{-N} + \ldots + a_{-1}\varepsilon^{-1} + a_0 + a_1\varepsilon + \ldots + a_n\varepsilon^n + \ldots $$

These form an ordered field in which $\varepsilon$ is an infinitesimal and can be used to extend any function expressible as a power series to a function $f(x+\varepsilon)$. This field was called the *supperreals* by Tall [17]. This is sufficient to deal with analytic functions (given by power series), which is essentially strong enough for combinations of standard functions in the calculus, but not for general functions in mathematical analysis.

The more general problem is to extend every real function $f$ to take on values in an extension field with infinitesimals.

We could begin by working with all sequences of real numbers $(x_n)$ and then operate with them term by term. Then a sequence such as $(1, 2, 3, ...)$ might be considered as an infinite number and its inverse would be $(1, \frac{1}{2}, \frac{1}{3}, ...)$ which would be an infinitesimal. The constant sequence $(k, k, k, ...)$ for any $k \in \mathbb{R}$ could then be identified with $k$, to let us embed $\mathbb{R}$ in the set of all such sequences. The problem is that such a system does not operate as an ordered field. For instance, even though we we might like to think of the sequence $(1, 2, 3, ...)$ as being bigger than any real number $k$, its initial terms might be less than $k$ and the $n$th term would only exceed $k$ once $n > k$.

The first step towards equivalence would be to say that two sequences $(a_n)$, $(b_n)$ are equivalent if they are equal for all but a finite number of terms. If we denoted the equivalence class containing $(a_n)$ by $[a_n]$, then we would have $[a_n] = [b_n]$ if and only if $a_n = b_n$ for all but a finite number of $n$.

This would give us a surprisingly good beginning to the problem, for if we let $\omega$ be the equivalence class of $(1, 2, 3, ...)$, then we would have $\omega > k$ for any real number $k$ because all but a finite number of its terms are greater than $k$ (by identifying $k$ with the sequence $(k, k, k, ...)$).

Furthermore $\omega + 1$ would be $(2, 3, ...)$, giving a situation in which every term of $(2, 3, ...)$ is 1 bigger than the corresponding term of $(1, 2, ...)$, so $\omega + 1 > \omega$, unlike cardinal infinities, where $\aleph_0 + 1 = \aleph_0$. The term $\omega^2$ would be far bigger, $1/\omega$ (the sequence $(1, \frac{1}{2}, \frac{1}{3}, ...)$) would be infinitesimal and $1/\omega^2$ would be a smaller infinitesimal still. We would even have a natural extension of a set $D$ to the larger set $*D$ consisting of all equivalence classes $[x_n]$ where $x_n \in \mathbb{R}$ and any function $f : D \to \mathbb{R}$ can be extended to $f : *D \to \mathbb{R}$ by defining $f([x_n])$ to be the equivalence class $[f(x_n)]$.

We still need to do more. As well as dealing with sequences that nicely tend to a limit, we need to deal with *every* sequence of real numbers, no matter how it is defined. For instance, the sequence $(0, 1, 0, 1, 0, ...)$ equals 0 on the set $O$ of odd numbers, but equals 1 on the set $E$ of even numbers. We need to assign it to an appropriate equivalence class. If we make the decision focusing on the odd numbers, it will be equivalent to 0, but if we make a decision focusing on the even numbers, it will be equivalent to 1. The consequence is that to define the equivalence relation fully, we must make a *choice*.

82

Making such a choice may seem strange at first, but it is only a technical device to make a decision in all cases so that we decide that every $[x_n]$ represents a specific finite or infinite quantity. Since the set of terms in the sequence $(x_n)$ is infinite, then at least one of the following three possibilities must hold:

  (i)  there is an infinite subsequence tending to $-\infty$.

  (ii)  there is an infinite subsequence tending to $+\infty$.

  (iii)  for some $A, B \in \mathbb{R}$, where $A < B$, there is an infinite number of terms between $A$ and $B$.

All three possibilities may occur, as in the case of the sequence $(a_n)$ given by $(-1, 2, \frac{1}{3}, -4, 5, \frac{1}{6}, ...)$ where $a_n = -n$ for $n = 3N - 2$, $a_n = n$ for $n = 3N - 1$, and $a_n = \frac{1}{n}$ for $n = 3N$, as $N$ increases through $1, 2, ...$ .

In general, if case (i) holds, then we may choose $[a_n]$ to be negative infinite by taking the decision set to be the set of $n$ for which a subsequence of terms tends to $-\infty$. (In the example, the decision set $(1, 4, 7, ...)$ gives $[a_n] = -\omega$. In case (ii), we may choose $[a_n]$ to be positive infinite. (In the example, the decision set $(2, 5, 8, ...)$ gives $[a_n] = \omega$. In case (iii), the terms have a subsequence tending to a finite number $L$ where $A \leq L \leq B$ and we may choose to make the decision on the set related to this subsequence, which gives $[a_n]$ equal to $L$ plus an infinitesimal. (In the example, the decision set is $(3, 6, 9, ...)$, $L = 0$ and $[a_n]$ is the infinitesimal $1/\omega$.)

The major problem is to choose all the decision sets in such a way that all the choices can be made in a consistent manner.

## 9. Making a serious choice

To make a coherent decision in all possible cases requires us to formulate a full collection of decision sets and say that a property $P(x)$ is true for a sequence $x = (x_n)$ if $P(x_n)$ is true for all $n$ in a particular decision set. If $S$ is chosen to be a decision set, we will say that $S$ is *decisive*.

First, we stipulate that a finite set cannot be decisive, while every cofinite set $S$ (a set consisting of the whole of $\mathbb{N}$ except for a finite number of elements) is necessarily decisive:

  (0)  If $S$ is finite, then $S$ is not decisive.

  (1)  If $\mathbb{N} \setminus S$ is finite, then $S$ is decisive.

Next, if we decide that $P(x)$ is true because $P(x_n)$ is true for all $n$ in a decision set $S$, then it may also be true in a larger set $T$. For the sake of coherence, $T$ should also be chosen to be decisive:

(2) If $S$ is decisive and $S \subset T \subset \mathbb{N}$, then $T$ is also decisive.

We further require that

(3) If $S$ and $T$ are decisive then $S \cap T$ is decisive.

Thus, the intersection of a pair of decisive sets should be decisive, as well. A collection of sets satisfying (0)–(3) is a *filter* on $\mathbb{N}$. It is relatively easy to construct a filter step by step. Just start with property (1) to include all subsets of $\mathbb{N}$ whose complements are finite. Then, if any new set $U$ is added, so must all sets of natural numbers that contain $U$, and any intersection of $U$ with a subset already in the filter. For example, if we start with the sets required by (1) and add the set $E$ of all even numbers, then we need to include any set containing the even numbers, as well as any subset of these sets formed by omitting a finite number of elements. The new set of sets now satisfies (0)–(3) and so it is again a filter.

The serious problem comes with expanding such a filter to satisfy the following additional requirement:

(4) For each subset $S$ of $\mathbb{N}$, one of the two sets $S$ and $\mathbb{N} \backslash S$ must be decisive.

This requires an infinite number of decisions to be made and seems impossible for a human being to accomplish in a finite lifetime. But that does not mean that we cannot imagine it happening.

A filter satisfying properties (1)–(4) is called a (nonprincipal) *ultrafilter*.

(Axiom (4) now renders (0) redundant as it follows from a combination of (1) and (4).)

The existence of such an ultrafilter is guaranteed by the axiom of choice, see Tarski [20]. The choice is not unique. For example, if we choose the odd numbers to be decisive, then for $a = [1, 0, 1, 0, ...]$ and $b = [0, 1, 0, 1, ...]$ we have $a > b$ and, if not, then by condition (4), the set of even numbers is decisive and we have $b > a$.

The set of equivalence classes is denoted by $^*\mathbb{R}$ and is called a field of hyperreal numbers.[10] The properties (1)–(4) guarantee properties that might be expected. For instance, comparing the element $\omega = [1, 2, ..., n, ...]$ with $k = [k, k, ..., k, ...]$ gives $n > k$ for all but a finite number of $n$, so by property (1), we have $\omega > k$ for all real numbers $k$ and hence $\omega$ is infinite. Similarly $1/\omega$ is a positive infinitesimal because $0 < 1/n < k$ for all $n > 1/k$ for any positive real number $k$.

---

[10] We say *a* field of hyperreals, rather than *the* field of hyperreals, because different choices are possible. If one assumes that the continuum hypothesis is true, then our ultrapower construction produces a unique ordered field up to isomorphism.

Such a system of hyperreals together with the extension of any subset $D$ to the subset

$$*D = \{[x_n] \in {}^*\mathbb{R} \mid x_n \in D\}$$

gives an extended map $f : {}^*D \to {}^*\mathbb{R}$ defined by $f([x_n]) = [f(x_n)]$.

## 10. Looking closely at Luzin's saw

Returning to the earlier difference of opinion between Luzin and Professor M., we now see that M. is correct in declaring that infinitesimals cannot be perceived nor made out of gypsum at their original size. However, the subsequent development of mathematical theory shows that he rejects the possibility of allowing students to give infinitesimals a mathematical meaning that can be perceived not only in imagination but also in a physical picture.

Let us imagine a finite sawtooth with $n$ equal steps from $(0, 0)$ to $(1, 1)$ as in Figure 4.

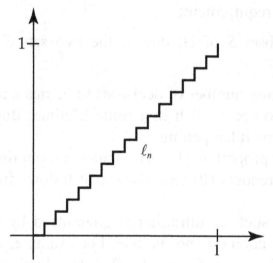

*Figure 4. Luzin's finite saw, parametrised as* $\ell_n : [0,1] \to \mathbb{R}^2$

This can be parametrised as

$$\ell_n : [0,1] \to \mathbb{R}^2$$

by tracing along it with a finger in time $t$ from 0 to 1 in which, for $k = 0, ...,$ $n-1$, the $k$th step begins at the point $(t, t)$ at $t = k/n$, moves up to $(t, t+\frac{1}{n})$ and then to $(t+\frac{1}{n}, t+\frac{1}{n})$ at time $t+\frac{1}{n}$.

The graph of $\ell_n(t)$ for $0 \le t \le 1$ is the $n$th finite Luzin saw drawn in the coordinate plane. His saw with infinitesimal teeth may be conceived as $\ell_N$ for infinite $N \in {}^*\mathbb{N}$. This is quite natural in ${}^*\mathbb{R}^2$, but as Luzin envisaged it, he thought of $N$ as being a *countable* set, and $N$ cannot be an infinite cardinal, because an infinite cardinal does not have an inverse $1/N$. However, this does not mean he cannot *imagine* infinitesimal quantities in an extension field, only that there is still work to be done to formulate a formally coherent structure (such as the hyperreals).

Professor M. is correct in asserting that the limit of $(\ell_n)$ is the straight line joining $(0, 0)$ to $(1, 1)$. However, the graph of $\ell_N$ in the extended $*\mathbb{R}^2$-plane has an infinite number $N$ of saw-teeth. The $k$th sawtooth (where $k$ may now be finite or infinite) starts at $(t, t)$ for $t = k/N$, moves up to $(t, t+\frac{1}{N})$ and then to $(t+\frac{1}{N}, t+\frac{1}{N})$ at time $t+\frac{1}{N}$.

When we magnify by the factor $N$ pointing at $(X,Y)$ where $X = k/N$, $Y = k/N$, we get a map $m : *\mathbb{R}^2 \to *\mathbb{R}^2$ in the form

$$m(x,y) = (N(x-X), N(y-Y)).$$

which gives

$$m(X,Y) = (0,0), \text{ and } m\left(X+\frac{1}{N}, Y+\frac{1}{N}\right) = (1,1)$$

so that the sides of the saw-tooth of size $1/N$ are magnified to unit lengths, giving the picture in Figure 5.

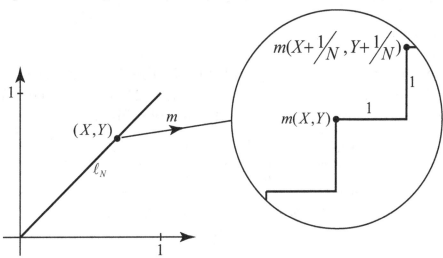

*Figure 5. Luzin's infinitesimal saw*

This reveals that the original line seen through perceptual human eyes (without magnification) does indeed look like M.'s diagonal line and, from M.'s viewpoint seeing only points in $\mathbb{R}$, the limit of the finite saw-teeth $(\ell_n)$ is indeed just the diagonal line. In the 'real' world of Cantor, with only the complete ordered field $\mathbb{R}$, Luzin would have to work a little harder to give an everywhere continuous nowhere differentiable curve. Rather than work with the Luzin sawteeth along the diagonal, let us look look at a sawtooth along the real line to get the first saw-tooth

$$s_1(x) = \begin{cases} x & \text{for } 0 \le x \le \frac{1}{2} \\ 1-x & \text{for } \frac{1}{2} \le x \le 1 \end{cases}, \quad s(n+x) = s(x) \text{ for } x \in \mathbb{Z}.$$

And then define successive saw-teeth as

$$s_n(x) = s_1(2^{n-1}x)/2^{n-1}.$$

These give successive teeth half the size of each previous one (figure 6). This variant of Luzin's finite saw-teeth has limit zero as $n$ increases:

$$\lim_{n \to \infty} s_n(x) = 0,$$

to give a straight line, just as M. declared it would be (Figure 6).

*Figure 6: The real limit of Luzin's sawteeth using M.'s mathematical analysis*

However, if we *add together* the saw-teeth to get

$$\mathrm{bl}(x) = \lim_{n \to \infty} \sum_{k=1}^{n} s_n(x),$$

then we get the blancmange function [18], identified by Takagi [16] in 1903, re-invented and generalized by van der Waerden [21] in 1930. This function is everywhere continuous and nowhere differentiable (Figure 7).

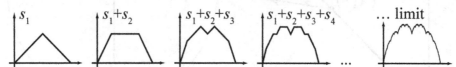

*Figure 7: The real limit adding Luzin's sawteeth*
*to give a real continuous, non-differentiable function*

The moral here is that the context in which one is working affects the nature of the mathematics. If one works in a context that only allows a complete ordered field $\mathbb{R}$, then M. is totally vindicated, along with Cantor and the mathematical culture of standard analysis. However, in the broader context of formal mathematics in extension fields of $\mathbb{R}$, infinitesimals *must* occur. Here the intuition of Luzin can be formally defined in an extended system (that he did not have at the time) in which the graph of $\ell_N$ has infinitesimal steps as he suggested. Furthermore, these steps can be represented in a physical drawing using a formally defined magnification, for all to see.

## 11. Conclusion

The approach outlined here allows the handling of sequences tending to zero or to infinity to be conceived in terms of infinitesimal and infinite quantities in a formal manner that is consonant with intuitive ideas of infinitesimals and infinite numbers in the calculus. The construction requires standard set theory together with the axiom of choice. Adjoining the axiom of choice does not introduce any contradictions in the sense that if standard set theory is consistent, so is the system when the axiom of choice is added.

While Professor M. and his modern counterparts believe that they are protecting the purity of mathematics by telling students that 'the actual infinity does not exist', this denies their students' right to imagine infinitesimal and infinite quantities in their mind's eye and to link their intuitive vision at some stage to a full formal approach.

Ideas of infinitesimals being generated by 'variables that tend to zero' were introduced by Cauchy and offer a meaning that is still used in applications today. Likewise 'variables that can be arbitrarily small' evoke a natural sense of dynamic limit processes appropriate for the calculus that can later be transformed either to standard arguments in mathematical analysis or infinitesimal methods based on the hyperreals. Meanwhile, as we saw in the structure theorem for any ordered extension field $K$ of $\mathbb{R}$, any such field can be imagined as an enriched number line with fixed infinitesimals that can be distinguished using a visual picture of a formal magnification. Moreover, the conceptions of infinitesimals as variables or as fixed quantities are different representations of the same underlying concept because *any* ordered field extension can be visualized as an enriched number line and magnified to see its infinitesimal quantities.

Most modern mathematicians now admit the axiom of choice, in the knowledge that it offers theoretical power without introducing contradictions that did not exist before. Is it not time to allow infinitesimal conceptions to be acknowledged in their rightful place, both in our fertile mathematical imagination and in the power of formal mathematics, enriched by the axiom of choice?

## References

[1] Cauchy, A. L.: *Cours d'Analyse de l'Ecole Royale Polytechnique*. Paris: Imprimérie Royale, 1821 (reissued by Cambridge University Press, 2009.)

[2] Cornu, B.: Limits, pp. 153-166, in *Advanced mathematical thinking*. Edited by D. O. Tall. Mathematics Education Library, 11. Kluwer Academic Publishers Group, Dordrecht, 1991.

[3] Courant, R.: *Differential and integral calculus*. Vol. I. Translated from the German by E. J. McShane. Reprint of the second edition (1937). Wiley Classics Library. A Wiley-Interscience Publication. John Wiley & Sons, Inc., New York, 1988.

[4] Donald, M.: *A Mind So Rare*. New York: Norton & Co., 2001.

[5] Kiepert, L.: *Grundriss der Differential-und Integralrechnung. I. Teil: Differentialrechnung*. Helwingsche Verlagsbuchhandlung. Hannover, 12th edition. 1912. XX, 863 S.

[6] Klein, F.: *Elementary Mathematics from an Advanced Standpoint. Vol. I. Arithmetic, Algebra, Analysis*. Translation by E. R. Hedrick and C. A. Noble [Macmillan, New York, 1932] from the third German edition [Springer, Berlin, 1924]. Originally published as *Elementarmathematik vom höheren Standpunkte aus*, Leipzig, 1908.

[7] Lakoff, G., Núñez, R.: *Where mathematics comes from. How the embodied mind brings mathematics into being*. Basic Books, New York, 2000.

[8] Luzin, N. N. (1931) Two letters by N. N. Luzin to M. Ya. Vygodskii. With an introduction by S. S. Demidov. Translated from the 1997 Russian original by Shenitzer. *Amer. Math. Monthly* 107 (2000), no. 1, 64–82.

[9] Medvedev, F. A.: *Nonstandard analysis and the history of classical analysis. Patterns in the development of modern mathematics* (Russian), 75–84, "Nauka", Moscow, 1987.

[10] Medvedev, F. A.: Nonstandard analysis and the history of classical analysis. Translated by Abe Shenitzer. *Amer. Math. Monthly* 105 (1998), no. 7, 659 – 664.

[11] Pinto, M. M. F., Tall, D. O.: Student constructions of formal theory: giving and extracting meaning. In O. Zaslavsky (Ed.), *Proceedings of the 23rd Conference of PME*, Haifa, Israel, 4, (1999), 65–73.

[12] Robinson, A.: *Non-standard analysis*. North-Holland Publishing Co., Amsterdam 1966.

[13] Robinson, A., Roquette, P.: On the finiteness theorem of Siegel and Mahler concerning diophantine equations. *J. Number Theory* 7 (1975), 121–176.

[14] Roquette, P.: Numbers and models, standard and nonstandard. *Math Semesterber* 57 (2010), 185–199.

[15] Roquette, P.: Nonstandard aspects of Hilbert's irreducibility theorem. In: Saracino, D.H., Weispfenning, B. (eds.) *Model Theor. Algebra, Mem. Tribute Abraham Robinson*, vol. 498 of Lect. Notes Math. pp. 231–275. Springer, Heidelberg, 1975.

[16] Takagi, T.: A simple example of the continuous function without derivative, *Proc. Phys.-Math. Soc,* Tokyo Ser. II 1 (1903), 176–177.

[17] Tall, D. O.: Looking at graphs through infinitesimal microscopes, windows and telescopes, *Mathematical Gazette*, 64 (1980), 22–49.

[18] Tall, D. O.: The blancmange function, continuous everywhere but differentiable nowhere, *Mathematical Gazette*, 66 (1982), 11–22.

[19] Tall, D. O., Katz, M.: A Cognitive Analysis of Cauchy's Conceptions of Continuity, Limit, and Infinitesimal, with implications for teaching the calculus (submitted for publication, 2011). Available on the internet at: http://www.warwick.ac.uk/staff/David.Tall/downloads.html.

[20] Tarski, A., Une contribution à la théorie de la mesure, *Fund. Math.* 15 (1930), 42–50.

[21] Waerden, B. L. van der: Fan einfaches Beispiel einer nichtdifferenzierbaren stetigen Funktion, *Math. Z.* 32 (1930), 474–475.

[22] Weber, K.: Traditional instruction in advanced mathematics courses: a case study of one professor's lectures and proofs in an introductory real analysis course, *Journal of Mathematical Behavior*, 23 (2004), 115–133.

# The didactical nature of some lesser known historical examples in mathematics

Kajsa Bråting

Uppsala University, Sweden

Nicholas Kallem

The University of Montana

Bharath Sriraman

The University of Montana

## 1. Introduction

In the field of history of mathematics the work of famous mathematicians, such as Cauchy, Newton and Leibniz, have been carefully studied which certainly have provided valuable knowledge regarding the development of mathematics. The results are well-documented in the literature. However, it may also be important to take into account the efforts of the less known (or sometimes unknown) mathematicians, contemporary to the famous mathematicians. In a didactical perspective, the study of the work of the less known mathematicians can provide valuable insights regarding mathematical thinking and conceptual development that often cannot be derived from the work of the well-known mathematicians. For instance, by studying the work of the less known mathematicians we get more information of the struggle behind famous mathematical results, which often include valuable didactical knowledge. Moreover, the general view of certain mathematical concepts at particular time periods can be better understood on the basis of not only the work of the leading mathematicians, but also on the basis of contemporary mathematicians working in the shadow of the famous mathematicians. Furthermore, in the work of the less known mathematicians one can sometimes find alternative solutions to a difficult mathematical problem, even before it was posed as a problem. However, these alternative solutions may not be as straightforward as the solutions documented in the modern textbooks. But they can provide valuable knowledge regarding mathematical thinking as well as the development in mathematics.

In this paper we try to highlight the didactical nature of some historical mathematical examples. We use examples from some less-known mathematicians in order to show how didactics of mathematics is naturally interweaved in the methods devised in the history of mathematics to solve difficult problems. In Sections 2 and 3 of this paper we consider some less known work from Islamic and Indian mathematics between the 11th and 16th centuries. Apparently, these mathematicians seem to have displayed an understanding of ideas such as limits, derivatives and integrals long before Newton and Leibniz developed the modern idea of calculus. For instance, we consider the 11th century Islamic mathematician ibn al-Haytham's (965-1040) work of calculating the volume of a parabola revolved around a line perpendicular to its axis. In fact, al-Haytham used a method similar to integration to correctly solve the problem. Moreover, within his calculations he actually derived the formulae of a sum of squares and a sum of the fourth powers. He did not have the machinery of Bernoulli numbers or polynomial computations but he used a similar algorithmic and recursive method. Al-Haytham accomplished this about 650 years before Newton an Leibniz began developing modern calculus. Moreover, we also consider the Islamic mathematician Sharaf al-Din al-Tusi who appeared to use methods similar to derivatives to find maximum values of functions already during the 12th century.

The so called "Kerala school", which consisted of Indian mathematicians between the 14th and 16th centuries, was developing methods similar to those in modern calculus. It appears that these mathematicians developed power series expansions for trigonometric functions without having the binomial expansion machinery that for instance Newton had at his disposal. In Section 3 of this paper we consider the Kerala shool's work with infinite series as well as the Indian mathematician Bhaskara II who addressed the idea of infinitesimals already in the 12th century. Moreover, it turns out that Bhaskara II used the idea of what is now known as Rolle's Theorem, that is 500 years before Rolle did.

In Section 4 we consider different views of some fundamental mathematical concepts based on examples from the 17th and 19th centuries. From the 17th century we consider a debate between the philospher Thomas Hobbes and the mathematician John Wallis. One of their issues dealt with whether there existed an angle between a circle and its tangent, the so called "angle of contact". Wallis claimed that "the angle of contact" was nothing, meanwhile, Hobbes argued that it was impossible that something that could be perceived in a picture could be nothing. It seems that one problem was that Hobbes and Wallis sometimes did not base their arguments on mathematical definitions. Instead their arguments were often based on intuitive and visual thinking. Another problem was that at least Hobbes in some cases did not clearly distinguish between mathematical objects and other objects.

A closely related issue to the debate between Hobbes and Wallis is the Swedish 19th century mathematician E.G. Björling's view of fundamental concepts in mathematical analysis. Björling lived during a time period when mathematical analysis underwent a significant shift from being considered as an empirical science based on time and space to being considered as a purely conceptual science (Jahnke, 1993, p. 267). It seems that Björling had a tendency to sometimes consider mathematical definitions as descriptions of entities rather than conventions. A typical example is that Björling considered particular examples of functions almost as if they were "existing independently of any definition". For instance, he argued that the function $f(x) = \frac{x}{|x|}$ (written with modern terminology) had two values at $x = 0$.

Meanwhile, the derivative did not exist, since the function representing the first derivtive jumped at $x = 0$. However, the derivative of the function $f(x) = |x|$ (written with modern terminology) existed and was equal to the two values $\pm 1$ (according to Björling).

## 2. Islamic mathematics

In this section we consider some work of the Islamic mathematicians ibn al-Haytham (965-1040) and Sharaf al-Din al-Tusi from the 11th and 12th centuries. As we are about to see, al-Haytham and al-Tusi were using methods similar to derivatives and integration.

### 2.1 Ibn al-Haytham's work on solids of revolution

The Islamic mathematician ibn al-Haytham was inspired by some of Archimedes' work on solids of revolution, and looked to solve the volume of a parabola revolved around a line perpendicular to its axis. Archimedes had previously studied the solid of revolution formed by revolving a parabola around a line parallel to its axis. To find this volume, ibn al-Haytham needed to introduce a formula for the sum of the fourth powers[1] (Katz, 1995, p. 165). He was successful in finding this formula. In fact, although ibn al-Haytham only meant to develop a formula for the sum of the fourth powers, his result could easily be extended for the sum of any integral powers.

Ibn al-Haytham began by considering a parabola described by the equation $x = k \cdot y^2$. He rotated the parabola around the line $x = k \cdot b^2$, where $b$ is the $y$-value of the parabola's vertex (see Figure 1)[2]. Ibn al-Haytham considered it to be the maximum height of the figure.

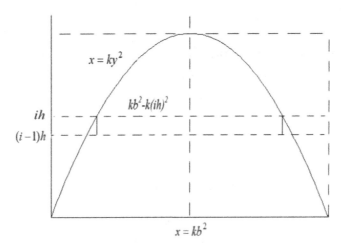

Figure 1: The Parabola

---

[1] The general method of Ibn- al Haytham described in this chapter is outlined by Katz. Here we fill in the detailed steps involved in evaluating the series described in Haytham's recursive formula in terms of modern notation. The steps involved in his discovery of the recursive formula are also unknown.
[2] Notice that the orientation of the xy-axis is somewhat arbitrary as is the location of the i[th] disk. The figure can be re-oriented with a change of co-ordinates to have the vertex align with the origin of the co-ordinate axes.

Using modern calculus methods, we can find the volume of the solid of revolution using integration. By adding up small slices of area, we can find the total area of the paraboloid. The general equation for the area of a slice of the paraboloid is

$$\pi \cdot r^2 = \pi \cdot (kb^2 - ky^2)^2,$$

where $r$ is the radius of the slice of the paraboloid. We integrate this area between $y = 0$ and $y = b$ in order to find the total area of the figure. The solution using modern notation is

$$\int_0^b A(y)dy = \int_0^b \pi * r^2 \, dy$$

$$= \int_0^b \pi * \left(kb^2 - ky^2\right) dy$$

$$= \int_0^b \pi \left(k^2 b^4 - 2k^2 y^2 b^2 + k^2 y^4\right) dy$$

$$= \pi \left( k^2 b^4 y - \frac{2k^2 y^3 b^2}{3} + \frac{k^2 y^5}{5} \right)\Bigg|_0^b$$

$$= \pi \left( k^2 b^5 - \frac{2k^2 b^5}{3} + \frac{k^2 b^5}{5} \right)$$

$$= \frac{8}{15} \pi * k^2 b^5.$$

We will now compare the modern solution with ibn al-Haytham's method from the 10th century. Al-Haytham begins by slicing the paraboloid into $n$ cylindrical disks of thickness $h = b/n$. The radius of the $i$th disk is $kb^2 - k(ih)^2$, so the volume of the $i$th disk is

$$\pi h(kb^2 - ki^2 h^2)^2 = \pi h(kh^2 n^2 - ki^2 h^2)^2 \qquad \text{(Since } h = bn\text{)}$$

$$= \pi h h^4 k^2 (n^2 - i^2)$$

$$= \pi k^2 h^5 (n^2 - i^2).$$

The volume of the paraboloid can be approximated by adding up the individual volumes of the $n$ disks that have been created. Note that the final disk has radius $kb^2 - k(b)^2 = 0$. Thus, the sum of the first $n - 1$ disks is necessary, since the $n$th disk has no radius and therefore no volume. So the approximation for the volume of the paraboloid is

$$\sum_{i=1}^{n-1} \pi k^2 h^5 (n^2 - i^2) = \pi k^2 h^5 \sum_{i=1}^{n-1} (n^2 - i^2) = \pi k^2 h^5 \sum_{i=1}^{n-1} (n^4 - 2n^2 i^2 + i^4)$$

The evaluation of the series gives us

$$\pi k^2 h^5 \left[ (n^4 - 2n^2(1)^2 + (1)^4) + (n^4 - 2n^2(2)^2 + (2)^4) + \ldots + (n^4 - 2n^2(n-1)^2 + (n-1)^4) \right]$$
$$= \pi k^2 h^5 \left[ (n-1)n^4 - 2n^2(1^2 + 2^2 + \ldots + (n-1)^2) + (1^4 + 2^4 + \ldots + (n-1)^4) \right]$$

We immediately notice two features of this series; the use of a sum of squares and the use of a sum of the fourth powers. Ibn al-Haytham actually derived the formulae of these series, which was later extended to a general formula for the sum of integral powers. His formula for the sum of the first $n$ integers raised to the $k$th power is

$$(n+1)\sum_{i=1}^{n} i^k = \sum_{i=1}^{n} i^{k+1} + \sum_{p=1}^{n}\left(\sum_{i=1}^{p} i^k\right)$$

or

$$\sum_{i=1}^{n} i^{k+1} = (n+1)\sum_{i=1}^{n} i^k - \sum_{p=1}^{n}\left(\sum_{i=1}^{p} i^k\right).$$

If we wish to find the sum of the fourth powers, then, we can make use of these equations. Letting $k = 3$, we have

$$(n+1)\sum_{i=1}^{n} i^3 = \sum_{i=1}^{n} i^4 + \sum_{p=1}^{n}\left(\sum_{i=1}^{p} i^3\right).$$

The notion $\sum_{i=1}^{p} i^3$ is recognizable as the sum of the first $p$ cubes. Ibn al-Haytham discovered that this is equal to

$$\left(\frac{p}{4} + \frac{1}{4}\right)p(p+1)p = \frac{p^4}{4} + \frac{p^3}{2} + \frac{p^2}{4}.$$

He also evaluated the sum of the squares, which he found to be

$$\sum_{i=1}^{p} i^2 = \left(\frac{p}{3} + \frac{1}{3}\right)p\left(p + \frac{1}{2}\right) = \frac{p^3}{3} + \frac{p^2}{2} + \frac{p}{6}.$$

So now we have

$$(n+1)\sum_{i=1}^{n} i^3 = \sum_{i=1}^{n} i^4 + \sum_{p=1}^{n}\left(\frac{p^4}{4} + \frac{p^3}{2} + \frac{p^2}{4}\right)$$

$$= \sum_{i=1}^{n} i^4 + \frac{1}{4}\sum_{p=1}^{n} p^4 + \frac{1}{2}\sum_{p=1}^{n} p^3 + \frac{1}{4}\sum_{p=1}^{n} p^2.$$

At this point, we can let $p = i$ without changing the value of the formula. Further simplifying, we get

$$(n+1)\sum_{i=1}^{n} i^3 = \sum_{i=1}^{n} i^4 + \frac{1}{4}\sum_{i=1}^{n} i^4 + \frac{1}{2}\sum_{i=1}^{n} i^3 + \frac{1}{4}\sum_{i=1}^{n} i^2$$

$$\frac{5}{4}\sum_{i=1}^{n} i^4 = (n+1-\frac{1}{2})\sum_{i=1}^{n} i^3 - \frac{1}{4}\sum_{i=1}^{n} i^2$$

$$\sum_{i=1}^{n} i^4 = \frac{4}{5}(n+\frac{1}{2})\sum_{i=1}^{n} i^3 - \frac{1}{5}\sum_{i=1}^{n} i^2$$

Thus, the formula for the sum of fourth powers depends on the sum of cubes and the sum of squares. The evaluation of the formula continues, giving

$$\sum_{i=1}^{n} i^4 = \frac{4}{5}(n+\frac{1}{2})(n+1)n - \frac{1}{5}(\frac{n^3}{3}+\frac{n^2}{2}+\frac{n}{6})$$

$$= \frac{4}{5}(n+\frac{1}{2})(\frac{n}{2}+\frac{1}{4})n(n+1)n - \frac{1}{5}(\frac{n}{3}+\frac{1}{3})n(n+\frac{1}{2})$$

$$= (\frac{n}{5}+\frac{1}{5})(n+\frac{1}{2})n(n+1)n - \frac{1}{3}(\frac{n}{5}+\frac{1}{5})n(n+\frac{1}{2})$$

$$= (\frac{n}{5}+\frac{1}{5})n(n+\frac{1}{2})\left[(n+1)n - \frac{1}{3}\right].$$

This was the value that ibn al-Haytham gave for the sum of the fourth powers, but it can be further simplified to

$$\frac{n^5}{5}+\frac{n^4}{2}+\frac{n^3}{3}-\frac{n}{30}.$$

Knowing the sum of the fourth powers, we can now approximate the volume of the paraboloid.

$$\pi k^2 h^5 \sum_{i=1}^{n-1}(n^4-2n^2 i^2+i^4) = \pi k^2 h^5\left[(n-1)n^4-2n^2(\frac{n^3}{3}+\frac{n^2}{2}+\frac{n}{6})+\frac{n^5}{5}+\frac{n^4}{2}+\frac{n^3}{3}-\frac{n}{30}\right]$$

$$\sum_{i=1}^{n-1}(n^4-2n^2 i^2+i^4) = (n-1)n^4-\frac{2n^5}{3}-n^4-\frac{n^3}{3}+\frac{n^5}{5}+\frac{n^4}{2}+\frac{n^3}{3}-\frac{n}{30}$$

$$= (n-1)n^4-\frac{7}{15}n^5-\frac{n^4}{2}-\frac{n}{30} = n^4(n-1-\frac{7}{15}n+\frac{1}{2})-\frac{n}{30}$$

$$= n^4(\frac{8}{15}n+\frac{1}{2})-\frac{n}{30} = \frac{8}{15}n^5+\frac{1}{2}n^4-\frac{n}{30}$$

Recall once again that $h = b/n$, so we can rewrite the volume as

$$\pi k^2 \frac{b^5}{n^5}\left(\frac{8}{15}n^5 + \frac{1}{2}n^4 - \frac{n}{30}\right) = \pi k^2 b^5\left(\frac{8}{15} + \frac{1}{2n} - \frac{1}{30n^4}\right).$$

The approximation for the volume using $n$ disks is

$$\pi k^2 b^5\left(\frac{8}{15} + \frac{1}{2n} - \frac{1}{30n^4}\right).$$

Ibn al-Haytham takes this one step further and notes that we can find a better approximation by making $n$ sufficiently large. In essence, he found that

$$\lim_{n\to\infty} \pi k^2 b^5\left(\frac{8}{15} + \frac{1}{2n} - \frac{1}{30n^4}\right) = \frac{8}{15}\pi k^2 b^5.$$

Of course, this was the same solution that was derived using modern methods. Ibn al-Haytham used a method similar to integration to correctly find the volume of a solid of revolution. He accomplished this about 650 years before Newton and Leibniz began developing modern calculus.

## 2.2 Al-tusi's work on cubic equations

The problem of finding volumes of solids of revolution was not the only instance of calculus used in medieval Islam. According to Berggren (1990, p. 306), there is strong evidence that some Islamic mathematicians had grasped the concept of using derivatives to find the maximum values of functions. In particular, Sharaf al-Din al-Tusi appeared to use the idea of derivatives to find extrema of functions over a given domain.

While solutions to certain forms of quadratic equations have existed since the time of ancient Babylon, cubic equations in general were still unsolved until Islamic mathematicians began studying them around the 11th century (Berggren, 1990, p. 304)[3]. The need for cubic equations arose from problems like doubling a cube and splitting a sphere into two parts (Katz, 2004, p. 173). According to Guilbeau (1930), ancient Greek mathematicians attempted to solve cubic equations, but were unsuccessful. However, Islamic mathematician Omar Khayyam addressed this problem again in the 11th century. He actually used a Greek method of intersecting conic sections to geometrically derive the solution to different types of cubic equations.

Sharaf al-Din al-Tusi was a Persian mathematician born in 1135 AD. His work on cubic equations primarily dealt with classifying cubics based on how many positive solutions the equation has. He wanted to be able to determine the number of positive solutions of a cubic by checking the value of the coefficients of the equation (Katz, 2004, p. 175). In al-Tusi's

[3] Berggren's (1990) description of al-Tusi's method is followed with details added to the steps involved in the calculation. The relatively modern (graphical) representation of the cubic function and simpler notation is meant to clarify the calculation delineated by Berggren. Kjeldsen (this volume) also presents the didactic possibilities of Al-Tusi's method in the context of student projects on the historical study of the theory of equations.

*Muadalat*, 5 cubic equations are listed that have positive solutions only if the coefficients of the equations fulfill certain requirements (Berggren, 1990, p. 306). One of the problems that al-Tusi examines in *Muadalat* is the cubic $x^3 + d = bx^2$, where $d$ and $b$ are positive. The solution to this problem illustrates the possibility that al-Tusi had knowledge of derivatives.

Sharaf al-Din al-Tusi begins by slightly manipulating the equation.

$$x^3 + d = bx^2$$
$$x^2(b-x) = d$$

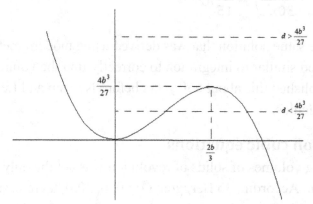

Figure 2: Graph of the cubic equation $x^2(b-x) = d$

Now al-Tusi states that the maximum local value of the function $f(x) = x^2(b-x)$ occurs at $x_0 = \frac{2b}{3}$. At this value;

$$x_0^2(b-x_0) = \frac{4b^2}{9}\left(b - \frac{2b}{3}\right) = \frac{4b^3}{27}.$$

So the maximum value of $x^2(b-x)$ is $\frac{4b^3}{27}$, and this gives a condition for the value of the coefficients. It is obvious that if $\frac{4b^3}{27}$ is the maximum value of the function, then $d$ cannot exceed this amount, since $x^2(b-x) = d$. Thus, to have any solutions, we must have $d \le \frac{4b^3}{27}$.

This also tells us that $x_0 = \frac{2a}{3}$ is the single root when $d = \frac{4b^3}{27}$. Further investigation reveals that $d < \frac{4b^3}{27}$ implies that there are two roots to the equation $x^2(b-x) = d$.

The main question that remains is how al-Tusi determined that a maximum occurs at $x_0 = \dfrac{2b}{3}$

. Berggren (1990, p. 309) suggests three proposed ideas for how al-Tusi found this result, each of which implicitly use derivatives. One proposed idea begins with us considering a more generalized form of the cubic equation.

Let $f(n) = -x^3 - bx^2 + cx$. Going back to the original cubic $x^3 + d = bx^2$, let $z$ be a solution to the cubic. Then $z^3 + d = bz^2$, and $z^3 < bz^2$ since $d$ must be positive. Then $z < b$ gives a boundary for the solutions of the equation. Namely, any solution to $x^3 + d = bx^2$ must be less than $b$.

Let $x$ be the point where $f(x) = -x^3 - bx^2 + cx$ is at its maximum and let $y$ be a number such that $x < y < b$. Then

$$[f(x) - f(y)] = -x^3 - bx^2 + cx + y^3 + by^2 - cy.$$

This result can be transformed into

$$[f(x) - f(y)] = (y^2 - x^2)(x+b) - (c - y^2)(y-x)$$
$$[f(x) - f(y)] = (y-x)\left[y + x)(x+b) - (c - y^2)\right]$$

.

Now al-Tusi looks for a condition to give $f(x)-f(y)>0$. For this to occur, we must have $(y+x)(x+b) - (c - y^2) > 0$. But since $y>x$, the condition $2x(x+b) \geq (c - x^2)$ is sufficient. What we have found is a condition so that $f(x)$ is greater than $f(y)$ while $y$ is greater than the value of $x$.

Now let us consider the reverse case. The value that maximizes $f(x)$ is still denoted by $x$. However, this time choose $y < x$. We still want to look for the case where $f(x)-f(y)>0$. Since $y < x$, though, we need to look for a condition such that $(y+x)(x+b) - (c - y^2) < 0$. Since $y < x$, the condition $2x(x+b) \leq (c - x^2)$ is sufficient.

If we look for a general value $y$ such that $f(x)-f(y)>0$, we find that for $x < y$ and $x > y$, the condition

$2x(x+b) = (c - x^2)$ is sufficient. Expanding this equation, we find the equation $3x^2 + 2xb - c = 0$. This is a condition that must hold to have $f(x)-f(y)> 0$. Now, having $f(x)-f(y)>0$ for local values of $y$ denotes that $x$ was a point at which a local maximum occurs, as we noted. The equation $3x^2 + 2xb - c = 0$ tells us where this local maximum occurs.

Going back to the general equation $f(x) = -x^3 - bx^2 + cx$, note that $f'(x) = -3x^2 - 2bx + c$. Setting this equal to zero to find critical points gives the equation $0 = -3x^2 - 2bx + c$. This is exactly what Sharaf al-Din al-Tusi accomplished in his analysis of the problem. Searching for a condition such that $x$ was a local maximum of the function allowed al-Tusi to find the derivative of the cubic polynomial.

The derivative is then used to find $x_0$ from above. We simply solve $0 = -3x_0^2 - 2bx_0$ to find that $x_0 = \frac{2}{3}b$. The solution $x_0 = 0$ is thrown away because it is not a maximum, as required for solution $x_0$. Sharaf al-Din al-Tusi used similar methods for solving other forms of cubic equations. While his use of the derivative appeared to be isolated given that the idea was not widely adopted by other mathematicians of the time, it is still quite incredible that he used some form of derivatives hundreds of years before Newton and Leibniz.

## 3. Indian mathematics

Goonatilake (1998, p. 135) asserts that calculus in India arose in the form of four problems: finding the value of pi, finding the instantaneous motion of a planet, finding "the 'position angle' of the ecliptic with any secondary to the equator," and finding the surface area and volume of a sphere. In this section we consider the 12th century Indian mathematician Bhaskara II, who was one of the first Indian mathematicians to present calculus-based solutions to problems. We also consider the Kerala-school

### 3.1 Bhaskara II

In 1150 AD, Bhaskara II introduced his *Siddhanta-siromani*. Within this treatise on astronomy, Bhaskara II touched upon some of the problems listed above. Bhaskara successfully solved the problem of instantaneous motion in this work, even introducing the term *Tatkalika Gati*, pertaining to the calculation of instantaneous motion. The solution to the problem makes use of small intervals to eventually find that $\delta \sin \theta = \cos \theta \delta \theta$. This is one of the first recorded uses of differential calculus. The problem is outlined by Datta and Singh (1984, pp. 96-99).

Bhaskara begins by considering the position of a planet located at point $P$. Point $Q$ will denote the location of the planet at a later time. The line $PT$ is the line tangent to the planet's orbit at P. Bhaskara states that if point $Q$ were sufficiently close to $T$, then arc $PQ$ would lie in the same direction as $PT$.

He notes that analyzing this problem in terms of larger intervals is not truly examining the instantaneous motion of the planet. In order to correctly solve the problem, he requires the use of a time interval called a *ksana*, which is an infinitesimal amount of time. Dattta and Singh mention that Bhaskara recognized this issue in solving the problem, but did not actually make us of infinitesimal amounts in the implementation of the solution.

Now to solve the problem, we first split the 90 degree arc in Figure 3 into $n$ parts, and let $A$ be the arc length of each one of the $n$ arcs. Bhaskara defines sine-differences as the set of values $R(\sin A - \sin 0), R(\sin 2A - \sin A), R(\sin 3A - \sin 2A), ... R(\sin nA - \sin((n-1)A)$. He terms these values *Bhogya khanda*. Inspection of the sine function reveals that the sine difference of each term decreases with successive terms, although the sine difference is never equal to zero.

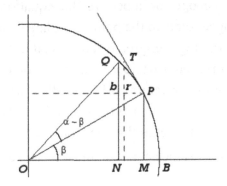

Figure 3: Bhaskara II's Orbit Problem

We now let $A$ equal the arc length $PQ$, the arc length that the planet traveled during a certain time interval. Let $R$ be the radius of the circular orbit. Let $\alpha$ be the angle $\angle BOQ$ and let $\beta$ be the angle $\angle BOP$. Then $R(\sin \alpha - \sin \beta) = QN - PM = Qb$. Now Bhaskara wishes to find the *Tatkalika Bhogya Khanda*, the instantaneous sine difference. Essentially this is the instantaneous movement of the planet at point $P$. Bhaskara's argument is that the *Tatkalika Bhogya Khanda* could be found if the movement of the planet were to extend along the tangent line $PT$ rather than the arc of $PQ$. Then the sine difference is easily found; it is equal to line $Tr$ in Figure 3. It can be shown that triangles $\triangle PTr$ and $\triangle PMO$ are similar. Therefore,

$$\frac{R}{PT} = \frac{R \cos \beta}{Tr}$$
$$Tr = PT \cos \beta$$

Now we take advantage of $Tr$ being a sine difference, as shown above.

$$Tr = R(\sin \alpha - \sin \beta).$$

Furthermore, $PT = R(\alpha - \beta)$. This takes advantage of the fact that $PT$ was stipulated to be the size of $A$, where $A$ was equal to the arc length $R(\alpha - \beta)$. Finally, using the results above, we obtain

$$(\sin \alpha - \sin \beta) = (\alpha - \beta) \cos \beta,$$

which is later interpreted as $\delta \sin \theta = \cos \theta \delta \theta$.

As Datta and Singh (1984) discuss, Bhaskara expanded upon his knowledge of calculus by demonstrating two important ideas. The first idea he demonstrated was that "the differential $dy/dx$ vanishes when the variable is at a maximum" (Goonatilake, 1998, p. 135). That is, the derivative is equal to zero when a maximum is found. The second idea uses the fact that "when a planet is either in apogee or in perigree the equation of the center vanishes" (Datta & Singh, 1984, p. 98). Apogee refers to the point in a planet's orbit when it is furthest from the center of the object it is orbiting. Perigree is the opposite; it is the point at which a planet is closest to the planet it is orbiting. Considering an equation describing the distance of a satellite from the center of the object it is orbiting, Bhaskara stated that the derivative of the distance equation must be equal to zero at the apogee and the perigree. If we choose one of these locations, and if a planet reaches this location at times $t_1$ and $t_2$, then during this time period the derivative of the distance equation must reach zero at some point in the interval $(t_1, t_2)$.

This was an early discovery of what is now known as Rolle's Theorem. Rolle's theorem is related to the mean value theorem of calculus, and it states that if a function $f$ is continuous on the closed interval $[a, b]$ and is differentiable on the open interval $(a, b)$ and $f(a)=f(b)$ such that $a \neq b$, then there exists some real number $c$ in the open interval $(a, b)$ such that $f'(c)=0$. In Figure 4, the value $c$ corresponds to the point $t_0$ where the derivative of the distance equation is equal to zero. While French mathematician Michele Rolle received credit for the theorem after working on it in 1691, Bhaskara II actually discovered Rolle's theorem around 500 years before Rolle did.

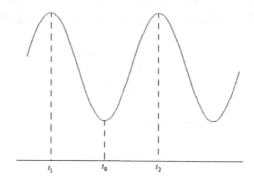

Figure 4: Bhaskara II's early development of Rolle's Theorem

## 3.2 The Kerala School[4]

Kerala, a region in the southwestern portion of India, was home to what is now known as the Kerala school of mathematicians between the 14th and 16th centuries (Katz, 2007, p. 480). These mathematicians, including Madhava and Nilakantha, developed elements of calculus that European mathematicians would not discover for several hundred years. The most notable work from the Kerala school regarded infinite series.

---

[4] This section is terse given the thorough explanations and details of the Kerala school found in the work of Rajagopal & Rangachari (1978), Raju(2001) and Ranjan (1990). Details of possible transmission of the work of the Kerala school is still a topic of research among historians of mathematics.

*Yuktibhasa* was written in the 16th century by Jyesthadeva, and it contained discussions on infinite series (Rajagopal & Rangachari, 1978). Several noteworthy formulas come from this document, including the series expansions for sine, cosine, the inverse tangent, and pi. Jyesthadeva attributes the series to Madhava, a Keralan mathematician who lived from around 1350 AD to 1425 AD (Katz, 1995, p. 169). Three of the series found by Madhava were

$$\theta = \tan\theta - \frac{\tan^3\theta}{3} + \frac{\tan^5\theta}{5} - \ldots$$
$$\sin\theta = \theta - \frac{\theta^3}{3!} + \frac{\theta^5}{3!} - \ldots$$
$$\cos\theta = 1 - \frac{\theta^2}{2!} + \frac{\theta^4}{4!} - \ldots$$

An infinite series for pi followed rather easily from the first series by simply letting $\theta = \frac{\pi}{4}$.

Since $\tan\frac{\pi}{4} = 1$, it follows that the series approximation for pi comes from the equation

$$\frac{\pi}{4} = 1 - \frac{1}{3} + \frac{1}{5} - \ldots.$$

Although these series are not found amongst the existing works from Madhava, the commentary on his work still exists in the form of Jyesthadeva's *Yuktibhasa* and Nilakantha's *Tantrasangraha* (Roy, 1990, p. 300).

The proof for the sine power series in the *Jyesthadeva* essentially divided a quadrant of a circle into small segments as seen in Figure 5. To find the formula for the power series, values for the differences between *y* values in successive segments are calculated, as well as values for the differences between *x* values of successive segments. Then allowing *n* to grow larger gives improved approximations for sine and cosine. In fact, allowing *n* to grow to infinity allowed Madhava and his followers to derive the sine and cosine power series, where the sine series comes from changes in *y* values and the cosine series comes from changes in *x* values between the *n* segments. In *The Mathematics of Egypt, Mesopotamia, China, India, and Islam: A Sourcebook*, Kim Plofker recognizes the relevance that these techniques had in the history of mathematics. She states that the methods involved in finding the sine and cosine power series "strongly recall elements in the seventeenth-century development of the infinitesimal calculus" (Katz, 2007, p. 493).

The Gregory series for pi was developed by James Gregory in 1671, and Newton and Leibniz receive most of the discovery of the sine, cosine, and inverse tangent series. Their work on series also occurred in the 1670's (Rajagopal & Rangachari, 1978, p. 89).

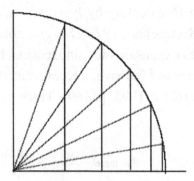

Figure 5: The partitioning of a quadrant of a
circle in the proof of the sine power series

## 4. Development of mathematical concepts

In this section we consider the view of fundamental mathematical concepts on the basis of historical examples from the 17th century and the 19th century. In these examples we point out some problems that arise when mathematical arguments are not based on mathematical definitions. In Section 4.1 we consider the 17th century debate between the philosopher Thomas Hobbes (1588-1679) and the mathematician John Wallis (1616-1703). Furthermore, in Section 4.2 we disucss the Swedish mathematician E.G. Björling's (1808-1872) efforts of understanding fundamental concepts in mathematical analysis.

### 4.1 The debate between Hobbes and Wallis

Hobbes and Wallis often discussed the relation between the finite and the infinite, or rather, if there is a relation between the finite and the infinite. It turned out that they had different ways of considering the nature of mathematical concepts. We will consider "The angle of contact" and "Torricelli's infinitely long solid" which are two examples that Hobbes and Wallis were discussing.

### 4.1.1 The angle of contact

"The angle of contact", which already occurred in Euclid's *Elements*, appeared to be an angle contained by a curved line (for instance a circle) and the tangent to the same curved line. A dispute between Hobbes and Wallis concerning "the angle of contact" was based on the following two questions:

1. Does there *exist* an angle between a circle and its tangent (see Figure 6)?
2. If such an angle exists, what is the size of it?

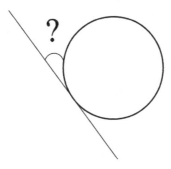

Figure 6: The angle of contact.

Wallis claimed that "the angle of contact" was *nothing*, whereas Hobbes argued that it was not possible that something which could be perceived from a picture could be nothing (Wallis, 1685, p. 71; Hobbes, 1656, pp. 143-144). In fact, this dispute originated from an earlier discussion between Jacques Peletier (1517-1582) and Christopher Clavius (1537-1612) (Peletier, 1563; Clavius, 1607).

According to Hobbes, it was not possible that something that actually could be perceived from a picture drawn on a paper could be nothing. Another reason why "the angle of contact" could not be nothing was the possibility of making proportions in a certain way between different "angles of contact". Hobbes claimed:

> [...] an angle of Contingence[5] is a Quantity[6] because wheresoever there is Greater or Less, there is also Quantity (Hobbes, 1656, pp. 143-144).

This statement was perhaps based on Eudoxos' theory of ratios, which is embodied in books V and XII of Euclid's *Elements*. Definitions 3 and 4 of book V states:

> DEFINITION 3. A ratio is a sort of relation in respect of size between two magnitudes of the same kind.

> DEFINITION 4. Magnitudes are said to have a ratio to one another which are capable, when multiplied, of exceeding one another (Heath, 1956, p. 114).

Hobbes, as well as Wallis, discussed the possibility of making proportions between different angles of contact on the basis of a picture similar to Figure 7 below. Hobbes' approach was to compare the "openings" (Hobbes' expression) between two different angles of contact. On the basis of Figure 7, Hobbes claimed that it was obvious that the "opening" between the small circle and the tangent line was greater than the "opening" between the large circle and the tangent line. That is, since one "opening" was greater than the other, the angle of contact must be a quantity since *"wherever there is Greater and Less, there is also quantity"* (see Hobbes' quotation above). From this Hobbes concluded that the angle of contact was a quantity (magnitude), and hence it could not be nothing.

---

[5] Hobbes used the term "the angle of Contingence", instead of "the angle of contact".
[6] Hobbes' term "quantity" can be interpreted as "magnitude", which is used in for instance Euclid's *Elements*.

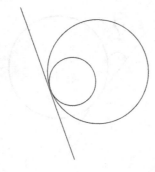

Figure 7: "The angle of contact" in proportion to another angle of contact.

Meanwhile, Wallis (1685) stressed that the "angle of contact" is of "no magnitude". He claimed that *"[...] the angle of contact is to a real angle as 0 is to a number"* (Wallis, 1685, p. 71). That is, according to Wallis it was not possible, by multiplying, to get the "angle of contact" to exceed any real angle (remember Definition 4 above). He pointed out that an "angle of contact" will always be contained in every real angle. However, at the same time he stressed that *"[...] the smaller circle is more crooked than the greater circle"* (Wallis, 1685, p. 91).

Today this is not a problem since we have determined that the answer to the question if the "angle of contact" exists is not dependent on pictures such as Figures 6 and 7 above. Instead the answer depends on which definition of an angle that we are using. In school mathematics an angle is defined as an object that can only be measured between two intersecting segments (Wallin et al., 2000, p. 93). According to such a definition the "angle of contact" is not an angle. However, an angle can be defined differently. For instance, in differential geometry an angle between two intersecting curved lines can be defined as the angle between the two tangents in the intersection point. According to such a definition "the angle of contact" exists and is equal to 0.

### 4.1.2 Torricelli's infinitely long solid

In 1642 the Italian mathematician Evangelista Torricelli (1608-1647) claimed that it was possible for a solid of infinitely long length to have a finite volume (Mancosu, 1996, p. 130). In modern terminology, if one revolves for instance the function $y = 1/x$ around the $x$-axis and cuts the obtained solid with a plane parallel to the $y$-axis, one obtains a solid of infinite length but with finite volume. Sometimes this solid is referred to as *Torricelli's infinitely long solid* (see Figure 8). The techniques behind the determintion of Torricelli's infinitely long solid were provided by the Italian mathematician Evangelista Cavalieri's (1598-1647) theory of indivisibles from the 1630:s (Mancosu, 1996, p. 131). However, the difference is that Torricelli was using *curved* indivisibles on solids of *infinitely long* lengths. Mancosu (1996, p. 131) discusses a debate between (among others) Hobbes and Wallis which was based on the issue if there could be a ratio between the finite and the infinite.

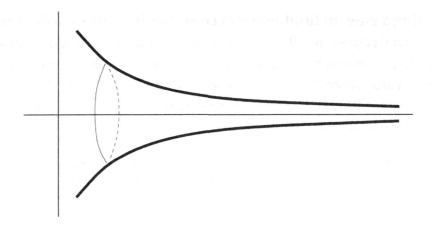

Figure 8: Torricelli's infinitely long solid.

Hobbes rejected the existence of infinite objects, such as "Torricelli's infinitely long solid", since *"[...] we can only have ideas of what we sense or of what we can construct out of ideas so sensed"* (Mancosu, 1996, pp. 145-146). He insisted that every object must exist in the universe and be perceived by the "natural light". Mancosu (1996, pp. 137-138) points out that many 17th century philosophers held that geometry provides us with indisputable knowledge and that all knowledge involves a set of self-evident truths known by "natural light". Hobbes stressed that when mathematicians spoke of, for instance, an "infinitely long line" this would be interpreted as a line which could be extended as much as one prefered to. He argued that infinite objects had no material base and therefore could not be perceived by the "natural light". According to Hobbes, it was not possible to speak of an "infinitely long line" as something given. The same thing was valid for solids of infinite length but with finite volume.

Meanwhile, for Wallis, "Torricelli's infinitely long solid" was not a problem as long as it was considered as a *mathematical object*. According to Mancosu, Wallis shared Leibniz' opinion that it was nothing more spectacular about "Torricelli's infinitely long solid" than for instance that the infinite series

$$\frac{1}{2} + \frac{1}{4} + \frac{1}{8} + \frac{1}{16} + \frac{1}{32} + \cdots$$

equals 1. If the new method led to the result that infinite solids could have finite volumes, then these solids existed within a mathematical context. Unlike Hobbes, it seems that Wallis (and Leibniz) made a distinction between mathematical objects and other objects. Perhaps one can also say that Wallis and Leibniz were *generalizing* the volume concept to not only be a measure on finite solids, but also a measure on solids of infinitely long lengths.[7]

---

[7]In (Bråting and Öberg, 2005) generalizations of mathematical concepts are discussed in more detail.

## 4.2 E.G Björling's view of fundamental concepts in mathematical analysis

In the history of mathematics the 19th century is often considered as a period when mathematical analysis underwent a major change. There was an increasing concern for the lack of "rigour" in analysis concerning basic concepts, such as functions, derivatives, and real numbers (Katz, 1998, 704-705). Mathematics was often connected to the intuition of time and space. Jahnke (1993) discusses that:

> "[…] there was a new emerging attitude among mathematicians during the mid 19th century whose aim was to erase the link between mathematics and the intuition of time and space (Jahnke, 1993, p. 268).

The definitions of many fundamental concepts in analysis were vague and gave rise to different views of not only the definitions, but also of the theorems involving these concepts. Laugwitz (1999) considers the 19th century as a "turning point" in the ontology as well as the method of mathematics. He argues that instead of using mathematics as a tool for computations, the emphasis fell on conceptual thinking (Laugwitz, 1999, p. 303).

In 1852 the Swedish mathematician E.G. Björling (1808-1872) wrote a survey of his view of some fundamental concepts in mathematical analysis. In fact, Björling's survey was the first half of a paper by Björling regarding power series expansions of complex valued functions. Björling pointed out that there seemed to exist different views of some of the most fundamental concepts in analysis which could effect the result of the main issue with his 1852 paper. Björling wrote;

> It goes without saying, that it has been necessary to return to some of the fundamental concepts in higher analysis, whose conception one has not yet generally agreed on […] It was, from my point of view, necessary, that I in advance – and before the main issue was considered - gave *my own* conception of these propositions' general applicability, then not only the base, which I have built, would be properly known, but also every misunderstanding of the formulation of the definite result would be prevented (Björling, 1852, p. 171).

Now, let us consider some examples from Björling's 1852 survey.

### 4.2.1 Examples from Björling's 1852 survey

It seems that Björling had a tendency to sometimes consider mathematical definitions as descriptions of entities rather than conventions. For instance, in the 1852 survey Björling defined a function as

> […] an analytical expression which contains a real variable $x$ (Björling, 1852, p. 171).

Björling certainly defined functions but sometimes he seems to consider the definition as describing something that already exists. As a consequence of his definition of a function, Björling considered every variable expression containing a variable that could be written up as a function. Furthermore, as we are about to see, for Björling a function did not need to be single-valued.

Let us consider three functions that Björling was discussing in his survey from 1852. In modern terminology these three functions would be expressed as;

$$f(x) = \frac{x}{|x|}, \quad g(x) = \frac{\sqrt{x} - \sqrt{a}}{x - a} \quad and \quad h(x) = |x|.$$

These functions are graphically represented in Figure 9.

f(x)                    g(x)                    h(x)

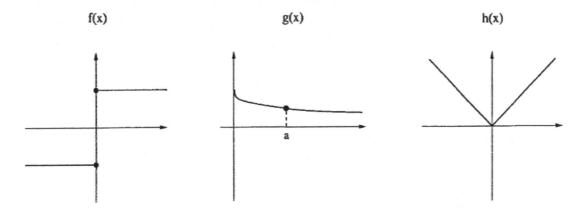

Figure 9: A graphical representation of the graphs $f$, $g$ and $h$.

Björling considered the function $f(x)$ as a function which attains the *two* values $\pm 1$ at

$$x = 0.$$

Another example of a function which Björling considered as two-valued is $y(x) = \sqrt{x}$. However, according to Björling this was another type of a two-valued function than $f(x)$ since $y(x) = \sqrt{x}$ has two values in more than finitely many points. Björling claimed:

> $\sqrt{x}$ has namely for each negative $x$-value ($= -y$) two specific values $\pm\sqrt{y}$ (Björling, 1852, p. 173).

Perhaps one could say that we have made it easier for us today since a function is always defined on a specific domain.

Furthermore, according to Björling, the derivative of $f(x)$ at $x = 0$ is *"undetermined"* since:

110

$D\left(\frac{x}{\sqrt{x^2}}\right)$ means namely that $\lim_{(\Delta=0)} \frac{F(\Delta)-F(0)}{\Delta}$, where $F(x)$ denotes $\frac{x}{\sqrt{x^2}}$. And since $F(\Delta)$, that is $\frac{\Delta}{\sqrt{\Delta^2}} = \pm 1$, as $\Delta$ is positive or negative, and $F(0)$ has the double value $\pm 1$; then obviously $D\left(\frac{x}{\sqrt{x^2}}\right)$ is highly undetermined (Björling, 1852, p. 177).[8]

Hence, for Björling the function $f(x)$ exists everywhere and attains two values at $x = 0$, but the derivative at the same point is "*undetermined*". However, the derivative of the function $h(x)$ at $x = 0$ exists and attains the two values $\pm 1$. According to Björling this is due to the fact that the function $h(x)$ attains only *one* value at $x = 0$. In general he stresses that

> [...] it is impossible for $DF(x)_{[x=x_0]}$ to be a finite and determined quantity unless $F(x_0)$ itself has one single finite value (Björling, 1852, p. 177).

According to Björling, the function $g(x)$ attains the one value $\frac{1}{2\sqrt{a}}$ at $x = a$, since

$$\lim_{\Delta \to 0} g(a + \Delta) = \frac{1}{2\sqrt{a}}.$$

In modern terminology we would say that Björling considers $g(a)$ as a removable discontinuity. Björling did not consider whether the derivative of $g(x)$ at $x = a$ existed or not, but probably (on the basis of similar examples in Björling, 1852) Björling would say that the derivative at $x = a$ was equal to $-\frac{1}{8a\sqrt{a}}$ since

$$\lim_{\Delta \to 0} \frac{g(a + \Delta) - g(a)}{\Delta} = -\frac{1}{8a\sqrt{a}}.$$

On the basis of the three functions $f(x)$, $g(x)$ and $h(x)$ above, it seems that Björling's approach was to investigate the behaviour of mathematical objects on the account of their "natural properties". At least one gets the impression that, for Björling, it was already presupposed that the expression written on the paper was a function and the task for Björling was to discover their exact properties. For instance, it seems that Björling tried to find answers in the graphs of the functions.

It is noticable that Björling lived during a time period when mathematics underwent a major change. One can discern the "old" mathematical approach as well as the "new" mathematical approach in Björling's work. For instance, Björling had an "old-fashioned" way of considering functions when he sometimes considered them as something that already existed and the definition worked as a description. At the same time, Björling made an important

---

[8] Björling uses the notation $\frac{x}{\sqrt{x^2}}$ instead of the modern notation $\frac{x}{|x|}$.

contribution to the development of uniform convergence in connection with Cauchy's theorem on the continuity of an infinite series.[9]

## Final remarks

In this paper we have studied mathematical results of some less-known mathematicians from different time periods in order to emphasize the didactical nature in historical examples. Apparently, the didactical nature in our examples can be seen from different perspectives. For instance, in the Islamic mathematican al-Tusi's work on finding maximum values of functions it turns out that he was using the idea of derivatives already in the 12th century. That is, al-Tusi's work provides, on the one hand, that the idea of derivatives can be grasped already in the 12th century. On the other hand, al-Tusi's work provides an alternative way (to the modern usage of derivatives) of finding maximum values of certain functions. Moreover, al-Tusi's work elucidates the complexity of finding maximum values of functions without the usage of the modern concept of derivative. A similar reflection can be noticed in al-Haytham's work of finding the volume of a solid of revolution. Al-Haytham's solution is correct but much more complicated than the modern way of using integrals. For instance, al-Haytham needed to introduce a formula for the sum of the fourth powers to find the volume.

Another example of a didactical view of this paper is the investigation of the Swedish 19th century mathematician E.G. Björling's efforts of understanding concepts in mathematical analysis. Björling's struggling of clarifying fundamental mathematical concepts can be seen as an example of the difficulty of working during a time period when mathematics (as a science) underwent a significant change. Laugwitz (1999) argues that during the mid 19th century mathematcs went from being considered more as "a tool for computations" to being considered as a study of abstract objects. One can discern both the "old" and the "new" mathematics in Björling's work. On the one hand Björling is concerned that there seemed to exist different definitions of some of the most fundamental concepts in mathematical analysis and tried to sort out this problem on the basis of his own survey from 1852. On the other hand, it seems that Björling had a tendency of sometimes consider mathematical definitions as descriptions of entities rather than conventions. The examples discussed in Section 4.2.1 in this paper reveals that Björling sometimes tried to "understand" the right behaviour of certain functions.

The 17th century debate between Hobbes and Wallis is another example of the difficulty of trying to "understand" mathematical objects on the basis of a picture and not on the basis of a definition. Hobbes insisted that [...] *every object must exist in the universe and be perceived by the "natural light"* (Mancosu, 1996, pp. 137-138). Of course, this became problematic in connection to concepts including "the infinitely large" or "the infinitely small". For instance, according to Hobbes, it was not possible that "Torricelli's infinitely long solid" (which was

---

[9] Björling's contribution to uniform convergence in connection with Cauchy's sum theorem can be found in (Bråting, 2007).

discussed in Section 4.1.2 in this paper) could be seen as something given since it had no material base and therefore could not exist (Mancosu, 1996, p. 147). Perhaps one may say that Hobbes in some cases did not clearly distinguish between mathematical objects and real objects. To be speculative, perhaps this kind of reasoning comes up in school mathematics in connection to the introduction of derivatives and integrals.

# References

Berggren, J. L. (1990). Review: Innovation and tradition in Sharaf al-din al-tusi's muadalat. *Journal of the American Oriental Society. 110(2)*, 304-309.

E.G. Björling (1852). Om det Cauchyska kriteriet på de fall, då functioner af en variabel låta utveckla sig i serie, fortgående efter de stigande digniteterna af variablen, *Kongl. Vetens. Akad. Förh. Stockholm.* 166-228.

Bråting, K & Öberg, A. (2005). Om matematiska begrepp – en filosofisk undersökning med tillämpningar. *Filosofisk Tidskrift.* 26, 11-17.

Bråting, K. (2007). A new look at E.G. Björling and the Cauchy sum theorem. *Archive for History of Exact Sciences.* 61, 519-535.

Clavius, C. (1607). *Euclidis Elementorum*, Frankfurt.

Datta, B., & Singh, A. (1984). Use of calculus in Hindu mathematics. *Indian Journal of History of Science. 19*, 95-104.

Goonatilake, S. (1998). *Toward a global science: Mining Civilizational Knowledge.* Bloomington, IN: Indiana University Press.

Guilbeau, L. (1930). The history of the solution of the cubic equation. *Mathematics News Letter. 5*, 8-12.

T. Hobbes (1656). *Six lessons to the professors of mathematics of the institution of Sir Henry Savile,* London.

Jahnke, H.N. (1993). Algebraic analysis in Germany, 1780-1840: Some mathematical and philosophical issues, *Historia Mathematica.* 20, 265-284.

Katz, V.J. (1995). Ideas of Calculus in India and Islam. *Mathematics Magazine. 68*, 163-174.

Katz, V.J. (1998). *History of mathematics: an introduction.* Addison-Wesley Educational Publishers, Inc.

Katz, V. J. (2004). *A history of mathematics: Brief edition.* Boston, MA: Pearson Education.

Katz, V. J. (Ed.). (2007). The mathematics of egypt, mesopotamia, china, india, and islam: A sourcebook. Princeton, NJ: Princeton University Press

Laugwitz, D. (1999). *Bernhard Riemann – Turning points in the conception of mathematics.* Birkhäuser, Boston.

P. Mancosu (1996). *Philosophy of mathematics and mathematical practise in the seventeenth century*, Oxford University Press.

J. Peletier (1563). *De contactu linearum.*

Rajagopal, C.T., & Rangachari, M.S. (1978). On an untapped source of medieval keralese mathematics. *Archive for History of Exact Sciences. 18*, 89-102.

Raju, C.K. (2001). Computers, mathematics education, and the alternative epistemology of the calculus in the yuktibhasa. *Philosophy East and West*, 51(3), 325-362 [Proceedings of the Eighth East-West Philosophers' Conference]

Roy, R. (1990). The discovery of the series formula for pi by Leibniz, Gregory, and Nilakantha. *Mathematics Magazine. 63*, 291-306.

Wallis, J. (1685). *A defense of the treatise of the angle of contact,* appendix to *Treatise of algebra,* London.

# The Brachistochrone Problem: Mathematics for a Broad Audience via a Large Context Problem

JEFF BABB
*Department of Mathematics and Statistics, University of Winnipeg, 515 Portage Avenue, Winnipeg, MB, Canada R3B 2E9 (E-mail: j.babb@uwinnipeg.ca)*

JAMES CURRIE
*Department of Mathematics and Statistics, University of Winnipeg, 515 Portage Avenue, Winnipeg, MB, Canada R3B 2E9 (E-mail: j.currie@uwinnipeg.ca)*

**Abstract.**

Large context problems (LCP) are useful in teaching the history of science. In this article we consider the brachistochrone problem in a context stretching from Euclid through the Bernoullis. We highlight a variety of results understandable by students without a background in analytic geometry. By a judicious choice of methods and themes, large parts of the history of calculus can be made accessible to students in Humanities or Education.

## Introduction

Each year the University of Winnipeg offers several sections of an undergraduate mathematics course entitled MATH-32.2901/3 *History of Calculus* (Babb, 2005). This course examines the main ideas of calculus and surveys the historical development of these ideas and related concepts from ancient to modern times. Students of Mathematics or Physics may take the course for Humanities credit; the course surveys a significant portion of the history of ideas and in fact is cross-listed with Philosophy. On the other hand, many students in Education take *History of Calculus (H of C)* to fulfill their Mathematics requirement; it is therefore necessary that the course offer solid mathematical content. Unfortunately, a significant fraction of these latter students are weak in pre-calculus material such as analytic geometry. Nevertheless, H of C welcomes the weak and the strong students together, and covers technical as well as historical themes.

In Stinner and Williams (1998) the authors enumerate the benefits of studying **large context problems (LCP)** in making science interesting and accessible. In their words 'the LCP approach provides a vehicle for traversing what Whitehead (1967) refers to as "the path from romance to precision to generalization"(p. 19).' For H of C, a useful LCP is the **Brachistochrone Problem:** the solution history of finding the curve of quickest descent. A focus on the brachistochrone motivates results ranging from Greek geometry, past the kinematics of Oresme and Galileo, through Fermat and Roberval to the Bernoullis, and the birth of the calculus of variations. By

116

careful selection of material, it is possible to find proofs accessible even to weak students, while still stimulating mathematically strong students with new content.

## Quickest Descent in Galileo

One of the topics considered by Galileo Galilei in his 1638 masterpiece, *Dialogues Concerning Two New Sciences,* is rates of descent along certain curves. In Proposition V of "Naturally Accelerated Motion", he proved that descent time of a body on an inclined plane is proportional to the length of the plane, and inversely proportional to the square root of its height (Galilei 1638/1952, p.212). Denoting height by H, length by L and time by T, we would write

$$T = kL/\sqrt{H} \qquad (1)$$

where k is a constant of proportionality. In fact, as we point out to students, this formulation is slightly foreign to the thought of Galileo; for reasons of homogeneity, he only forms ratios of time to time, length to length, etc. He therefore says that the ratio of times (a dimensionless quantity) is proportional to the ratio of lengths, inversely proportional to the ratio of square roots of heights. Galileo proves (1) in a series of propositions starting with the "mean speed rule":

> 'The time in which any space is traversed by a body starting from rest and uniformly accelerated is equal to the time in which that same space would be traversed by the same body moving at a uniform speed whose value is the mean of the highest speed and the speed just before acceleration began.' (Proposition I, Galilei 1638/1952, p.205).

It is insufficiently well-known that this rule had been proven geometrically by Nicole Oresme three centuries earlier! Oresme's very accessible geometric derivation is presented early in H of C. (Babb, 2005)

Moving closer to the question of quickest descent, with his Proposition VI, Galileo established the law of chords:

> 'If from the highest or lowest point in a vertical circle there be drawn any inclined planes meeting the circumference, the times of descent along these chords are each equal to the other'. (Galilei 1638/1952, p.212)

Galileo's proofs are given in a series of geometric propositions. Unfortunately, many of our students would identify with the complaint that Galileo places in the mouth of Simplicio:

> 'Your demonstration proceeds too rapidly and, it seems to me, you keep on assuming that all of Euclid's theorems are as familiar and available to me as his first axioms, which is far from true.' (Galilei 1638/1952, p. 239)

Happily, in an earlier part of H of C dealing with Greek mathematics, some geometrical rudiments are established. In particular, students see a proof of Thales' theorem – an angle inscribed in a semi-circle is a right angle – using similar triangles. This allows the following demonstration of the law of chords:

Figure 1: The law of chords

Let a circle of diameter D have a chord of length L and height H inscribed as shown in Figure 1. Let the descent time along the chord be T. By similar triangles L/H = D/L, so that $L^2/H = D$. Then by (1),

$$T = kL/\sqrt{H} = k\sqrt{D} \qquad (2)$$

which is indeed constant. Q.E.D.

The law of chords can also be demonstrated by deriving expressions for the velocities and descent time, and noting that the expression for descent time is independent of the upper point of the chord along the circular arc. The expression for descent time may be obtained by considering the component of the force of gravity along the inclined chord. Alternatively, Nahin (2004, pp. 202-206) derives the descent time by applying the principle of conservation of mechanical energy. Our brief geometrical derivation is rather close to the spirit of Galileo's proof, and is accessible to students without a physics background.

Lattery discusses an interesting approach suggested by Matthews in 1994 for leading students towards a derivation of the law of chords. Students are asked to consider the following thought experiment:

> 'Suppose a ball is released at some point A on the perimeter of a vertical circle and rolls down a ramp to point B, the lowest point on the circle … The ramp may be rotated about point B. For what angle will the time of descent along chord AB be the least? '
> (Lattery, 2001, p. 485)

This way of posing the problem helps students to greater appreciate the surprising result. It also allows the option of discussing the distinction between sliding and rolling motion. H of C is a mathematics course and has no laboratory component. Nevertheless, (particularly for students with weak physical intuition) it is useful to be able to observe the results studied in various descent problems. As experimentalists know, however, the design and operation of physics demonstration apparatus can be as much art as science. For this reason, we have chosen to use physically realistic computer simulations to illustrate various theorems. A computer program in MAPLE allows freedom in the choice of curves studied, as well as the possibility of speeding, slowing or freezing demonstrations. Figure 2 shows two screen snapshots from the program. Portability is one more advantage of this approach, to go along with flexibility of use.

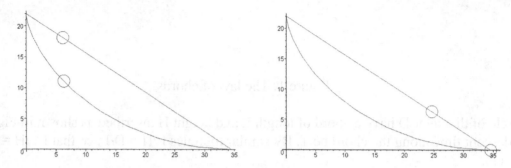

Figure 2: Screen snapshots from *MAPLE* race: cycloid vs. straight-line ramp

Details of the simulation are closely related to Jakob Bernoulli's solution of the brachistochrone problem, and are detailed in a later section.

With his Proposition XXXVI, Galileo proved that the descent time from a point on the lower quadrant of a circle to the bottom is quicker along two consecutive chords than along a direct chord. He began his proof by a clever application of the law of chords, but then completed it by a fairly involved geometric argument. In H of C, the proof is completed using a shorter method proposed by Erlichson (1998), based on conservation of mechanical energy.

**Three Curves**

Three curves of major interest to the mathematicians of the seventeenth century were the cycloid, the isochrone and the brachistochrone. (See, for example, Eves, 1990, p. 426.) The definitions of these curves are *kinematic*; as students learn in H of C, the acceptance of curves defined via motion was part of a mathematical revolution in the seventeenth century. A *cycloid* is the curve traced by a point on the circumference of a circle as the circle rolls, without slipping, along a straight line. A *brachistochrone* from point A to point B is a curve along which a free-sliding particle will descend more quickly than on any other AB-curve. (It is thus an optimal shape for components of a slide or roller coaster, as we inform our students.) An *isochrone* is a curve along which a particle always has the same descent time, regardless of its starting point. A surprising discovery was that these three curves are one and the same!

Galileo may have been the first to consider the problem of finding the path of quickest descent. This is suggested by the initial statement of his Scholium to Proposition XXXVI:

'From the preceding it is possible to infer that the path of quickest descent from one point to another is not the shortest path, namely, a straight line, but the arc of a circle'. (Galilei 1638/1952, p.234)

Many researchers, such as Stillman Drake and Herman Goldstine, have concluded that Galileo incorrectly claimed that a circular arc is the general curve of quickest descent (Erlichson 1998, Erlichson 1999). However, Erlichson (1998, p.344), argues that Galileo restricted himself to descent paths that used points along a circle. Galileo's claim is based on an argument that the descent time for a particle along a twice-broken path is less than along a twice-broken path, and that the descent time along a multiply-broken path would be even less. According to Nahin (2004, pp. 208-209), Galileo's claim is correct, but his reasoning was flawed; Erlichson (1998) noted that Galileo's method of proving Proposition XXXVI holds for descent from rest, but fails to generalize to situations in which a particle is initially moving.

In fact, as we have mentioned, the brachistochrone is not the circle; it is a segment of an inverted cycloid. The cycloid was discovered in the early sixteenth century by the mathematician Charles Bouvelles (Cooke, 1997, p. 331). In the 1590s, Galileo conjectured and empirically demonstrated that the area under one arch of the cycloid is approximately three times the area of the generating circle. That the area is exactly three times that of the generating circle was proven by Roberval in 1634 and by Torricelli in 1644 (Boyer and Merzbach, 1991, p. 356). Roberval constructed the tangent to the cycloid in 1634 (Struik, 1969, p. 232). According to Cooke (1997, p. 331), constructions of the tangent to the cycloid were independently discovered circa 1638 by Descartes, Fermat and, and slightly later by Torricelli. In 1659, Christopher Wren also determined the length of a cycloidal arch, showing it to be exactly four times the diameter of the generating circle (Stillwell, 2002, p. 318). In 1659, Huygens discovered that the cycloid is a solution to the isochrone or tautochrone problem; he showed that a particle sliding on a cycloid will exhibit simple harmonic motion with period independent of the starting point (Stillwell, 2002, p. 238). Huygens published his discovery of the cycloidal pendulum in *Horologium oscillatorium* in 1673 (Boyer and Merzbach, 1991, p. 379). The cycloidal pendulum also features in Newton's Principia (Gauld, 2005). In 1696, Johann Bernoulli demonstrated that a brachistochrone is a cycloid (Erlichson, 1999).

Most of these discoveries concerning the cycloid are inaccessible to students with no calculus background. Remarkably, Roberval's historical construction (Struik, 1969, pp. 234-235) of a tangent to the cycloid is quite accessible to students, as it uses only the parallelogram law for vector addition. The result is presented in H of C:

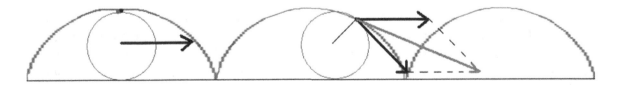

Figure 3. Finding a tangent to the cycloid

The result can almost be presented "without words". (See Figure 3.) Follow the path of the "tracing" point on the generating circle of a cycloid. The motion of this point at a given instant

120

has a horizontal component, corresponding to the horizontal motion of the center of the circle; it also has a component normal to a radius of the circle, since the circle rolls. These components are of equal magnitude, since the circle rolls without slipping. The resultant of the motions is found by the parallelogram law, and is the tangent to the cycloid.

It was early in 1696 that Johann Bernoulli solved the problem of finding the curve of quickest descent; he showed that the brachistochrone was a cycloid.  Later, in June of that year, he posed the problem in the journal *Acta Eruditorum*

> ' PROBLEMA NOVUM, ad cujus Solutione Mathematici invitantur.
> " Datis in plano verticali duobus punctis *A* et *B,* assignare mobili *M* viam *AMB,* per quam gravitate sua descendens, et moveri incipiens a puncto *A,* brevissimo tempore perveniat ad alterum punctum *B.* "' (Woodhouse 1810, pp.2 – 3)

An English translation is as follows:

> 'If two points A and B are given in a vertical plane, to assign to a mobile particle M the path AMB along which, descending under its own weight, it passes from the point A to the point B in the briefest time.' (Smith, D.E. 1929, p.644)

The problem was also solved by Jakob Bernoulli, Leibniz, L'Hôpital and Newton.  Newton's solution was published anonymously in the *Philosophical Transactions of the Royal Society* in January, 1697. Solutions by Johann Bernoulli, Jakob Bernoulli, Leibniz and Newton were published in *Acta Eruditorum* in May, 1697.  According to Stillwell (2002, p. 239), the most profound was Jakob Bernoulli's solution, which represented a key step in the development of the calculus of variations.  The historical development of what became the calculus of variations is closely linked to certain minimization principles in physics, namely, **the principle of  least distance, the principle of least time**, and ultimately, **the principle of least action**. (See Kline, 1972, pp. 572-582.) To understand Johann Bernoulli's solution of the brachistochrone problem, students in H of C are led through Fermat's **principle of least time**: light always takes a path that minimizes travel time.

## Principle of Least Time

An accessible application of the principle of least time is in deriving the **law of reflection**: if a ray of light strikes a mirror, then the angle of incidence equals the angle of reflection. This law was first noted by Euclid in the fourth century BCE (Ronchi, 1957, p. 11), and was explained using a principle of least distance by Heron of Alexandria in the first century CE (Cooke, 1997, p. 149). In the H of C course, the law of reflection is derived geometrically, as per Heron. Since the speed of light (in a fixed medium) is constant, this is equivalent to a derivation from the principle of least time.

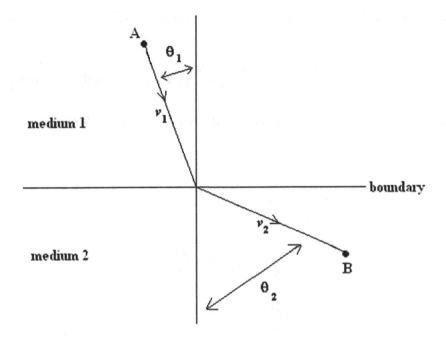

Figure 4: Snell's law of refraction

The **law of refraction** states that when a ray of light crosses the boundary between transparent media, it experiences a change in direction characterized by the relation

$$\sin \theta_1 / \sin \theta_2 = k$$

in which $\theta_1$ is the angle of incidence, $\theta_2$ is the angle of refraction and $k$ is a constant dependent on the nature of the two media. (See Figure 4.) This law was discovered experimentally by the Dutch physicist Willebrord Snel circa 1621; in English it is known as **Snell's law**. Snel noticed that if the first medium is less dense than the second, then $k > 1$; that is, upon entering the second medium, the light ray bends toward the normal to the boundary. (Nahin, 2004, p. 103)

In the mid-seventeenth century, Fermat demonstrated that Snell's law of refraction may be derived from the principle of least time. In the H of C course, such a derivation is given using an abbreviation of the approach outlined by Nahin: Consider a ray of light crossing the boundary between transparent media. For $i = 1, 2$ let $v_i$ denote the speed of light in medium $i$. Referring to Figure 5, let $T$ denote the transit time for a light ray travelling from point $A$ in medium 1, through point $B$ at the boundary, to point $C$ in medium 2. Then

$$T = (length\ of\ AB)/v_2 + (length\ of\ BC)/v_2 = \sqrt{{h_1}^2 + x^2}\Big/v_1 + \sqrt{{h_2}^2 + (d-x)^2}\Big/v_2$$

To obtain the path of least time, it is necessary to determine $x$ so that the transit time $T$ is minimized.

$$dT/dx = x\big/v_1\sqrt{h_1^2 + x^2} - (d-x)\big/v_2\sqrt{h_2^2 + (d-x)^2} = \sin\theta_1/v_1 - \sin\vartheta_2/v_2$$

The necessary condition for a minimum, namely that $dT/dx = 0$, yields the requirement that

$$\sin\theta_1 / \sin\theta_2 = v_1 / v_2$$

Thus, Snell's media-dependent constant is $k = v_1 / v_2$.

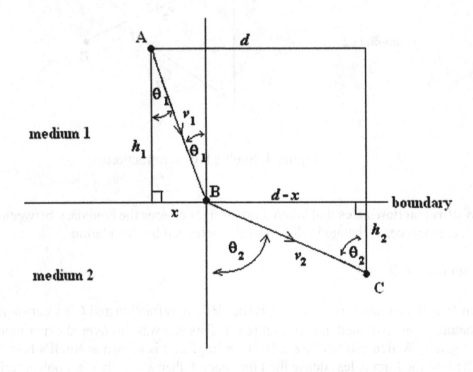

Figure 5: Derivation of Snell's law

It should be noted that Fermat achieved the minimization using his **method of adequality**, which is comparable to differentiation; however, he had to introduce some approximations, since he could not apply his adequality method directly to expressions involving square roots. Nahin (2004, p. 127-134) gives a detailed presentation of Fermat's solution.

**Johann Bernoulli's Solution to the Brachistochrone Problem**

In the brachistochrone problem, an ideal particle traverses an AB-curve under gravity. The traversal time will be determined if we can fix the speed of the particle at each point along its path. The principle of conservation of mechanical energy implies that if the particle starts at rest, and the vertical drop from A to a point is $y$, then the particle will acquire a speed at the given point of

$$v = (2\,g\,y)^{1/2} \qquad\qquad\qquad (3)$$

This speed is independent of whether the particle has dropped vertically, moved along an inclined line, or followed some more complicated path. The brachistochrone problem thus becomes the following:

> A particle moves from A to B in such a way that whenever its vertical drop from A is $y$, its speed is given by (3). Find the AB-curve with the shortest traversal time.

Johann Bernoulli solved the problem via a brilliant thought experiment. Consider a non-uniform optical medium which becomes increasingly less dense from top to bottom. If light enters from above, its speed becomes faster and faster as it moves down. By a judicious varying of the density, light may be constrained to travel through this medium in a manner satisfying (3). However, by the principle of least time, in any situation, light will always travel along a path with the shortest traversal time. We therefore see that if A is located at the top of this non-uniform medium, and B at the bottom, the path taken by light travelling from A to B is the brachistochrone!

This leaves the question of how light will travel through our non-uniform medium. Consider a light ray travelling through two transparent media, from point A in medium 1 (upper) to point B in medium 2 (lower). Let $\theta_1$ denote the angle of incidence and $\theta_2$ the angle of refraction. Let $v_1$ and $v_2$ denote the speed of light in the respective media. Suppose that medium 2 is less dense than medium 1, so that $v_2 > v_1$. Then, since

$$\sin\theta_1\,/\,\sin\theta_2\ =\ v_1\,/\,v_2\ <\ 1$$

$\theta_2 > \theta_1$, and the light ray bends away from the normal to the boundary. (See Figure 4.) Note, also, that

$$\sin\theta_1\,/\,v_1\ =\ \sin\theta_2\,/\,v_2\ =\ \text{constant}$$

Now, consider a similar situation with a light ray traveling downward through many layered transparent media, with each medium less dense than the layer above it. The speed of light increases in the successive media as it progresses through deeper layers and the ray of light bends further away from each successive normal to the boundary at point of contact. (See Figure 6.)

124

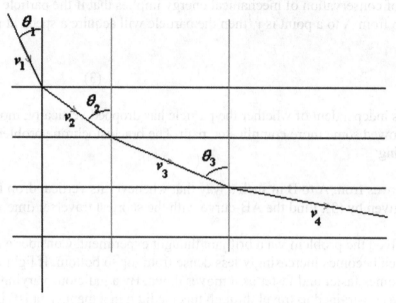

Figure 6: Refraction through multiple layers

Note also, that:

$$\sin \theta_1 / v_1 \;=\; \sin \theta_2 / v_2 \;=\; \sin \theta_3 / v_3 \;=\; \dots \;=\; \text{constant}$$

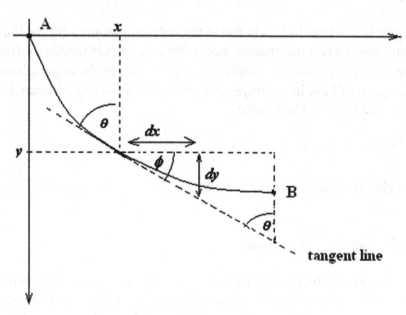

Figure 7: Johann Bernoulli's proof

Letting the number of layers increase without bound and the thickness of each layer decrease towards zero, the path of the light ray becomes a smooth curve. (See Figure 7.) At each point along the curve

$$\sin \theta / v = c \qquad\qquad\qquad (4)$$

Bernoulli thus realized that a particle falling along the curve of quickest descent from A to B must satisfy both equations (3) and (4). From the triangle in Figure 7

$$\sin \theta = \cos \phi = 1 / \sec \phi = 1 / [1 + \tan^2\phi]^{1/2} = 1 / [1 + (dy/dx)^2]^{1/2}$$

Thus

$$\sin \theta / v = \text{constant} = 1 / (v [1 + (dy/dx)^2]^{1/2})$$

where $v = (2gy)^{1/2}$. This yields the following nonlinear differential equation:

$$y [1 + (dy/dx)^2] = k \qquad\qquad\qquad (5)$$

where $k$ is some constant. Algebraic manipulation of the differential quantities, $dx$ and $dy$, yields the equation

$$dx = dy [y/(k-y)]^{1/2} \qquad\qquad\qquad (6)$$

which Bernoulli recognized as a differential equation describing a cycloid. In the translated words of Johann Bernoulli:

> 'from which I conclude that the Brachystochrone is the ordinary Cycloid'
> (Struik 1969, p .394 translation of Johann Bernoulli 1697)

Note that in Bernoulli's May 1697 paper in *Acta Eruditorum*, the usual labelling of $x$ and $y$ coordinates is reversed. Equation (6) may be further manipulated to obtain parametric equations for $x$ and $y$; for details, see the excellent accounts by Simmons (1972), Erlichson (1999) and Nahin (2004).

## Jakob Bernoulli's Solution to the Brachistochrone Problem

Although Johann Bernoulli's solution to the brachistochrone problem impresses us with its elegance, it is tailored to a very specific application. Jakob Bernoulli's more methodical approach generalizes, and in fact became the basis of the calculus of variations. In fact, neither of the Bernoullis' solutions uses calculus explicitly in the foreground. Each of the brothers, by a different method, sets up a differential equation, and having found this equation, declares the problem solved. (Struik, 1969, pp. 392-399). Their quickness at this early date to recognize a differential equation of the cycloid is striking!

Since the calculus is in the background, Jakob Bernoulli's solution to the cycloid may be outlined to H of C students. We give our abbreviated presentation of his proof below. It relies on the use of similar triangles, some (typical) hand-waving regarding infinitesimals, and the concept of *stationary points* of functions.

A *stationary point* of a function is one at which the function's rate of change is zero. In the ordinary calculus, we recognize that local extrema of a function occur at stationary points. Jakob Bernoulli extended this idea to the brachistochrone problem. Since the brachistochrone minimizes descent time, the rate of change of descent time must be zero with respect to infinitesimal variation of the brachistochrone path. Consider Figure 8.

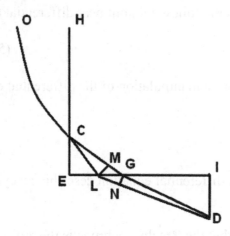

Figure 8: Jakob Bernoulli's solution of the brachistochrone problem

Let curve $OCGD$ be a small section of the brachistochrone. Letting $y$ measure vertical drop from $O$, choose units so that a particle moves along the curve with instantaneous speed $\sqrt{y}$ at any point. We consider $CG$ to be so short that a particle moves along $CG$ with constant speed $\sqrt{|HC|}$ where $|HC|$ denotes the length of $HC$. Similarly, we assume a constant speed $\sqrt{|HE|}$ on $GD$.

Vary the path by moving $G$ an infinitesimal distance horizontally, to $L$. As the brachistochrone is stationary, the descent time along $OCLD$ must also be minimal. Add construction lines $ML$ and $NG$ such that triangles $\Delta CML$ and $\Delta DNG$ are isosceles. The descent times along $CM$ and $CL$ are thus equal, as are descent times along $GD$ and $ND$. As the total descent times along $OCGD$ and $OCLD$ are to be equal, the descent time along $MG$ must equal that along $LN$, and

$$|MG|/\sqrt{|HC|} = |LN|/\sqrt{|HE|}. \qquad (7)$$

As we are dealing with infinitesimal distances, we may consider *ML* to be an arc of a circle centered at *C*, and *LMG* to be a right angle. By similar triangles, then, $|MG| / |LG| = |EG| / |CG|$. If we let *x* measure horizontal distance, and *s* arc length along the brachistochrone, this can be rewritten as

$$|MG| / |LG| = dx / ds \text{ on segment } CG.$$

By an analogous argument,

$$|LN| / |LG| = dx / ds \text{ on segment } GD.$$

Dividing by $\sqrt{y}$ on each of segments *CG* and *GD*, and applying equation (7), we find that

$$dx / \sqrt{y} ds = k \qquad\qquad (8)$$

with the same constant *k* on both segment *CG* and segment *GD*. We conclude that equation (8) holds everywhere on the brachistochrone. Jakob Bernoulli recognized this as a differential equation of the cycloid.

Note that various mysteries involving infinitesimals take place; segment *CG* is only short, while *LG* is infinitesimal. Also, isosceles triangle $\Delta CML$ contains two right angles, and this is essential to the argument with similar triangles. Again, Bernoulli interchanged *x* and *y*, which can be confusing to a modern reader.

### Computerized Demonstration of the Brachistochrone

It may be difficult for students to grasp the nature of the minimization problem involved in finding the brachistochrone. We are not finding the tangent or area of some particular *given* curve, as is usually the case in calculus. Instead, we must search among *all possible hypothetical curves* to find that which allows least time descent. The nature of the problem also precludes physical demonstration; we may build an apparatus to demonstrate sliding descent on a particular curve, or we may "race" physical beads on two or three particular curves; however, it is hard to conceive how one would physically demonstrate beads descending on enough curves to allow students to conceptualize descent on an *arbitrary* curve. Here the computer comes to the rescue:

Using the symbolic programming language *MAPLE*, we simulate the descent under gravity of a particle along a curve as follows:
1. Curves are given parametrically: $x = x(s)$, $y = y(s)$, $a \le s \le b$.
2. A given curve is systematically sampled at $n + 1$ points $s_i = a + i(b - a)/n$, $i = 0, 1, ..., n$.
3. The curve is modeled by straight-line segments, from $(x_i, y_i)$ to $(x_{i+1}, y_{i+1})$, $i = 0, 1, ..., n - 1$.

4. The motion of a particle under gravity down the digitized curve is modeled in a straight-forward way: If the particle enters the straight-line segment from $(x_i, y_i)$ to $(x_{i+1}, y_{i+1})$ moving at speed $v_0$, its acceleration under gravity will be $a = g \sin \theta$, where $g$ is the acceleration (downward) due to gravity, and $\sin \theta = (y_i - y_{i+1})/\sqrt{(x_i - x_{i+1})^2 + (y_i - y_{i+1})^2}$. The length of the segment is $D = \sqrt{(x_i - x_{i+1})^2 + (y_i - y_{i+1})^2}$. At time $t$ after entering the segment, the particle has moved distance $d = v_0 t + \frac{1}{2} a t^2$ along the segment. The time $T$ spent traversing the segment is then the solution of $D = v_0 T + \frac{1}{2} a T^2$, so that the particle enters the next segment with a speed of $v_0 + aT$.

5. The positions of particles on several digitized curves can thus be worked out as parametric functions of time. The *MAPLE* command *animate*() is then capable of presenting trajectories of two or more of these particles simultaneously, as they each move along their underlying curves. Our code as currently implemented contrasts the motion of particles along two desired curves.

We have found that $n = 11$ already gives very smooth-looking approximations. Using our software, we "race", for example, a particle on a cycloid arc against a particle on a straight-line ramp (see Figure 2), or a particle on a circular arc. We can in fact race along any (parametric) curve suggested by students. The same software will illustrate Galileo's law of chords. After students have seen Johann Bernoulli's solution to the brachistochrone, they can appreciate the analogy between modern "digitization" and Bernoulli's layers.

## Conclusion

We have traced the thread of quickest descent problems and the brachistochrone from Galileo, through Fermat and Roberval, to the Bernoullis and the dawn of the calculus of variations. We have spelled out in detail a selection of mathematical results which we have presented to our H of C students. These results include mathematical content ranging in difficulty from geometry, through vectors, to differential equations. A themed unit on the cycloid shows students at very different levels the strong interplay between mathematics and physics: Geometry informs optics (Fermat), optics informs kinematics (Johann Bernoulli), and kinematics informs geometry (Roberval).

Of particular interest is Johann Bernoulli's beautiful *Gedankenexperiment*, whereby a falling particle becomes a ray of light, moving through media arranged without regard for the possibility of actual physical construction. Again, the purely mathematical and hypothetical nature of the frictionless bead in the brachistochrone problem motivates our software demonstrations to students. The digital sampling of curves in our *MAPLE* code echoes the (finite) layering of media by Johann Bernoulli. A careful examination of the arguments of the Bernoullis introduces students to the interesting philosophical and technical issues related to infinitesimals/differentials in mathematics and physics.

# References

Babb, J.: 2005, 'Mathematical Concepts and Proofs from Nicole Oresme: Using the History of Calculus to Teach Mathematics', *Science and Education* 14(3-5), 443-456.

Boyer, C.B. & Merzbach, U.C.: 1991, *A History of Mathematics*, 2^nd edition, John Wiley and Sons, New York.

Cooke, R.: 1997, *The History of Mathematics: A Brief Course*, John Wiley and Sons, New York.

Erlichson, H.: 1998, 'Galileo's Work on Swiftest Descent from a Circle and How He Almost Proved the Circle Itself Was the Minimum Time Path', *American Mathematical Monthly* 105, 338-347.

Erlichson, H.: 1999, 'Johann Bernoulli's brachistochrone solution using Fermat's principle of least time', *European Journal of Physics* 20, 299-344.

Eves, Howard W. 1990, *Introduction to the History of Mathematics* 6^th *edition,* Thomson, Toronto

Galilei, G.: 1638/1952, *Dialogues Concerning Two New Sciences,* translated by H. Crew and A. de Salvio, Encyclopaedia Britannica, Chicago.

Gauld, C.: 2005, 'The Treatment of Cycloidal Pendulum Motion in Newton's *Principia*' in *The Pendulum: Scientific, Historical, Philosophical and Educational Perspectives*, Springer, The Netherlands, pp.115-125.

Kline, M.: 1972, *Mathematical Thought from Ancient to Modern Times*, Oxford University Press, New York.

Lattery, M.J.: 2001, 'Thought Experiments in Physics Education: A Simple and Practical Example', *Science and Education* 10: 485-492.

Nahin, P.J.: 2004, *When Least is Best: how mathematicians discovered many clever ways to make things as small (or as large) as possible*, Princeton University Press, Princeton, NJ.

Ronchi, V.: 1957, *Optics: The Science of Vision* (reprinted in 1991), Dover Publications, New York.

Simmons, G.F.: 1972, *Differential Equations with Applications and Historical Notes*, McGraw-Hill Book Company, New York.

Smith, D.E.: 1929, *A Source Book of Mathematics, Volume Two* (reprinted in 1959), Dover Publications, New York.

Stillwell, J.: 2002, Mathematics and Its History, Springer-Verlag, New York.

Stinner, A. & Williams H.: 1998, 'History and Philosophy of Science in the Science Curriculum', International handbook of science education, Springer, New York.

Struik, D.J. (editor): 1969, *A Source Book in Mathematics 1200-1800*, Harvard University Press, Cambridge.

Whitehead, A.:1967, *The aims of education.* New York: The Free Press.

Woodhouse, R.: 1810, *A History of the Calculus of Variations in the Eighteenth Century*, Chelsea Publishing Company, Bronx, New York (reprint of *A Treatise on Isoperimetrical Problems and the Calculus of Variations*, originally published 1810).

References

# Chopping Logs:
# A Look at the History and Uses of Logarithms

*Rafael Villarreal-Calderon[1]*
*The University of Montana*

Abstract:
Logarithms are an integral part of many forms of technology, and their history and development help to see their importance and relevance. This paper surveys the origins of logarithms and their usefulness both in ancient and modern times.

Keywords: Computation; Logarithms; History of Logarithms; History of Mathematics; The number "e"; Napier logarithms

## 1. Background

Logarithms have been a part of mathematics for several centuries, but the concept of a logarithm has changed notably over the years. The origins of logarithms date back to the year 1614, with John Napier[2]. Born near Edinburgh, Scotland, Napier was an avid mathematician who was known for his contributions to spherical geometry, and for designing a mechanical calculator (Smith, 2000). In addition, Napier was first to make use of (and popularize) the decimal point as a means to separate the whole from the fractional part in a number. Napier was also very much interested in astronomy and made many calculations with his observations and research. The calculations he carried out were lengthy and many times involved trigonometric functions (RM, 2007). After many years of slowly building up the concept, he finally developed the invention for which he is most known: logarithms (Smith, 2000).

In his book (published in 1614) *Mirifici Logarithmorum Canonis Descriptio* (Description of the wonderful canon of logarithms), Napier explained why there was a need for logarithms,

> Seeing there is nothing…that is so troublesome to Mathematicall practise, nor that doth more modest and hinder Calculators, than the Multiplications, Divisions, square and cubical Extractions of great numbers, which besides the tedious expense of time, are for the most part subject to many errors, I began therefore to consider in my minde by what certaine and ready Art I might remove those hindrances. (Smith, 2000)

During Napier's time, many astronomical calculations required raw multiplication and division of very large numbers. Sixteenth-century astronomers often used prosthaphaeresis, a method of

---

[1] rafael.villarreal-calderon@umontana.edu; rafaelvillarreal2000@hotmail.com
[2] The word "logarithm" was coined by Napier from the Greek "logos" (ratio) and "arithmos" (number) [9].

obtaining products by using trigonometric identities like $\sin\alpha \cdot \sin\beta = \frac{1}{2}[\cos(\alpha - \beta) - \cos(\alpha + \beta)]$ and other similar ones that required simple addition and subtraction (Katz, 2004). For example, if one were to multiply 2994 by 3562, then $\sin\alpha$ would be 0.2994 (the decimal is placed so that the value of $\alpha$ can be used later) and $\sin\beta$ would be 0.3562—these would make $\alpha \approx 17.42$ and $\beta \approx 20.87$ (values obtainable in a table). Next, $\alpha$ and $\beta$ values would be inserted into the equation, again a table would be used, and simple subtraction and division by 2 would occur—the result would yield ~0.10665158. By moving the decimal the same number of times that it was moved in order to accommodate the trigonometric equation (eight places to the right), the answer becomes 10,665,158 (approximating the actual 10,664,628). Because this answer is an estimate, the desired number of accurate digits would be dependent on the values initially given to $\alpha$ and $\beta$. Performing such calculation tricks, astronomers could reduce errors and save time (Katz, 2004).

In addition to prosthaphaeresis, Napier also knew about other methods for simplifying calculations. Michael Stifel, a German mathematician, developed in 1544 a relationship between arithmetic sequences of integers and corresponding geometric sequences of 2 raised to those integers (Smith, 2000): $\{1, 2, 3, 4,\ldots, n\}$ and $\{2^1, 2^2, 2^3, 2^4,\ldots, 2^n\}$. Stifel wrote tables in which he showed that the multiplication of terms in one table correlated with addition in the other (Katz, 2004). For example, to find $2^3 \cdot 2^5$, one would add 3+5 (terms in the arithmetic sequence) and the answer could then be inserted back into the geometric sequence to obtain $2^3 \cdot 2^5 = 2^{3+5} = 2^8 =256$. These tables were limited, however, in their calculating ability; Napier's approach to using logarithms, on the other hand, allowed the multiplication of any numbers through the use of addition (Katz, 2004).

To define logarithms, Napier used a concept that is rather different from today's perception of a logarithm.[2] Since astronomers at the time often handled calculations requiring trigonometric functions (particularly sines), Napier's goal was to make a table in which the multiplication of sines could be done by addition instead (Katz, 2004). The process consisted of having a line segment and a ray, where a point was made to move on each (from one extreme end to the other). The starting "velocity" for both points was the same, but the difference began as one point moved uniformly (arithmetically on the ray) and the other moved geometrically such that its velocity would be proportional to the distance left to travel to the endpoint of the line segment. Using this mental model, Napier defined the distance traveled by the arithmetically moving point as the logarithm of the distance remaining to be traveled by the point moving geometrically (Cajori, 1893). In Napier's words, "the logarithm of a given sine is that number which has increased arithmetically with the same velocity throughout as that with which radius began to decrease geometrically, and in the same time as radius has decreased to the given [number]" (Katz, 2004). A detailed account of the process can be seen below in the *Calculation Techniques* section. Clearly, Napier's definition differed from the modern concept of just having a base raised to the corresponding exponent.

It took Napier about 20 years to actually assemble his table of logarithms (Katz, 2004), but shortly after publishing his book, Napier was visited by the English mathematician Henry Briggs (Smith, 2000). A professor of geometry in London, Briggs was impressed with Napier's work,

My lord, I have undertaken this long journey purposely to see your person, and to know by what engine of wit or ingenuity you came first to thing of this most excellent help in astronomy, viz. the logarithms; but, my lord, being by you found out, I wonder nobody found it out before, when now known it is so easy. (Cajori, 1893)

They both discussed the convenience of setting the logarithm of 1 equal to 0 (rather than the original 10,000,000) and setting the logarithm of 10 at 1. In this way, the more familiar form of the logarithm was born, and a common property like log (xy) = log x + log y could be used to make a new table. Napier died in 1617, so Briggs began to do the calculations to construct the table (Katz, 2004). Briggs did not convert Napier's logarithms to the new common logarithms, however. Instead he set out to calculate successive square roots to obtain the logarithms of prime numbers, and used these to calculate the logarithms of all natural numbers from 1 to 20,000 and from 90,000 to 100,000. Although he did use algorithms to obtain the roots, the amount of work needed to calculate all those logarithms is nonetheless astounding. To calculate the logarithm of 2, for instance, he carried out forty-seven successive square roots (Smith, 2000). In addition, all of the calculations for logs were carried out to 14 decimal places (Cajori, 1893). An example of the calculations needed for this task is shown below in the *Calculation Techniques* section. Finally, in 1624, Briggs published his tables in his *Arithmetica Logarithmica*. The logarithms of the numbers between 20,000 and 90,000 were calculated by the Dutchman Adrian Vlacq, who published the complete table from 1 to 100,000 in 1628 (Cajori, 1893).

*Below is a page from Briggs's Arithmetica Logarithmica (MatematikSider, 2007).*

The way logarithms were viewed changed over time, and today's notation for a logarithm was developed by Leonhard Euler in the late 1700s. He related exponential and logarithmic functions by defining $\log_x y = z$ to hold true when $x^z = y$ (Smith, 2000). This definition proved very useful and found multiple applications. A classic example of a practical application of logarithms is the slide rule. In 1622 the Englishman William Oughtred made a slide rule by placing two sliding logarithmic scales next to each other. The slide rule could replace the need to look up values in a logarithm table by instead requiring values to be aligned in order to perform the multiplication, division, and many other operations (depending on the model). Up until the 1970s, with the incoming of electronic calculators, the slide rule was widely used in the fields of science and engineering (Stoll, 2006). A look at how the slide rule could be used for calculations is shown below in the *Calculation Techniques* section.

Although the common logarithm has many practical uses, another logarithm is widely used in fields ranging from calculus to biology. The natural logarithm is of the form $\log_e a = n$. The base of a logarithm could be any number larger than 1, but the use of $e$ brings on various advantages (Lowan, 2002). The definition of $e$, the limit of $(1+1/n)^n$ as n approaches infinity, might seem a bit awkward at first, but it turns out that $e$ not only turns up frequently in nature, but it also makes natural logarithms have the simplest derivatives of all logarithmic systems (Evans, 1939). Various solutions to applied mathematical problems can be expressed as powers of $e$: the flow of electricity through a circuit, radioactive decay, bacterial growth, etc. (Lowan, 2002). The natural logarithm arose from modifications of Napier's logarithms made by John Speidell, a mathematics teacher from England. In 1622 he published the book *New Logarith* with logarithms of tangents, sines, and secants in a format that showed natural logarithms (except that he had omitted decimal points). For example, he gave log 10= 2302584, which would be written today as $\log_e 10 = 2.302584$ (Cajori, 1893). As an interesting note, the Napier log of x would be equivalent to the expression $10^7\log_{1/e}(x/10^7)$ in modern terms (Smith, 2000).

## 2. Calculation Techniques

> *Napier's definition of logarithms* (see Cairns, 1928, and Cajori, 1893, and Katz, 2004, and Pierce,1977)*:*

Given a ray and a line segment, the point $G$ moves along the ray and the point $H$ moves along the line segment.

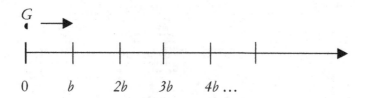

$G$ moves at a constant velocity by traveling $b$ distance in equal time intervals (along an increasing arithmetic sequence).

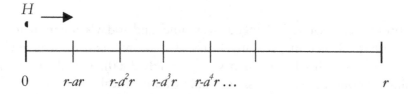

$H$ moves towards $r$ in equal time intervals from 0 to $r\text{-}ar$, $r\text{-}ar$ to $\text{r-a}^2\text{r}$, $\text{r-a}^2\text{r}$ to $\text{r-a}^3\text{r}$, etc. Napier made $r = 10{,}000{,}000$ and $a$ be less than 1 (but very close to 1).

He made the line segment (from 0 to $r$) be the "sine of 90°", and the distance from $r$ to $H$ the sine of the arc with the distance traveled by $G$ as its logarithm. Thus, Napier had log $10^7$ = 0.

Under this system, the notion of using bases with corresponding exponents did not apply.

In a calculus sense, Napier's logarithms could be seen as measures of "instantaneous" velocities. For example, the velocity of $H$ could be $V_H = \Delta d/\Delta t = d(r\text{-}x)/dt$, where x is the distance remaining to be traveled by $H$ to reach $r$. Similarly, the velocity of $G$ would be $V_G = dy/dt$, where y is the distance traveled by $G$ (this velocity is constant).

To obtain the definition of a Napier logarithm in modern calculus terms: $d(r\text{-}x)/dt = x$, since the velocity of $H$ is proportional to the distance remaining to be traveled by $H$ to reach $r$. So, $dr/dt - dx/dt = x$, and since $r$ is a constant ($10^7$): $0 - dx/dt = x \rightarrow 1/(\text{-}dx/dt) = 1/x \rightarrow \text{-}dt/dx = 1/x \rightarrow \int\text{-}dt = \int 1/x\,dx \rightarrow \text{-}t = \ln x + c.$ Since both $G$ and $H$ start at the same velocity, when $t = 0$ then $x = r$, thus $0 = \ln r + c \rightarrow c = \text{-}\ln r$, therefore, $\text{-}t = \ln x - \ln r$.

Point $G$ progresses in an arithmetical fashion, and its velocity is $dy/dt$. Having established that its velocity is constant and that it is equal with $H$'s velocity at $t = 0$, then $dy/dt = r$ so $dy = rdt \rightarrow \int dy = \int r\,dt \rightarrow y = rt$.

Finally, to relate $x$ and $y$: $\text{-}t = \ln x - \ln r \rightarrow t = \ln r - \ln x \rightarrow t = \ln (r/x) \rightarrow y = r \ln (r/x)$

By his definition, the Napier log $x = y$ is Naplog $x = r \ln(r/x) = 10^7 \ln(10^7/x)$

Napier did not use the notion of $e$ in calculating his logarithms, but this perspective helps to see the connection between logarithms, calculus, and the usefulness of $e$ and the natural logarithm.

➢ Using Napier's Logs in calculations (see Katz, 2004):

To use his logs in calculations, Napier had to note that Naplog $10^7 = 0$.
If $j/p = w/z$, then Naplog $(j)$ – Naplog $(p)$ = Naplog $(w)$ – Naplog $(z)$.
If $f/q = q/m$, then Naplog $(f)$ – Naplog $(q)$ = Naplog $(q)$ – Naplog $(m)$ and 2Naplog $(q)$ = Naplog $(f)$ + Naplog $(m)$
And if $f/q = m/k$, then Naplog $(f)$ + Naplog $(k)$ = Naplog $(q)$ + Naplog $(m)$.

Using these properties he established, conforming to his logarithms, a triangle could be solved by reference to his tables.

Example: using the law of sines, $\sin\theta / t = \sin\delta / d$ for triangle

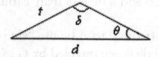

So to find $\delta$ the properties are applied,

Naplog $(\sin\delta)$ = Naplog $(\sin\theta)$ + Naplog $(t)$ – Naplog $(d)$

Referring back to his tables, Napier could calculate $\delta$ by simple addition and subtraction.

➢ Briggs's logarithms (see Cairns, 1928 and Henderson, 1930):

Briggs adapted Napier's logs to fit log 10 = 1 instead, thus giving birth to today's common logarithms. By taking successive square roots, Briggs concluded, for example, that if
$\sqrt{10} \approx 3.162277$, then log 3.162277 = 0.5
$\sqrt{\sqrt{10}} \approx 1.77828$, then log 1.77828 = 0.25
$\sqrt{\sqrt{\sqrt{10}}} \approx 1.33352$, then log 1.33352 = 0.125, etc.

To find the logarithms of prime numbers Briggs used the following method:
To find log 2, he noticed that if he raised 2 to a certain power, the number of digits in the result gave an approximation for log 2 (because of the properties of using logarithms with base 10); the log of a number with $x$ number of digits is between $x-1$ and $x$. For example, $2^8 = 256 \rightarrow 2 < \log 256 < 3$.
He then noted that $x$ and $x-1$ could be divided by the exponent to which 2 was raised to get an approximation of the log of 2:

| | | |
|---|---|---|
| $2^{10} = 1024$ | $\rightarrow 3 < \log 1024 < 4$ | so $0.3 < \log 2 < 0.4$ |
| $2^{20} = 1048576$ | $\rightarrow 6 < \log 1048576 < 7$ | so $0.3 < \log 2 < 0.35$ |
| $2^{40} \approx 1.1 \times 10^{12}$ | $\rightarrow 12 < \log 2^{40} < 13$ | so $0.3 < \log 2 < 0.325$ |
| $2^{60} \approx 1.2 \times 10^{18}$ | $\rightarrow 18 < \log 2^{60} < 19$ | so $0.3 < \log 2 < 0.3167$ |
| $2^{80} \approx 1.2 \times 10^{24}$ | $\rightarrow 24 < \log 2^{80} < 25$ | so $0.3 < \log 2 < 0.3125$ |
| $2^{100} \approx 1.3 \times 10^{30}$ | $\rightarrow 30 < \log 2^{100} < 31$ | so $0.3 < \log 2 < 0.31$ |

…and so forth until Briggs obtained log 2 to 14 decimal places. Once he calculated the logs for other prime numbers, he followed the rules of logarithms: for example, log 10 = log (2·5) = log 2 + log 5. Until his tables covered the logarithms of 1-20,000 and 90,000-100,000.

➢ Slide rule calculations (see Stoll, 2006):

The slide rule works by simplifying multiplications and divisions into logarithmic scale additions or subtractions. Slide rules basically print fit scales into a ruler-type setup and by just sliding a cursor against another scale, long operations can be done quickly. One could get away with using a slide rule without really understanding logarithms, but to make one, the following rules are essential:

$\log xy = \log x + \log y$         $\log (x/y) = \log x - \log y$
$\log x^y = y\log x$                   etc.…

Briggs's logarithms allowed long operations like $10478 \cdot 97503$ to become
$\log 10478 + \log 97503 = 4.020278 + 4.989018 = 9.009296$, then
antilog $9.009296 = 10478(97503) \approx 1{,}021{,}636{,}000$.

➤ Natural logarithms:

Logarithms with base $e$ unavoidably spring up in calculus (which was developed a little after Napier's death). To see how these logs are essential to obtain certain integrals: let $f(x) = \log_e x = \ln x$

$$f'(x) = \lim_{h \to 0} [f(x+h) - f(x)] / h = \lim_{h \to 0} [\ln(x+h) - \ln(x)] / h = \lim_{h \to 0} [\ln((x+h)/x)] / h$$

$$= \lim_{h \to 0} [\ln(1 + h/x) / h] \, [(1/x)/(1/x)] = \lim_{h \to 0} [1/x][\ln(1 + h/x)] / [h/x]$$

$$= \lim_{h \to 0} [1/x][x/h][\ln(1 + h/x)] = \lim_{h \to 0} [1/x][\ln(1 + h/x)^{x/h}] = [1/x]\lim_{h \to 0} \ln(1 + h/x)^{x/h}$$

$$= [1/x]\lim_{x/h \to \infty} \ln(1 + h/x)^{x/h} = [1/x] \ln e = 1/x$$

Thus, $d(lnx)/dx = 1/x$, and $\int 1/x \, dx = \ln x + c$.

The definition of $e$, $\lim_{n \to \infty}(1 + 1/n)^n = e$, allows the above demonstration to hold.

## 3. Conclusions & Implications

Today's concept of logarithms might make it seem strange that logarithms really developed out of comparing velocities of arithmetically and geometrically moving points. Napier's idea took him decades to fully develop and conclude, and the work of Briggs helped simplify and enhance a useful mathematical invention. What today seems like a simple base to exponent relationship really has a long history of work and improvements. The natural logarithm further helps us see the connection between the labors of a Scottish mathematician (and many others) with calculus and all its modern applications in math, science, and technology.

Napier's invention of the logarithm has surely left an important mark in the history of mathematics. The applications derived from the calculations he and others developed, still have relevance today. Although slide rules are now obsolete, the principles that allow them to work are not. The story of the development of logarithms is a good example of the effects that mathematical discoveries and inventions can have on society and the technological world.

In writing this paper I have learned a great deal about these calculation aids. But perhaps more importantly, I have realized that figuring out mathematical operations and tricks certainly takes significant amounts of effort, time, and devotion. Today, we often take for granted those symbols and explanations that are neatly compiled into math and science textbooks. It is easy to forget that every equation encases a story: frustration, fascination, arduous work, friendly collaborations, disappointment, and the occasional serendipity. Mathematics is not just about numbers, but it is also about the people whose work gives us the luxury and pleasure of understanding.

## Acknowledgements

Special thanks to Dr. Bharath Sriraman of the University of Montana.

## References

Cairns, W. D. (1928). Napier's logarithms as he developed them. *The American Mathematical Monthly. 35*, 64-67.

Cajori, F. (1893). *A history of mathematics.* New York, NY: Chelsea Publishing Company.

Evans, J. E. (1939).Why logarithms to the base e can justly be called natural logarithms. *National Mathematics Magazine. 14*, 91-95.

Henderson, J. (1930). The methods of construction of the earliest tables of logarithms. *The Mathematical Gazette. 15*, 250-256.

Katz, V. J. (2004). *The history of mathematics: Brief version.* Boston, MA: Pearson Education.

Logaritmer- Henry Briggs. Retrieved April 28, 2007, from Vestergaards Matematik Sider Web site: http://www.matematiksider.dk/briggs.html

Lowan, A. N. (2002). Logarithm. In *AccessScience@McGraw-Hill* [Web]. from http://www.accessscience.com, DOI 10.1036/1097-8542.389000

Pierce, R. C. (1977). A brief history of logarithms. *The Two-Year College Mathematics Journal. 8*, 22-26.

RM. (2007). Biography: Napier, John (1550–1617). In *AccessScience@McGraw-Hill* [Web]. Helicon Publishing . Retrieved January 11, 2008, from http://www.accessscience.com.weblib.lib.umt.edu:8080/content.aspx?id=M0090803

Smith, D. W. (2000). From the top of the mountain. *Mathematics Teacher. 93*, 700-703.

Stoll, C. (2006, May). When slide rules ruled. *Scientific American*, 80-87.

# The history of mathematics as a pedagogical tool: Teaching the integral of the secant via Mercator's projection

*Nick Haverhals[1] & Matt Roscoe[2]*
*Dept of Mathematical Sciences*
*The University of Montana*

Abstract: *This article explores the use of the history of mathematics as a pedagogical tool for the teaching and learning of mathematics. In particular, we draw on the mathematically pedigreed but misunderstood development[i] of the Mercator projection and its connection to the integral of the secant function. We discuss the merits and the possible pitfalls of this approach based on a teaching module with undergraduate students. The appendices contain activities that can be implemented as an enrichment activity in a Calculus course.*

Keywords: conformal mapping; history of mathematics; integrals; Mercator projection; rhumb lines; secant function; undergraduate mathematics education

## Introduction

There is no shortage of research advocating the use of history in mathematics classrooms (Jankvist, 2009). Wilson & Chauvot (2000) lay out four main benefits of using the history of mathematics in the classroom. Its inclusion "sharpens problem-solving skills, lays a foundation for better understanding, helps students make mathematical connections, and highlights the interaction between mathematics and society" (Wilson & Chauvot, 2000, p. 642). Bidwell (1993) also recognizes the ability of history to humanize mathematics. His article opens with description of mathematics instruction treated as an island students perceive as "closed, dead, emotionless and all discovered" (p. 461). By including the history of mathematics, "we can rescue students from the island of mathematics and relocate them on the mainland of life that contains mathematics that is open, alive, full of emotion, and always interesting" (p. 461). Marshall & Rich (2000) argue that the history of mathematics can be a facilator for the reform called for by the NCTM.

In addition to the benefits mentioned above, Jankvist (2009) identifies more gains that can be had by using the history of mathematics. Among them are increased motivation (that can be found in generating interest and excitement) and decreased intimidation - through the realization that the mathematics is a human creation and that its creators struggled as they do. Jankvist (2009) also mentions history as a pedagogical tool that can give new perspectives and insights into material and even can serve as a guide to the difficulties students may encounter as they learn a particular mathematical topic. Marshall & Rich (2000) conclude their article by saying:

---

[1] nicolas.haverhals@umconnect.umt.edu
[2] roscoem@mso.umt.edu

To sum up, history has a vital role to play in today's mathematics classrooms. It allows students and teachers to think and talk about mathematics in meaningful ways. It demythologizes mathematics by showing that it is the creation of human beings. History enriches the mathematics curriculum. It deepens and broadens the knowledge that students construct in mathematics class. (p. 706)

This is by no means a comprehensive summary of the research advocating the use of history in mathematics. All of the research cited above, particularly Jankvist (2009), provides many more sources. Bidwell (1993) also mentions three ways of using history in the classroom. The first is an *anecdotal display*, which features the display of pictures of famous mathematicians or historical facts in the classroom. The second is to *inject anecdotal material as the course is presented*. Here, Bidwell is referring to making historical references to coursework while it is being covered. Barry (2000), however, cautions against letting the use of history limited to the use of anecdotes. The third use mentioned is to *make accurate developments of topics a part of the course*. This third use best describes the remaining contents of this article.

*Background of this research*
The current research evolved out of an assignment given to two of the authors (N. Haverhals & M. Roscoe) in a graduate level history of mathematics class at the University of Montana[ii]which mutated into the study that is currently being reported. The remainder of the article is a reporting of this research.

**Methodology**
The study sought to investigate the merits of employing a historical approach through the teaching and learning of the topic of the integral of the secant, a topic that is common to most second semester calculus courses at both the high school and university level. The integral of the secant played a key role in the development of the Mercator map in the 16th and 17th centuries. The map was a critical tool during the age of discovery due to the fact that it was a conformal projection of the globe onto the plane, that is, it projected the globe in such a manner as to preserve angles (at the cost of distorting lengths and areas). This property allowed mariners to navigate across large expanses of featureless ocean by following compass bearings that the map provided.

In preparation for the investigation, the authors conducted a review of pertinent literature. In particular, we sought material treating the subject of the Mercator projection that was easily translated into an educational setting where the integral of the secant is taught *through* the historical reenactment of its discovery. Furthermore, we wanted to construct a unit that could be realistically included in a traditional calculus course. Since these courses typically allow for little divergence from firmly established traditional content, we decided that the unit had to be brief, able to be employed in a single class meeting.

After reading a number of articles and several educational units which dealt with the role of the integral of the secant in the Mercator projection we set out to design and create our own activity. It was decided that two documents would be produced. The first document consisted of a "take home" primer on the Mercator projection (which can be found in Appendix 1). This document "set the stage" for the investigation. In it we gave a brief description of the problem of conformal projection and motivated the historical need for such a map during the age of discovery. We included new terms such as "rhumb line", "loxodrome" and "conformal" as well as introduced the key historical figure in the development of the map, namely, Gerhardus Mercator. We also included an example of the important role of the conformal map by demonstrating how a seaman's bearing changes for a non-conformal plane projected map leading to errors in navigation.

The second document that we produced was conceived as the "in-class" exploration of the integral of the secant (Appendix 2). In this document, we hoped to lead students through a "historical reenactment" of the discovery of the integral of the secant motivated by a desire for mathematical description of the Mercator projection. The document first asked students to reason about the horizontal scaling of latitudes and then went on to describe the "mechanical integration" that was carried out by Edward Wright which determined the vertical conformal scaling. Students were asked carryout and compare the accuracy of two such approximating integrations. A proof of the closed form of the integral was provided with several missing steps and students were asked to complete the traditional proof. Finally, a number of extensions to the in class exploration asked students to investigate the way that distance is distorted by the projection.

A sample of 16 undergraduate students consisting of 9 males and 7 females participated in the study. The students were all mathematics majors who had completed their calculus sequence. Students were given the "take home" document one week before being asked to complete the "in-class" document. A period of two hours was scheduled for the in-class portion. Students completed the exercise in groups of two. The authors of the study circulated about the classroom, answering questions. At the end of the period, the completed documents were collected.

One week after participation in the classroom investigation into the historical account of the integral of the secant, two groups of four students each were chosen for separate case study analysis. One group of four students was interviewed to probe for affective reaction to the educational activity. These students were asked the following questions.

1. Describe what you learned in the activity on the historical approach to the integral of the secant.
2. How was the activity different from a typical mathematics class?
3. Here is the calculus textbook that we use here at the University of Montana. This is the presentation for the integral of the secant. How does it differ from the historical presentation of the integral of the secant that was presented last week?
4. Did the activity change the way that you view mathematical discovery?

5. Did the activity change the way that you view learning mathematics?
6. Would you say that you were more or less motivated to complete the traditional proof of the integral of the secant after having placed its discovery in a historical context?
7. Does including mathematics history make mathematics more meaningful? How?

Student response to these questions was audio recorded and transcribed for analysis. A second group of four students was shown a false physical model of the Mercator projection (chosen because of its commonality in supposedly "explaining" the projection). These students were asked to disprove the physical model using the knowledge that they had acquired through participation in the educational activity on the Mercator projection. Specifically, these students were shown the following:

A common misconception about the Mercator projection involves a physical model where the globe is projected onto a cylinder tangent to its radius through "illumination" of the globe from its center. Use the figure at the right to find the vertical stretching factor to determine whether or not this physical model gives rise to the Mercator projection.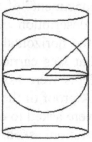

Student response to this prompt was audio recorded and transcribed for analysis.

**Framework**
While the use of history in mathematics classrooms is widely supported, it is probably safe to say that implementation is not so widely seen. Man-Keung Siu (2007) provides the following list of 16 *unfavorable factors* that contribute to the lack of history in mathematical classes:

(1) "I have no time for it in class!"
(2) "This is not mathematics!"
(3) "How can you set question on it in a test?"
(4) "It can't improve the student's grade!"
(5) "Students don't like it!"
(6) "Students regard it as history and they hate history class!"
(7) "Students regard it just as boring as the subject mathematics itself!"
(8) "Students do not have enough general knowledge on culture to appreciate it!"
(9) "Progress in mathematics is to make difficult problems routine, so why bother to look back?"
(10) "There is a lack of resource material on it!"
(11) "There is a lack of teacher training in it!"
(12) "I am not a professional historian of mathematics. How can I be sure of the accuracy of the exposition?"
(13) "What really happened can be rather tortuous. Telling it as it was can confuse rather than to enlighten!"

(14) "Does it really help to read original texts, which is a very difficult task?"
(15) "Is it liable to breed cultural chauvinism and parochial nationalism?"
(16) "Is there any empirical evidence that students learn better when history of mathematics is made use of in the classroom?"

The list was compiled by Siu for the purpose of collecting the views of mathematics educators. The authors took used their experience in preparing and administering their Mercator map activity to address each of these factors. The list was divided into groups of related items and these sub-lists form the next four sections.

## A Philosophical Response to Three Unfavorable Factors

Many of Siu's (2007) unfavorable factors for the use of history of mathematics in classroom teaching are tied to philosophical questions concerning the nature of mathematics and mathematics instruction. That is, in response to the query of, "Why don't you use the history of mathematics in your classroom?" teachers often disclose personal beliefs about mathematics and how it should be taught. Specifically the following list of three of Siu's unfavorable factors fit this description:

(1) "I have no time for it in class!"
(2) "This is not mathematics!"
(9) "Progress in mathematics is to make difficult problems routine, so why bother to   look back?"

By stating that there is no time for a historical approach to the teaching of mathematics in the classroom, teachers reveal personal beliefs that the history of mathematics is peripheral to other content matter in the subject which are given higher priority in classrooms where time is a limited commodity. This is especially the case in the modern American setting where student performance in mathematics on state and federally mandated tests is directly tied to school funding which places direct pressure on mathematics teachers to produce students who are computationally proficient in arithmetic, geometry, algebra and the like.

The statement "this is not mathematics" is a rejection of the history of mathematics as traditional mathematical classroom content. Here, the personal philosophy of mathematics might be seen as one which draws a clear line between that which is *history* and that which is *mathematics* thereby promoting a vision of mathematics that is at once highly specialized while also strictly compartmentalized from other areas of study.

Finally, the statement equating progress in mathematics with making "difficult problems routine" is a firm expression of a philosophy of mathematics which can be best described as one which seeks to avoid the complexities associated with the historical development of the subject in favor of routines, algorithms and memorized procedures.

While many authors have written about the role of personal philosophies in the teaching of mathematics (Thom, 1973; Hersh, 1986; Ball, 1988; etc), perhaps Paul Ernest's (1988) framework of philosophies of mathematics provide the most succinct and streamlined approach to the subject. Ernest identifies three psychological systems of beliefs about mathematics each with components addressing the nature of mathematics, the nature of mathematics learning and the nature of mathematics teaching.

Ernest identifies the *instrumentalist view*. Here the conception of mathematics is one of an "accumulation of facts, rules and skills to be used in the pursuance of some external end" (Ernest, 1988, p. 2). Mathematics is then thought of as a useful collection of unrelated rules and facts. The teacher's role is then envisioned as an *instructor* who promotes skills mastery and correct performance in his or her students through strict adherence to curricular materials. The student of mathematics fulfills a role characterized by compliant behavior leading to the mastery of mathematical content, namely, the rules, skills and mathematical procedures presented by the teacher.

Ernest secondly describes the *Platonist view* of mathematics. Here the conception of mathematics is one of a "static but unified body of certain knowledge" (Ernest, 1988, p.2) which is discovered (not created) by humans through mathematical investigation. Thus mathematics is inherent to the world in which we live. It is a "universal language" which exists independently from human knowledge or awareness of the subject. The teacher's role is then envisioned as an *explainer,* tasked with the promotion of conceptual understanding in his or her students as well as a presentation of mathematics as a unified system of knowledge. The student then learns mathematics through reception of mathematical knowledge. Proficiency is demonstrated through student presentation of knowledge possession, usually taking the form of variations of the same sorts of problems presented by the teacher during instruction.

Finally, Ernest describes the *problem solving view* of mathematics. Mathematics is conceived of as "a dynamic, continually expanding field of human creation and invention, a cultural product" (Ernest, 1988, p.2). Here mathematics is seen as a process rather than product, a means of inquiry rather than a static field of knowledge. As a human created body of knowledge, mathematics is envisioned as uncertain and open to refutation and revision. The teacher's role is then taken as *facilitator* and is tasked with the confident presentation of problems. The student then learns mathematics through the act of problem solving, actively constructing knowledge through investigation. Proficiency in mathematics is equated with autonomous problem solving and even problem posing.

Placing the historical approach to the integral of the secant into this philosophical framework it seems apparent that our approach to this common calculus topic seems most strongly associated with the problem solving view of mathematics. The activity was presented to the student group with little more than an introduction concerning the problem of mapping a spherical globe onto a planar map. The questions posed were largely open-ended and lacked any algorithmic approach.

The role of the teacher (here, the authors) was one of facilitator. Students were expected to construct their own knowledge through active investigation and group collaboration.

Perhaps more notable is the fact that the presentation of the discovery of the integral of the secant, as a necessary component of a conformal projection of the globe, can be thought of as a historical argument for the problem solving view of mathematics. Indeed, the first map presented by Mercator was produced without the aid of the integral – Mercator produced his map through geometric construction (Rickey, 1980). The map was improved upon by Wright through "mechanical integration" and the use of tables of values of the secant taken at one minute intervals (Sachs, 1987). Finally, the actual exact value for the integral of the secant was discovered by Henry Bond through the keen observation that Wright's sums seemed to agree with tables of values of

$$\ln\left|\tan\left(\frac{\theta}{2}+\frac{\pi}{4}\right)\right|,$$

which can be shown to equal

$$\ln\left|\sec\theta+\tan\theta\right|,$$

the value of the integral of the secant that is presented in modern calculus textbooks today. Certainly this presentation of the subject presents a notion of mathematics that is "dynamic", "continuously expanding" and "open to refutation and revision" as Ernest's problem solving approach describes.

If we place each of the three of Siu's unfavorable factors listed above into Ernest's framework of mathematical philosophies it seems apparent that these objections are most closely aligned with the instrumentalist view of mathematics. "I have no time for it in class" seems to imply a classroom where the teacher's role is taken as instructor (note the use of "I" instead of "we"). "This is not mathematics" seems to reject historical lessons on the basis that they do not promote any specific "skill". Finally, the instrumentalist approach is especially apparent in the last comment that identifies "progress in mathematics" as making "difficult problems routine" which presents a truly procedural philosophy of mathematics and mathematics instruction.

While philosophical debate over the true nature mathematical knowledge continues, educators from both Platonist and problem-solving perspectives level criticism directed at the instrumentalist approaches to mathematics education. Indeed, Thompson (1992) notes that none of the philosophical models of mathematics education have "been the object of more criticism by mathematics educators than the model following most naturally from an instrumentalist perspective" (p.136). Critics of the approach argue that computational proficiency is not necessarily a measure of mathematical understanding and point to studies that document impoverished notions of mathematics by students who display satisfactory performance on routine tasks (Schoenfeld, 1985). Proponents of the problem solving view also object that the instrumentalist approach denies the student the opportunity to "construct" their own mathematical knowledge thereby disallowing the student true understanding of the structure in mathematics which is discovered through active investigation.

It seems evident that these three unfavorable factors for the use of a historical approach to the integral of the secant are actually subtle philosophical arguments concerning the nature of mathematics and mathematics instruction. When placed within Ernest's framework of philosophies of mathematics it is apparent that the incorporation of such an activity into instruction on the topic most closely aligns with the problem-solving view of mathematics, while the reasons not to incorporate such an activity align most closely with the instrumentalist view of mathematics. Perhaps the debate is best concluded through deictic example by imagining a classroom where historical approaches to mathematics are strictly forbidden. In such a world, the student would come away from a mathematics lesson with little notion of where mathematics comes from or how it is developed. There would be no sense of mathematics driven by both practical necessity and human curiosity, both of which play into the story surrounding the integral of the secant. Finally, there would be little appreciation for those that have given us the wealth of knowledge that we now enjoy or biographical inspiration to further the science.

If we have no time for history in mathematics instruction then we have abandoned crucial sources of inspiration and insight. If history of mathematics is not mathematics then mathematics is without a story: alien to the student, not of this world. And if mathematics is meant to "make difficult problems routine" then we should expect our students to excel only in that which is "routine" which certainly will not equip them with the tools to adapt in a changing world.

## Student Responses to Unfavorable Factors

Several of Siu's (2007) unfavorable factors for the use of the history mathematics in classroom teaching relate to teacher's beliefs regarding student's opinions about the use of such materials in the classroom. The following four, in particular, fit this description:

(5) "Students don't like it!"
(6) "Students regard it as history and they hate history class!"
(7) "Students regard it just as boring as the subject mathematics itself!"
(8) "Students do not have enough general knowledge on culture to appreciate it!"

Each of these reasons for not incorporating the history of mathematics into mathematics instruction proceeds from the standpoint of the student and argues against its incorporation into the mathematics classroom on two fronts.

The first three factors (5, 6, and 7) seem to argue that the inclusion of the history of mathematics in the mathematics classroom has a negative (or negligible) outcome on student motivation in the subject. Students who do not like or even "hate" the history of mathematics are likely to be unmotivated and even repelled by its inclusion in the classroom. If historical approaches to mathematical topics, such as the integral of secant, are "just as boring" as a more traditional approach then, it is argued, such approaches are perceived as a waste of a teacher's valuable classroom and preparation time. These three factors seem to argue that the benefits of historical approaches to mathematical topics do not outweigh the costs that such approaches require of the teacher in terms of research, planning and implementation.

The last factor (8) is, perhaps, more severe than the first three. For here there is a tone of cultural superiority on the behalf of the teacher. The student is perceived as culturally deficient in their ability to perceive and understand mathematics when it is placed in a historical context. There is a tone of "teacher knows best" what is "good for the student" in terms of the lessons of history.

Our study, which placed the integral of the secant in a historical context by examining the development of the Mercator projection map, found evidence which dispels the factors provided by Siu which are outlined above. During the implementation of the unit, students displayed intense curiosity in the mathematics behind the projection. With some instructional guidance, all student groups were able to successfully finish the unit in the two hour classroom time that was allotted for the experiment.

In a follow up to the activity, four students were chosen at random for interview which was conducted one week after the classroom meeting in which the experiment had been conducted. Each student was asked the following questions:

1. Describe what you learned in the activity on the historical approach to the integral of the secant.
2. How was the activity different from a typical mathematics class?
3. Here is the calculus textbook that we use in this mathematics department. This is the presentation for the integral of the secant. How does it differ from the historical presentation of the integral of the secant that was presented last week?
4. Did the activity change the way that you view mathematical discovery?
5. Did the activity change the way that you view learning mathematics?
6. Would you say that you were more or less motivated to complete the traditional proof of the integral of the secant after having placed its discovery in a historical context?
7. Does including mathematics history make mathematics more meaningful? How?

Transcripts from these interviews were analyzed for evidence for or against the merits of Siu's factors outlined above. All four interviewees were found to respond favorably to the historical approach to the integral of the secant. Consider the following response from student 1 to question 4:

> I think that it's unbelievable, first of all. That people find these connections...and the fact that these guys did it without the tools that I have now. I mean, Mercator doing this, not perfectly, but pretty good, pretty good, having an idea, um, I just think that it's really cool that they know there's an answer. That these guys are so intelligent that they know something's up...and through their own intuition and

through their own work they get there…and that has to be the greatest feeling ever for these guys. So, it gives more respect to anyone that has discovered something that we use or even something that we don't use in our Calc books…it definitely gave me more respect for these guys. It's unbelievable that they did these things…(Student 1)

Certainly the response to the question displays a sense of wonder at the use of mathematics in the Mercator projection. Words such as "cool" and "unbelievable" and "respect" are used in describing the historical discovery of the integral in the making of the map. In response to question 3, student 1 comments:

And, again, you can show me this and I am going to accept it, 'cause it's in my Calc book and we have no choice but to accept it and memorize it…but that is completely different than starting with this [points to historical approach to integral of secant activity]…starting with integration and ending with the natural log of the secant of x plus the tangent of x. So, for me and the way that I think and the way that I enjoy school, it was helpful, and I can even imagine seeing myself start out learning about integration with this example (Student 1)

Again, student 1 responds favorably to the activity placing it in the category of activities that the student "enjoys" in school and calling it "helpful". Student number 2 also expressed a positive reaction to the activity. In response to question 5 the student comments:

I found it that it made me feel that my work was more important than it usually is. And the fact that usually when you do a problem, you get an answer, and think you're done, but, there really is no point to it that you see…I mean…when you're taking…you're doing integration by parts, it's like, okay, what are we ever going to use this for? And so, you do all this work and you never see really ever where it applies…they'll try to do stuff and…I mean it's really, really basic and it doesn't really apply, but, if they could take examples and show where its used, the historical context, it makes it feel as if you're kind of working along side of those people when they were actually doing the work hundreds of years ago…you went through and saw what they did, and so it gives a level of importance that isn't usually ever there…you know…that was valuable. (Student 1)

Here the student contrasts "traditional" approaches to common calculus topics (integration by parts) with the historical approach to the integral of the secant and describes how the historical approach lends a "level of importance" to an otherwise mundane mathematical topic. Student 3 in response to question 7 echoes this sentiment:

It's nice to be able to first learn about the secant and then they'll show you what it's used for…then it makes a lot more sense because you have been exposed to it already and you're already kind of familiar to it. It's nicer to see people apply it to their life and situations. (Student 3)

Here we see the characterization of the historical approach as adding a real world "applied" aspect to instruction which is positively characterized as "nice" by the student. Finally, student 4, in response to questions 1 and 7, comments that:

> I liked the historical context. It helps me put things in perspective. It's cool that people were applying integration before integration was codified. It also illustrated the closer and closer estimations using smaller rectangles better than my Calculus study of Riemann sums... understanding how anything, especially math, is related to real world problems and solving them makes me more motivated to understand the methodology and consider broader applications of the problem solving technique. (Student 4)

And so, our analysis of student response to the historical approach to the integral of the secant is unanimous in its approval of the educational technique. Rather than Siu's suggestion that they hate it, find it boring and non-motivating, our study group reported that they "enjoy" it and characterize the approach as "cool" and "useful" commenting that they are "more motivated" and "interested" in mathematics which is given an added "level of importance" when it is couched in a historical context. Furthermore, our data shows that students *do* have enough cultural maturity to appreciate the approach. There is a sense that the accomplishment of a conformal map was an "unbelievable" achievement won through great intelligence with "the tools that they had" before the advent of calculus through mechanical integration. There is evidence of an understanding that these early map makers were "applying integration before integration was codified" which displays the student's ability to imagine a mathematical culture before the invention of the calculus. Finally, there is a sense that the student is "working along side of those people when they were actually doing the work hundreds of years ago" which certainly indicates a level of cultural respect and admiration.

## A Logistical Response to Unfavorable Factors

As expected from a list as comprehensive as Siu's, a number of the factors that discourage teachers from employing the history of mathematics deal with very practical matters. This list of unfavorable factors relating to practicality is divided into two groups: logistics and preparation.

The following list is comprised of the factors the authors would describe as logistical in nature. These are factors that might discourage even those who are inclined to include the history of mathematics in their teaching. Each factor in the list will be addressed individually, from the perspective of the authors and through the lens of creating and implementing the teaching module. The list of logistical factors, determined by the authors, is as follows:

(3) "How can you set question on it in a test?"
(4) "It can't improve the student's grade!"
(13) "What really happened can be rather tortuous. Telling it as it was can confuse rather than to enlighten!"
(14) "Does it really help to read original texts, which is a very difficult task?"

150

The first factor on the list, (3), is one that most teachers would probably agree needs to be addressed before using the history of mathematics in their classes. Rarely is anything presented by a teacher deemed unimportant. However, it is generally necessary to test students on material in order for them to see it as important. At the same time, asking students to be accountable for historical and/or biographical information is likely to re-enforce the sorts of ideas responsible for factor (2).

The trick, then, is to create assessment questions that use the skills developed in the course of using the history of mathematics. This can come in a number of forms. The first and most obvious is the form that assessment generally comes in. Often, when new material is presented in a class, the teacher will lecture for some period of time and then leave the students to practice the skills taught for the remainder of the class and on the assigned homework. However, this practice usually only matches a small portion of the lecture time – when the teacher presents examples, usually occurring at the end of the lecture. Typically the practice problems assigned does not match a bulk of the lecture – the part when the teacher explains, justifies or proves the technique or material to be taught. Thus if this portion of lecture time is spent presenting (explaining, justifying or proving) the material from a historical perspective, little need be changed in the way that students are assessed.

This response to factor (3) may easily be criticized, however, on the grounds that it still promotes the sort of "drill and kill" mentality that the inclusion of the history of mathematics is largely meant to discourage. If the goal for including the history of mathematics in the classroom is to get away from this mentality and promote the development of other skills (such as problem solving) then assessment should reflect this aim. For this, the preparers of the historical content need to get creative.

For our study, a problem very much related to the Mercator projection was chosen and given to four students during a follow-up interview a week after the study. The students were asked to evaluate the validity of a commonly used physical characterization of the Mercator projection. The illustration (see methodology section above) was described as follows: students were asked to imagine a semi-transparent globe sitting snuggly in a cylinder with a light bulb glowing in the center of it. Light shines through the globe and "shadows" are cast by land masses on the globe. These shadows become the placement of the land masses on the cylinder, which is then sliced and laid flat to form the map. While this characterization does share some properties with the Mercator projection (the poles of the globe can never be projected and the stretching increases with increases in latitude), it is actually a different (non-conformal) projection. As the students were able to deduce, the factor of stretching in this alternative representation is a tangent function which is not equal to

$$\ln|\sec\theta + \tan\theta|,$$

the factor of stretching found in the exploration.

While the students definitely needed some nudging to get them started, once the ball was rolling all four were able to deduce that the representation would not yield the Mercator projection. The authors feel that a question such as this would make for a legitimate test question. Granted, the students did need encouragement and some may not have known where to start if the saw it on a test they had to work out on their own. However, the authors believe that this is due in large part to the fact that the students are rarely asked to perform this type of task. If mathematics was taught from more of a historical perspective, they would be more accustomed to problem-solving and therefore would be more flexible in their thinking. It is worth noting that the first part of the Mercator exploration asked the students to find the factor of stretching for arbitrarily chosen latitude. The potential assessment question had the students do the same thing from a different perspective – so it indeed was assessing a skill they used in the activity.

The second item on this sub-list of unfavorable factors, (4), deals with students' grades. Like before, the way in which using the history of mathematics in the classroom affects students' grades depends on how it is used. If it is used simply as an alternative lecturing format with little or no change in assessment then it is possible that students see no benefit in terms of grade. However, the history of mathematics can be used as a guide for how students learn mathematics. This can be seen in the difficulties students have in learning particular mathematical ideas (Moreno-Armella & Waldegg, 1991; Jankvist, 2009) and in students' conceptions of mathematical proof (Bell, 1976; Almeida, 2003). By using history as a guide for how students learn, it is possible that instruction could be improved.

This can even be taken a step further. Fawcett (1938/1966) describes a high school geometry class which, it could be argued, was set up in a fashion that mimics the historical development of Euclidean geometry. Under the guidance and supervision of the teacher and through class discussion and consensus, the experimental geometry class created their own textbooks consisting of definitions, axioms and theorems. The main goal of the experiment was to improve the students' knowledge of mathematical proof, a goal that was achieved. It should be noted, however, that the students also outperformed a control class on a standardized geometry test administered state-wide. This was despite the fact that the experimental class covered less material than the control class. Some of this uncovered material showed up on the standardized test, but the students in the experimental class were flexible enough to deal with material new to them.

The last two unfavorable factors in this section, (13) and (14), are quite similar and will be addressed together. Basically, they are both speaking to the fact that dealing with historical mathematics can be quite difficult. Much of the time, this is true. While modern day mathematicians can often handle the mathematical content associated with historical mathematical documents, other barriers to understanding exist. One stumbling block stems from the fact that the first solution to a problem is rarely the most elegant or straightforward to understand, as mentioned in factor (13). Language (terminology) and notation are two other major obstacles, referred to in (14).

The authors believe, however, that the module provided serves as an example that these concerns can be addressed. Although the authors made every effort to make the activity historically accurate, much of the historical difficulty was described, rather than recreated. Students were asked to mimic the process of mechanical integration (before they likely realized that was what they were doing) used historically but to a far less accurate, but more user friendly, degree. Based on student responses this served the intended purpose, as each group was able to recognize that smaller intervals gave better approximations. This was a necessary insight to understand the link between the Mercator projection and the integral of the secant. Also, students were told about the "lucky accident" that resulted in the discovery of the closed form for the integral of secant; they were not expected to find it on their own. Relieving the students of unnecessary difficulty does not mean they are left with nothing to do on their own. As is mentioned in the module, the original proof for the validity in question was extremely laborious and difficult. The students were then guided through an alternative (and later) historical proof – one that allowed the use of methods familiar to the students from their pre-calculus and calculus classes.

The amount of editing of historical material virtually eliminates the factor (14) from the students' perspective. The only original material that made it into the final teaching module was quotes carefully chosen to provide historical context (or humor, as the case may be). Factor (14) is not yet eliminated from the content-preparer's perspective. However, this will be addressed in the next section.

## A Response to Unfavorable Class Preparation Factors

Factor (14) raises a completely different issue from the teacher's point of view. If the students can be shielded from difficult to read original texts, are not the teachers responsible for doing the shielding? Not completely, as was seen by the authors in the preparation of the teaching module. This issue is tied into the next three factors that Siu mentions. They are:

(10) "There is a lack of resource material on it!"
(11) "There is a lack of teacher training in it!"
(12) "I am not a professional historian of mathematics. How can I be sure of the accuracy of the exposition?"

The bulk of the material that made its way into the activity came from journal articles or other teaching modules relating to the topic. In this way, the authors were not responsible for dealing with the difficult task of reading original material. Rather, they were free to concentrate on preparing the material in such a way as to be appropriate for their students.

This speaks to factor (10) as well. In preparing the module, the authors found more than enough resource material on the topic. It is true that not all of the material was deemed suitable by the authors for their targeted students. However, the materials found did provide enough for a complete, coherent teaching module to be put together. It should be noted that the authors acknowledge the possibility that factor (10) has not been interpreted as Siu intended. It is possible that what is being referred to is a lack of ready-to-use materials that can be implemented by teachers with little or no modification. The authors can not speak to this concern directly. Although some of the materials used were indeed designed to be used without modification, as mentioned, none were deemed appropriate for the students who were to see it. This was of no concern to the authors, however, because the creation of the module was an end in and of itself. Appropriate, ready-to-use materials were not sought. Instead, enough material was collected to complete the activity and that is all. It is unclear whether or not a completed module that was appropriate for the students in question could have been found.

The experience the authors had while completing this activity also helps dispel factor (11). While the module was originally meant to be part of a history of mathematics course the authors were taking at the time, no skills were explicitly taught in the class that lent themselves to its creation. What was gained from the class, however, was an appreciation for and interest in the history of mathematics. This new motivation, coupled with the authors' existing mathematical skills, was sufficient to see them through to the completion of the project. As the activity was designed for students taking Calculus II, the authors feel that it (or something similar) could have been created by any teacher with a grasp of Calculus II material and the desire and interest to do so. Thus, in general, a lack of training in the history of mathematics need not be a deterrent for those teachers who wish to use it in their classrooms.

The last factor to be addressed in this section, (12), that will be addressed in this section is related to (11). One may get the feeling that since he or she lacks training in the history of mathematics, they may be ill-equipped to judge the accuracy of sources. The authors were able to alleviate this concern through the use of articles from reputable scholarly journals. The use of such journals assures the readers (content-producers) that the materials have been peer-reviewed. That way, the burden of verification is placed on professionals and the teachers preparing the material can concentrate on making it appropriate for and useful to their students.

**A Response to the Final Two Unfavorable Factors**
The authors thought that the last two factors did not relate closely with the others and will be addressed briefly here. They are:

(15) "Is it liable to breed cultural chauvinism and parochial nationalism?"
(16) "Is there any empirical evidence that students learn better when history of mathematics is made use of in the classroom?"

The first of these, (15), speaks to the potential that the history of mathematics has to create a classroom setting that is not agreeable to the teacher. It is possible that the history of mathematics could be used to create a narrow view of the development of mathematics. This narrow view, in turn, may lead to the impression that a select group of peoples alone were responsible for (and therefore good at) mathematics. The can easily be avoided by the careful inclusion of mathematics from many different cultures. While the contributions of the ancient Greek and later European cultures are well known, they are not the only wells from which to draw. The articles referenced by Katz (1995) and Wang (2009) serve as examples of articles that describe methods developed by other cultures.

The last unfavorable factor, (16), is on to which the authors can not respond. To their knowledge, there is no convincing empirical evidence that students learn better when the history of mathematics is used in the classroom. However, the student responses gathered from this article suggest that students would welcome the inclusion of the history of mathematics – and more enthusiastic students generally make for better learners.

## Conclusion

Siu has done the mathematics educational community a service by playing the role of devil's advocate in maintaining a list of popular reasons why teachers do not use historical approaches in mathematics education. His list provided the framework for analysis of this educational experiment on the historical approach to the integral of the secant in the development of the Mercator projection map.

We found that several of Siu's unfavorable factors could be characterized as subtle philosophical statements regarding the nature of mathematics and mathematics instruction. When viewed within the framework of Ernest's (1988) philosophies of mathematics it is apparent that these objections most closely align with an instrumentalist view of mathematical knowledge and mathematical instruction. This view equates computational proficiency with mathematical understanding and is subject of much criticism in denying true mathematical understanding. In contrast, the historical approach employed in this study placed problem solving at the heart of instruction. By taking a historical approach to the subject, students learn that the closed form of the integral of the secant was "needed" to mathematically explain the Mercator projection. A historical approach allows for crucial sources of inspiration, insight and motivation which are missing from strict instrumentalist approaches, seen in this light, any argument against historical approaches can be seen as an argument in favor of an impoverished notion of mathematics.

A number of Siu's unfavorable factors were characterized as teacher statements regarding negative student predispositions to historical approaches in the mathematics classroom. Analysis of our data disproves these notions. Student response to the activity was universally positive thus affirming the approach from a student standpoint and dispelling misapplied characterizations commonly held by teachers.

There were unfavorable factors that were seen as logistical concerns. Our unit demonstrates that each of these concerns can be overcome. We were able to creatively "set a question on a test" to the historical approach. We feel that, in the area of problem solving, the historical approach does "help student's grades" by endowing them with a richer and more meaningful understanding of the process of mathematical meaning making. Finally, student difficulty in confronting historical text can be alleviated by careful and thoughtful presentation that is at once historically accurate while educationally streamlined toward an intended goal, in this case, an understanding of the integral of the secant.

In terms of Siu's unfavorable classroom preparation factors, our study has shown to dispel many of the commonly espoused concerns. We encountered ample resources that aided the creation of the educational unit. No special teacher training was required. Lastly we appealed to reputable journals to insure accuracy of historical exposition, thus, an educator need not be a professional historian of mathematics in order to create educational materials which teach mathematical concepts from a historical standpoint.

While we acknowledge the concerns of cultural chauvinism and parochial nationalism raised by Siu, we feel that an evenhanded approach to historical topics in mathematics education may lead to quite the opposite outcome. Here the "historically educated" student of mathematics might come to an awareness of the great cultural and national diversity that has contributed to the development of the subject.

Finally, in response to Siu's assertion that there is a lack of empirical evidence that supports historical approaches to mathematics in terms of improving student understanding we stand by the fact that student response to the unit pointed to greater interest and enthusiasm in the subject, which, we assert, are prerequisites to deep and meaningful learning in mathematics.

**References**

Almeida, D. (2003). Engendering proof attitudes: Can the genesis of mathematical knowledge teach us anything? *International Journal of Mathematical Education in Science and Technology, 34*(4), 479 – 488.

Ball, D. (1988). Unlearning to teach mathematics. *For the Learning of Mathematics*, 8(1), 40-48.

Barry, D.T. (2000). Mathematics in Search of History. *Mathematics Teacher, 93*(8), 647- 650.

Bell, A. W. (1976). A study of pupils' proof-explanations in mathematical situations. *Educational Studies in Mathematics, 7*(1/2), 23 – 40.

Bidwell, J. K. (1993). Humanize Your Classroom with The History of Mathematics. *Mathematics Teacher*, 86, 461-464.

Ernest, P. (1988). *The impact of beliefs on the teaching of mathmatics.* Paper presented at ICME VI, Budapest, Hungary.

Fawcett, H. P. (1966). *The nature of proof: A description and evaluation of certain procedures used in senior high school to develop an understanding of the nature of proof.* New York, NY: AMS Reprint Company.

Hersh, R. (1986). Some proposals for revising the philosophy of mathematics. In T. Tymoczko (Ed.), *New directions in the philosophy of mathmatics* (pp. 9-28). Boston: Birkhauser.

Jankvist, U. T. (2009) A characterization of the "whys" and "hows" of using history in mathematics education. *Educational Studies in Mathematics, 71*(3), 235 – 261.

Katz, V. J. (1995). Ideas of calculus in Islam and India. *Mathematics Magazine, 68*(3), 163-174.

Marshall, G. L., & Rich, B. S. (2000). The Role of History in a Mathematics Class. *Mathematics Teacher*, 93 (8), 704-706.

Moreno-Armella, L., & Waldegg, G. (1991). The Conceptual Evolution of Actual Mathematical Infinity. *Educational Studies in Mathematics, 22*, 211–231.

Rickey, V. F. & Tuchinsky, P. M. (1980). An application of geography to mathematics: history of the integral of the secant. *Mathematics Magazine*, 50(3), 162-166.

Sachs, J. M. (1987). A curious mixture of maps, dates, and names. *Mathematics Magazine*, 60(3), 151-158.

Schoenfeld, A. (1985). *Mathematical problem solving*. San Diego, CA: Academic Press, Inc.

Siu, M.-K. (2007). No, I don't use history of mathematics in my class. Why? In F. Furinghetti, S. Kaijser, & C. Tzanakis (Eds.), *Proceedings HPM2004 & ESU4* (revised edition, pp. 268–277). Uppsala: Uppsala Universitet.

Sriraman, B., Roscoe, M., & English, L. (2010). Politicizing Mathematics Education: Has Politics gone too far? Or not far enough? In B. Sriraman & L. English (Eds), Theories of Mathematics Education: Seeking New Frontiers (pp. 621-638), Springer, Berlin/Heidelberg.

Thom, R. (1973). Modern mathematics: Does it exist?. In A. G. Howson (Ed.), *Developments in mathematical education: Proceedings on the Second International Congress on Mathematics Education* (pp. 194-209). Cambridge: Cambridge University Press.

Thompson, A. G. (1992). Teachers' beliefs and conceptions: a synthesis of research. In D. A. Grouws Ed*., Handbook of research on mathematics teaching and learning* (pp. 127-146). New York: Macmillan.

Wang, Y. (2009). Hands-on mathematics: Two cases from ancient Chinese mathematics. *Science & Education*, *18*(5), 631 – 640.

Wilson, P.S., & Chauvot, J. B. (2000). Who? How? What? A Strategy for Using History to Teach Mathematics. *Mathematics Teacher*, 93(8), 642-645.

## Appendix 1
## Student take home

### Mercator's World Map
### A Historical Approach to the Integral of the Secant

Suppose that you are tasked with navigating a ship that is to travel from a point in Europe to the "New World" recently discovered across the Atlantic Ocean. How would you navigate the vessel using only 16th century technology? Most mariners during the age of discovery steered their ships along lines of constant bearing using a magnetic compass. A path of constant bearing on the globe is called a "rhumb" line named for the Spanish *rumbo* meaning "way" or "direction". This concept of a path of constant bearing was later named a loxodrome from the Latin *loxos* signifying "slant" and *drome* signifying "running". So, most mariners of the 16th century travelled paths across the ocean that we know call loxodromes.

On the globe a loxodrome intersects all north-south lines of constant longitude at the same angle. Parallels, or lines of constant latitude, are therefore loxodromes because they intersect all north-south lines of constant longitude (meridians) at right angles. Early sailors, cartographers and later mathematicians realized that these paths of constant bearing became spiral-like curves whenever the direction chosen was not due east or west. This effect is due to the fact that as a rhumb line moves north the distance separating meridians grows closer and thus the line must turn away from the pole to maintain the heading.

**Figure 1**: *Two views of a typical rhumb line, a path of constant bearing, on a globe. All rhumb lines, except paths of constant latitude, create spiral paths on the globe differing only in slope.*

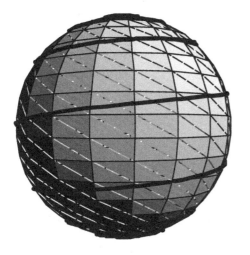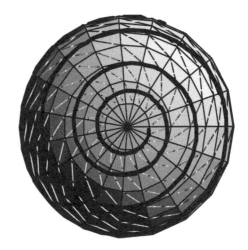

The spiraling nature of lines of constant bearing created the need for a special kind of map in which a sailor could draw a line from his present location to his objective and measure the bearing by determining the angle that is formed by the path and the meridians that are crossed in route. Such a map was presented to the world in 1569 by Gerhardus Mercator and is today known as the Mercator projection. The map signified a gigantic improvement over previous plane projection maps and is still widely used in navigation today.

**Figure 2**: *A Mercator Projection Map*

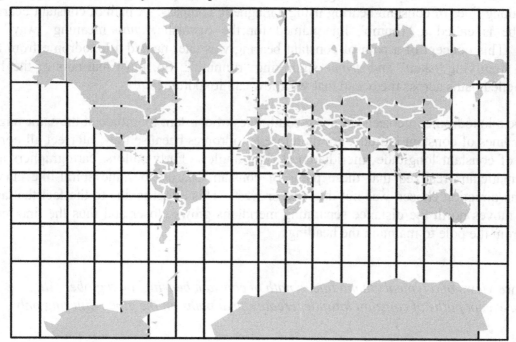

**Figure 3**: *A Plane Projection Map*

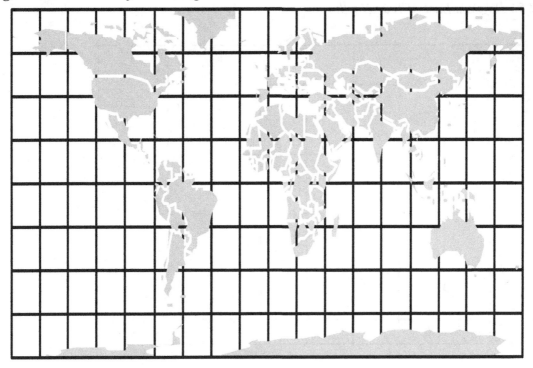

Close inspection of the two maps reveals that the plane projection has evenly spaced lines of constant latitude. In contrast, the distance between lines of constant latitude grows as a function of distance from the equator in Mercator's version.

In order to better understand the effect of Mercator's special scaling, consider a path on the globe that carries a seaman from Colon, Panama to Land's End, England. Using a magnetic compass (or the North Star) a sailor can successfully make such a trip by following a rhumb line that leaves Colon at a bearing of approximately 56° from true north. Such a path of travel over the globe then crosses all meridians at this same angle and thus scribes a spiraling loxodrome across the surface of the globe. If we were to plot the path of such a journey on both the Mercator and plane projection maps we would find that only on the Mercator projection would such a journey actually cross all meridians at an angle of 56° from true north thus correctly directing the sailor to his home (figure 4). On the plane projection map such a journey crosses all meridians at an angle of 60° from true north (figure 5). If we plot a course that leaves Colon at 60° from true north we find that our sailor is erroneously directed to France as indicated on the Mercator map (figure 6).

**Figure 4**: *Our Seaman's Journey on the Mercator Projection Map: Directs Seaman to a Bearing of 56 Degrees East of True North*

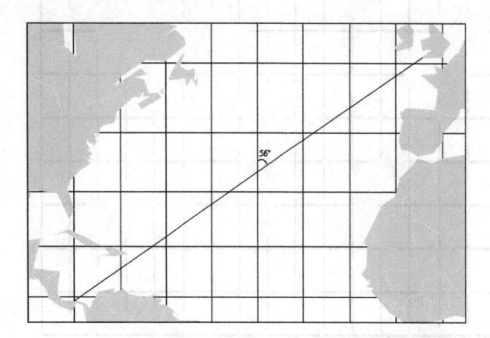

**Figure 5**: *Our Seaman's Journey on the Plane Projection Map: Directs Seaman to a Bearing of 60 Degrees East of True North*

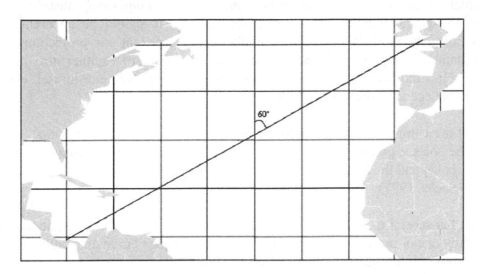

**Figure 6**: *Our Seaman's Journey on the Mercator Projection Map: Bearing 60 Degrees and Bearing 56 Degrees East of True North*

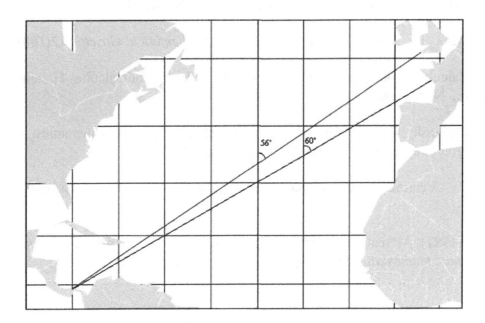

So, it becomes apparent that the Mercator projection provides the seaman with a much more useful tool where a line of constant compass direction corresponds to a straight line on the map which making it possible for a 16th century seaman to determine the correct line of bearing to follow in order to arrive at the intended destination.

From a mathematical point of view, Mercator's projection is *conformal,* meaning that the projection from the globe onto the plane preserves angles. It should be apparent that the projection does not preserve distances. It is a mathematical fact that any projection from the sphere onto the plane cannot preserve both of these quantities, but that is another story best saved for another day. How did Mercator decide on his special scaling? Mercator himself comments on this scaling in the legend of the map of 1596:

> In view of these things, I have given to the degree of latitude from the equator towards the poles, a gradual increase in the length proportionate to the increase of the parallels beyond the length which they have on the globe, relative to the equator. (Sachs, 1987)

Mercator created his special map using a compass and straight edge but mathematicians of the era challenged "any one or more persons that have a mind to engage" to mathematically describe the scaling that produced the successful map. (Rickey, 1980)

## References

Alexander, J. (2004). Loxodromes: a rhumb way to go. *Mathematics Magazine*, 77(5), 349-356.

Carslaw, H. S. (1924). The story of Mercator's map. *The Mathematical Gazette*, 12(168), 1-7.

Comap. (N.D.). Calculus of the Mercator map. Derive Lab Manual for Calculus. Houghton Mifflin Co., New York, NY.

Rickey, V. F. & Tuchinsky, P. M. (1980). An application of geography to mathematics: history of the integral of the secant. *Mathematics Magazine*, 50(3), 162-166

Sachs, J. M. (1987). A curious mixture of maps, dates, and names. *Mathematics Magazine*, 60(3), 151-158.

Tuchinsky, P. M. (1987). Mercator's world map and the calculus. In *UMAP modules in undergraduate mathematics and its applications.* Lexington, MA: COMAP.

## Appendix 2
## Mercator in class activity

**Mercator's World Map**
**A Historical Approach to the Integral of the Secant**

Mercator wrote, "In making this representation of the world we had…to spread on the plane the surface of the sphere in such a way that the positions of places shall correspond on all sides with each other both in so far as true direction and distance are concerned and as concerns correct longitudes and latitudes…With this intention we have had to employ a new proportion and a new arrangement of the meridians with reference to the parallels…It is for these reasons that we have progressively increased the degrees of latitude towards each pole in proportion to the lengthening of the parallels with reference to the equator." (Rickey, 1980)

Using the figure provided below determine the function that governs, "The lengthening of the parallels with reference to the equator." That is, given a parallel at latitude $\theta$ determine the function $f(\theta)$ that tells us how the latitude lines must be stretched horizontally in order to appear equal in length to the equator.

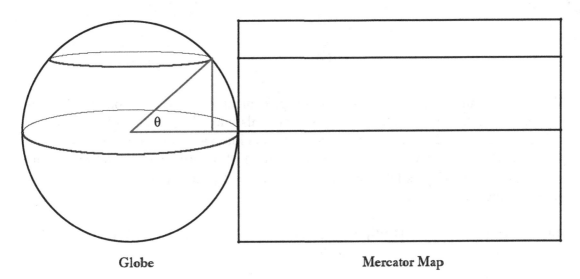

Globe                    Mercator Map

In the previous quote Mercator comments, "...It is for these reasons that we have progressively increased the degrees of latitude towards each pole in proportion to the lengthening of the parallels with reference to the equator..." Mercator determined this vertical scaling through compass constructions. It was not until 1610 that Edward Wright, a Cambridge professor of mathematics and a navigational consultant to the East India Company, described a mathematical way to construct the Mercator map which produced a better approximation than the original. In 1599 he published *Errors in Navigation Detected and Corrected*. Wright argued that in order to preserve angles on the Mercator projection, the vertical scaling factor had to be the same as the horizontal scaling factor. To visualize this phenomena imagine a 45° angle drawn on a small portion of a globe. Recall that when this region gets projected to the plane, it gets stretched in the horizontal direction by an amount that depends on the latitude. Notice what happens. The angle as projected is no longer 45°. In order for the angle to be preserved, a stretch must occur in the vertical direction that matches the horizontal stretch.

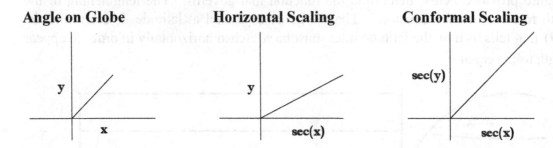

Wright also realized that the correct interval of placement of a parallel on the Mercator projection was the result of the addition of any subintervals into which it could be divided. To this end, Wright made a table of secants taken at a common interval, added these results and then multiplied by the interval widths to determine the location of a particular parallel on the map. So, if the location of the 60th parallel is desired and an interval width of 10' is used then Wright would have performed the following:

| Table of Secants | Multiply by Interval Width | Location on Mercator Map |
|---|---|---|
| Secant 10' = 1.0154 | $10° \times 8.0954 = 80.954$ | Place the 60th parallel at a location that is 80.954' north of the equator. |
| Secant 20' = 1.0642 | | |
| Secant 30' = 1.1547 | | |
| Secant 40' = 1.3054 | | |
| Secant 50' = 1.5557 | | |
| Secant 60' = <u>2.0000</u> | | |
| Total     = 8.0954 | | |

Use Wright's method to determine the location of each of the following parallels using an interval length of 5 degrees.

| Latitude on the Globe | Location on Mercator Projection |
| --- | --- |
| 15' | |
| 30' | |
| 45' | |
| 60' | |
| 75' | |
| 90' | |

You should notice that the location of the 60th parallel that you just calculated is different than the one that was calculated in the example that proceeded. Which placement produces a more accurate map? How do you know?

What difficulty did you encounter in determining the placement of 90' Latitude? What is this location on the Globe? What are the implications for the Mercator Map?

Historically, Wright's table of secants had an interval width of one minute or one sixtieth of a degree. Describe mathematically, using modern notation, the process that Wright is carrying out in determining the vertical scaling of the Mercator projection. Is Wright's method exact? How could it be improved?

As you have probably discovered, the exact mathematical explanation for Wright's technique in developing the vertical scaling for the Mercator projection hinges on a closed form for the integral of the secant. In Wright's time this result was still some 50 years from being discovered. However, with Wright's charts at his disposal, Henry Bond in the 1640s had a very lucky accident. Bond, who fancied himself a teacher of navigation and mathematics, compared Wright's table to a table of values in which the tangent function was composed with the natural logarithm. This led him to conjecture that the closed form for the integral of the secant equaled

$$\ln\left|\tan\left(\frac{\theta}{2}+\frac{\pi}{4}\right)\right|,$$

which can be shown to equal

$$\ln\left|\sec\theta+\tan\theta\right|.$$

The first proof of the integral of the secant was provided in 1668 by James Gregory. Edmund Halley commented on the proof, "The excellent Mr. James Gregory in his *Exercitationes Geometricae*, published Anno 1668, which he did not, without a long train of consequences and complication of proportions, whereby the evidence of the demonstration is in a great measure lost, and the reader wearied before he attain it." (Rickey, 1980). And so we avoid this proof and instead offer guidance through a proof offered by Isaac Barrow. Complete the missing steps in the proof

$$\int \sec\theta\, d\theta \quad =$$

$$=$$

$$= \quad \int \frac{\cos\theta}{(1-\sin\theta)(1+\sin\theta)}\, d\theta$$

$$=$$

$$=$$

$$=$$

$$= \quad \frac{1}{2}\int \frac{\cos\theta}{1-\sin\theta} + \frac{\cos\theta}{1+\sin\theta}\, d\theta$$

$$=$$

$$=$$

$$= \quad \frac{1}{2}\Big[-\ln|1-\sin\theta| + \ln|1+\sin\theta|\Big] + c$$

$$=$$

$$= \quad \frac{1}{2}\ln\left|\frac{(1+\sin\theta)^2}{(\cos\theta)^2}\right| + c$$

$$=$$

$$= \quad \ln|\sec\theta + \tan\theta| + c$$

The angle $\theta$ in the previous integral assumes a radian measure. If $\theta$ is measured in degrees then a change of variables will yield the following:

$$\int_0^\theta \sec\theta \ d\theta = \frac{180}{\pi} \ln|\sec\theta + \tan\theta|$$

Use this result to determine the exact location of each of the following parallels.

| Latitude on the Globe | Location on Mercator Projection |
|---|---|
| 15' | |
| 30' | |
| 45' | |
| 60' | |
| 75' | |
| 90' | |

How do these placements compare to those found earlier in the exercise?

EXTENSIONS

1.  How do distances vary on the Mercator map relative to latitude? Consider travelling parallel to the equator at various latitudes on the map. Consider travelling perpendicular to the equator at various latitudes on the map.

2.  How do areas vary on the Mercator map relative to latitude? What happens to the area of a region as one moves farther from the equator? Give examples.

Notes

Reprint of Haverhals, N., & Roscoe, M. (2010). The history of mathematics as a pedagogical tool: Teaching the integral of the secant via Mercator's projection. *The Montana Mathematics Enthusiast*, Vol. 7, nos.2&3, pp.339-368, 2010© Information Age Publishing

---

[i] The mathematics behind the Mercator map has nothing to do with the way the map ended up being used for political purposes. A number of critical theorists who have no idea of the mathematics behind the map run around saying "the map was purposefully made that way" . Gerhardus Mercator (1512-1594) created the map for navigational purposes with the goal of preserving conformality, i.e., angles of constant bearing crucial for plotting correct navigational courses on charts . In other words a line of constant bearing on a Mercator map is a rhumb line on the sphere. Conformality as achieved by Mercator with his projection came at the price of the distortion that occurred when projecting the sphere onto a flat piece of paper. The history of the map is also linked to the limitations of the Calculus available at that time period, and the difficulty of integrating the secant function (see Carslaw, 1924). Mercator himself comments, "…It is for these reasons that we have progressively increased the degrees of latitude towards each pole in proportion to the lengthening of the parallels with reference to the equator…" Mercator determined this vertical scaling through compass constructions. It was not until 1610 that Edward Wright, a Cambridge professor of mathematics and a navigational consultant to the East India Company, described a mathematical way to construct the Mercator map which produced a better approximation than the original (Sriraman, Roscoe & English, 2010).

[ii] Math 606 was a topics course in the history of mathematics taught by Professor Bharath Sriraman in Spring 2009. One of the assignments in the course was to take Carslaw's (1924) paper and rewrite in such a way as to make it readable by modern students. In an effort to get more mileage out of the work to be done, the students asked if it would be possible to turn the assignment into something that could be used in the future – namely an activity designed for use in a calculus classroom. Dr. Sriraman allowed for the change and arranged for the activity to be completed by undergraduate students who had completed Honors Calculus II. He also encouraged us to use the opportunity to perform some research.

170

# Topics in History and Didactics
# of Geometry and Number

# Euclid's Book on the Regular Solids:
# Its Place in the *Elements* and Its Educational Value

Michael N. Fried
*Ben Gurion University of the Negev*

## Introduction

In his commentary on the first book of the *Elements*, Proclus tells us no less than four times that one of two aims of the entire work is the investigation of the cosmic figures, the five regular solids.[1] This is a remarkable claim, and it is one not easy for modern readers of the *Elements* to swallow. In his own commentary on the *Elements*, written a millennium and a half after Proclus, Thomas Heath says in this regard Proclus is "obviously incorrect," and Heath goes on to explain that, "It is true that Euclid's *Elements* end with the construction of the five regular solids; but the planimetrical portion has no direct relation to them, and the arithmetical no relation at all; the propositions about them are merely the conclusion of the stereometrical division of the work" (Heath, 1956, I, p.2). Heath's skepticism[2] cannot be dismissed causally. It is indeed difficult to see how Proclus can put aside the immense wealth of other mathematical work in the *Elements* and single out so pointedly one of its shortest books, Book XIII on the regular or Platonic solids—Book XIII, in this view, must take precedence over Book X with its 115 propositions on incommensurables, or Book V which sets out the theory of proportion, or Books VII-IX on the theory of numbers, not to speak of Books I, III, IV, and VI, which make up every modern student's own elements of geometry. And Heath is right to cast doubt on Book XIII as in any way culminating the deductive structure of the *Elements*: the propositions in Book XIII do rely widely on propositions from other parts of the *Elements*, but not from all parts, as Heath points out, and not in a very striking way.[3]

---

[1] *In Eucl.* (Friedlein). pp. 68, 70, 71,74 (see also pp. 82-83). The English translations are taken from Morrow (1970); however, I will always use Friedlein's page numbers, which, being included also in Morrow's text, will allow the reader to follow the Greek text or Morrow's translation.

[2] The skepticism is also shared by Morrow (1970, p. l)

[3] Mueller is less dismissive of Proclus' claim, even though Mueller, like Heath, places great weight on deductive structure in judging the relative importance of parts of the *Elements*. Mueller writes: "...although from the point of view of deductive structure the remark [by Proclus] is a gross exaggeration, one can see how book XIII might have led Proclus to make it. For in book XIII Euclid makes direct use of material from every other book except the

174

According to Proclus, though, there are two aims behind the *Elements*, as I mentioned at the outset. The second of these refers explicitly to learners, to *manthanontes*, namely, that with Euclid's work at hand learners will have a treatment of the elements before them and the means to perfect their understanding of the whole of geometry (*In Eucl.* p.71). On the face of it, this seems to be a different kind of aim than the first. Heath seems to take it more seriously than that concerning the Platonic solids, which he attributes only to Proclus' Platonic loyalties. He claims that Proclus himself had difficulty reconciling the two aims and that "To get out of the difficulty…" (p.2) Proclus delegated one aim to learners and the other to the subject. But, if this really is a difficulty for Proclus, it is hard to see how it is resolved by referring one aim to the learner and the other to the subject; can learners perfect their knowledge about the whole of geometry and yet exclude its subject? It is more likely that not only did Proclus not see any difficulty here but that he also saw these two aims as essentially complementary.

As for that, it is of obvious importance that Proclus did associate Euclid with Plato, which he does explicitly just when he first tells us that Book XIII was the aim of the *Elements*.[4] There is no surprise here of course since Proclus was one of the last heads of the Platonic academy— about Proclus' Platonic loyalties, Heath is certainly right. But, keeping in mind not only that the academy was a place of learning but also that education itself was at center of Plato's thought, Proclus' assertion that "Euclid belonged to the persuasion of Plato…," meant, in effect, that Euclid's *Elements* had a place among the studies of the academy, that it had educational value, or, to use the far more subtle Greek term, that it contributed to *paideia*. So the thesis I will maintain in this paper is that to understand how Book XIII can be conceived as the aim of the *Elements*, one must view it in terms of *paideia*—and this means not so much what facts or skills

---

arithmetic books and book XII; and XIII is ultimately dependent on the arithmetic books because book X is. In this sense the treatment of the regular solids does constitute a synthesis of much of the *Elements*, a culmination of the Euclidean style in mathematics. However, the significance of the Elements lies less in its final destination than in the regions traveled through to reach it" (Mueller, 1981, p. 303). Of course the books left out are not insignificant, if one wants to judge Book XIII by its place in the deductive structure of the entire work. The argument about Book X is also moot, because, as will be discussed in more detail in the next section, the propositions in Book X are not applied in an absolutely necessary way, that is, they are not needed for constructing of the solids, for determining that there can be no more than five solids, nor for proving many of the geometrical interrelationships among the solids. In all fairness to Mueller, however, I should point out that he himself admits that the reason Book X is brought into the discussion of Book XIII is not completely clear (see Mueller, 1981, pp. 270-1).

[4] *In Eucl.* p. 68: "Euclid belonged to the persuasion of Plato and was at home in this philosophy; and this is why he thought the goal of the *Elements* as a whole to be the construction of the so-called Platonic figures".

it affords the learner, but what view of the world it gives the learner and, with that, what it does to the learner.

The paper will have two parts. Since some sense of the mathematics in the book is necessary to understand anything about it,[5] the first part of the paper will look at the structure and the character of Book XIII itself. As for its character, I will try to make that clear by looking in some detail at the dodecahedron in XIII.17. The second part of the paper will then look at how this book of the *Elements*, given the particular nature and the treatment of its subject, contributes to *paideia*. I shall try and make the case that educative value of Book XIII lies in its attention to wholeness, or rather in the way it turns readers' own attention to wholeness and forces them to reflect on the whole of the *Elements*. In the course of doing this, I will also make some general remarks about mathematics education in the classical world.

## I. Book XIII of the *Elements*: On the Cosmic Figures

Book XIII can be divided into four parts. The first part spans propositions 1 to 6; the second, propositions 7 to 12; the third, propositions 13 to 17; while the last part contains just the single proposition, proposition 18.

Propositions 1-6 making up the first part of Book XIII, contain aspects the extreme and mean ratio (*akron kai meson logon*), what is popularly called the "golden section." These propositions, with one exception, function as lemmas used in the rest of the book,[6] particularly in propositions 16 and 17 in which the icosahedron and dodecahedron are constructed. The exception is XIII.2, which states that, "If the square on a straight line be five times the square on a segment of it, then, when the double of the said segment is cut in extreme and mean ratio, the greater segment is the remaining part of the original straight line."[7] It is the converse of XIII.1: "If a straight line be cut in extreme and mean ratio, the square on the greater segment added to the half of the whole is five times the square on the half." The existence of an exception, any exception,

[5] This is not always considered obvious (see Unguru & Fried, 2007)
[6] Noted also by Heath (1956) and Mueller (1981).
[7] With occasional and minor modifications by myself, all English translations of Euclid's text come from Heath (1956).

weakens the claim of course that these are indeed just lemmas. One way out of the difficulty is to claim, as Heath does, that proposition 2 is "probably not genuine" (Heath, 1956, III, p.441); Heiberg too has doubts at least about parts of the proposition. But if proposition 2 was not a later addition, and Euclid truly intended it to be in the book, then its function must be only to complete XIII.1, as if to emphasize that that proposition and, by implication, the other propositions on the extreme and mean ratio as well form part of the subject of the book, and not merely provide tools for it; for as tools, we should only require only those actually used. Viewed this way, Euclid would be underlining that the regular solids are born in realm of proportion; the proportion determined by the extreme and mean ratio, moreover, has a special place being a continued proportion (*sunexēs analogia*) whose three terms are the segments of a line and the whole line. And it is at least suggestive that Plato calls a continued proportion, which "…makes itself and the terms it connects a unity in the fullest sense…," the "most beautiful of bonds" (*desmōn kallistos*).[8]

Before leaving this first part, the last proposition of the set, XIII.6 should be mentioned. It states: "If a rational straight line be cut in extreme and mean ratio, each of the segments is the irrational straight line called apotome." Like XIII.2, this may also have been a later interpolation[9]; however, despite some details, it is thematically appropriate in connecting the material from the previous propositions to the development in Book X of the kinds and relationships among incommensurable lines.[10] We might as well, at this point, define the terms in this proposition since they will come up again later. The terms, "rational" (*rētē*), "irrational" (*alogos*), and "apotome" (*apotomē*) all come from Book X.[11] A line set out as given for comparison is *rational* as are all lines commensurate with it in length or in square: thus with p as the assigned line, q is rational if p:q or sq.p:sq.q (square on p: square on q) have the same ratio

---

[8] *Timaeus*, 31c, translation by M. Cornford (1975), with some minor modification by myself.
[9] Heath (1956, III, 451) writes: "It seems certain that this proposition is an interpolation."
[10] It is generally agreed that both Book X and Book XIII are derived from the work of Theaetetus (c.417-369 B.C.E). This is said both in the Sudas and also in the first *scholium* for Book XIII, where Theaetetus is given credit particularly for the investigation of the octahedron and icosahedron.
[11] A very enlightening guide to Book X is Taisbak (1982).

as a whole number to a whole number.[12] A line, which is not rational relative to p is *irrational*. An *apotome*, defined much later in Book X, is simply the difference between two rational lines that are commensurate in square only. Thus, for example, the difference between the diagonal of a square and its side is an *apotome*. Proposition XIII.6 says, then, that if AB is divided at a point C in extreme and mean ratio, that is, if AC:CB::CB:AB, then both AC and CB are *apotomes*.

The way in the segments of a line divided according to the extreme and mean ratio are classified and compared in XIII.6 and the other propositions in the first part of Book XIII hints, perhaps, at the way the sides of the cosmic figures will be compared and classified later in the constructions and especially in the final part, proposition 18, where all of the sides of the various figures are brought together and described in one diagram. But in propositions XIII.1-6 that hint is at best vague and visible at all only on hindsight. In the next set of propositions, propositions XIII.7-12, the prefigurement of the constructions, however, is more cogent.

Propositions XIII.7-11 all concern, one way or another, properties of regular pentagons or of the decagons derived from them, while proposition 12 concerns an equilateral triangle inscribed in a circle, telling us, in particular, that  square on the side of the triangle is triple the square on the radius of the circle. Indeed, except for XIII.7, all of these propositions consider the regular figures inscribed in circles, just as cosmic figures, the regular solids, are inscribed in spheres. Now, it is true that, like the first set of propositions on the extreme and mean ratio, these propositions serve as lemmas for the constructions of the regular solids. Mueller (1981), for example, refers to all of propositions XIII.1-12 as "consist[ing]…of lemmas for the principal propositions 13-17" (p.251). Yet, consider the enunciations of propositions 11 and 16:

XIII.11: If in a circle which has its diameter rational an equilateral pentagon be inscribed, the side of the pentagon is the irrational straight line called minor.[13]

XIII.16: To construct an icosahedron and comprehend it in a sphere like the aforesaid figures; and to prove that the side of the icosahedron is the irrational straight line called minor.

---

[12] For this reason, the diagonal of a square relative to the side of the square is rational! One can see from this that there is great difference between what we call rational/irrational numbers and what Euclid calls rational/irrational magnitudes.

[13] The *minor* (*elassōn*) is another kind of *irrational* line defined in Book X (in X.76). Taisbak (1982) uses the term "the lesser," which is a direct translation of *elassōn* (see pp.49ff).

178

Admittedly, XIII.11 is needed in the demonstration of XIII.16. That said, however, it is hard to imagine that the similarity between these two propositions was not meant to be noticed, especially when one considers the interest in analogy in other part of Greek mathematics (see Fried & Unguru, 2001; Fried, 2003). Thus, propositions XIII.1-12, what I have framed as the first two parts of Book XIII, prepare us for the constructions of the regular solids not only by providing necessary lemmas, that is, prepare us in a deductive sense, but also it prepares us, one might say, in a thematic sense. With that, let us move on to the third part of Book XIII, the constructions themselves.

To start, it is worth noting that, except for the tetrahedral,[14] the definitions of the regular solids are not given in Book XIII, but in Book XI.[15] One arrives to Book XIII, therefore, already knowing what the solids are: the job of Book XIII, rather, is the construction and ordering of the solids. With that, the constructions appear in the book as follows. The tetrahedral composed of four equilateral triangles, three triangles meeting at each vertex, is the first solid constructed in XIII.13. Next, the octahedron composed of eight equilateral triangles, four triangles meeting at each vertex, is constructed in XIII.14. The cube is constructed in the next proposition. The icosahedron composed of twenty equilateral triangles, five triangles meeting at each vertex, is constructed in XIII.16. Finally, in XIII.17 the dodecahedron composed of twelve regular pentagons is constructed. The form of the propositions in every case follows same general pattern as in XIII.16, quoted above: each proposition requires the figure be constructed, inscribed in a sphere, and its side compared to the diameter of the sphere. The constructions themselves, however, are completely distinct—there is no master scheme for the constructions; there are five procedures for the five solids. To get a feeling for how the propositions in this group proceed, let

---

[14] In Book XIII, the word "pyramid" (*puramis*) is used. This is defined in Book XI, but only in a very general way, as a figure contained by planes and constructed from one plane to one point. Propositions concerning pyramids with triangular bases appear in Book XII. When Euclid constructs the tetrahedral in XIII.13, then, the reader, it seems, is simply expected to understand that a pyramid contained by four equilateral triangles is meant. This is consistent with the claim that one comes to Book XIII already knowing what the regular solids are.

[15] The definitions are precise; yet they do not allow one to visualize the solids easily. For example, the dodecahedron is defined as a "solid figure contained by twelve equal equilateral, and equiangular pentagons." The constructions in Book XIII also leave to the reader much of the work of visualizing the solids, but enough guidance is given to make that possible.

us look at XIII.17. We shall not look at all the details of the demonstration, only its overall strategy.

Proposition XIII.17 concerns, as said above, the construction of the dodecahedron. Its statement is as follows:

> To construct a dodecahedron and comprehend it in a sphere, like the aforesaid figures, and to prove that the side of the dodecahedron is the irrational straight line called apotome

The first part of the demonstration, the construction itself, begins with another solid, the previously constructed cube[16]: the dodecahedron is, literally, built on the cube. So let two faces of the cube be ΑΒΓΔ and ΓΒΕΖ (see fig. 1).[17] Let ΝΞ and ΘΜ, meeting at Ο, join the midpoints of the sides of face ΓΒΕΖ, and let ΘΛ and ΗΚ, meeting at Π, join the midpoints of the sides of face ΑΒΓΔ. Divide the lines ΟΝ and ΟΞ in extreme and mean ratio at points Ρ and Σ such that ΡΟ and ΟΣ are the greater segments; similarly divide ΠΘ in extreme and mean ratio at point Τ, such that ΤΠ is the greater segment. From Ρ, Σ and Τ, let the lines ΡΥ, ΣΦ, ΤΧ be set up perpendicularly to the faces of the cube, and let them be equal to the equal lines ΡΟ, ΟΣ, and ΤΠ. Join the segments, ΥΒ, ΒΧ, ΧΓ, ΓΦ, ΦΥ. Then the pentagon ΥΒΧΓΦ is equilateral, lies in one plane, and is equiangular.

Proving these three facts, in that order, takes up the next three parts of the demonstration. Without going through the details, suffice it to say that the proofs rely crucially on the propositions concerning extreme and mean ratio from the first part of the book and on those concerning the pentagon from the second part, as well as on propositions from other parts of the *Elements*, from Book XI, the first book in the *Elements* on solid geometry, from Book I, and Book VI. For example, I.47, the "Pythagorean theorem" is used in proving that the pentagon is equilateral. This is important to point out in light of what was said above in the introduction: Book XIII depends on propositions from other parts of the *Elements*; however, to say that I.47,

---

[16] Indeed, Euclid refers not to *a* cube but *the* cube previously mentioned (*proeirēmenos kubos*), presumably that in XIII.15.

[17] Figures 1 and 2 contain only the relevant components of Euclid's diagram for XIII.17.

for example, was proven for the sake of this particular juncture in XIII.17 hardly seems reasonable. But let us return to the proposition.

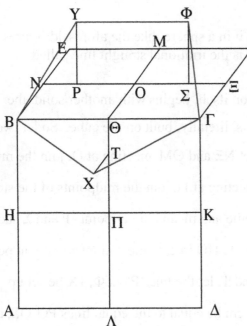

Fig. 1. Proposition XIII.17: construction

From Euclid's point of view, having proved that ΥΒΧΓΦ is equilateral, planar, and equiangular, he has sufficiently demonstrated how the construction is done. He only adds, "Therefore, if we make the same construction in the case of each of the twelve sides (*pleurai*) of the cube,[18] a solid figure will have been constructed which is contained by twelve equilateral and equiangular pentagons, and which is called a dodecahedron." In other words, the reader, the learner, is not only asked to complete the details of how the construction is to be completed for the other sides of the cube (the construction given being with respect to the one side ΒΓ), but also to imagine the shape of the finished figure (see fig. 2).[19] While these tasks are not impossibly difficult, given the head start Euclid has already provided, they are also not effortless. They require an act of the imagination more than a concrete execution with, say, compass and straight edge; they are

---

[18] That is, the edges of the cube.

[19] A comparison might be made here to the constructions ending Book I of Apollonius' *Conica* (for example, Conica, I.52). Having provided the vertex and base circle, Apollonius asks us there to "imagine a cone" (*noesthō kōnos*), that is, to complete the picture.

distinguished in this way from constructions such as that of the equilateral triangle, the first regular figure, which opens the very first book of the *Elements*.

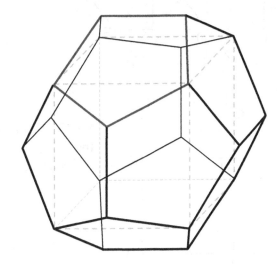

Fig. 2. The completed dodecahedron

The next part of the proposition is to show how the dodecahedron can be inscribed in a sphere. Again, Euclid relies on the previous construction for the cube. From the midpoint $\Psi$ of $Y\Phi$, $\Psi O$ is drawn and produced to where it meets the diameter of the cube at $\Omega$ (see fig. 3). In fact, $\Omega$ is the midpoint of the diameter and therefore the center of the sphere circumscribing the cube. Then $\Omega Y$, which join $\Omega$ to a vertex of the dodecahedron, is joined. It is then shown that the square on $\Omega Y$ is equal to three times the square on half

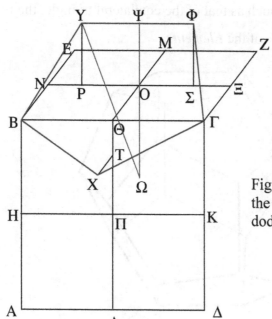

Fig. 3. Proposition XIII.17: finding the sphere circumscribing the dodecahedron

the side of the cube. But the square of the radius of the sphere circumscribing the cube was also shown to be equal to three times the square on the half-side (or the square on the diameter of the sphere is three times the square on the side of the cube). Therefore, the sphere that passes through the points of the dodecahedron not shared with the cube, such as Υ and Φ, also passes through those points which are shared by the two figures, such as points Β and Γ.

Finally, the side of the dodecahedron must be shown to be an apotome. In brief, Euclid shows that since the side of the dodecahedron, ΥΦ, is twice the segment ΡΟ, and Ρ divides ΝΟ in extreme and mean ratio, it follows that if the side of the cube, which is equal to ΝΟ, were divided in extreme ratio, the greater segment would be equal to ΥΦ. Therefore, ΥΦ is, by XIII.6, an apotome with respect to the side of the cube. But since three times the square on the side of cube is equal to the square on diameter of the sphere, the side of the cube is rational with respect to the rational diameter.[20] Therefore, ΥΦ is an apotome (with respect to the diameter of the circumscribing sphere). It is fair to ask why is it important to show that the side of the dodecahedron is an apotome. It is not hard to see why one should want to know that the ratio of

---

[20] The diameter is rational by definition, being the given line to which the others are compared.

the side of the dodecahedron to the side of the cube is the extreme and mean ratio, for, together with the fact that the square on the diameter of the sphere circumscribing the cube is three times the square on the side of the cube, this fact allows one to construct the side of the dodecahedron given the side of its circumscribing sphere. But why must we know that it is an apotome? I think the only reasonable answer is that Euclid wanted to show how the ordering of the solids fits with the other great taxonomic book in the *Elements*, namely, Book X.[21]

It might appear a simpler claim just to say that Book XIII presented an opportunity for Euclid to apply the material from Book X. Ockham's razor might cut the matter here if the Book XIII ended with proposition 17. But proposition 18 shows that a taxonomy of the solids and their sides was truly on Euclid's mind. Proposition XIII.18, which I have deemed distinct part of Book XIII, comprises two completely independent parts.[22] The first part of this proposition is a synopsis of the relations between the sides of the various solids and the diameter of the sphere circumscribing them. It is unique in the *Elements* in that while it contains geometrical arguments and there are claims proved along the way, its main job is not out to prove a particular claim or solve a particular problem; what it does is "to set out and bring into comparison with one another" (*ekthesthai kai sunkrinai pros allēlas*) various lines that have already been determined. The sides of the five constructed figures are laid out all together in a half circle whose diameter is the diameter of the sphere circumscribing them. Thus, in figure 4, AB is the diameter of the sphere, AZ is the side of the tetrahedral, BZ is the side of the cube, BE is the side of the octahedron, MB is the side of the icosahedron, and BN is the side of the dodecahedron. The point Γ divides AB into half; Δ divides AB such that AΔ twice ΔB or two thirds AB; Λ is such that ΓΛ=ΛK where K is determined by the intersection of the circle with HΓ drawn from the extremity of AH, which is equal to AB and at right angles to it; and N divides BZ, the side of the cube, in the extreme and mean ratio.

---

[21] The point remains even if one fully accepts that Theaetetus was behind both Book X and Book XIII. Euclid was, nevertheless, the editor, if not more than that. He was not obliged to preserve any connection between the theory of irrational lines and the regular solids that might have been in Theaetetus' own work (assuming that a written work existed), unless he himself deemed that connection essential for the flow of the work.

[22] The merging of these two parts—the comparison of the sides of the regular figures and the proof that there cannot be more than five such figures—was undoubtedly a later modification. Support for this comes from the fact that the enunciation refers only to the first part, while the second begins with its own formal enunciation.

184

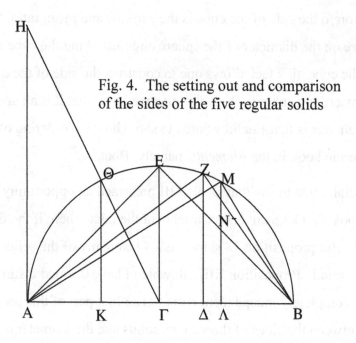

Fig. 4. The setting out and comparison
of the sides of the five regular solids

This first part of XIII.18 concludes with a proof that BM, the side of the icosahedron is greater than BN the side of the dodecahedron. But before that, Euclid establishes that the sides of the tetrahedral, cube, and octahedral are rational (in the sense of Book X) while those of the icosahedron and dodecahedron are irrational, the one being an irrational "minor" while the other an apotome. None of this is new. It has all been established in the constructions themselves. As in the diagram, the cosmic figures are brought together in the text and compared as parts of a cosmos. One is brought together with the other like the clicking of glasses at a meeting of friends:

> …of parts of which the square on the diameter of the sphere contains six, the square on the side of the pyramid[23] contains four, the square on the side of the octahedron three, and the square on the side of the cube two. Therefore the square on the side of the pyramid is four thirds of the square on the side of the octahedron, and double of the square on the side of the cube; and the square on the side of the octahedron is one and a half times the square on the side of the cube. The said sides, therefore , of the three figures, I mean the pyramid, the octahedron and the cube, are to one another in rational ratios. But the remaining two, I mean the side of the icosahedron and the side of the dodecahedron, are not in rational ratios either to one another or to the aforesaid sides; for they are irrational, the one being minor and the other an apotome.

---

[23] As noted above, Euclid has no other special term for the tetrahedral; it is just called the pyramid.

The second part of XIII.18 shows that there can be no other regular solids than those constructed. This Euclid shows quite simply using the fact proved in Book XI (prop. 21) that a solid angle is contained by (three or more) plane angles which together must be less than four right angles. Therefore, the solid angles of a regular solid can be contained by three, four, or five equilateral triangles, three squares, or three pentagons. Except for a final lemma that only clarifies a point in the last proof and is probably an interpolation anyway, this second part of XIII.18 ends the book. In what sense does it end the book and in what sense does it belong together with the first part of XIII.18? Both questions can be answered at once. The second part of XIII.18 is, in a way, a limit of possibility, a *diorismos*, like that at the end of *Elem.* I.22 where the construction of a triangle given three lines is limited by the condition that any two of the lines together be greater than the third. But in another way it is utterly different. It is not a bound for an unlimited number of possibilities; rather, it shows that the five figures constructed in propositions 13-17 form a totality. The first part of XIII.18 shows the relations between the cosmic figures with respect to a single sphere that circumscribes them: now, we know that that sphere circumscribes a true cosmos, an ordered whole (as in the sense of the Greek, *kosmos*). Obviously there is more one can say about the regular solids, as is proven by the added pseudo-Euclidean books, Book XIV (by Hypsicles) and Book XV. But, unlike any other topic in the *Elements*, this one treated in Book XIII can claim a set of *objects* that cannot be further extended. There is no other book in the *Elements* that presents a *whole* in this sense and a true ending.

This display of wholeness, I believe, is the key to understanding the special status Proclus attaches to Book XIII as a *telos* and to understanding its special educative value. But to complete that argument, we need to consider first what being educative might have meant for a thinker like Proclus, and that means, among other things, considering what place mathematics occupied in education in Classical times.

## II. Mathematics Education in Classical Times and the Educational Contribution of Book XIII

It is natural from a modern perspective to expect a continuous educational nexus leading to the kind of mathematical work one finds in Archimedes' *On the Sphere and Cylinder,* or Apollonius' *Conica,* or Euclid's *Elements*: one wants to see a program, or at least, a pattern of mathematical education from K-12 to undergraduate to graduate studies. Yet, between very rudimentary mathematical training and studies in higher mathematics at institutions of advanced learning, such as the *Museum* in Alexandria and the *Academy* in Athens, there appears to be a gap.[24] Indeed, given the sophistication and level of such mathematical works as those of Archimedes, Euclid, and Apollonius, it is surprising to discover that the ordinary education of youth, at least in $4^{th}$ and $5^{th}$ century Athens, seems to have included very little mathematics of any weight at all. Thus, Ian Mueller (1991) observes that despite an apparent common ability to perform calculations such as and 2000/10 and 3×700,[25] "...it appears that the average Athenian citizen knew remarkable little arithmetic from our point of view and that he did not acquire his knowledge in school. But even if he did learn arithmetic at school, we have no right to assume he learned any geometry, astronomy, or music theory, despite the fact that we have plenty of evidence associating these subjects with the intellectual heights of fifth-century culture" (p. 88).[26]

---

[24] Morgan (1999) writes: "Among non-literary elements of education, our ignorance of mathematics and mathematical education in the Classical period constitutes a problem in a class of its own" (p.52). For example, one of the only texts referring to older students' studying geometry is a remark by Teles, quoted by Stobeus to the effect that teachers of arithmetic and geometry (and riding!) are among chief plagues of students' life (see Marrou, 1982, p.176; Freeman, 1912, p.160). Nevertheless, the scarcity of texts about mathematics education other than the most elementary kind may be a sign that, generally, there truly was a gap between basic numeracy and advanced work: the question, then, is what filled the gap.

[25] Mueller is relying here on passages from Aristophanes' *Wasps* and Plato's *Hippias Minor*, respectively.

[26] Thomas Heath is more generous in his estimate of children's arithmetical education. He says, "The main subjects [of elementary education] were letters (reading and writing followed by dictation and the study of literature), music and gymnastics; *but there is no reasonable doubt that practical arithmetic (in our sense), including weights and measures, was taught along with these subjects* [emphasis added]" (Heath, 1981, vol. I, pp. 18-19). However, whether or not mathematics was included in the basic education of Athenian youth in fact, if we consider the accounts of basic Athenian education by Protagoras in Plato's *Protagoras* (325e-326c) and Glaucon in the *Republic* (522b), we must accept that neither saw mathematics as an *obvious enough* component of elementary education to mention it in their descriptions; for them, it seems, "the three R's" of education were Reading, Rhythm, and wRestling! Further indication of this, at least from the pre-Hellenistic age, comes from vase-paintings of school scenes connected with mathematics: those connected with mathematics are rare and do not show anything that hints

One possible form of education filling the gap between elementary and advanced mathematics education was education in the rhetorical tradition, that tradition beginning with the Sophists in 5th century B. C. E and arriving, finally, to a point of great technical perfection and sophistication by the end of the Hellenistic period. By that time, and certainly by late antiquity, rhetorical education had become the predominant form of education in the classical world:[27] we must imagine this, then, as the basic education of citizens in the classical world, certainly of the intellectual elite, including mathematicians, and certainly from the Hellenistic period onward. The structure and vocabulary of ancient mathematical texts reveal the influence of rhetorical education. In late classical mathematical works, such as those of Pappus, one can see the influence of this source of *paideia* in the particular shape of those works (Bernard (2003a, 2003b). Such works were written by people trained to write rhetorical texts that inspire rhetorical practice. A text written with this background "...therefore functions as a kind of *trap* for its reader or its listener...Mathematical texts, that is, texts that are *mathemata* in the true sense, '*learning matters*', also share in this particular form" (Bernard, 2003b, p.409). Like the rhetorical texts they knew so well, it reasonable to think that writers of these mathematical texts might also have thought of them as models for imitation and sources for invention.

Without denying this influence of the rhetorical tradition on writers of mathematical texts, one ought to be cautious about putting too much weight on it in elucidating the ancient study of mathematics. For one, mathematics does not appear to have been of great interest to teachers of rhetoric. In a famous passage from the *Antidosis*, Isocrates, in what may be concession to Plato,[28]is willing to see mathematics as useful to learners, allowing them to grasp and learn "...more easily and more quickly those subjects which are of more importance and of greater value" (*Antid.*, 265); thus, mathematics is useful, but not more than that. And in the *Panathenaicus* he repeats this position, but says explicitly that studies such as geometry, astronomy, and eristics are not the sort that should be carried into adulthood (*Panath.*, 27-28), when, we might add, mathematics would be pursued at an advanced level. Much later,

---

at anything but the simplest kind of mathematics; for example, although there wax tablets with multiplication exercises preserved, pupils are not seen writing in the mathematical scenes (see the vase-paintings of school scenes collected in Beck, 1975).

[27] Marrou (1982) writes, "Hellenistic culture was above all things a rhetorical culture..." (p.195).

[28] See Jaeger (1945), III, pp.147-148.

Quintilian also accepts that some training in mathematics might be beneficial to orators by awakening their thinking and sharpening their intellect as well as helping them avoid appearing inept and unlearned (*indoctus*) in court; however, he too emphasizes that it is learning appropriate only for the young (*teneris aetatibus*).[29]

The place of higher mathematical studies is better placed in the tradition of philosophical education, particularly but not exclusively[30] in the stream of that tradition associated with Plato (where indeed it has typically been placed). Besides the central role mathematical studies assumes in the *Republic* and *Laws*, mathematics is prominent in "Socrates' school" lampooned in Aristophanes' *Clouds* (e.g. 180-200); whatever mathematics was in fact taken up among those inspired by Socrates, he was certainly *perceived* by his fellow Athenians to be deeply interested in it. And although Plato does not present Theaetetus, who was probably behind much of the work in Book XIII, as being taught by Socrates in the dialogue named after him, Plato does make him out to be something like an ideal learner in Socrates' circle.[31]

The association between mathematical learning and the philosophical tradition is evident also in later antiquity. Proclus' own education provides a good example of that. Marinus, his biographer and successor, tells us that having taken great delight in rhetoric and achieving renown in the subject he was advised by Athena:

> …to devote himself up to philosophy, and to attend the Athenian schools. So he said farewell to rhetoric, and to his other former studies, and first returning to Alexandria, he attended only what philosophical courses were there given. To begin his study of Aristotle's philosophy he attended the instruction of the Younger Olympiodorus, whose reputation was very extensive. *For mathematics, he trusted himself to Heron, a very*

---

[29] See *Institutio Oratoria*, Book I, 3.34-3.49.

[30] That mathematics formed part of the program adopted by Sophists is apparent in the activities of Hippias of Elis (see Plato, *Hippias Minor*, where Hippias is referred to as one interested in the mathematical subjects (367d), and *Hippias Major*, where Hippias is said to teach these subjects (285c). See also *Protagoras*, 318d)

[31] Another context for mathematical studies besides the rhetorical and philosophical educational traditions was that of sciences like architecture in which mathematics was applied directly. Vitruvius, for example, in describing the education of an architect says that an architect, among other things, should be learned in geometry (*eruditus geometria*) (*De Architectura*, I. 3). But while he emphasizes mathematics only as a tool for the architect, Vitruvius, unlike Isocrates, shows a genuine appreciation of those rare people who are true mathematicians, and he lists, as examples, Aristarchus, Philolaus, Archytas, Apollonius, Eratosthenes, Archimedes and Scopinas (I.17). The role advanced mathematics played in this kind of education needs to be investigated further, but it is beyond the present paper.

*pious person, who possessed and practiced the best methods of his art* [emphasis added] (§9)[32]

So from this remark two things are clear: first, Proclus' studies of mathematics begin when he begins his philosophical studies, and second, those mathematical studies, together with his studies of Aristotle, are considered themselves philosophical studies. That mathematical studies may be thought of as not just aligned with philosophical studies, but, in some sense, being philosophical studies, implies that the gap between rudimentary and advanced mathematics might be an illusion based on a faulty expectation: one wants to think that the proper propaedeutic for higher mathematics is mathematics of a similar but simpler type. The case may be more like that of, say, art history, where one does necessarily see students beginning with primary school arts and crafts and progressing through a sequence of classes of on art of increasing sophistication until reaching the level of an academic course in a university art program; one expects rather that the preparation for the latter comprises classes in general literature and history.

The idea of a general preparation, or rather a general way of being, that forms the constant foundation of one's thinking and one's works is close to the classical idea of *paideia*. As an educational ideal, *paideia* was hardly the monopoly of philosophy; the rhetorical tradition claimed that for itself quite as much, if not more. Both aspired to genuine knowledge and a perspective on how one should live. In the rhetorical tradition, a component of *paideia* entailed knowledge of a certain corpus of literature; however, it meant most of all having the skills and presence of mind allowing one to speak and act in an intelligent way, one might say in a *cultured* way. "Culture," in fact, just as "education" itself, is a frequent translation of *paideia*, and the Latin translation of *paideia* came to be, revealingly enough, *humanitas* (see Marrou, 1982, p.218). Similarly, in the philosophical tradition as represented by Plato, to possess *paideia* means leading a philosophical way of life. As Jaeger puts it in his great three volume work (Jaeger, 1945) entirely dedicated to defining the notion of *paideia*, "[Plato's] *philosophos* is not a professor of philosophy, nor indeed any member of the philosophical 'faculty', arrogating that title to himself because of his special branch of knowledge (*texnudion*)... Although...he uses the

---

[32] Translation by K. S. Guthrie (1925).

word to imply a great deal of specialized dialectical training, its root meaning is 'lover of culture', a description of the most highly educated or cultured type of personality…[Plato's *philosophos*] is averse to all petty details; he is always anxious to see things as a whole; he looks down on time and existence from a great height" (vol. II, p.267). In either case, learning, *paideia*, is not something done when one is a youth and then put aside. Rather it accompanies one for an entire life, one's true unassailable possession; thus, as Jaeger points out (1945, II, p.70), when Megara was sacked, the philosopher Stilpon, having been offered compensation, replied only that nothing of his own was lost since no one had taken away his *paideia,* that he still had his words and his knowledge (*paideian gar mēdena exenēnochenai, ton te logon echein kai tēn epistēmēn*).[33]

This then is the more compelling reason why mathematics, although having some place in the rhetorical tradition and undoubtedly influenced by it, cannot be considered part of that tradition's view of *paideia*: for mathematics, in that view, is not to be embraced as an essential part of one's life—indeed, as we have seen, it is to be eventually abandoned, like many other things of youth. While the philosophical tradition does not necessarily see mathematics as the final end of learning—certainly not in Plato—it does recognize the power of mathematics to make one "see things as a whole" and point one to a higher life; this gives mathematics a place in one's thinking life, in one's *paideia*. The true mathematician, that is, one for whom mathematics is part of *paideia*, is the kind of mathematical learner that Proclus must have in mind when he says that Book XIII is one of the two aims of the *Elements*.

So, what is it that such a learner gains from Book XIII? The main thing a learner gains is what we have already discussed at the end of the last section, namely, that one is given an image of a whole in a way that more clear than in any other book of the *Elements*. This is so in two ways: first, that the five solids are only five and, therefore, form a totality; second, that each figure is itself a kind of totality, whose likeness to a sphere is underlined by each being circumscribed within a sphere. In this connection, one ought to take into consideration how these figures could

---

[33] The story of Stilpon is told in Diogenes Laertius, "The Life of Stilpon," (Book II, 115). Jaeger points out (II, p.70) that this remark by Stilpon is a restatement of the phrase by Bias of Priene, one of the seven sages, which is, in its lapidary Latin version, *omnia mea mecum porto*.

turn one's attention towards the cosmic figures described in the *Timaeus*: the tetrahedron, octahedron, cube, and icosahedron corresponding to the elements fire, air, earth, and water, while the dodecahedron serving by itself as an image of the all (*to pan*), or, more concretely, corresponding to the field of stars where the constellations lie. It is naturally far beyond the scope of this paper to look at the *Timaeus* in any deep way; but let it suffice to say that the dialogue is very much a dialogue about wholeness: words referring to the whole (*to holon* or the adjective *holos*) or the all occur in it with striking frequency. Proclus, who also wrote a commentary on the *Timaeus*, refers to the *Timaeus* often in his commentary on the *Elements*. For example, in describing the contribution of mathematics to the theory of nature (*phusikēn theōrian*) he writes:

> It reveals the orderliness (*eutaxian*) of the ratios according to which the universe is constructed (*dedēmiourgētai to pan*) and the proportion that binds things together in the cosmos, making, as the *Timaeus* somewhere says, divergent warring factors into friends and sympathetic companions. It exhibits the simple and primal causal elements as everywhere clinging fast to one another in symmetry and equality, the properties through which the whole heaven was perfected (*ho pas ouranos eteleōthē*) when it took upon itself the figures appropriate to its particular region; and it discovers, furthermore, the numbers applicable to all generated things and to their periods of activity and of return to their starting-points, by which it is possible to calculate the times of fruitfulness or the reverse for each of them. All these I believe the *Timaeus* sets forth, using mathematical language throughout in expounding its theory of the nature of the universe [literally, the "whole"] (*peri tēs phuseōs tōn holōn theōrian*). (*In Eucl.* pp.22-23)

The connection between the regular solids and the construction of the elements described in the *Timaeus* can also be seen in the scholiast's remarks where the identification of the tetrahedron, octahedron, cube, and icosahedron with fire, air, earth, and water, and the dodecahedron with the all (*tōi panti*) is repeated.[34] It is not altogether clear when that scholium was written, but there is some reason to believe it was written prior to Proclus;[35] nevertheless, it shows that not only Proclus related the material in Book XIII to what is spoken of in the *Timaeus*. Recalling the *Timaeus* is important also in the way it presents wholes coming to be, namely, as, in some way, being constructed (the nature of which is, of course, one of the points where Neo-Platonism is

---

[34] See Euclid (1969-77), vol. V., p. 309.
[35] The particular scholium cited here is from Heiberg's *Schol Vat. series P*, which means there is a good chance it was written before Theon's time in the 4th century (see Heath, 1956, I, pp. 64-69).

not clearly in line with Platonism). In the *Timaeus* itself the four solids corresponding to the elements are constructed in a heuristic way, as if only to show that the elements can be constructed rationally the way the *demiourgos* constructs the entire universe, the whole itself. The dodecahedron is not constructed in the *Timaeus*, though it is, as we have seen, in the *Elements*. What one sees then in Book XIII is the precise completion of those constructions, how they might be realized in detail. It ought to be recalled, however, that that detail is meant only to be enough to allow the learner to imagine the completion: the geometrical constructions, as we described in the last section, are a kind of prompt for them to open the eyes of their imaginations.[36]

But construction in the *Timaeus* is central not only to the conception of the individual cosmic figures, but also of cosmos considered as a whole, whether or not Plato took that construction literally or metaphorically—and this, needless to say, was a point hotly debated even in ancient times, where, interestingly enough, the comparison was sometimes made with the geometers constructions which were done for purposes of instruction (*didaskalias xarin*), as Aristotle mentions in *De Caelo* (279b).[37] Be that as it may, what construction shows is how parts fit together, or, to use the more suggestive Greek word, how they demonstrate *harmonia*. Indeed it is the fitting together into a whole that makes a cosmos a cosmos. Its order is one of organization and place, of *taxis*, rather than, as Jacob Klein (1985, esp. pp. 30-34) has pointed out, one of *lex*, of law, which marks the modern sense of an ordered universe. It is this kind of order that is exemplified so well in proposition XIII.18 setting out the relationships of the five regular solids. As suggested in that discussion, this is not the only proposition which relates one kind of mathematical object to another—what proposition does not?—but, with the final part of XIII.18 showing that there are no other regular solids, that the five form a totality, the first part becomes a *taxis* indeed of a whole.

Interestingly enough, this brings us to Proclus' second claim regarding the aim of the *Elements*, namely, that the work presents learners with a thorough treatment of the elements of geometry

---

[36] One might make a comparison here to how Socrates, at the start of the *Timaeus*, asks to see the image created in the previous day's talk to come alive (*Tim.* 19b): learners, reading Book XIII, must bring the cosmic figures to life.

[37] Aristotle makes the remark, it should be said, to show that this comparison does not save Plato from the apparent contradiction between a kosmos that is indestructible yet created (*aphtharton men ... genomenon de*).

and the means to perfect their understanding of the whole of geometry (*In Eucl.* p.71). Now, we have already addressed the point that learners are those who want to cultivate their *paideia* and therefore are not easily separated from other mathematicians. But what ought to be said in additions that learners' perfecting their understanding of the whole of geometry may not be very far from their contemplating a cosmos, like that suggested by the material of Book XIII. There is no fine line between the *taxis* of the material of geometry and its discursive means. Thus Proclus' description of the formal aspects of geometry, of synthesis and analysis, of definitions, hypotheses, and demonstrations, flows seamlessly from his description of the material of geometry moving from the simple to the complex, from points to bodies, and from the complex back to the simple (*In Eucl.* pp.57-58). In this, he may be betraying his NeoPlatonic rather than strictly Platonic outlook. For in struggling with the question of wholeness of The Soul with respect to individual souls, Plotinus seriously considers the image of geometrical science, even though, it must be said, he ultimately rejects it:

> Is the individual soul a "part" in the sense that a theorem of a science is called a "part" of this science taken in its totality, since the science continues to exist despite this division into theorems? This division consists only in the enunciation and actualization of each of its parts. Each theorem contains accordingly the total science in potency and the total science does not exist one whit the less (*Enneads*, IV, 3, in O'Brien, 1964)

The point is that if contemplating a cosmos such as that represented by the five regular figures can be related to the organized whole of the *Elements* as a science, Proclus' two aims of the *Elements* really become one: seeing the whole portrayed by the science becomes at once seeing the whole of the science.

## Summary

In this paper, I have argued that to understand how Proclus could see Book XIII as one of the principal aims of the *Elements,* Euclid's work should be placed in the context of *paideia* as Proclus might have understood it, that is, within philosophical educational tradition. Thinking of the *Elements* in this way forces one to try and see not how the work supplies a mathematical toolbox (though it certainly does that) but how it can turn its reader, the learner, towards a view of the world. In this light, Book XIII, with its focus on the construction of the regular solids, their mutual relations, and the way they form a totality, presents an ordered cosmos much as does Plato's *Timaeus,* which has always been associated with the regular solids and from which they have come to be called Platonic solids. The manner in which wholes are constructed in Book XIII, both the individual solids and their *taxis*, also turns one towards the wholeness of the entire treatment of geometry in the *Elements*. This makes, in fact, the two goals of the *Elements* stated by Proclus—to investigate the cosmic figures *and* to provide an orderly presentation of geometry from which the learner may perfect an understanding of the science as a whole—plausible, complementary, and, perhaps, ultimately one.

## References

Beck, F. A. (1975). *Album of Greek Education*. Sydney, Australia: Cheiron Press.

Bernard, A. (2003a). Sophistic Aspects of Pappus's Collection. *Archive for History of Exact Sciences,* 57(2), 93-150.

Bernard, A. (2003b). Ancient Rhetoric and Greek Mathematics: A Response to a Modern Historiographical Dilemma. *Science in Context*, 16(3), 291-412

Cornford, M. (1975). *Plato's Cosmology*. Indianapolis, IN.: Bobbs-Merrill Company, Inc.

Euclid (1969-77.) *Euclidis Elementa*, post Heiberg edidit E. Stamatis, 5 vols. Stuttgart: B. G. Teubner.

Freeman, K. (1912). *Schools of Hellas*. London: MacMillan and Co. Limited.

Fried, Michael N. & Unguru, Sabetai (2001). *Apollonius of Perga's* Conica*: Text, Context, Subtext*. Leiden, The Netherlands: Brill Academic Publishers.

Fried, Michael N., (2003). The Use of Analogy in Book VII of Apollonius' Conica. *Science in Context, 16*(3), pp.349-365.

Gutherie, K. S. (1925). Proclus' Biography, Hymns and Works. New York: Platonist Press. (Transcribed by Roger Pearse for the Website www.tertullian.org/fathers/marinus 01 life of proclus.htm . Accessed April 2, 2008)

Heath, T. H. (1981). *A History of Greek Mathematics*, 2 volumes. New York: Dover Publications, Inc.

Heath, T. L. (1956). *The Thirteen Books of Euclid's Elements. Translated from the text of Heiberg with introduction and commentary*, 3 vols. New York: Dover Publications, Inc., 1956)

Isocrates (1962). *Isocrates in Three Volumes, vol.2*, G. Norlin (transl.). Cambridge: Harvard University Press (Loeb Classics).

Jaeger, W. ,1945, *Paidea: The Ideals of Greek Culture*, 3 vols., Gilbert Highet (trans.), New York: Oxford University Press.

Klein, J. (1985). The World of Physics and the "Natural" World. In R. B. Williamson and E. Zuckerman (Eds.), *Jacob Klein: Lectures and Essays* (pp. 1-34) Annapolis, MD.: St. John's College Press.

Marrou, H. I. (1982). *A History of Education in Antiquity*. George Lamb (trans.). Madison, Wisconsin: University of Wisconsin Press.

Morgan, T. J. (1999). Literate Education in Classical Athens. *Classical Quarterly, 49*(1), 46-61.

Morrow, G. R. (1970*). Proclus: A Commentary on the First Book of Euclid's Elements*. Princeton: Princeton University Press.

Mueller, I. (1981). *Philosophy of Mathematics and Deductive Structure in Euclid's Elements*. Cambridge, MA: MIT Press.

Mueller, I. (1991). Mathematics and Education: Some Notes on the Platonic Program. In *ΠΕΡΙ ΤΩΝ ΜΑΘΗΜΑΤΩΝ*, special edition of *Apeiron, XXIV*(4), 85-104.

O'Brien, E. (1964). *The Essential Plotinus*. New York: New American Library.

Proclus (1967). *Procli Diadochi in primum Euclidis Elementorum librum commentarii*, (G. Friedlein). Hildesheim.

Taisbak, C. M. (1982). *Coloured Quadrangles: A Guide to the Tenth Book of Euclid's Elements*. Copenhagen: Museum Tusculanum Press.

Unguru, S. and Fried, M. N. (2007). Apollonius, Davidoff, Rorty, and Zeuthen: From A to Z, or, What Else Is There? *Sudhoffs Archiv, 91*(1),1-19.

# BOOK X OF *THE ELEMENTS*: ORDERING IRRATIONALS

*Jade Roskam[1], The University of Montana*

**Abstract:** Book X from The Elements contains more than three times the number of propositions in any of the other Books of Euclid. With length as a factor, anyone attempting to understand Euclidean geometry may be hoping for a manageable subject matter, something comparable to Book VII's investigation of number theory. They are instead faced with a dizzying array of new terminology aimed at the understanding of irrational magnitudes without a numerical analogue to aid understanding. The true beauty of Book X is seen in its systematic examination and labeling of irrational lines. This paper investigates the early theory of irrationals, the methodical presentation and interaction of these magnitudes presented in The Elements, and the application of Euclidean theory today.

Keywords: Book X; Euclid; Euclid's Elements; Geometry; History of mathematics; Irrational numbers

---

## 1. BACKGROUND

Book X of Euclid's *The Elements* is aimed at understanding rational and irrational lines using the ideas of commensurable and incommensurable lengths and squares. Unfortunately, a lack of documentation of the early study of incommensurables leads to speculation on its exact origin and discoverer. Wilbur R. Knorr in a 1998 article from The American Mathematical Monthly dates original knowledge, but not necessarily understanding, of irrational quantities to the Old Babylonian Dynasty Mesopotamians. The mathematical tables of these peoples, dating back to 1800-1500 BC, supposedly demonstrate knowledge of the fact that some values cannot be expressed as ratios of whole numbers. However, many sources disagree with Knorr's article and attribute original knowledge of irrational magnitudes to the school of Pythagoras around 430 BC (Fett, 2006; Greenburg, 2008; Robson, 2007; Posamentier, 2002). Given the most well-known accomplishment to come from the Pythagoreans, the Pythagorean theorem, it seems inevitable that this group of people would discover irrational values in the form of diagonals of right triangles. Take for example the length of the hypotenuse of an isosceles triangle with side lengths 1. This gives one of the most studied irrational quantities, $\sqrt{2}$. Prior to this inexorable discovery, the Pythagoreans viewed numbers as whole number ratios and therefore could not incorporate irrational quantities into their theory of numbers. Irrationals, considered to be an unfortunate discovery and the result of a cosmic error, were treated as mere magnitudes inexpressible in numerical form (Fett, 2006; Greenburg, 2008). These ideas were continued during the writing of The Elements, and would remain until the Islamic mathematician al-Karaji

---

[1] Jade.Roskam@umontana.edu

translated Euclidean terminology into irrational square roots of whole numbers approximately 13 centuries after Euclid wrote (Berggren, 2007).

The Pythagoreans attitude toward irrationals stunted any studying of the magnitudes beyond the incommensurability of a square's side to its diagonal. Fortunately, the superstition surrounding irrationals did not reach Plato's camp. Theodorus, a student of Plato, and one of Theodorus' own students, Theaetetus, took it upon themselves to study irrational magnitudes at length and put forth the first known theory of irrational lines (Knorr, 1975). Theodorus is cited as the first to produce varying classes of incommensurable lines through arithmetic methods argues Knorr (1975). However, Theodorus' discoveries were limited to specific cases, like lines cut in extreme-and-mean ratio, and he was unable to generalize his findings. It was his student, Theaetetus, who is generally considered as the first to put forth an organized, rigorous theory of irrationals, a work that started intuitively with his master but one that Theodorus ultimately could not prove (Knorr, 1975; 1983). The assembled findings of Theodorus and Theaetetus were published by Plato in a dialogue titled after the younger mathematician. Unfortunately much of Theaetetus has been lost over time and the little that is known about Theaetetus' early theory of irrationals comes from Eudemus, a student of Aristotle. Eudemus lived between the times of Plato and Euclid and is credited as having passed the early theory of irrational lines to Euclid's generation to be examined in full force in Book X of The Elements (Knorr, 1975; Euclid, 2006). If it was not for Plato's Theaetetus and the accounts from Eudemus, we may very well have attributed the entirety of the ideas of commensurable and incommensurable magnitudes to Euclid (Knorr, 1983).

Theaetetus is the one credited with having classified square roots as those commensurable in length versus those incommensurable (Knorr, 1983; Euclid, 2006). The three main classes of irrational magnitudes are the medial, binomial, and apotome. The medial line is defined as the side of a square whose area is equal to that of an irrational rectangle. The binomial and apotome oppose one another, as the binomial is formed by the addition of two lines commensurable in square only and the apotome is defined as the difference between two lines commensurable in square only. Each class of magnitude will be discussed in more detail later. Theaetetus is also said to have tied each class of magnitude with a unique mean: he medial is tied to the geometric mean, the binomial to the arithmetic, and the apotome to the harmonic mean (Euclid, 2006). However, these terms may just have been a replacement by Eudemus to tie irrational lines to Euclidean means, rather than the original correlations Theaetetus may have used (Knorr, 1983). The history behind the advancement of irrationality theory cannot exclude Euclid from its discussion. It was Euclid who generalized the idea of commensurable and incommensurable to squares, and also ordered the binomial and apotome irrational lines into six distinct classes each (Knorr, 1983). Most of the post-Euclidean advancement of the theory of irrational lines is found in propositions 111-114 of Book X which are generally considered to have been additions due to the lack of contiguity between these and the previous properties of irrationals presented. It is important to note that Book X details a theory of irrational *magnitudes* and not a theory of irrational *numbers* (Grattan-Guinness, 1996). Theaetetus' original theory of irrationals may have included numbers, but Euclidean theory deals solely with irrational lines and geometric lengths. The six classes of binomial and apotome are now more easily understood using algebra as the ordering of irrational magnitudes is explained through solutions of a general quadratic formula. The basis of this development is somewhat controversial. Knorr (1975) attributes some of the "geometric algebra" to Theodorus. Most sources believe this understanding of geometry through algebra originated in the 8th Century through the vast advances made by many Islamic mathematicians in the area of algebra (Gratten-Guinness, 1996; Berggren, 2007). Some now argue that much, if not most, of The Elements is actually algebra disguised as geometry (Gratten-Guinness, 1996). However, as will be discussed later, this idea is a hindrance to understanding Euclidean theory. While using the solutions to a general quadratic is a good way to help understand how each order of binomial and apotome is derived, it inherently ignores all irrationals that are not in the form of a square root and treats irrationals as values rather than magnitudes (Burnyeat, 1978).

## 2. EUCLID ON IRRATIONALS

At the start of Book X Euclid (2006) provides definitions for commensurability and rationality. For commensurability Euclid states that "magnitudes are said to be commensurable which are measured by the same measure, and those incommensurable which cannot have any common measure" and that "Straight lines are commensurable in square when the squares on them are measured by the same area, and incommensurable in square when the squares on them cannot possibly have any area as a common measure" (p. 693). Euclid (2006) then moves to rationality which he defines as:

Let then the assigned straight line be called rational, and those straight lines which are commensurable with it, whether in length and in square, or in square only, rational, but those that are incommensurable with it irrational….And then let the square on the assigned straight line be called rational, and those areas which are commensurable with it rational, but those which are incommensurable with it irrational, and the straight lines which produce them irrational (p. 693). To simplify, given a rational length (or number), all lengths (numbers) that have common measure with the rational and/or with the square of the rational are also rational. Those lengths that do not have a common measure with the given line are irrational. Squaring a rational length produces a rational area, and those areas that are commensurable with the rational area are rational and those incommensurable with the rational area are irrational. If an area is irrational, the length that was squared to create the irrational area is also irrational.

In total, there are 13 distinct irrational straight lines. In addition to the medial, Euclid sets up six orders of binomials and six orders of apotomes. The Elements also defines a subgroup of irrational lines that can be constructed from the thirteen distinct irrationals which include first and second order bimedial lines, first and second order apotome of a medial line, major, and minor, the first four of which will be discussed briefly.

According to Euclid (2006), a medial is formed when a rectangle contained by two rational straight lines commensurable in square only is irrational and the side of the square equal to it is irrational. The side of the square is called the medial (X. 21)[1].

Book X. Proposition 21

In the diagram below, lines *AB*, *BC* are assumed to be rational lengths that are commensurable in square only. That
is to say, the square on *AB* and the square on *BC* have the ratio of a whole number to a whole number, but lengths *AB*, *BC* do not have a common measure. Now construct the square *AD* such that *AB* = *BD*.

[1] Note that for ease, I will denote propositions from <u>The Elements</u> by (Book. Proposition Number). For instance, Proposition 47 from Book I will be cited as (I. 47).

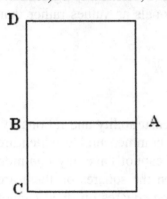

Then the square *AD* is rational since $AD = AB^2$ and *AB* is rational. We know that *AB* and *BC* are incommensurable in length, which implies that *BD*, *BC* are also incommensurable in length. Note that

$$BDBC = BD*ABBC*AB = ADAC$$

Since *BD*, *BC* are incommensurable, this implies that *AD*, *AC* are also incommensurable. But we know that *AD* is rational, so *AC* must be irrational. Since *AC* is an irrational area, a square with equal area will also be irrational and, by definition, will have a side of irrational length. This irrational side length is known as a medial.

Binomials on the other hand are formed when two rational straight lines commensurable in square only are added together, making the whole irrational. The following is adapted from <u>The Elements</u> (X. 36):

Let lines *x*, *y* be rational and commensurable in square only, meaning that nothing measures both *x* and *y*, but $x^2$ and $y^2$ have a common measure. It is proposed that their sum, *x* + *y*, will be irrational and, as per <u>The Elements</u>, called a binomial.

(i) Since *x* is commensurable in square only with *y*, *x* and *y* are incommensurable in length. Therefore, since

$$xy = x2x*y$$

(*ii*) It follows that $x^2$ and $x*y$ are also incommensurable.
But since $x^2$ and $y^2$ have a common measure, $a$, then

$n*a=x2$

$m*a=y2$
where $m$, $n$ are integers.
By substitution,
$x2+ y2=n*a+m*a=a*(n+m)$

(*iii*) So $a$ measures $x^2$ and $a$ measures $(x^2 + y^2)$, which implies that $x^2$ and $x^2 + y^2$ are commensurable.
(*iv*) It is obvious that $x*y$ is commensurable with $2*(x*y)$.
(*v*) Since (*iii*) $x^2$ and $x^2 + y^2$ are commensurable, (*iv*) $x*y$ and $2*(x*y)$ are commensurable, but (*ii*) $x^2$ and $x*y$ are incommensurable, it follows that $x^2 + y^2$ and $2*(x*y)$ are incommensurable. From this, we see that $(x^2 + y^2 + 2*(x*y))$ must be incommensurable with $(x^2 + y^2)$. Rearranging the first term, $(x + y)^2$ and $(x^2 + y^2)$ are incommensurable. Since $x$, $y$ are rational, then $x^2$, $y^2$ are also rational and it follows that $x^2 + y^2$ is rational. This implies that $(x + y)^2$ is irrational, and therefore $(x + y)$ is irrational.

Euclid defines an apotome in proposition 73 of Book X as the remainder of two rational straight lines, the less subtracted from the greater, which are commensurable in square only. It is, in essence, the counterpart of the binomial. Euclid's proof that the apotome is irrational follows the same logical steps as those used to prove the irrationality of the binomial. We start will the same basic assumption, that lines $x$, $y$ are rational and commensurable in square only. It is proposed that the apotome, $x - y$, is irrational. Steps (*i*) through (*v*) are identical to the proof of Proposition 36. For the apotome, note that

$x2+ y2=2*x*y+x-y2$

Since (*v*) $x^2 + y^2$ and $2*(x*y)$ are incommensurable, it follows that $x^2 + y^2$ and $(x - y)^2$ are also incommensurable.
But since $x$, $y$ are rational by construction, $x^2 + y^2$ must be rational. This implies that $(x - y)^2$ is irrational, from which it follows that $x - y$ is irrational. Thus we have proven that if a rational straight line is subtracted from a rational straight line, and the two are commensurable in square only, the remainder will be irrational.

It was stated earlier that Theaetetus tied the three known types of irrationals at the time to unique means: the medial with the geometric, the binomial with the arithmetic, and the apotome with the harmonic. The first two of these pairings follow somewhat simply. The medial is tied to the geometric mean, which can be found using the following general formula.
$G$x, x, $\ldots$, x$=n$x$*$x$*\ldots*$x

A medial is defined as the length of the side of a square whose area is equal to that an irrational rectangle formed by two rational lines commensurable in square only. Using our above diagram, the square on the medial, we will call it *MN* for simplicity, is equal to the area of rectangle *AC*. Algebraically,

$MN2=AC=AB*BC$

$MN= AB*BC$

Medials can therefore be represented symbolically as the geometric mean

$x*y$

for two given rational magnitudes $x$ and $y$ commensurable in square only.

As stated earlier, the binomial is defined as the sum of two rational straight lines commensurable in square only and is closely related to the arithmetic mean, of which the following is the general formula.

$Ax, x, \ldots, x=1n*(x+x+\ldots+x)$

It is obvious how the representation of the binomial $(x + y)$ is closely linked to this formula. However, the coupling of the apotome and the harmonic mean is more complex. To explain, the harmonic mean of two numbers, $x$ and $y$, is

$2*x*yx+y$

If you consider the propositions X.112-4, you can see that if a rational area has a binomial for one of its sides, the other side will be an apotome commensurable with the binomial and of corresponding order. Using our knowledge of the general form of an apotome and a binomial, we can see that this area would be

$x+y*x-y= x2-y2$

With $x*y$ representing a medial area and the above equation for the given rational area, we see that

$2*x*yx2- y2*x-y$

with $(x - y)$ representing the basic form of an apotome. Again, this seems like a stretch given the ease with which the binomial and medial are tied to their respective means. It should be noted that this relationship between the apotome and the harmonic mean is explained in the commentary by Woepcke in an Arabic translation of Book X of The Elements (Euclid, 2006). Whether this was Theaetetus' original reasoning for pairing the apotome and harmonic mean is unknown. Again, these algebraic explanations are not the original work of Euclid, but theories imposed upon his work by later mathematicians. This is important to note because Euclidean theory pertained solely to irrational magnitudes and not to irrational numbers. Since most of Theaetetus' originally theory is lost, it cannot be determined conclusively whether the Platonic mathematician described the above relationships. The ties between three types of Euclidean quantities and three Aristotelian quadrivium is seen elsewhere in The Elements. According to Ivor Gratten-Guinness' 1996 article, the three types of quantities Euclid addresses, number, magnitude, and ratio, correlate to arithmetic, geometry and harmonics, respectively. These relationships certainly follow more readily than the irrational magnitudes to the corresponding means, and the latter associations may have been formed in response to the former.

## 3. ORDERING IRRATIONALS

A class is defined as a set of objects connected in the mind due to similar features and common properties (Forder, 1927). All magnitudes in the class of binomials are of the form $(x + y)$ where $x, y$ are lines commensurable in square only. All binomials share common features, which will be discussed later. The same is true of apotomes. All are of the form $(x - y)$ where $x, y$ are lines commensurable in square only, and they share common properties. These represent two of the three classes proposed in Theaetetus' early theory of irrationals. Within each of these classes, Euclid defines six orders, or sub-classes, of each. Each member of a sub-class contains all the properties common to the class as a whole, but has different properties from members of other sub-classes (Forder, 1927). Theaetetus is credited with ideas of the medial, binomial, and apotome, but he makes no reference to the six orders of binomials and apotomes listed in The Elements. Therefore it was up to Euclid's discretion on how to best order the sub-classes. The difficulty in Euclidean theory of irrationals lies in the overlap of properties between sub-classes. As will be discussed in detail, the six orders of each class are paired into three groups, with one from each pair representing commensurability and one from each pair representing incommensurability. The struggle arises from what is most important in the class, the commensurability or the pairing with another sub-class. Despite the algebraic understanding of Euclid's irrationals making the pairing of sub-classes easier to follow, Euclid chose to first break each class of irrational line into commensurable versus incommensurable, and then pair the members in each.

Euclid defines each of the six orders of binomial and apotome in Definitions II and III, respectively, of Book X and the introduction to Book X provides an algebraic understanding of how each type is derived. To clarify the definitions given by Euclid, we will represent a binomial using the general form $(x + y)$ with $x$ being the greater segment and $y$ being the lesser segment and an apotome using the general form $(x - y)$ with $x$ being the whole and $y$ being the annex (or what is subtracted from the whole).

Consider the general quadratic formula
$x2\pm2*a*x*p\pm b*p=0$

where $p$ is a rational straight line and $a$, $b$ are coefficients. Only positive roots of this equation will be considered as $x$ must be a straight line. Those roots include

$x=p*(a+a2-b)$

$x=p*(a-a2-b)$

$x=p*(a2+b+a)$

$x=p*(a2+b-a)$

First, consider the expressions for $x_1$ and $x_1*$. Suppose $a$, $b$ do not contain any surds. That is to say, they are either integers or of the form $mn$, where $m$, $n$ are integers. If this is the case, either

(*i*) $b=m2n2*a2$

Or

(*ii*) $b{\neq}m2n2*a2$

If (*i*), then $x_1$ is a first binomial and $x_1*$ is a first apotome. Euclid defines the first order in Definitions II, for the binomial, and III, for the apotome, in Book X as:
Given a straight line and a binomial/apotome…the square on the greater term/whole [*x*] is greater than the square on the lesser/annex [*y*] by the square on a straight line *commensurable* (emphasis added) in length with the greater/whole…the greater term/whole commensurable in length with the rational straight line set out then the entire segment is known as a first binomial/first apotome (p. 784, 860). This wordy definition is translated by Charles Hutton in his 1795 two volume edition of *A Mathematical and Philosophical Dictionary* to more comprehensible terminology: the larger term, $x$, is commensurable with a rational and is thus a rational itself and $x^2 - y^2 = z^2$, where $z$ is commensurable in length with $x$, so $z$ must also be rational. Using this new definition and numerical examples provided by Islamic mathematician al-Karaji in the 10th Century, we can understand better what Euclid was representing geometrically (Berggren, 2007). For instance, $3+\sqrt{5}$ would be considered a first binomial and $3-\sqrt{5}$ would be considered a first apotome. The greater term (3) is rational and

$32- (5)2=9-5=4=22$

If (*ii*), then $x_1$ is a fourth binomial and $x_1*$ is a fourth apotome.
The fourth order of binomial and apotome are defined as opposing the first order. Euclid's definition for the fourth order of each class of irrational states:
If the square on the greater term/whole [*x*] be greater than the square on the lesser/annex [*y*] by the square on the straight line *incommensurable* (emphasis added) in length with the greater/whole, then if the greater term/whole be commensurable in length with the rational straight line set out then the entire segment is called a fourth binomial/fourth apotome (p. 784, 860). Much like the first binomial and first apotome, the greater term, $x$, will be rational. However, unlike the first order, the square root of the difference of the squares of the two terms,

$z$, will be incommensurable with $x$, meaning that $z$ will not have a rational ratio to $x$. Take, for example 4-√3. The greater term (4) is a rational number, and

42−324=16−34=134

which is not a rational ratio.

Now look at the possibilities for the $x_2$, $x_2$* expressions. If we stick to our supposition that $a$, $b$ do not contain surds, then either

($i$) $b=m2n2−m2*a2$

Or

($ii$) $b{\neq}m2n2−m2*a2$

If ($i$), then $x_2$ is a second binomial and $x_2$* is a second apotome.
Like the first order, the second order of binomials/apotomes has the square root of the difference of squares of the two terms ($z$) commensurable with the greater term/whole ($x$). The difference between the first and second order is that the lesser term/annex ($y$) is the segment that is commensurable with the rational straight line set out. This indicates that the lesser term, $y$, is rational, and that the ratio of the square root of the difference of the squares of the two terms, $z$, and the greater term, $x$, is a rational ratio expressible in whole numbers. Again, a wordy definition easily explained with actual values, like √18±4. The lesser term (4) is a rational number and

182−4218=18−1618=218=19 =13

which is a rational ratio expressed in whole numbers.

If ($ii$), then $x_2$ is a fifth binomial and $x_2$* is a fifth apotome.
The fifth order of binomial and apotome is a combination of the second and third order definitions. Like the second order, the lesser of the two terms ($y$) is commensurable with the rational straight line set out. However, the square root of the difference between the two terms ($z$) is incommensurable in length with the greater term/whole ($x$). This means that again the lesser term is rational and that the ratio of the square root of the difference of the squares of two terms, $z$, and the greater term, $x$, is not a rational ratio. For instance, √6±2. 2 is a rational number and

62−226=6−46=26=13 =13

which is not a rational ratio.

To obtain the final two orders of binomial and apotome, we must consider the case where

206

*a=mn*

where *m, n* are integers. To abbreviate, let $\lambda=mn$. Therefore

$x=p*(\lambda+\lambda-b)$
$x=p*(\lambda-\lambda-b)$
$x=p*(\lambda+b+\lambda)$
$x=p*(\lambda+b-\lambda)$

"If $\lambda-b$ in $x_1$, $x_1*$ is not surd but of the form (*mn*), and if $\lambda+b$ in $x_2$, $x_2*$ is not surd but of the form (*mn*), the roots are comprised among the forms already shown" (X. Introduction). To explain, in our original equations for $x_1$, $x_1*$, $x_2$, and $x_2*$, *a* was assumed to be rational (containing no surds) and $a2\pm b$ would then be irrational. In our new equations, we define *a* as being irrational. The above quote states that if $\lambda\pm b$ is rational (containing no surds) then we well again obtain the 1st, 2nd, 4th, and 5th order binomials or apotomes. The original $x_1$, $x_1*$ and the new $x_2$, $x_2*$ are taking a rational magnitude plus (binomial) or minus (apotome) an irrational to obtain the 1st and 4th orders. The original $x_2$, $x_2*$ and newly formed $x_1$, $x_1*$ start with an irrational magnitude and add (binomial) or subtract (apotome) a rational magnitude, forming the 2nd and 5th orders of binomial and apotome. Therefore, the only case that needs to be investigated is the case where an irrational magnitude is added or subtracted from another irrational magnitude.

If $\lambda-b$ in $x_1$, $x_1*$ is surd, then either
(*i*) $b=m2n2*\lambda$

Or

(*ii*) $b\neq m2n2*\lambda$

If (*i*), then $x_1$ is a third binomial and $x_1*$ is a third apotome.
In the case of the third order of each type of irrational, we again have a connection to the language describing the first order. The square on the greater term/whole (*x*) is greater than the square on the lesser term/annex (*y*) by the square on a straight line commensurable with the greater/whole. However, in this order neither of the terms, *x* or *y*, are commensurable with the rational straight line set out. In terms of real numbers, both *x* and *y* must be irrational and ratio of the square root of the difference of the squares of the terms, *z*, and the greater term, *x*, is a rational ratio expressible in whole numbers. To explain by example, look at the third order binomial ($\sqrt{24}+\sqrt{18}$) or the third order apotome ($\sqrt{24}-\sqrt{18}$). Both terms are irrational and

*242-18224=24-1824=624=14=12*

with *12* being a rational ratio expressible in whole numbers.

If (*ii*), then $x_1$ is a sixth binomial and $x_1$* is a sixth apotome.
Much like the third, in a sixth order of binomial and apotome neither the lesser term (*y*) nor the greater term (*x*) are commensurable in length with the rational straight line set out, but the square on the greater/whole is greater than the square on the lesser/annex by the square on a straight line incommensurable in length with the greater term/whole. This translates to the sixth apotome having the form of two irrational terms with the ratio of the square root of the difference of the squares of the two terms, *z*, to the greater term, *x*, being an irrational ratio. Look at the sixth order binomial/apotome $\sqrt{6}\pm\sqrt{2}$. Both terms are irrational and

$$62-226=6-26=46=26$$

which is not a rational ratio.

Table 9.1 summarizes the six orders of binomial and apotome.

Euclid also designates two orders of bimedial lines and two orders of apotome of medial straight lines. Bimedial lines are the sum of two medial lines which are commensurable in square only. Proposition 37 demonstrates how to construct a first bimedial line (two medial lines commensurable in square only and containing a rational rectangle can be added together). Constructing a second bimedial line is discussed in proposition 38, where by all the same conditions as the first bimedial apply, accept that the two medial lines form a medial rectangle instead of a rational rectangle. An apotome of a medial is defined as the difference between two medial lines, the lesser of which is commensurable with the whole in square only. If a rational rectangle is contained with the square of the whole, then the remainder is a first apotome of a medial straight line (X. 74). If a medial rectangle is contained with the square of the whole, the remainder is known as a second apotome of a medial straight line (X. 75). An obvious connection can be drawn between bimedial lines and apotome of medial lines. All four types are constructed by manipulating two medial lines, with the first orders of each referring to a contained rational rectangle and the second orders of each having a medial rectangle contained by the two medial lines. The name apotome of a medial is fitting in an obvious way: the line is formed by the difference of two medials (*x* − *y*). What is confusing is the naming of the bimedial. With the connection between the bimedial and apotome of a medial mentioned above and the definition of the bimedial as the sum of two medial lines (*x* + *y*), it is interesting that Euclid did not use the more obvious title of binomial of a medial line. It is possible that the original terminology was binomial of a medial line and through translation was shortened to bimedial, but this is mere speculation.


## 4. PROPERTIES AND INTERACTIONS

One of the most fascinating things about the three main types of irrational lines is studying the ways that they interact with each other. One example of this is the algebraic representation of

General formulas (applying to the roots):

$$x_1 = p\cdot\left(a+\sqrt{a^2-b}\right)$$
$$x_1^* = p\cdot\left(a-\sqrt{a^2-b}\right)$$
$$x_2 = p\cdot\left(\sqrt{a^2+b}+a\right)$$
$$x_2^* = p\cdot\left(\sqrt{a^2+b}-a\right)$$

| Order | Definition of Binomial (Apotome) — Square on the greater segment (whole) greater than the square on the lesser segment (annex) by the square on a straight line ___ in length with the greater term (whole); ___ be commensurable in length with the rational straight line set out Rational Term | | Algebraic Interpretation of the Roots of General Quadratic Formula | Meaning | Real Values / Example |
|---|---|---|---|---|---|
| First | commensurable | greater segment (whole) greater term is rational | If $a$, $b$ do not contain surds and $x_1$ is a binomial and $x_1^*$ is an apotome | Difference of the squares of the two terms is a square number | $3 - \sqrt5$ |
| Second | commensurable | lesser term (annex) lesser term is rational | If $a$, $b$ do not contain surds and $b = \frac{m^2-m^2}{n^2-m^2}\cdot a^2$; $x_2$ is a binomial and $x_2^*$ is an apotome | Square root of the difference of the squares of terms is a rational ratio to greater term | $\sqrt{18} - 4$ |
| Third | commensurable | neither segment (neither) neither term is rational | If $\sqrt{(a^2-b)}$ is surd, $a=\sqrt{\frac{m}{n}}$, $b=\frac{m^2}{n^2}\cdot\lambda$; $x_1$ is a binomial and $x_1^*$ is an apotome | Square root of the difference of the squares of terms is a rational ratio to greater term | $\sqrt{24} - \sqrt{18}$ |
| Fourth | incommensurable | greater segment (whole) greater term is rational | If $a$, $b$ do not contain surds and $b\neq\frac{m^2}{n^2-m^2}\cdot a^2$; $x_1$ is a binomial and $x_1^*$ is an apotome | Square root of the difference of the squares of terms is not a rational ratio to the greater term | $4 - \sqrt3$ |
| Fifth | incommensurable | lesser term (annex) lesser term is rational | If $a$, $b$ do not contain surds and $b\neq\frac{m^2}{n^2-m^2}\cdot a^2$; $x_2$ is a binomial and $x_2$ is an apotome | Square root of the difference of the squares of terms is not a rational ratio to greater term | $\sqrt6 - 2$ |
| Sixth | incommensurable | neither segment (neither) neither term is rational | If $\sqrt{(a^2-b)}$ is surd, $a=\sqrt{\frac{m}{n}}$, $b\neq\frac{m^2}{n^2}\cdot\lambda$; $x_1$ is a binomial and $x_1^*$ is an apotome | Square root of the difference of the squares of terms is not a rational ratio to greater term | $\sqrt6 - \sqrt2$ |

binomials and apotomes.  Binomials can be understood as a process of addition represented by $(x + y)$. The opposite is true of the apotome which is represented as $(x - y)$. These in turn have a product of $(x^2 - y^2)$. It is obvious that there are numerous relations that these lines hold with each other, and yet for all of their similarities, each of the categories of irrational lines are mutually exclusive.  Euclid goes so far as to say, "The apotome and the irrational straight lines following it are neither the same with the medial straight line nor with one another" (X. 111).  However, these lines are not just mutually exclusive categories, but are also unique in their division into parts.  Proposition 42 demonstrates that if $AB$ is a binomial, then there is only one point $C$ between $A$ and $B$ such that $AC$, $BC$ are rational and commensurable in square only.  This proves that for a given binomial, there is only one way to separate its length into greater and lesser segments.  The same is proven of a first bimedial (X. 43) and a second bimedial (X. 44). Likewise, from a given apotome, only one length can be subtracted such that both segments are rational and commensurable in square only (X. 79).  Again, Euclid goes on to prove in propositions 80 and 81 the uniqueness of first and second apotome of a medial lines.

We must first look at the major properties of medials, binomials, and apotomes before we can delve into the interactions between these lines.

Common to all of the types of irrational lines is that fact that lines commensurable with the given length are of the same type and order where applicable (X. 23(medial), 66(binomial), 67(bimedial), 103(apotome), 104(apotome of a medial)).  Unique to medials are the ideas that rectangles contained by medial lines commensurable in length is medial (X. 24), that rectangles contained by medial lines commensurable in square only are either rational or medial areas (X. 25), and that the difference between two medial areas will never be a rational area (X. 26).  In maybe the most important proposition of book X, Euclid proves that an infinite number of unique irrational lines arise from a medial line (115).  Interestingly, he chose to make this the last proposition in the book, possibly with the hopes that future students would use this property of medials to further investigate the theory of irrationals, possibly coming up with new unknown forms of irrational lines or a new classification system.  There are a few propositions that deal with only binomials or apotomes, but these are usually taken in sets with the ensuing propositions using a bimedial or apotome of a medial, and thus will be discussed later.  However, propositions 48-53 do deal strictly with binomials, in that each describes how to find a binomial of particular order.  Propositions 85-90 perform the same action for orders of apotome.

Why Euclid chose to classify apotomes and binomials such that the first and fourth, second and fifth, and third and sixth orders were paired is not explained.  We can note that the first three orders deal with commensurable lengths between the differences in squares explained above and the greater segments while the last three orders have greater terms being incommensurable with the difference in squares.  We can also note each of the pairings are based on which term (greater, lesser, or neither) are rational.  From this, it is plausible to assume Euclid's ordering is first based on the commensurability of given aspects of the line, and second based on which part of the given line is rational, leading to the two fold classification system seen today.  Whether this was Euclid's reasoning or not, it does appear that he was not particularly concerned with functional order throughout the elements.  For example, the first time a reader is introduced to a

210

line cut in extreme-and-mean ratio is in the beginning of Book II. Yet it is not until Book X that the properties of such a line (with greater length is an apotome and lesser length a first apotome) are explained and not until Book XIII that this type of line is applied, which will be discussed in more detail later. It is also interesting to note Euclid devotes books VII, VIII, and IX to investigating numbers and number theory, but certain properties of numbers appear in many of the other ten books, including addressing ratios of numbers in Book V well before putting forth a theory of numbers (Grattan-Guinness, 1996).

Despite what may or may not be a flawed ordering system, the vast majority of Book X is devoted to exploring the interactions between the classes of irrational lines. It should be noted that for each property of binomial lines, the same property is proven just thirty-seven short propositions later for apotome lines. Starting with propositions 54 and 91, Euclid proves that if a rectangle is formed by a rational line and a first order binomial or apotome, the "side" or diagonal of that rectangle will be a binomial or apotome. As I mentioned before, the propositions describing the interactions of irrational lines often come in sets. Just as 54 and 91 prove the above statements, propositions 55, 56 and 92, 93 prove a similar situation occurs with bimedials and apotome of medial lines. An area formed by a rational and a second order binomial has for its side a first order bimedial (X. 55). For a rational and a third order binomial, the second bimedial is the diagonal (X. 56). Switching "apotome" for binomial and "apotome of a medial" for bimedial, we have the statements of propositions 92 and 93. We learn that if a rectangle is formed with rational length and area equal to a binomial squared, the width of the rectangle will be a first order binomial in proposition 60. The likewise is true of apotomes (X. 97). Propositions 61-62 and 98-99 are devoted to proving a similar statement: that if a straight line *AB* is a first bimedial or apotome of a medial (or second order for proposition 99), and a rational straight line *CD* is the side of rectangle *CE* such that the area of *CE* is equal to the square on *AB*, then the other side of rectangle *CE*, side *CF*, is a second binomial or apotome (third order for 99). Finally, as stated previously, propositions 112-113 prove that if a rational area has a binomial for one of its sides, the other side will be an apotome commensurable with the binomial and of corresponding order, with 114 proving that if a binomial and apotome that are commensurable and of the same order form the length and width of a rectangle, the diagonal will be rational.

## 5. MODERN IMPLICATIONS

Euclid's' dialogue on irrational lines is not restricted to Book X. Indeed he puts forth an important application of the apotome in Book XIII. Proposition 6 states that if a line is cut in extreme-and-mean ratio (first introduced in II. 11), then the greater segment will be an apotome and the lesser segment a first apotome. This one proposition has enormous implications for the theory of irrational magnitudes. The golden ratio, one of the most applicable and well-studied areas of math, is created by a line cut in extreme-and mean ratio. This is an important topic to understand due to the vast number of properties held by objects that contain this ratio. One example is the logarithmic spiral which is formed through the construction of both golden rectangles, whose sides, when taken in proportion, equal the golden ratio, and the golden triangle, whose angles are 72°, 72°, 36°. Logarithmic spirals are seen throughout nature. Ram horns, elephant tusks, nautilus shells, pine cones, sun flowers and many other living things grow in accordance with the golden ratio (Fett, 2006). This proportion is said to be the most aesthetically pleasing to look at, which is why many great paintings and sculptures contain the golden ratio. The Parthenon in Athens, which not only houses sculptures containing the golden ratio but in fact can be inscribed in a golden rectangle, and five of Leonardo da Vinci's works, including two of his most famous "Madonna on the Rocks" and "Mona Lisa", are also said to contain the golden ratio (Fett, 2006). Many plastic surgeons still use the golden ratio several times over to construct what is believed to be a universal standard of beauty (Fett, 2006).

Each of the five Platonic solids, the only existing solids to have identical and equilateral faces and convex vertices, incorporates the golden ratio in its construction. The tetrahedron, octahedron and icosahedron are based on equilateral triangles, while the cube and dodecahedron are based on the square and pentagon. These shapes are discussed in detail in Book XIII of The Elements after the introduction of the line cut in extreme-and-mean ratio. Of particular interest are the dodecahedron and the icosahedron. Exodus of Cnidus, who lived after Theaetetus, is credited with having first discovered the irrationality of a line divided in extreme and mean ratio after working with the problem of inscribing a regular pentagon in a given circle (Knorr, 1983). The pentagon is actually formed by three golden triangles, and the ratio of the shorter side to the longer is equal to the golden ratio. This implies the construction of the icosahedron is dependent upon the golden ratio. The golden ratio is also present in the calculation of surface area and volume of the dodecahedron as well as the volume of the icosahedron (Fett, 2006). Since Theaetetus is credited with first discovering the icosahedron, Book XIII, along with Book X, is firmly based in the Athenian's work. In fact, M. F. Burnyeat quotes B. L. Van der Walden in his 1978 journal article as saying "The author of Book XIII knew the results of Book X, but…moreover, the theory of Book X was developed with a view to its applications in Book XIII. This makes inevitable the conclusion that the two books are due to the same author…Theaetetus".

The golden ratio is also seen in the comparison of sequential values in the also well-studied Fibonacci sequence. The Fibonacci numbers are defined by the recursive formula

$$f_{n+1} = f_n + f_{n-1}$$

for $f_n > 2$ with $f_1 = 1$ and $f_2 = 1$ (Fett, 2006; Posamentier, 2002; Rosen, 2005). When comparing the $n^{th}$ Fibonacci number with the $(n-1)^{th}$, the ratio will approach the golden ratio as $n$ increases. Not surprisingly given its relationship with the golden ratio, the Fibonacci sequence is often found in the growth of natural objects. For example, the number of spirals in plants that grow in a phyllotaxis pattern will always be a Fibonacci number (Rosen, 2005). Another sequence related to the both the Fibonacci numbers and the golden ratio is the Lucas numbers. These are defined using the same recursive formula as the Fibonacci numbers and still begins the sequence with 1, but $\ell_2 = 3$ instead of 1 (Posamentier, 2002; Rosen, 2005). Interestingly enough, the same relationship exists between Lucas numbers and the golden ratio as the Fibonacci numbers and the golden ratio (Posamentier, 2002).

## 6. CONCLUSIONS

It is unfortunate that so little is known about the early theory of irrationals or who advanced our understanding to what it is today. What is also regrettable is the lack of progress we have made since the days of Euclid. On the positive, we can be thankful for the meticulous systematic presentation of irrational magnitudes and their properties and interactions demonstrated in Book X of The Elements. While this chapter of Euclidean geometry has not been developed to the degree that most of his work has, the multitude of Book X leaves us with plenty of information on irrationals.

We know that there are 13 types of irrational lines, with each category being mutually exclusive. We also know that for every irrational line, there is only one way to divide the line to meet the criteria of its category and order, proving the uniqueness of each. Maybe most importantly, we know from proposition 115 that there are infinitely many irrational lines. Euclid provides us with the application of this theory in Book XIII, showing how the pentagon and icosahedron utilize irrational magnitudes in the construction of each. This application of the extreme-and-mean ratio has led to significant discoveries in the area of aesthetics, art, and music through the apotome known as the golden ratio.

Perhaps this information was enough to satisfy mathematicians throughout history. Or possibly our Euclidean understanding of irrationals is complete. We highly doubt the latter, but believe so much time is spent simply trying to understand the already burdensome theory of irrational lines that little is left for the advancement of the theory. Many mathematicians have devoted time to aiding future students in understanding the classification of commensurable and incommensurable magnitudes. Hopefully this will eventually lead to a more readily comprehensible theory, a base step from which a more innovative, improved theory of irrational lines can be developed.

N. Sirotic and R. Zazkis (2005) conducted a research project to find out how much we retain of the Euclidean theory of irrational lengths. They asked a group of college students studying to be secondary teachers if it was possible to locate $\sqrt{5}$ on a number line. The results were somewhat frightening. Less than 20% of the participants used a geometric construction to find $\sqrt{5}$ on the number line, most of those having used the Pythagorean Theorem, approximately 65% used some sort of decimal approximation in varying degrees of exactness, and an abysmal 15% either did not answer, or worse, argued it was not possible to find exactly where $\sqrt{5}$ falls on a number line. Most of those who argued it was impossible reasoned that since $\sqrt{5}$ was irrational, the decimal approximation in infinite and non-repeating and that was why it cannot be accurately positioned. However, these same participants believed a repeating infinite decimal, like 13, could be placed in its exact position, but could not explain why whether the decimal repeated or not made a difference. This means that 80% of those future secondary teachers could not think past our understanding of decimal approximations to use a well-known and highly practiced idea (the Pythagorean Theorem) to find where $\sqrt{5}$ falls on a number line. To these people, it seems the number line is really a *rational* number line and irrational numbers cannot be placed exactly since "because [the decimal] never ends we can never know the exact value" (Sirotic, 2005). This is an unfortunate side effect of Theodorus' and al'Karaji's work to aid students in understanding irrationals. The original understanding of irrational lines using geometry is lost to the more easily comprehensible geometric algebra presented in almost all current editions of The Elements. While the use of algebra is integral to helping students understand this dense topic, a return to irrational magnitudes' geometric roots appears to be just as important for students to gain a true understanding of Book X.

## ACKNOWLEDGMENTS

Special thanks to Steve Williams for initial assistance in researching this vast subject and to Dr. Bharath Sriraman of the University of Montana for his valuable input on various drafts and for challenging me with such a complex topic.

## REFERENCES

Burnyeat, M. F. (1978) The Philosophical Sense of Theaetetus' Mathematics. ISIS: The Official Journal of the History of Science Society. 69(4), 489-513

Euclid (2006) The Elements. (Sir T. L. Heath, Trans.) New York: Barnes & Noble Publishing, Inc. (Original work written around 300 BC).

Fett, B. (2006) An In-Depth Investigation of the Divine Ratio. The Montana Mathematics Enthusiast. 3(2), 157-175.

Forder, H. G. (1927) The Foundations of Euclidean Geometry. London: Cambridge University Press.

Grattan-Guinness, I. (1996) Numbers, Magnitudes, Ratios, and Proportions in Euclid's *Elements*: How Did He Handle Them? Historia Mathematica. 23, 355-375.

Greenburg, M. J. (2008) Euclidean and Non-Euclidean Geometries (4th ed.). New York: W. H. Freeman and Company.

Hutton, C. (1795) A Mathematical and Philosophical Dictionary. London

Imhausen, A., Robson, E., Dauben, J., Plofker, K., and Berggren, J. L. (2007) The Mathematics of Egypt, Mesopotamia, China, India, and Islam: A Sourcebook. (V. Katz, eds.) Princeton: Princeton University Press.

214

Knorr, W. R. (1975) <u>The Evolution of the Euclidean Elements: A Study of the Theory f Incommensurable    Magnitudes and Its Significance for Early Greek Geometry</u>. Dordrecht, Holland: D. Reidal Publishing        Company.

Knorr, W. (1983) Euclidean Theory of Irrational Lines. <u>Bulletin of the American Mathematical Society</u>. 9(1), 41-69.

Knorr, W. (1998) 'Rational Diameters' and the Discovery of Incommensurability. <u>The American Mathematical Monthly</u>. 105(5), 421-429

Kyburg, H. Jr. (1997) Quantities, Magnitudes, and Numbers. <u>Philosophy of Science</u>. 64, 377-410.

Posamentier, A. S. (2002) <u>Advanced Euclidean Geometry</u>. New York: Key College Publishing.

Rosen, K. H. (2005) <u>Elementary Number Theory and Its Applications</u> (5th ed.). Boston: Pearson Addison Wesley.

Sirotic, N. and Zazkis, R. (2007). Irrational numbers on the number line – where are they? <u>International Journal of Mathematical Education in Science and Technology</u>. 38(4), 477-488.

# The Origins of the Genus Concept
# in Binary Quadratic Forms

*Mark Beintema*[1]          &          *Azar Khosravani*[2]
*College of Lake County, Illinois*          *Columbia College Chicago*

ABSTRACT: We present an elementary exposition of genus theory for integral binary quadratic forms, placed in a historical context.
KEY WORDS: Quadratic Forms, Genus, Characters
AMS Subject Classification: 01A50, 01A55 and 11E16.

**INTRODUCTION:** Gauss once famously remarked that "mathematics is the queen of the sciences and the theory of numbers is the queen of mathematics". Published in 1801, Gauss' *Disquisitiones Arithmeticae* stands as one of the crowning achievements of number theory. The theory of binary quadratic forms occupies a large swath of the *Disquisitiones*; one of the unifying ideas in Gauss' development of quadratic forms is the concept of genus. The generations following Gauss generalized the concepts of genus and class group far beyond what Gauss had done, and students approaching the subject today can easily lose sight of the basic idea.

Our goal is to give a heuristic description of the concept of genus – accessible to those with limited background in number theory – and place it in a historical context. We do not pretend to give the most general treatment of the topic, but rather to show how the idea originally developed and how Gauss' original definition implies the more common definition found in today's texts.

1, 2

Mark Beintema
Department of Mathematics
College of Lake County
19351 W. Washington Ave.
Grayslake, IL 60030
(847) 543-2913
markbeintema@clcillinois.edu

Azar Khosravani
Science and Math Department
Columbia College Chicago
600 S. Michigan Ave.
Chicago, IL 60605
(312) 344-7285
akhosravani@colum.edu

**BASIC DEFINITIONS:** An integral binary quadratic form is a polynomial of the type $f(x,y) = ax^2 + bxy + cy^2$, where $a$, $b$, and $c$ are integers. A form is *primitive* if the integers $a$, $b$, and $c$ are relatively prime. Note that any form is an integer multiple of a primitive form. Throughout, we will assume that all forms are primitive. We say that a form $f$ represents an integer $n$ if $f(x,y) = n$ has an integer solution; the representation is *proper* if the integers $x$, $y$ are relatively prime. A form is *positive definite* if it represents only positive integers; we will restrict our discussion to positive definite forms.

The *discriminant* of $f = ax^2 + bxy + cy^2$ is defined as $\Delta = b^2 - 4ac$. Observe that $4af(x,y) = (2ax + by)^2 - \Delta y^2$. Thus, if $\Delta < 0$, the form represents only positive integers or only negative integers, depending on the sign of $a$. In particular, if $\Delta < 0$ and $a > 0$ then $f(x,y)$ is positive definite. Moreover, $\Delta = b^2 - 4ac$ implies that $\Delta \equiv b^2 \pmod 4$. Thus we have $\Delta \equiv 0 \pmod 4$ or $\Delta \equiv 1 \pmod 4$, depending on whether $b$ is even or odd. Moreover, we will write $(Z/\Delta Z)^*$ to denote the multiplicative group of congruence classes which are relatively prime to $\Delta$.

We say that an integer $a$ is a quadratic residue of $p$ if $x^2 \equiv a \pmod p$ has a solution. When discussing quadratic residues, it is convenient to use Legendre symbols. If $p$ is an odd prime and $a$ an integer relatively prime to $p$, then $\left(\dfrac{a}{p}\right)$ is defined as follows:

DEFINITION: $\left(\dfrac{a}{p}\right) = \begin{cases} 1 & \text{if } x^2 \equiv a \pmod p \text{ has a solution} \\ -1 & \text{otherwise} \end{cases}$

This notation allows us to concisely state some well-known facts about quadratic residues; here $p$, $q$ are distinct odd primes:

i) $\left(\dfrac{-1}{p}\right) = (-1)^{(p-1)/2}$  ii) $\left(\dfrac{2}{p}\right) = (-1)^{(p^2-1)/8}$

iii) $\left(\dfrac{p}{q}\right)\left(\dfrac{q}{p}\right) = (-1)^{(p-1)(q-1)/4}$  iv) $\left(\dfrac{a}{p}\right)\left(\dfrac{b}{p}\right) = \left(\dfrac{ab}{p}\right)$

Item (iii) is called the Quadratic Reciprocity Law; discovered independently by Euler and

Legendre, the first correct proof appeared in Gauss' *Disquisitiones*. Items (i) and (ii) are known as the First and Second Supplements to Quadratic Reciprocity and were proved by Euler (1749) and Legendre (1785) respectively.

More generally, let $m = p_1 p_2 \cdots p_k$, and let $a$ be any positive integer. The Jacobi symbol is defined as $\left(\dfrac{a}{m}\right) = \left(\dfrac{a}{p_1}\right)\left(\dfrac{a}{p_2}\right)\cdots\left(\dfrac{a}{p_k}\right)$. Observe that if $a$ is a quadratic residue modulo $m$, then $\left(\dfrac{a}{m}\right) = 1$, but the converse is not true. The Jacobi symbol has many of the same basic properties as the Legendre symbol; in particular the four results above are valid when $p$ and $q$ are replaced by arbitrary odd integers. The Jacobi symbol also satisfies $\left(\dfrac{a}{m}\right)\left(\dfrac{a}{n}\right) = \left(\dfrac{a}{mn}\right)$. The reciprocity law for Jacobi symbols was also proved by Gauss [**7**, Art 133], and can be stated as follows: If $m$ and $n$ are odd integers, then $\left(\dfrac{m}{n}\right) = \left(\dfrac{n}{m}\right)$ if either of $m, n \equiv 1 \pmod 4$ and $\left(\dfrac{m}{n}\right) = -\left(\dfrac{n}{m}\right)$ if $m \equiv n \equiv 3 \pmod 4$.

**HISTORICAL BACKGROUND:** The earliest investigations concerning the representation of integers by binary quadratic forms were due to Fermat. In correspondence to Pascal and Marsenne, he claimed to have proved the following:

THEOREM 1:
1. Every prime number of the form $4k + 1$ can be represented by $x^2 + y^2$.
2. Every prime number of the form $3k + 1$ can be represented by $x^2 + 3y^2$.
3. Every prime number of the form $8k + 1$ or $8k + 3$ can be represented by $x^2 + 2y^2$.

These results motivated much later research on arithmetic quadratic forms by Euler and Lagrange. Beginning in 1730, Euler set out to prove Fermat's results; he succeeded in proving (1) in 1749 (as well as the more general Two-Square Theorem), and made significant progress on the other two [**1**]. In a 1744 paper titled *Theoremata circa divisors numerorum in hac forma* $paa \pm qbb$ *contentorum*, Euler recorded many examples and formulated many similar conjectures (presented as theorems). It was in this paper that he also established many basic facts about quadratic residues. His most general result along these lines was the following:

THEOREM 2: Let $n$ be a nonzero integer, and let $p$ be an odd prime relatively prime to $n$. Then $p \mid x^2 + ny^2$, $\gcd(x, y) = 1$ $\Leftrightarrow$ $\left(\dfrac{-n}{p}\right) = 1$.

In 1773, Lagrange published the landmark paper "*Recherches d'arithmetique*", in which he succeeded in proving Fermat's conjectures concerning primes represented by the forms $x^2 + 2y^2$ and $x^2 + 3y^2$. The same paper contains a general development of the theory of binary quadratic forms, treating forms of the type $f = ax^2 + bxy + cy^2$. Lagrange's development of the theory is systematic and rigorous – it is here that he introduces the crucial concepts of discriminant, equivalence, and reduction. One of the first results is a connection between quadratic residues and the representation problem for general quadratic forms:

THEOREM 3: Let $m$ be a natural number that is represented by the form $ax^2 + bxy + cy^2$. Then $\Delta = b^2 - 4ac$ is a quadratic residue modulo $m$.

One of Lagrange's primary innovations was the concept of equivalence of forms (although the terminology is due to Gauss). We say that two forms are *equivalent* if one can be transformed into the other by an invertible integral linear substitution of variables. That is, $f$ and $g$ are equivalent if there are integers $p, q, r$, and $s$ such that $f(x, y) = g(px + qy, rx + sy)$ and $ps - qr = \pm 1$. It can be shown (e.g. see [6] or [11]) that equivalence of forms is indeed an equivalence relation. Moreover, equivalent forms have the same discriminant and represent the same integers (the same is true for proper representation). Gauss later refined this idea by introducing the notion of proper equivalence. An equivalence is a proper equivalence if $ps - qr = 1$, and it is an improper equivalence if $ps - qr = -1$. Following Gauss, we will say that two forms are in the same class if they are properly equivalent. Using these ideas, we obtain the following partial converse of Theorem 3:

THEOREM 4: Let $p$ be an odd prime. Then $p$ is represented by a form of discriminant $\Delta$

if and only if $\left(\dfrac{\Delta}{p}\right)=1$.

*Proof:* Let $f = ax^2 + bxy + cy^2$ represent $p$, say $p = ar^2 + brs + cs^2$. Because $p$ is prime, we must have $\gcd(r, s) = 1$. Hence, we can write $1 = ru - st$ for integers $t$, $u$. If $g(x, y) = f(rx + ty, sx + uy)$, then $g$ is properly equivalent to $f$ and thus has discriminant $\Delta$. Moreover, by direct calculation we have $g = px^2 + b'xy + c'y^2$. Thus, $\Delta = b'^2 - 4pc'$ and so $b'^2 \equiv \Delta \pmod{p}$.

Next, suppose that $m^2 \equiv \Delta \pmod{p}$. We can assume that $m$ has the same parity as $\Delta$ (replacing $m$ by $m + p$ if necessary). Writing $m^2 - \Delta = kp$, and recalling that $\Delta \equiv 0$ or $1 \pmod{4}$, we have $kp \equiv 0 \pmod{4}$. Thus the form $px^2 + mxy + (k/4)y^2$ has integer coefficients and represents $p$

Once we have partitioned the set of binary quadratic forms into equivalence classes, the next logical step is to choose an appropriate representative for each class. This naturally leads another of Lagrange's innovations, the concept of reduction. A primitive positive definite form $ax^2 + bxy + cy^2$ is said to be *reduced* if $|b| \leq a \leq c$ and $b \geq 0$ if either $|b| = a$ or $a = c$. Lagrange showed that every primitive positive definite form is properly equivalent to a unique reduced form, and that there are only there are only finitely many positive definite forms with a given determinant $\Delta$ [**6, 11**]. We write $h(\Delta)$ for the number of classes of primitive positive definite forms of discriminant $\Delta$. Thus, $h(\Delta)$ is the number of reduced forms of discriminant $\Delta$.

In the special case where $h(-4n) = 1$, the only reduced form of discriminant $-4n$ will be the form $x^2 + ny^2$. In this case, $p = x^2 + ny^2 \iff \left(\dfrac{-n}{p}\right) = 1$. This situation is in fact quite rare – Gauss conjectured that the only values of $n$ for which $h(-4n) = 1$ are $n = 1, 2, 3, 4$, and $7$. The conjecture was proved by Landau in 1903. More generally, we call $\Delta$ a *fundamental discriminant* if it cannot be written as $\Delta = k^2\Delta_0$, where $k > 1$ and $\Delta_0 \equiv 0$ or $1 \pmod 4$. Gauss conjectured that if $\Delta < 0$ is a fundamental

220

discriminant then $h(\Delta) = 1$ only for $\Delta = $ -3, -4,  -7, -8, -11, -19, -43, -67, -163.  This was proved in 1952 by Heegner [12].

**GENUS THEORY:**  We say that two primitive positive definite forms of discriminant $\Delta$ are in the same *genus* if they represent the same values in $(Z/\Delta Z)^*$.  Recall that equivalent forms represent the same integers and so must be in the same genus.  Thus, the concept of genus provides a method of separating reduced forms of the same discriminant according to congruence classes represented by the forms.  In his table of reduced forms, Lagrange showed forms grouped according to the congruence classes represented by the forms.   For this reason, many authors credit the original idea of genus to Lagrange.   Some authors have even attributed the idea to Euler [10].   However, Gauss is the first to explicitly discuss the concept of genus.   More importantly, he is the first to put it to use.

Before presenting Gauss' definition of genus, a few remarks concerning notation and terminology are in order.   Throughout most of the *Disquisitiones Arithmetica*, Gauss assumes forms have even middle coefficient – that is, he mostly considers forms of type $ax^2 + 2bxy + cy^2$.  (Forms with odd middle coefficient are called "improperly primitive", and are treated separately.)  Instead of discriminants, he uses the *determinant* of the form, defined as $D = b^2 - ac$.  Note that the discriminant $\Delta$ satisfies $\Delta = 4D$.

The following result, found in Article 229 of *Disquisitiones Arithmetica*, is the foundation of genus theory.  The proof is paraphrased slightly from the original text.

THEOREM 5:   Let $F$ be a primitive form with determinant $D$ and $p$ a prime number dividing $D$: then the numbers not divisible by $p$ which can be represented by the form $F$ agree in that they are either all quadratic residues of $p$, or they are all nonresidues.

*Proof*:  Let  $m = ag^2 + 2bgh + ch^2$ and $m' = ag'^2 + 2bg'h' + ch'^2$.  Then

$$mm' = [agg' + b(gh' + hg') + chh']^2 - D(gh' - hg')^2.$$

Thus $mm'$ is a quadratic residue mod $D$, and hence is also a quadratic residue mod $p$ for any $p$ dividing $D$. It follows that $m$, $m'$ are either both residues, or both are non-residues mod $p$. That is, if $m$ and $m'$ are both represented by $F$, then $\left(\dfrac{m}{p}\right) = \left(\dfrac{m'}{p}\right)$ $\square$

From the relation $\Delta = 4D$ we get two important observations: First, any odd prime that divides $D$ also divides $\Delta$. Moreover, if $p$ is an odd prime, then $\Delta$ is a residue mod $p$ if and only if $D$ is. Thus Theorem 5 still holds if the word determinant is replaced by discriminant. Henceforth, we will revert to the more common practice of using discriminants.

The argument used to prove Theorem 5 also shows that if $8 \mid D$ or $4 \mid D$, then the product of two numbers represented by $F$ will be a quadratic residue mod 8 or a quadratic residue mod 4, respectively. Hence if $8 \mid D$, then exactly one of the following is true: all numbers represented by $F$ are $\equiv 1 \pmod 8$, or all are $\equiv 3 \pmod 8$, or all are $\equiv 5 \pmod 8$, or all are $\equiv 7 \pmod 8$. Likewise, if $4 \mid D$, but $8 \nmid D$, then all numbers represented by $F$ are $\equiv 1 \pmod 4$, or all are $\equiv 3 \pmod 4$.

These observations are then used to classify forms according to *characters*. Let $p_1, p_2, ..., p_k$ be the odd prime divisors of $D$. Define $\chi_i = Rp_i$ if the numbers represented by $F$ are quadratic residues of $p_i$, and $\chi_i = Np_i$ if the numbers represented by $F$ are quadratic non-residues of $p_i$. We define one additional character, $\chi_0$, which will be an ordered pair $a, b$ chosen from the list $\{(1,4), (3,4), (1,8), (3,8), (5,8), (7,8)\}$, where all numbers $m$ represented by the form $f$ satisfy $m \equiv a \pmod b$. For example, we write $\chi_0 = 1,4$ to indicate that all numbers represented by the form are congruent to 1 mod 4. Finally, the *complete character* for a form is then defined as: $\chi_0; \chi_1, \chi_2, ..., \chi_k$. Two forms then said to be in the same *genus* if they have the same complete character. In Article 231, Gauss discusses the possibilities for $\chi_0$ based on the prime factorization of the determinant, as well as the number of potential complete characters in each case. In each case, the number of potential complete characters is a power of 2.

Let $p_1 p_2 \cdots p_k$ be all of the odd primes dividing $\Delta$. We summarize the results in the table below:

| $\Delta$ | Possible $\chi_0$ | Number of potential complete characters |
|---|---|---|
| $\Delta = 8 \cdot 2^r \cdot p_1 p_2 \cdots p_k$ $(r \geq 0)$ | 1,8 3,8 5,8 7,8 | $2^{k+2}$ |
| $\Delta = 4 \cdot p_1 p_2 \cdots p_k$ | 1,4 3,4 | $2^{k+1}$ |
| $\Delta = p_1 p_2 \cdots p_k \equiv 1 \pmod 4$ | 1,4 | $2^k$ |

**Table 1**

EXAMPLE:  Let $\Delta = -55$; then $\chi_0 = 1,4$ and there are four reduced forms:

$$f_1 = x^2 + xy + 14y^2, \quad f_2 = 2x^2 + xy + 7y^2$$
$$f_3 = 2x^2 - xy + 7y^2, \quad f_4 = 4x^2 + 3xy + 4y^2$$

$f_1$ represents 1, and 1 is a residue for any prime $p$, so the complete character for $f_1$ is 1,4; $R5, R11$. $f_2$ and $f_3$ each represent 2, which is a non-residue mod 5 and mod 11, so the complete character for each of these forms is 1,4; $N5, N11$. Finally, $f_4$ represents 4, which is a residue modulo any odd prime $p$. Thus the complete character for $f_4$ is $R5$, $R11$.

It follows that there are two genera, each with two proper equivalence classes:

| Complete Character | Reduced Forms |
|---|---|
| 1,4; $R5, R11$ | $f_1 = x^2 + xy + 14y^2, \quad f_4 = 4x^2 + 3xy + 4y^2$ |
| 1,4; $N5, N11$ | $f_2 = 2x^2 + xy + 7y^2, \quad f_3 = 2x^2 - xy + 7y^2$ |

Note that $f_2, f_3$ are equivalent, so they must be in the same genus. However, they are *not* properly equivalent since $f_3 = f_2(-x, y)$. Thus they represent two distinct elements within the genus.

Observe also that in the example above, there were four possible complete characters, but only two actually defined a genus. In Articles 261 and 287, Gauss shows that the number of genera is always exactly half the number of possible complete characters and must always be a power of 2. For odd, non-square discriminants, this is easy to see: Let $m$ be an odd integer represented by a form $f$ of odd discriminant $\Delta$, and let $p$ be an odd prime dividing $\Delta$. If $Rp$ is a character, then $\left(\dfrac{m}{p}\right) = 1$, whereas if $Np$ is a character, then $\left(\dfrac{m}{p}\right) = -1$. Replacing the characters by their respective Legendre symbols and multiplying, we get $\left(\dfrac{m}{p_1}\right)\left(\dfrac{m}{p_2}\right)\cdots\left(\dfrac{m}{p_k}\right) = \left(\dfrac{m}{\Delta}\right)$, where $\left(\dfrac{m}{\Delta}\right)$ is the Jacobi symbol and $\Delta = p_1 p_2 \cdots p_k$. By reciprocity we have $\left(\dfrac{m}{\Delta}\right) = (-1)^{(m-1)(\Delta-1)/4}\left(\dfrac{\Delta}{m}\right)$. Since $m$ is odd and $\Delta \equiv 1 \pmod 4$, we have $\left(\dfrac{m}{\Delta}\right) = \left(\dfrac{\Delta}{m}\right)$. Finally, since $m$ is represented by $f$, we have $\left(\dfrac{\Delta}{m}\right) = 1$ by Theorem 3. Thus, for $m$ represented by $f$, the product of the characters is always 1; if $k-1$ of the characters are known, the $k$-th is also determined. It follows that there must be $2^{k-1}$ complete characters.

Reciprocity plays a critical role in the argument above, and this is no accident. In Article 261, Gauss shows that at least half the possible complete characters cannot belong to a genus – this fact serves as the basis of his second proof of the Quadratic Reciprocity [7, Art 262].

The argument above (or Theorem 3) shows that if $m$ is represented by a form of odd discriminant $\Delta$, then $\left(\dfrac{\Delta}{m}\right) = 1$. Gauss' Theorem 5 then allows us to extend this relationship to elements of $(Z/\Delta Z)^*$. That is, $\chi(\overline{m}) = \left(\dfrac{\Delta}{m}\right)$ is a well-defined map from $(Z/\Delta Z)^*$ to $\{\pm 1\}$. This is a homomorphism since $\left(\dfrac{\Delta}{m}\right)\left(\dfrac{\Delta}{n}\right) = \left(\dfrac{\Delta}{mn}\right)$. Moreover, this is the *unique* homomorphism $\chi : (Z/\Delta Z)^* \to \{\pm 1\}$ such that $q \in \ker(\chi)$ if and only if $q$ is represented by a form of discriminant $\Delta$. A famous result of Dirichlet guarantees that there are infinitely many primes in an arithmetic progression, provided the first term and common difference are relatively prime. Thus, each element of $(Z/\Delta Z)^*$ can be represented as $\bar{q}$, for some odd prime $q$ not dividing $\Delta$. From this, it follows that the the condition $\chi(\bar{q}) = \left(\dfrac{\Delta}{q}\right)$ for odd primes $q$ determines $\chi$ uniquely.

Let $\Delta \equiv 0, 1 \pmod 4$ be a discriminant. The *principal form* is defined by

$$x^2 - \frac{\Delta}{4} y^2 \quad if \quad \Delta \equiv 0 \pmod 4$$

$$x^2 + xy + \frac{1-\Delta}{4} y^2 \quad if \quad \Delta \equiv 1 \pmod 4$$

The class and genus containing the principal form are called the *principal class* and *principal genus*, respectively. Note that the principal form has discriminant $\Delta$ and is reduced. When $\Delta = -4n$, the principal form is $x^2 + ny^2$. Many fundamental properties of genus can be described in terms of the homomorphism $\chi$ and the principal form:

THEOREM 6: Given a negative integer $\Delta \equiv 0, 1 \pmod 4$, let $\chi$ be the homomorphism of Theorem 4, and let $f$ be a form of discriminant $\Delta$.

  i) For an odd prime not dividing $\Delta$, $\bar{p} \in \ker(\chi)$ if and only if $p$ is represented by one of the $h(\Delta)$ forms of discriminant $\Delta$.
  ii) $\ker(\chi)$ is a subgroup of index 2 in $(Z/\Delta Z)^*$
  iii) The values in $(Z/\Delta Z)^*$ represented by the principal form of discriminant $\Delta$ form a subgroup $H \subset \ker(\chi)$
  iv) The values in $(Z/\Delta Z)^*$ represented by $f(x, y)$ form a coset of $H$ in $\ker(\chi)$.
  v) For odd $\Delta$, $H = \{x^2 \mid x \in (\mathbf{Z}/\Delta \mathbf{Z})^*\}$

Part (i) of the theorem is a restatement of Theorem 3: $\chi(\bar{p}) = \left(\dfrac{\Delta}{p}\right) = 1$ if and only if $p$ is represented by some form of discriminant $\Delta$. Part (ii) states that exactly half the congruence classes in $(Z/\Delta Z)^*$ are represented by some form of discriminant $\Delta$; for odd $\Delta$, this follows from our argument that exactly half of all possible complete characters actually result in a genus. Parts (iii) and (iv) get to the heart of genus theory; since distinct cosets are disjoint, different genera represent disjoint classes in $(Z/\Delta Z)^*$. That is, we can now describe genera in terms of cosets $kH$ of $H$ in $Ker(\chi)$. We could then define a genus to consist of all forms of discriminant $\Delta$ that represent the values of $kH$ mod $\Delta$. Note that this definition could be used to show that each genus contains the same number of classes [9, Art. 252].

EXAMPLE: Recall that there were four reduced forms of discriminant $\Delta = -55$:

$$f_1 = x^2 + xy + 14y^2, \quad f_2 = 2x^2 + xy + 7y^2$$
$$f_3 = 2x^2 - xy + 7y^2, \quad f_4 = 4x^2 + 3xy + 4y^2$$

There are $\Phi(55) = 55(1 - \frac{1}{5})(1 - \frac{1}{11}) = 40$ elements in $(Z/55Z)^*$. Of these 40 elements, exactly 20 are represented by a form of discriminant -55.

Since $f_1(x,0) = x^2$, the principal form $f_1 = x^2 + xy + 14y^2$ represents all of the squares:

$$H = \{1, 4, 9, 14, 16, 26, 31, 34, 36, 49\}$$

Thus the set of classes in $(Z/55Z)^*$ represented by $f_1, f_4$ is $H$, which is easily verified to be a subgroup of $(Z/55Z)^*$. Also note that $f_2(0, y) = 7y^2$, so the set of classes represented by $f_2, f_3$ can be written as $7H = \{2, 7, 8, 13, 17, 18, 28, 32, 43, 52\}$.

Of special interest are those discriminants $\Delta$ such that each genus contains exactly one class; in this situation, the primes that are represented by a form of discriminant $\Delta$ are determined by congruence conditions mod $\Delta$. (See [2] for details.)

**COMPOSITION OF FORMS:** The theory of composition is intricately linked to that of genus. Composition of forms was first investigated by Legendre and Lagrange, but the theory was brought to fruition by Gauss, who discovered a remarkable group structure. Gauss' exposition is long and technical, and is one of the most difficult parts of the *Disquisitiones*. However, the main result – that classes of binary quadratic forms of fixed discriminant form an abelian group under the operation of composition – is justly celebrated as one of the milestones of 19th century mathematics. Mathematicians following Gauss were able to streamline the theory considerably.

Gauss showed that any two forms of the same discriminant can be composed in such a way that composition is a well-defined operation on (proper) equivalence classes of forms. For simplicity, we present a version of the operation developed by Dirichlet [2, 3] which is based on a case singled out by Gauss for special consideration [7, Art 242]. We say that $f_1 = a_1 x^2 + b_1 xy + c_1 y^2$ and $f_2 = a_2 x^2 + b_2 xy + c_2 y^2$ are *concordant* (the terminology is due to Dedekind [3]) if the following conditions hold:

$$\text{i) } a_1 a_2 \neq 0 \qquad \text{ii) } b_1 = b_2 \qquad \text{iii) } a_1 \mid c_2 \text{ and } a_2 \mid c_1$$

If two concordant forms have the same discriminant, say $b^2 - 4a_1 c_1 = b^2 - 4a_2 c_2$, then $a_1 c_1 = a_2 c_2$, and so $c_1 / a_2 = c_2 / a_1$. We then define the composition of two concordant forms $f_1, f_2$ of discriminant $\Delta$ as $f_1 * f_2 = a_1 a_2 x^2 + bxy + cy^2$, where $b = b_1 = b_2$ and $c = c_1 / a_2 = c_2 / a_1$. Dirichlet showed that given two equivalence classes of forms $C_1$, $C_2$, it is always possible to find concordant forms $f_1, f_2$ with $f_1 \in C_1$ and $f_2 \in C_2$.

Suppose that $f_1 = a_1 x_1^2 + bx_1 y_1 + a_2 c y_1^2$ and $f_2 = a_2 x_2^2 + bx_2 y_2 + a_1 c y_2^2$ are concordant forms. Then setting $X = x_1 x_2 - c y_1 y_2$ and $Y = a_1 x_1 y_2 + a_2 y_1 x_2 + b y_1 y_2$, we have $(a_1 x_1^2 + bx_1 y_1 + a_2 c y_1^2)(a_2 x_2^2 + bx_2 y_2 + a_1 c y_2^2) = a_1 a_2 X^2 + bXY + cY^2$ (by direct calculation). Using this identity and the definition of composition given above, we quickly deduce that $f_1 * f_2$ represents $m_1 m_2$ whenever $f_1$ represents $m_1$ and $f_2$ represents $m_2$. The following theorem summarizes the main properties of composition [7, Art 242]:

THEOREM 8 [Gauss]: For a fixed discriminant $\Delta$, the set of equivalence classes of primitive positive definite forms comprise an abelian group under the operation of composition. The identity of this group is the class containing the principal form. The class containing the form $ax^2 + bxy + cy^2$ and the class containing its "opposite" $ax^2 - bxy + cy^2$ are inverses.

This group is called the *class group*, and has cardinality $h(\Delta)$. The proof is long and technical, as might be expected; the results themselves represent an unprecedented level of abstraction for their time. Soon after discussing composition of classes, Gauss defines *duplication*: let $K$ and $L$ be proper equivalence classes of forms of discriminant $D$. If

$K * K = L$, then we say that $L$ is obtained by duplication of $K$. In Article 247, Gauss points out that the duplication of any class lies in the principal genus; in Articles 286-287 he shows the converse, stating that

"*it is clear that any properly primitive class of binary forms belonging to the principal genus can be derived from the duplication of some properly primitive class of the same determinant*".

This fact is often referred to in the literature as the Principal Genus Theorem. While the statement is made rather casually (not even stated as a formal theorem), Gauss nonetheless describes it as "*among the most beautiful in the theory of binary forms*". (See [12] for a discussion of the many generalizations of this result.)

We conclude with a description of Gauss' proof of the Principal Genus Theorem. To demonstrate how duplication of any class is in the principal genus, Gauss defines composition of genera, and in doing so describes another group structure. In Article 246, he shows that if $f, f'$ are primitive forms from one genus, and if $g, g'$ are primitive forms from another genus, then the compositions $f * g$ and $f' * g'$ will be in the same genus. He then explains how one can determine the genus of $f * g$ using the characters for $f, g$ respectively. First, he gives a multiplication table for the characters $\chi_0$; then he describes multiplication of characters $\chi_i, \chi_i'$ as $Rp_i$ if $\chi_i = \chi_i'$ and as $Np_i$ if $\chi_i \neq \chi_i'$.

The characters of $f * g$ are then the products of $\chi_i, \chi_i'$, $i = 0,1,...,k$. If the discriminant $\Delta$ is odd, we can illustrate this by replacing the characters by their respective Legendre symbols. Let $\Delta = p_1 p_2 \cdots p_k$ be odd, and let $f, g$ come from the genera $G_1, G_2$ respectively. Suppose that $m$ is represented by $f$ and that $n$ is represented by $g$, so the total characters of the forms can be described as $\left(\frac{m}{p_1}\right)\left(\frac{m}{p_2}\right),\cdots,\left(\frac{m}{p_k}\right)$ and

$\left(\frac{n}{p_1}\right)\left(\frac{n}{p_2}\right),\cdots,\left(\frac{n}{p_k}\right)$ respectively. Then $G_1 * G_2$ is the genus with total character

$\left(\frac{mn}{p_1}\right)\left(\frac{mn}{p_2}\right),\cdots,\left(\frac{mn}{p_k}\right)$. Note that the principal genus always represents 1, which is a quadratic residue modulo any prime; that is, $\left(\dfrac{1}{p_i}\right) = 1$ for all $i$. Thus the principal genus

$G$ is the genus in which all the characters have value 1. On the other hand, if $G_i$ is any other genus and $m$ is an integer represented by $G_i$, the characters for $G_i * G_i$ will be

$$\left(\frac{m^2}{p_1}\right)\left(\frac{m^2}{p_2}\right),\ldots,\left(\frac{m^2}{p_k}\right) = 1, 1,\ldots, 1.$$ Hence $G_i * G_i = G$. Moreover, it follows that the

genera form a group of order 2, whose identity is the principal genus.

## BIBLIOGRAPHY

1.  M. Beintema and A. Khosravani, "Binary Quadratic Forms: A Historical View", Mathematics and Computer Education, **40** (2006), p. 226-236.

2.  D.A. Cox, *Primes of the Form* $x^2 + ny^2$, Wiley-Interscience, John Wiley and Sons, New York, 1989.

3.  L.E. Dickson, *Theory of Numbers Vol. III: Quadratic and Higher Forms*, Chelsea, New York, 1952.

4.  L. Euler, *Oeuvres Vol II*, Gauthier-Villars and Sons, Paris, 1894.

5.  P. de Fermat, *Oeuvres Vol II*, Gauthier-Villars and Sons, Paris, 1894.

6.  D. Flath, *Introduction to Number Theory*, Wiley, New York, 1989.

7.  C.F. Gauss, *Disquisitiones Arithmeticae*, Springer-Verlag, New York, 1986.

8.  J.L. Lagrange, "Recherches d'arithmetique", *Oeuvres III*, Gauthier-Villars, Paris, 1867.

9.  A.M. Legendre, *Theorie des Nombres*, Paris, 1830; reprint, Blanchard, Paris, 1955.

10.  F. Lemmermeyer: "The Development of the Principal Genus Theorem", ArXiv Mathematics e-prints, math/0207306, 2002.

11.  W. Scharlau and H. Opolka, *From Fermat to Minkowski, Lectures on the Theory of Numbers and Its Historical Development*, Springer-Verlag, New York, 1985.

12.  J.P. Serre, $\Delta = b^2 - 4ac$, Math. Medley **13** (1985), pp. 1-10.

# Where are the Plans?- A socio-critical and architectural survey of early Egyptian Mathematics

Gabriel Johnson & Bharath Sriraman
The University of Montana

Rachel Saltzstein
Volcano Vista High School, Albuquerque, New Mexico

The majority of the mathematics taught in secondary schools was invented in the ancient world, and this fact opens the curriculum up to organic opportunities to teach mathematics from a socio critical historical perspective. Teachers and education theorists have observed a fascination on the part of students of all ages with the culture of ancient Egypt, which makes a study of the mathematics of ancient Egypt a particularly attractive topic for the public school mathematics classroom. How is one to approach this topic given the controversy surrounding the Egyptians and their place in the history of mathematics? This controversy is discussed by Joseph (1991) but most explicitly examined in Bernal's (1987) still controversial, multi-volume work *Black Athena,* in which he defends his thesis that there are two competing models of the genesis of Ancient Greek language and mathematics, one that is Eurocentric or "Aryan", and the other, Bernal's (1987) "revised ancient model," that is essentially Egyptian and Semitic. Bernal (1987) presents evidence that the Eurocentric model emerged during the late 17$^{th}$ through 19$^{th}$ centuries and was motivated by skin color racism that developed during the era of African slavery and European and American imperialism.

Much of Bernal's (1987) evidence of racism comes in the form of quotes from scholars who worked during this time period, and these quotes are ideal for a critical discussion in class. There is even a section of the first volume of *Black Athena* that deals with the mathematical elements of the Great Pyramid, which many modern scholars believe was built not just with knowledge of "Pythagorean" triples, but with knowledge of the golden ratio, a mathematical find attributed to the Greeks. Discussion of the Great Pyramid controversy should not only include an examination of the intricacies of the mathematics involved in this engineering feat but also racist quotes from the 19$^{th}$ century, without which our secondary school curriculum falls into what Bernal (1987) calls the *status quo*. Bernal (1987) expresses the view that the persistence of the Eurocentric model is caused by academics who maintain the *status quo* and if that model persists into the secondary schools, change to Bernal's revised ancient model will be further delayed.

A discussion of *Black Athena* demands a discussion of racism and specifically of racism against Jews and people of African descent. This discussion might not seem particularly germane to a mathematics classroom by conservative members of a community, or even by success-driven parents who care more about their children's careers than about exposing their children to social justice issues. Thus teaching about the math of ancient Egypt is suggested to those interested in a socio-cultural and historical approach to the mathematics. The specifics of the architectural splendors discussed via the use of the elementary ideas in mathematics makes such an approach accessible to middle and high school teachers, and even those using critical theory in the discussion of history at universities.

Many modern scholars of ancient Egypt, such as Imhausen (2003), seek caution when discussing the evidence. They begin the conversation with a discussion of the paucity of source material extant from this ancient civilization. Imhausen, an Egyptologist by training, often cautions readers about the lack of papyri found, then poses the question, if we were to walk into a library containing all the books published in one country during the last 2,000 years and arbitrarily grab two books, what would we really know about the extent of the mathematical knowledge of that period? When scholars make the leap to say that everything Egyptians knew about math is contained in the Rhind and Moscow papyri, they commit an offense Bernal

230

(1987) calls "archaeological positivism." That is, they assume they can infer absence, which in reality is impossible to prove. A discussion of the problems of finding source material when dealing with ancient cultures should preface any classroom discussion of Egyptian mathematics, and can do much to invalidate the dismissive attitudes of other scholars with regard to the mathematics found in the papyri.

One common criticism of Egyptian scribes is that they did not think of numbers in an abstract manner, only in concrete situations. They did, however, come up with procedures that can be translated into modern formulas for finding the areas of plane figures and solids like the truncated pyramid. This is an easy topic to bring into a classroom, as these area formulas are usually taught at multiple junctures in the elementary and secondary mathematics curriculum. The dialogs found in the papyri give clues that the Egyptians might have thought more abstractly, for example, the text explaining the procedure for finding the volume of the truncated pyramid begins with something like 'take any truncated pyramid, for example, one with these dimensions.' A class could read the original text and discuss if it is evidence of abstract thinking on the part of the Egyptian scribe, since it does appear they knew their procedure worked in general for any truncated pyramid. This discussion is germane to the mathematics curriculum, and teaches a controversy that is part of the Eurocentric model.

The ancient pyramids of Egypt are still in many ways a mystery to mankind even after centuries of study. The pyramids were primarily built as burial tombs for Kings, known as Pharaohs, in Egypt during the Ages of the Pyramid (2635-1780 BC). The aim of the consequent sections is to see how were these great geometric structures built?, and why has mankind not been able to replicate the construction plans and methods? We explore the many theories about methods and mathematics used to construct specifically The Great Pyramid, better known as the Pyramid of Khufu. In doing this, one hopes to ascertain exactly how these ancient Egyptian architects built such an exquisitely perfect structure.

Many famous scholars and engineers of the past studied in Egypt and provided the world with much of the basis for mathematics and geometry today. Just to name a few, Euclid, Archimedes, and Pythagoras all studied in Egypt. Euclid studied in Alexandria, Egypt during the reign of Ptolemy I (323–283 BC). Euclid is often referred to as "The Father of Geometry" and wrote such works as The Elements, which is still used as a text for Euclidean Geometry courses today. It is also believed that Archimedes studied in Alexandria, Egypt during his youth. Archimedes was a Greek mathematician and engineer who discovered Archimedes Principle, which is the basis for hydrostatics. Archimedes also designed many machines and weapons of warfare, which gained him notoriety as an engineer including the Archimedian Screw and catapults. Variations of the Archimedean Screw are used in many applications still today. Pythagoras most famous for the formulating the Pythagorean Theorem is also believed to have visited Egypt during his travels as a youth in search of knowledge. It is widely accepted among historians that the Egyptians knew of the application of the Pythagorean Theorem but did not actually formally present or formulate the idea. An example of this is that it is believe the "rope stretchers in Egypt used a rope with knots tied at units of 3,4, and 5 to create right angles for building.

So if such great scholars as Euclid, Archimedes, and Pythagoras studied in or at least visited Egypt to gain knowledge, they must have believed this area of the world had knowledge of mathematics and engineering that was useful for the future. Since the pyramids have been studied for centuries and the human mind still has not been able to exactly replicate the construction methods of the Egyptians, then is there more to be learned from this ancient civilization? Did the Egyptians have some superior knowledge or tools that allowed them to build this great structure, or did they just have more knowledge than originally believed? Understanding the construction methods, techniques, and mathematical ideas used to build the pyramids just might lead to breakthroughs in engineering and mathematical concepts.

One cannot understand how The Great Pyramid was built without first knowing the colossal measurements of The Great Pyramid. Many individuals in history studied the measurements and construction methods of The Great Pyramid. It is even rumored that the Greek Philosopher Thales traveled to Egypt in the 6th century BC and was able to accurately calculate the height of The Great Pyramid by measuring the shadow of the pyramid at the same time of day that his shadow was equal to his height. However, the earliest complete account came from Herodotus.

Herodotus called by many "The Father of History" traveled to Egypt around 450 BC to study Egypt. Herodotus indicated that each of the four "perfectly triangle faces of the pyramid were covered with polished limestone (Tompkins, p.2). Herodotus also hypothesized that it took 100,000 slaves twenty years to construct the pyramid.

Strabo, a Pontine geographer, voyaged up the Nile River in 24 BC wrote forty-seven books about the History of Egypt. Unfortunately, most of his work was lost. The only remaining information we have is a description and measurements of an entry way on the north side of The Great Pyramid made of hinged stone which could be raised and lowered and looked identical to the surrounding masonry (Tompkins p.3)

It would not be until 1638 that John Greaves, a mathematician and astronomer traveled to Egypt. His main goal was to find clues in the Great Pyramid about the circumference of the earth and identify the ancient unit of measurement used by The Great Pyramid's builders. Greaves surveyed The Great Pyramid with greater accuracy than anyone since the builders. Greaves measured the height of the Pyramid to be 499 feet high, within 12 feet of being correct. He also measured the base of the Pyramid to be 693 feet; however, due to the large piles of rubble at the base of the Pyramid he underestimated the length of the base by approximately 70 feet. Greaves returned to England and published his work in 1646 as "Pyramidographia: Or a Description of the Pyramids of Aegypt". Sir Issac Newton unaware that Greaves survey was flawed used Greaves results to come to the conclusion that the Great Pyramid had been built with two distinct measurements the "profane" cubit, about 21 inches long, and the "sacred" cubit, about 25 inches long (Jackson, Stamp, p125-126).

No remarkable discovery or interest in the Pyramids was rekindled until Napolean Bonaparte set sail to conquer Eqypt as a means to gain India and work domination. Napolean took along with him 175 "savants, erudite French civilians, who supposed to have a knowledge of Egyptian antiquities and 35,000 soldiers. Napolean and his army defeated 10,000 Mameluke horseman at the Great Pyramid. The "savants" explored and measured the Pyramid at this time. The "savants" explored the interior of the Pyramid but found nothing of real interest; however outside the Pyramids they found a long flat area on which the Pyramid had originally been established. They also found two trenches carved 10 by 12 feet approximately 20 inches into the bedrock at two corners of the Pyramid. This gave them two concrete points to measure the base of the Pyramid. Edme-Francois Jomard, one of the savants, was able to take a series of measurements giving a base length of 230.902 meters or 757.5 feet. Jomard also measured the height of the Pyramid by measuring down each step to equal 144 meters or 481 feet. Using trigonometry, he obtained an angle of slope of 51°19'14" and an apothem of 184.722 meters. Jomrad remembered according to Strabo, the apothem of the Pyramid was supposed to be one stadium long (Tompkins, p. 45). He knew that a stadium was 600 Greek feet which is 185.5 meters. Jomrad also learned from reading classics of his time that a stadium of 600 feet was held to be 1/600th of a geographical degree. He then calculated that a geographical degree at the mean latitude of Egypt was 110,827.78 meters. Dividing this figure by 600 resulted in a measure of 184.712 meters (Tompkins, p46). It seemed to Jomrad that the Egyptians had worked out basic units of measure like the cubit and stadium from the size of the earth and displayed this knowledge in the Pyramid structure. Jomrad discovered that some Greek authors indicated that the perimeter at the base of the Pyramid was supposed to measure one half a minute of longitude. This means 480 times the base of the Pyramids is equal to one geographical degree. Jomrad calculated that base 230.8 meters by dividing the 110,827-meter degree by 480. So Jomrad wanted to compare the cubit to these measures. Jomrad read that according to Herodotus 400 cubits was equal to 600 feet. Jomrad divided the apothem by 400 and came up with 0.4618 meter. This actually turned out to be the present Egyptian cubit. Jomrad had skeptics but he continued to assert what he thought to be true. He even pointed out that Herodotus, Plato, and Diodorus and others had called Egypt the "birthplace" of geometry (Tompkins p.48).

Jomrad's colleagues Gratien Le Pere and Colone Coutelle re-measured the base Pyramid and found it to be 2 meters longer than Jomrad. They also re-measured the height and found that Jomrad's angle of inclination was too small and thus his apothem too short.

In the early 1800's, Colonel Howard-Vyse , a British Guards officer, began exploring the Pyramid. Since, the Middle Ages when the Arabs had taken the limestone from the outercasing to use for other buildings, the perimeter of the Pyramid had been littered with debris and rubble. The sand had also covered the corners that the French had recently cleared. Howard-Vyse decided to clear a space on the north side of the

Pyramid. What he discovered were two limestone blocks on the lowest level of the Pyramid in their original form and at their original location. The limestone was so precisely carved that it allowed Howard-Vyse and his help to measure the angle of the slope the block that would be the angle at which the Pyramid was originally built. The blocks were 5 feet high, 12 feet long, and 8 feet wide and displayed an angle of approximately 51°51'. This angle is slightly steeper than the angle measured by the French. Now that Howard-Vyse had the angle of inclination and the length of the base as 763.62 feet measured by Frenchmen Coutelle and Le Pere, it was possible for him to calculate by using trigonmetry the perpendicular height to the point where the missing capstone should have been. He arrived at the solution of 485.5 feet or 147.9 meters. In 1840 Howard-Vyse returned to England and produced "Operations Carried on at the Pyramids of Gizeh in 1837".

In the mid-1800's, John Taylor, an English poet, essayist, and gifted mathematician as well as an amateur astronomer, began comparing and compiling accounts of the many who had visited and studied the Pyramid. Taylor made a scaled model of the Pyramid and analyzed the measurements from a mathematical point of view. Taylor began drawing and redrawing the Pyramid according to the measurements of Howard-Vyse. It seemed odd to Taylor that the Egyptians chose to build the Pyramid with an angle of inclination of 51°51' instead of choosing an angle such as 60°. Taylor noticed that Herodotus' reported that the Egyptian priests gave him information about the surface of each face from which Taylor concluded that the faces of the Pyramid had been built equal in area to the square of the Pyramid's height. Taylor also discovered that if the perimeter of the Pyramid was multiplied by 2 times the height, he arrived with the number 3.144. This number being very close to $\pi$ which is 3.14159. This meant that the height of the Pyramid seemed to be in relationship to the perimeter of its base as the radius of a circle is to its circumference (Tompkin, p.70). Taylor was curious to understand why $\pi$ was represented in the Pyramid. He theorized that the perimeter of the Pyramid might have been constructed to represent the circumference of the Earth at the equator and the height might be representative of the distance from the Earth's center to a pole. This theory supported Jomrad's research. Taylor also was confident that the Egyptians had not used a unit of measure such as the British foot because this unit of measure did not fit the height or the base exactly. So Taylor explored for a unit of measure a proportion that contained $\pi$ and fit the geometry of the Pyramid in whole numbers. Taylor came up with 366:116.5. Taylor noticed that 366 was obviously close to the number of days in a year and thought maybe the Egyptians designed the perimeter of the Pyramid to represent the solar year (Tompkins p.72). Taylor then converted the perimeter of the Pyramid to inches and noticed that it was nearly 100 times 366. He also noticed that if he divided the base of the Pyramid by 25 inches he got 366. This led Taylor to believe that the Egyptians used a unit very close to the British foot but not exactly (25 of these units appeared to be = a cubit). At the same time Sir John Herschel, a Britian astronomer discovered that a unit half a hair's length longer than the British inch was a unit based on the actual size of the earth. It turned out that 25 of these units were the actual cubit that Taylor had discoverer in his calculations. Taylor could not understand how such an ancient civilization could a vast knowledge of the planet's shape, size, and motion. Taylor concluded that the Egyptians must have built the Pyramid by Divine Revelation much the same way Noah built the Ark. Taylor endured must criticism for his belief.

Piazzi Smyth, a mathematician near the same time as Taylor, did not mock Taylor's beliefs or reasoning. Piazzi Smyth actually supported Taylor's ideas. In 1864, Taylor died and Smyth decided to travel to Egypt to measure the Pyramid in hopes to see whether Taylor's theories of the $\pi$ relations and the cubit used to supposedly build the Pyramid were true. In 1864, Smyth and his wife set sail for Egypt with more accurate scientific instruments than had ever been used to measure the Pyramid. Smyth used a 500-inch cord, along with theodolites, sextants, and telescopes to measure elevations and take measurements of the outside of the Pyramid. From the Summit of the Pyramid Smyth calculated the latitude of the Pyramid to be 29°58'51" very close to 30° (Tompkins p.86). Smyth later explained the difference in the latitude by the gradual shifting of latitude registered at Greenwich as 1.38" per century. Smyth then subtracted 26°17' (angle of the descending passage inside the Pyramid) from the Pyramid's latitude of 30°. He came up with an angle of 3°43'. Smyth then calculated that the alpha Draconis would have been located 3°43' from the pole at its lower culmination in 2123 BC and again in 3440 BC. Supporting the later date, Smyth concluded that Pyramid had taken place at midnight of the equinox of 2170 BC when alpha Draconis was at the meridian below the pole, another star Alcyone of the Pleiades would have been crossing the meridian above the pole (Tompkins p.87). In other words, the Egyptians could have aligned the Pyramid with these constellations at that specific time.

Smyth was still intrigued to know if Taylor's claim that the proportion of π was part of the structures dimensions. Smyth found fragments of casing stones like Howard-Vyse had used to measure the angle of inclination of the structure. The angle of inclination checked out to be 52°. Smyth in an effort to refine the angle, used an accurate altitude azimuth circle with he had borrowed from Professor Lyon Playfair to measure the angle of inclination while holding the stones silhouetted in the sky. Smyth obtained an angle of 51°49' while Sir John Herschel measured 51°52'15.5" from casing stones. Smyth took the mean of these two measures obtaining 51°51'14.3". Smyth also took the mean of the French base measurement of 763.62 feet and Howard-Vyse's base measurement of 764 feet and obtained 763.81 feet. These mean measurements produced a value of π in the Pyramid's proportions of 3.14159+. Smyth employed Taylor's reasoning of the base being divided by 366 units essentially equal to the number of days in a year to explain why π was present in the Pyramid's proportions. In order to be exact the perimeter of Pyramid should have measured 365.24 Pyramid inches that meant that each side of the Pyramid should have been 9140.18 British inches. The measurements of Howard-Vyse and the French were about 2 feet too long. So Smyth knew he needed to dig up the sockets at the corner of the Pyramid and measure the base. Smyth had to return to Scotland but it so happened that two engineers from Glasgow, Messrs, Inglis and Aiton, were passing through Egypt at this time. The engineers agreed to take the measurements for Smyth and forward the results to him. The engineers measured the side of the Pyramid to be 9110 British inches. Smyth decided that the true length must be the average of the engineers' measurements and Howard-Vyse's measurement of 9168 British inches. Smyth obtained 9140 British inches, which resulted in a year of 365.2 days instead of the exact value of 365.24 required by Smyth's theory. Smyth also attributed these ideas of such an advanced ancient society to Divine Wisdom in his work, *Life and Work at the Great Pyramid of Jeezeh during the Months of January, February, March, and April 1865 A.D.* Smyth like Taylor, underwent extreme scrutiny for his theories and ideas related to Divine Wisdom. Smyth and the many who had measured and studied the Pyramids in the past had discrepancies with their results because the measurements had been taken with great piles of rubble still at the base of the Pyramid.

To solve the controversy over the Pyramid's dimension, William Petrie set out to design and build instruments more advanced sextants, theodolites and vernier to take measurements than even Smyth himself had used. This task would take William Petrie approximately 20 years to accomplish. William Petrie's son, William Flinders Petrie, became impatient with his father and decided he would travel to Egypt to measure the Pyramid. William Flinders Petrie, a talented professional surveyor, set of for Egypt in 1880 at the age to 26 with his father's instruments to measure the Pyramid. Petrie first set out to create a precise triangulation over the hill of Giza. He included points all around all three pyramids and the surrounding temples that were in the Giza complex. Petrie was able to figure with the help of his father's theodolite the layout of the three pyramids at Giza to within a quarter of an inch. Petrie considered the base length of the Pyramid to be measured in reference to the edges of the pavement some 20 inches higher on the Pyramid than the sockets measured by Smyth. Petrie's measurements came up with a base length of 9069 British inches as compared to Smyth's base measurement/calculation of 9140 British inches. Petrie showed by his calculations and measurements that Egyptian builders had used the royal cubit of 20.63 British inches, which translates into a base measurement of 440 cubits and a height of 280 cubits. Petrie appeared to have validated Taylor's theory that the Pyramid was a representation of the Earth by obtaining a value of π at 3.14285. However, Petrie seemingly nullified the theory of Smyth that the perimeter of the Pyramid was a representation of the exact number of days in a year when Petrie came up with a value of 362.76 days.

David Davidson, a structural engineer from Leeds, England, was the next to investigate the Pyramid. Davidson noticed that Petrie had failed to take into account in his measurements a hollowing of the core masonry of each side of the Pyramid which was observed in a an aerial photograph taken by Bridgadier P.R.C. Groves, the British prophet of air power. The photograph was taken during Petrie's lifetime. Davidson used the hollowing feature of the masonry to confirm Smyth's theories. The theoretical length of each side of the base calculated by Smyth gave values of 9131.5 Pyramid inches which is 9141.1 British inches. Davidson revised Petrie's figure to take into account the hollowing effect and came up with 9141.4. Davidson indicated that the hollowing effect led to three basic lengths on the base of the Pyramid that showed three different lengths of year. The lengths included: An outer or shortest length, from corner to corner, bypassing the hollowing, a second, slightly longer, which included part of the indentation of the four hollowed faces at the base; and a third which included the entire angle within each hollow face (Tompkins p.111). The three measurements could be the equivalent to the three lengths of year known to modern science as: the solar, the sidereal, and the anomalistic year according to Davidson. Davidson like Smyth and Taylor had many skeptics and faced great criticism. Pyramidologist continued to study and try to understand the Pyramid into the 1920's while most dismissed Davidson, Smyth, and Taylor's theories.

So all these dimensions beg the question: What did the geometry of the Pyramid of Khufu look like. Below are several drawings and pictures of the Great Pyramid.

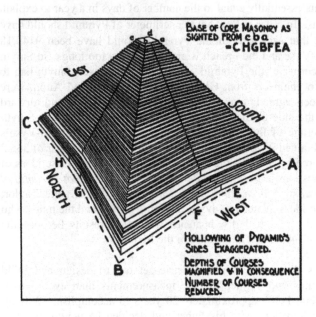

*From: http://www.world-mysteries.com*

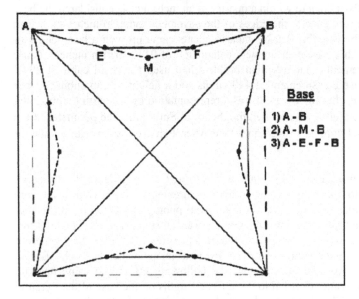

According to this model:

$\tan(\phi) = 4/\pi$ (where $\phi$ is the side slope angle).

This results in $\phi = 51.854$ (deg).

This is very close to the measured value of 51.87 (deg)*.

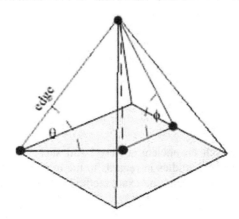

The angle of elevation of the edge of a side ($\theta$),
can be calculated from this equation:

$\tan(\theta) = 4/(\pi * \sqrt{2}\ )$

The result is $\theta = 42.00$ (deg)*.

Again, this is very close to the measured
value of 42.01 (deg).

*Side slope data taken from ***The Pyramids of Egypt*** by I.E.S. Edwards.

http://www.world-mysteries.com

The Pyramid of Khufu was also believed to have been built to include the Golden Section. Herodotus alluded to this when he indicated that the temple priest had told him the Pyramid was built is such a way that the are of each of the faces is equal to the square of the height (Tompkins p.190). If the apothem of one of the side faces equal to 356 cubits is divided by half of the base known to be 440 cubits then we see that 356/220 is equal to 1.618. The Golden Ratio is $(1+\sqrt{5})/2$ which is 1.618 or commonly called phi ($\varphi$).

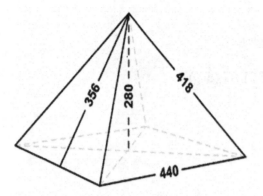

[From http://www.world-mysteries.com]

So, we are still stuck with the questions of how did such an ancient society build such an enormous structure as the Great Pyramid. There are many theories and studies in regards to the number of laborers needed, the task of how the Egyptians moved such huge stones, and the exact methods of construction the structure of the Pyramid.

Egyptologists agree that the builders of the Pyamid would have had to clear the sand and gravel off of the bedrock prior and level the area for the base of the Pyramid. R.L. Engelback, a student of Petrie, maintains the theory that to level the site the Egyptians surrounded area of construction with low banks of mud from the Nile and then filled the area with water and leveled the area by cutting any think that protruded above the water level (Tompkins, p.220). Although the Great Pyramid is built on a hillside of limestone, surrounding the entire foundation area with water would have been very difficult although it could have been done for one side fairly easily and this is likely what Engelback is referring to. Egyptologists have shown that the builders of the Pyramid actually used part of the hill that the Pyramid is built of by carving the natural limestone rock into the block shape of the rest of the Pyramid.

The Pyramid is aligned with the four points of the compass. The Egyptian builders also likely used the merkhet and bay to determine the exact position of astronomical North by sighting the position of a determined star when it was rising and setting with help of an artificial horizon consisting of a vertical limb wall (Siliotti, p.42). The merkhet is a tool that has a gnomon like stick with a plumb bob attached to the right-angled end, The bay is basically a stick with a V –notch in the top. Ludwig Borchardt, Egyptian archaeologists in the early 1900's, and J. P. Laur believe that when the Pyramid was laid out to be build that the first straight side would have been drawn or shown by repeatedly observing the rising and setting of circumpolar stars most likely alpha Darconis (Tompkins, p. 220) As I discussed earlier, Smyth also had this same theory.

The Great Pyramid of Khufu contains approximately 2.3 million blocks of stone, but were did all this stone come from? Archaeologists have established that the bulk of the limestone for the Great Pyramid came from the Mokattam quarries or Tura quarries several miles across the Nile River and the Giza Plateau itself. They would have quarried some of the limestone in Tura because it is of very high quality and was like used for casing the Pyramid. The limestone of Tura would have been loaded on barges and floated down the Nile and unloaded on a causeway that led to the Pyramid. The remains of canals that had to be dug to assist the barges transporting the block to the causeway can still be seen. The causeway built from the Nile to the Pyramid is also visible in aerial photographs although it is buried under sand today. The blocks quarried from the Giza Plateau would have been roughly a ramp's length from the Pyramid itself. The limestone in the Giza Plateau also serves as the foundation for the Great Pyramid. The granite used in the in the King's Chamber, the interior of the Pyramid, came from the Aswan quarry. The Aswan quarry is approximately 500 miles up the Nile from Giza. It is proposed that this stone was also floated down the Nile of reed barges. The workers had access to copper tools as early as the First Dynasty and would have used copper saws and chisels to quarry the limestone. It is believed that they used moistened quartz sand to quarry the granite.

The Great Pyramid of Khufu contains an estimated 2.3 million blocks and is believed to have been built in a span of 23 years. This great structure obviously took an enormous workforce. Herodotus claimed the pyramid took 100,000 men twenty years to build. However, it is widely accepted that Herodotus meant that 25,000 men would be working in three-month increments instead of 100,000 men at one time. Egyptologists generally accept that 20,000-30,000 worked on the Great Pyramid. However, the figures range from 100,000 to 4,000 workers. A team led by Mark Lehner, an American archeologist, who began studying the in Egypt in the early 1970's, theorized that it took as few as 4000 workers. These figures generally include quarrying the stone, transporting the stone form the quarries to the structure, lifting the blocks up the structure and placing the blocks, and masons to finish the blocks. Although these are the main functions needed to build the Great Pyramid there were many other needs such as blacksmiths to make and sharpen copper tools, workers to carry water to builders and to lubricated sled roads.

One of the first questions that pops into an engineer or architect's mind when trying to understand how something was built, is where are the plans for the project? How was a structure such as the Great Pyramid of Khufu constructed to such great precision without a drawn plan? There have not been any records found of actual plans for the Great Pyramid, although there are a few theories of the plans they might have employed. There are two main plans and both involve a theory that the Egyptians employed a grid system to communicate the construction plans of the pyramid with what they believed to be about 10,000 illiterate workers.

The first plans suggest that the Egyptians builders used a six-square grid plan. It is proposed that the Pyramid was cross-sectioned at the Northern Baseline, the Prism Point, and the Well among other un-named points were used to divided the plan in six-square grids. However, the Great Pyramid was built on a cliff of limestone and the ancient Egyptians did not have the modern survey equipment that we do today to layout the building on the ground. So, how did they actually lay the building out? One theory suggests the technique was similar to how they had laid out temples. The Egyptians had been known to layout plans beside structures on the ground. Today on the eastern side of the Pyramid, there is a level area, large and flat enough to layout at least the northern half of the Pyramid's six square grids and thus set the angle of the Pyramids rise with the 20-40-60 cubit additions to the three grid-coordinates (Romer, p.336). These plans would have been translated from the eastern plateau up to the Pyramid.

Another such theory was formally introduced by Ole Jorgen Byrn, a Professor of Architecture at Norwegian University of Science and Technology, in September of 2010. Byrn who noticed that aerial photographs taken in the 1920's show the Pyramid is not set on a square base. He states this has been largely ignored by researchers. Byrn suggests that what the Pyramid actually creates is a Diamond Matrix. Basically Byrn says that by subtracting one royal cubit from baseline allows the baseline to be related to a six-square grid system. Byrn shows that if one royal cubit is subtracted from half of the baseline then the baseline is 219 royal cubits which when divided by 3 (half of the grid) equal exactly 73 royal cubits. Since the grid predicts a distance from each level or mastaba base to the surface of the pyramid removing 1 royal cubit and adds 1/7 th of a royal cubit for each step in the Pyramid thus creating a "Diamond Matrix" as shown below. Byrn writes, "The Apex point of a true pyramid could not have been reached without the precision and tolerance embedded in a simple and redundant 3-d grid with the precision system separated from the building of constituent parts."(Byrn, p.142) The full Diamond Matrix is shown below:

238

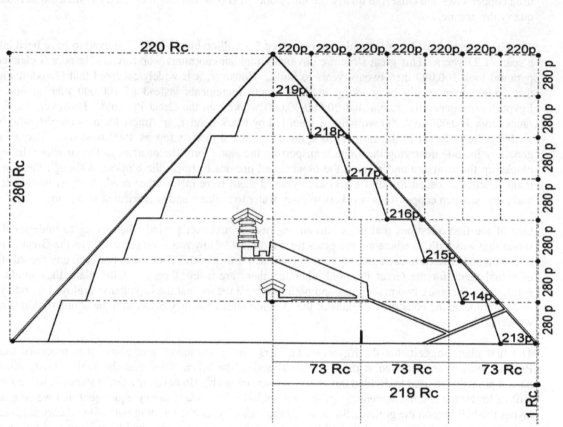

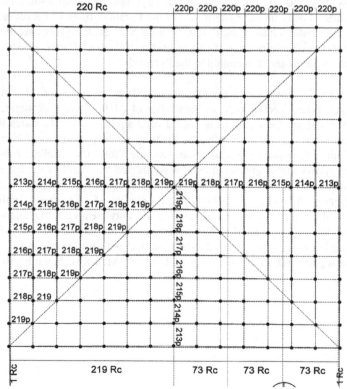

238

Byrn also refutes Romer's six-square grid plan. Byrn writes:

"Romer's grid, however, was probably not used as a grid for the construction of the pyramid itself for three reasons. First, the gird, when transferred to the Great Pyramid lacks any relation to an existing built structure; it has too few grid points, and it cannot be divided into smaller fractions." (Byrn, p.141).

Basically, Byrn suggests that Romer's grid cannot be converted to a 3-d construction grid as his is. There are many more theories about how the massive blocks of limestone and granite were lifted into place to actually build the Pyramid. These theories range from crane type structures to different ramp construction theories.

Herodotus' wrote in the 5th century BC that the upper part of the Pyramid was finished first, then the middle part, and last of all the base part. Herodotus also indicated that casing blocks were lifted step by step on pieces of wood. Herodotus' description is very vague but could have some validity.

Diodorus Siculus in the 1st century BC maintained that the stone blocks were transported up earthen ramps to build the Pyramid. Commander F.M. Barber, an American naval commander and engineer stationed in Egypt in the early 1900's thought that the stone blocks could not have been lifted without steel cranes. He reverted to the idea that the Egyptians must have used ramps.It is widely accepted among Egyptologists that ramps were used to construct the Pyramid ,but most believe that some form of leverage device must have been used finish the construction. Ramps used to construct some pyramids have been found to contain wood strip or logs like railroad ties, which were lubricated with water to make pulling the sleds load with stone up the ramps. The sleds used in these processes have also been found at some pyramid sites.

The most common ramp theories are a large straight ramp as shown below.

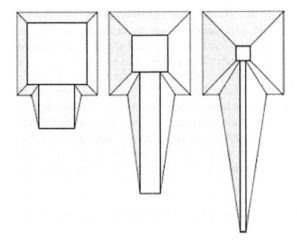

(http://en.wikipedia.org/wiki/Egyptian_pyramid_construction_techniques)

The other ramp theories include zig-zag ramps, ramps utilizing part of the incomplete structure, and spiraling ramps as shown below.

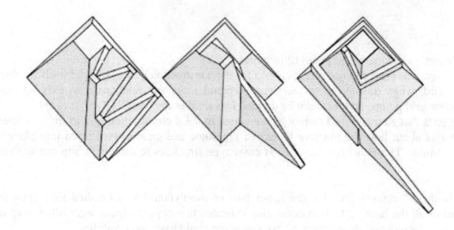

(http://en.wikipedia.org/wiki/Egyptian_pyramid_construction_techniques)

Spiraling ramps would have spiraled around the outside rim or the inside of the Pyramid. Most Egyptologist maintain the idea that such ramps would have worked great to a certain height, but would have reached a point would have to be built so long and steep to construct the Great Pyramid that they would not longer be feasible. Most hold the idea that the ramps would have been feasible up to 200 feet high. Evidence of ramps has been found at other Egyptian pyramids, but there is not evidence of an external ramp at the edge of The Great Pyramid.

Jean-Pierre Houdin formulated the internal ramp hypothesis in 1999. Actually Jean-Pierre's father, an architect, came up with the theory that the Great Pyramid was constructed with the use of an internal ramp. His father brought the idea to Jean-Pierre, also an architect, who became so interested in the idea that he quit his job and began studying to prove the theory true. Houdin believed that the builders had used an external ramp to construct the first 200 feet of the height of the Pyramid then used the internal ramp to build further. He also believed the stones used to construct the external ramp were then re-used to build the remainder of the pyramid thus leaving no trace of an external ramp. In 2005, Houdin teamed up with a group of engineers from Dassault Systemas, a French 3-D software company, who used computer aided designs to test Houdin's theory. The theory appeared proven or at least plausible according to the computer software. Houdin's theory for movement of the block along the internal ramp that each part of the internal ramp ended in a notch of sorts where a crane type structure would be located to lift the block with the aid of men up to the next ramp. As with other theories there were critics; however in 1986 a French team performed a micro-gravimetric analysis of the structure and in one of their plotting there is what appears to be a visible internal ramp.

Another recent theory formulated by Paul Hai in 2006 re-visits Herodotus's research and more specifically the statement:

"The pyramid was built in steps, battlewise, as is called, or according to others, altarwise. After laying the stones for the base, they raised the remaining stones to their places by means of machines formed of short wooden planks." (http://www.haitheory.com/)

Hai writes, "This pulley surrounds its load, which is a Pyramid block, and then as it is hoisted the pulley step-walks the Pyramid's stepped layers with a mechanical advantage of 2.8 (MA=2.8)".
This is known today as a three-wheel step trolley.

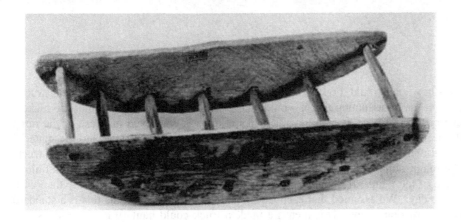

Above is a picture of a "Petrie Rocker". (http://www.haitheory.com/pdfs/Paul%20Hai's%20article.pdf)

There are many variations of these theories and even theories that the block were cast by a form of limestone concrete, but as a whole the primary theories and studies revolve around ramps and lifting techniques.

Since the main theories revolve around ramps and lift techniques, we decided to concentrate the investigation on the ramp theories. The two theories that revolve around a frontal ramp include mainly a ramp built to the top of the Pyramid as the Pyramid was being constructed and a ramp built to the 200 feet up the ramp where some other ramp such as a spiral internal or external ramp would have been employed.

So we investigate the volume of material that would be needed to build a ramp to the top of the Pyramid at an 8% slope. It is difficult for even a truck to pull loads up grades greater than 6 % today and most scholars believe if ramps were employed by the Pyramid builders they were not any steeper than 8%. We calculated this volume of material by using the average end area method. First, we found the horizontal length (= 6025.2 feet) and actual length (=6044.4 feet) of the ramp by using trigonometry. Second, we divided the ramp into 100 feet increments using the horizontal length. We proposed that the path of ramp was 15 feet because many of the blocks used to construct the Pyramid are up to 10 feet wide. Third, we let the side slopes of the Pyramid be equal to a 1.5:1. This means that for every 1 foot in rise the slope has 1.5 feet of run. So the width of the ramp increased by 24 feet every 100 feet of length. We began calculating the volume of the Pyramid using the average end area method. This is a common method used in engineering today to calculate earthwork volumes. The calculations were performed in the following way:

A1= Area at 0+00 beginning of the ramp
A2= Area at 1+00 or 100 feet from the beginning of the ramp

V1  = (A1 + A2)/2  X 100 feet = Volume of the area between 0+00 to 1+00

The volumes are calculated until the top of the Pyramid is reached from 0+00 to 60+00. The volumes were then added together and a total volume of the ramp was found to be 26,403,555.6 cubic yards. Since the actual Pyramid inhabits some of this volume, we had to calculate the volume of the Pyramid and subtract half of it from the ramp volume. We calculated the volume of the Pyramid simply by using the formula: BASE LENGTH * BASE WIDTH * HEIGHT * 1/3. So the volume of the Pyramid was as follows:

440 cubits X 440 cubits X 280 cubits X 1/3 = 18,069,333.3 cubits=3,405365.3 cub. yards.

So the actual volume of the ramp is= 26,403,555.6 cubic yards – (3,405365.3)/ 2 cubic yards= 24,700,873 cubic yards.

So, the next question was how long would it take employing modern construction techniques to construct a ramp of this size. So we began by assuming that it would take a tandem dump truck 5 minutes to make a round trip from the quarry to the ramp. It is believe that if a ramp was employed the base of the ramp would have essentially been at the area where stoned was quarried in the Giza Plateau and the material believed by many to build ramps was the left over chips from quarries. So, five minutes to load a truck, haul the material to the ramp, dump it and return to the quarry is a quick round trip but will definitely illustrate the extreme size of the ramp. If the workers work on the ramp for 10 hours per day then each truck can haul 120 loads a day. We assumed that each truck can haul 10 cubic yards as this is a standard measurement for most construction today. This means a tandem truck could haul 10 cubic yards X 120 trips = 1200 cubic yards per day. We also assumed that 30 trucks would be used to haul the material so a total of 1200 cubic yards X 30 trucks = 36,000 cubic yards would have been moved each day. If working 300 days/year at total of 10,800,000 cubic yards would have been moved in one year. So how long would it have taken to construct the ramp at this rate? In order to calculate this, we took the volume of the ramp and divided it by the volume of material moved in a day. So the volume is:

24,700,873 cubic yards/36,000 cubic yards = 686.1 days = 2.29 years

This is quite a while considering using modern equipment.

Since many theories are centered on the idea that ramps were used but frontal ramps were only used to a height of 200 feet we decided to see if the volume of such a ramp was of a reasonable magnitude to construct in a timely fashion. The volume of this ramp was calculated in the by the same method as above. The volume of the ramp with out subtracting the portion of the Pyramid contained in the ramp was 1,992,222.2 cubic yards.

The volume of the Pyramid contained in the ramp was calculated by first calculating the length of the Pyramid covered with the ramp. This calculation was performed by using trigonometry as follows:

tan 52° = 200/x where x is the horizontal =156.3 feet. So the base of the Pyramid is 2345.9 feet from the base of the ramp.

So from our calculations of volume done previously we know the width of the ramp at 2350 feet, 2400 feet, and 2500 feet from the base of the ramp. The widths are 579 feet, 591 feet, and 615 feet respectively. Next the height of the Pyramid at 2400 feet and 2500 feet. The heights were calculated by using trigonometry once again as follows:

tan 52° = y/50 where y is the height at 2400 , y=64 feet
tan 52°= y/150 where y is the height at 2500, y=192 feet

The cross-section of the Pyramid can be shown by using a 1.5:1 slope and extending it from the base width to the height just calculated. The volume of the Pyramid under the ramp is calculated by using the average end area from these cross-sections as has been shown above.
The volume of the Pyramid under the ramp is equal to 202,133.2 cubic yards. The total volume of the ramp itself is 1,992,222.2 cubic yards-202,133.2 cubic yards = 1,790,089 cubic yards. Using the same techniques as above it would take 50 days to construct such a ramp.

The Pyramid of Khufu, also known as the Great Pyramid because it is the largest ever built, is an absolutely amazing structure. The Pyramid of Khufu like most Egyptian Pyramids was built as a tomb for Khufu and is believed to have been built in approximately 23 years corresponding with the reign of Khufu. The Great Pyramid was built with extraordinary accuracy with joints to with a 1/50th of an inch. So how was this great structure built.

Based on this research and exploration, we believe that Egyptians did employ ramps to build the Great Pyramid. We do not however believe they used a frontal ramp built to the top of the Pyramid. As can be seen from the calculations, even with modern techniques a ramp to the top of the Pyramid would have taken over two years to construct. Most Egyptologists believe it would have taken almost 10 years to build such a ramp with manpower alone. This prediction is very probable according to my calculations with modern techniques. We believe like many Egyptologists, that the builders of the Great Pyramid used a frontal ramp to a height of 200 feet. The calculations show that it would have take approximately 50 days to construct a ramp to 200 feet. The large ramp would have taken approximately 14 times as long. It appears by these calculations that the ramp to 200 feet could have been built in approximately 215 days with manpower. This is quite a feat but with lots of manpower it is probable.

After constructing the lower portion of the Pyramid up to 200 feet, the Egyptians would have had to employ some different method of erecting the Pyramid. Without computer software and visits to the Great Pyramid, it is very hard to get a true idea of how this was performed. However, we are certain that it would have been performed by some kind of spiral type ramp whether it be an internal or external ramp, that remains to be decided. It seems based on this research that the internal ramp would allow for safer and easier construction. One side of the Pyramid could be essentially stair stepped to close to the top and then the ramp replaced with blocks as coming back out. However, as Houdin hypothesized there were probably a series of internal ramps where each ramp led to a point where some kind of lifting device would lift the blocks to a different level with another ramp. This would have kept the slope of the ramp to a minimum and thus made the blocks easier to pull. Either way, in order to place the blocks as close as they are together the Egyptians must have had some lifting device or ramming type device to move the blocks so close together as worker would need lots of room to pull the blocks. We also believe there is some validity to Herodotus's writing that indicate blocks were pulled or lifted up the side of the Pyramid. It appears to me that Herodotus's statement,

"The pyramid was built in steps, battlewise, as is called, or according to others, altarwise. After laying the stones for the base, they raised the remaining stones to their places by means of machines formed of short wooden planks." is referring to the casing blocks. The stones laid for the base seems to refer to the blocks inside the Pyramid, but then the Pyramid was covered with limestone casing blocks. If the actual structure of the Pyramid was already built, then casing blocks would have been pulled up the Pyramid to cover the outside probably using a device such as a "Petrie Rocker".

It is highly probable that the ancient builders also used some form of grid system to ensure the structure remained true to the form proposed. Bryn's theory of a diamond matrix used to construct the structure is a simple yet concise plan that could have been communicated easily and efficiently to a work force. It is very probable that the ancient builders used ramps to construct other structures like the Sphinx since ramps have been found at the sites of other pyramids and the Egyptians had been building pyramids for approximately 80 years before the Great Pyramid.

Although no one is still exactly sure how The Great Pyramid was constructed, it is certain that the ancient builders were very skilled with quarrying stone to exact dimension and masonry techniques. The Egyptians also had a good knowledge of geometry and astronomy based on the dimensions of the Pyramid and it relation to the Earth and stars.

References

Bernal, M. (1987). *Black Athena: Afro-asiatic Roots of Classical Civilization, Volume I: The Fabrication of Ancient Greece, 1785-1985*. Rutgers University Press.

Brier, B.,& Houdin, J.P (2008). *The Secret of The Great Pyramid*. New York, Harper Collins.

Bryn, O. J. (2010). "Ole Jorgen Bryn." *Nordic Journal of Architectural Research* , 135-143.

Hai, P. "Raising Stones" Accessed January 12, 2011 from http://www.haitheory.com/

Imhausen, A. (2003). *Ägyptische Algorithmen. Eine Untersuchung zu den mittelägyptischen mathematischen Aufgabentexten*[Egyptian Algorithms. A Study of Middle Egyptian Mathematical Problem Texts], Wiesbaden: Otto Harrassowitz

Jackson, K., & Jonathan S. (2003). *Building The Great Pyramid*. Toronto, Ontario, Firefly Books.

Joseph, G.G., (1991). *The Crest of the Peacock*. Princeton University Press.

Lehner, M (1997). *The Complete Pyramids Solving the Ancient Mysteries*. New York, Thames and Hudso..

*Mystic Places The Great Pyramid* . Accessed November 29, 2010 from http://www.world-mysteries.com/mpl_2.htm.

Romer, J.(2007). *The Great Pyramid*. Cambridge, UK, Cambridge University Press.

Siliotti, A. (1997). *Guide to the Pyramids of Egypt*. Vercelli, Italy, White Star.

Tompkins, P.(1971) *Secrets of the Great Pyramid* . New York, Harper and Row.

Wikipedia (2011) Accessed December 2, 2010 *Egyptian Pyramid Construction Techniques* URL: http://en.wikipedia.org/wiki/Egyptian_pyramid_construction_techniques

# History of Mathematics in Mathematics Education

# CLASSIFYING THE ARGUMENTS & METHODOLOGICAL SCHEMES FOR INTEGRATING HISTORY IN MATHEMATICS EDUCATION[1]

**Constantinos Tzanakis**

*Department of Education, University of Crete,*

*Rethymnon 74100, Greece*

*tzanakis@edc.uoc.gr*

**Yannis Thomaidis**

*Regional Administration of Education, Central*

*Macedonia, Thessaloniki, Greece*

*gthom54@gmail.com*

## ABSTRACT

*The ICMI Study volume "History in Mathematics Education", published in 2000, includes a comprehensive list of arguments for integrating history in Mathematics Education and methodological schemes of how this can be accomplished. Recently Jankvist distinguished between using "history-as-a-goal" and using "history-as-a-tool" to classify the above arguments. Independently, Grattan-Guinness distinguished between "history" and "heritage" in the hope that –among other things - this will help to understand better which history is expected to be helpful and meaningful in Mathematics Education. We attempt to connect these two conceptual "dipoles", aiming to provide in this way a finer and deeper classification of the arguments and methodological schemes for integrating history in Mathematics Education that will serve as an appropriate theoretical framework. These ideas are illustrated by outlining their application in a specific example.*

Keywords: *History, Heritage, History–as–a-tool, History–as-a-goal, anchoring of meta-issues, complementarity.*

---

[1]This is an elaborated version of the paper presented at the 7th European Conference on Research in Mathematics Education (CERME 7), 9-13/2/2011, Rzeszów, Poland.

248

## 1. INTRODUCTION AND THE BASIC IDEAS

In the last four decades, there is a worldwide growing interest in integrating the history of mathematics (HM) in mathematics education (ME). Several attempts have been made, educational material has been produced, empirical research has been conducted, methodological schemes have been invented & implemented and arguments for this integration have been put forward to refute possible objections and/or to enhance the interest of the ME community in this direction. For a long time, there were no coherent theoretical ideas and framework to place, see and compare all these activities and empirical evaluation of the effectiveness of a historical dimension in ME has remained limited (Jankvist & Kjeldsen in press, §1). A serious attempt in this direction is the comprehensive ICMI Study volume (Fauvel & van Maanen 2000). In particular, it presents a comprehensive list of the arguments **for** integrating the HM in ME (the *whys*)[2] and the general ways of **how** to accomplish this task (the *hows*)[1], in the sense that the ***whys*** correspond to *tasks* that one attempts to accomplish and the ***hows*** correspond to *methodological approaches* one could follow (Tzanakis & Arcavi 2000[3]).

Since then, further important work followed, however:

(a) Jankvist (2009a, b, c, d) has reconsidered the *whys* and the *hows*. He made the distinction between two broad ways (or purposes as he calls them) to introduce the HM in ME, namely in his terminology:

- ***History as a tool*** focusing mainly on the *internal issues* **of** mathematics, that is issues *within* mathematics, or ***inner issues*** of Mathematics (for brevity called *in-issues* from now on), and

- ***History as a goal*** focusing mainly on issues **about** mathematics, that is issues from a ***meta-***

---

[2] In Jankvist's (2009a) terminology.

*perspective*, or *meta-perspective issues* in Mathematics (for brevity called *meta-issues* from now on) He attempted to classify the *whys* using this distinction (Jankvist 2009a §§2.3, 8.1, 8.5; 2009b §1.1; 2009c §§2, 3), which constitutes the first **conceptual dipole**, in the context of the present paper.

Moreover, he attempted to classify the *hows* by distinguishing possible implementations into three broad types: ***illumination approaches, modules approaches*** and ***history-based approaches*** and to connect them to the above conceptual dipole (Jankvist 2009a §§2.4, 2.6, 2009c §§6, 9).

We should note here that the term "dipole" is used, instead of a more directly interpretable one, like "pair", in order to emphasize the deep interconnections between the two concepts, which reflect better their complementary character described below.

(b) Independently, Grattan-Guinness (2004a, b) introduced in a more general context the distinction between "***history***" and "***heritage***", to interpret mathematical activities and their products, in an effort to clarify existing conflicts and tensions between a mathematician's and a historian's approach to mathematical knowledge. Grattan-Guiness (2004b) gives several examples by contrasting the general characteristics of the two concepts, which constitute our second **conceptual dipole**. It could be an important tool to revisit the issue of "which history is appropriate to ME?" (see e.g. Barbin 1997) and is close to a similar distinction of attitudes towards the development of mathematics that has been introduced by Rowe, the one adopted by "cultural historians" and the other adopted by "mathematical historians" (Rowe 1996, pp.3, 5).

As we will argue, within each *dipole*, the two "*poles*" are **complementary** to each other,

---

[3]In the same work, a list of possible objections against this integration is given (*op. cit.* p.203), subsequently enlarged and enriched in Siu 2006, §2).

in the sense that they are **mutually exclusive** for a **simultaneous** use, but none of them, taken alone, can lead to a sufficiently wide and deep enough understanding of what (a specific piece of) mathematics is; instead **both are indispensable** for understanding mathematics as a cultural endeavour and human intellectual activity, either didactically or/and epistemologically. Here, the term "complementary" is used in a way close to that introduced by N. Bohr to describe the microphysical reality and subsequently was raised to a general epistemological principle, used as a tool to understand reality (see Bohr 1934 ch.2, 1958 ch.2 & 3, Pauli 1950/1994, Jammer 1974, ch.4). In this connection we note that "mutually exclusiveness" should not be understood in its strictly **logical** sense, but rather in Bohr's broader sense, according to which

"...the coupling between phenomena and their observation, forces us to adopt a new mode of description designated as *complementary* in the sense that any given application of ...concepts precludes the simultaneous use of other ...concepts which in a different connection are equally necessary for the elucidation of the phenomena" (Bohr 1934, p.10)[4].

In the context of the present paper, "complementarity" refers to the fact that within each dipole, each pole focuses and puts emphasis on a particular aspect of what mathematics and its development are. However, none of these emphases is sufficient for that if taken alone, and an extreme version of each one of them leaves no space for the other.

Appreciating the complementary nature of the concepts within each dipole may help to smooth out existing strong tensions (e.g. between "cultural" and "mathematical" historians mentioned in Rowe 1996 and similarly in Grattan-Guinness 2004b, especially p.8) and to refute more convincingly objections against the integration of the HM in ME. In fact, the key idea and aim of this paper is to **classify** the ICMI Study *whys* and *hows* in a **finer** way, by projecting them onto the 2X2 grid formed by the two dipoles introduced above, thus getting "… a clear[er] idea about **why** history should be used in a given situation; i.e. what *whys* the use of history should fulfil…" and which are suitable options of *hows* (Jankvist 2009c p.256). Connecting the two dipoles in this way may contribute to clarify their relevance to ME and provide a finer conceptual framework to integrate HM into ME. However, this **2-dimensional classification** is **tentative**, possibly subject to modifications, as **specific** realizations & examples of the *whys* and the *hows* are examined how they fit into it.

The paper is structured as follows: A description of each *dipole* is given in section 2, with reference to distinctions pertinent to the *History-Heritage* dipole, found in Tzanakis & Arcavi 2000. Section 3 presents the ICMI Study *whys* and *hows*, Jankvist's three broad types of implementations, and the list of possible objections mentioned in footnote 3. An attempt is made to describe some of the *whys* in a more refined way, so that their classification using the above mentioned 2X2 grid is clearer. Section 4 presents the **classification tables** with brief reference to indicative examples that support these classifications, or reveal some possibly controversial aspects of them. The emphasis in this paper is on the presentation of the general theoretical ideas. Specific examples should be explored in detail to check the validity of these ideas, their usefulness in actual implementation and their efficiency to better understand different aspects of which and how HM could be integrated in ME and for what purpose. Nevertheless, in section 5 we outline the application of these ideas to a specific example, as an illustration of these ideas; logarithms and related concepts to be taught in upper high school.

---

[4]E.g. Understanding biological systems requires a holistic view, whereas, understanding their biochemical processes needs a reductionist approach: Both are indispensable for a sufficiently wide and deep understanding of life phenomena,

Detailed analysis of other examples will be given in future publications. Some final remarks and possible further connections of the theoretical ideas of this paper with other similar ideas in the existing literature are given in section 6.

## 2. THE TWO CONCEPTUAL DIPOLES

**2.1** As already mentioned, Jankvist introduced two broad ways in which HM could be helpful and relevant to ME: *History-as-a-tool* and *History-as-a-goal*, which are intimately connected with issues within mathematics (what he calls *in-issues*) and with issues that concern mathematics itself (what he calls *meta-issues*). In his own words,

"*History-as-a-tool* concerns the use of history as an assisting means, or an *aid*, in the learning [or teaching] of mathematics…. in this sense, history may be an aid both …"[5] "as a motivational or affective tool, and … as a cognitive tool …"[6] More precisely, "…[w]hen restricting ourselves to the use of history as a tool, we may (sub) categorize such uses into at least three different types: history as a motivational and/or affective tool; history as a cognitive tool (e.g. the idea of epistemological obstacles); and the role of history in what may be referred to as the evolutionary arguments (the recapitulation argument or historical parallelism)."[7] "[It] concerns… inner issues, or *in-issues*, of mathematics [that is] issues related to mathematical concepts, theories, disciplines, methods, etc. — the internal mathematics"[8].

On the other hand,

---

but it is impossible to put **absolute** emphasis on the one, without "destroying the other (Bohr 1958, chs.2).

[5]Jankvist 2009b §1.1.

[6]Jankvist 2009d, p8.

[7] Jankvist & Kjeldsen (in press), §2.

[8]Jankvist 2009c, p240.

"*History-as-a-goal* does not serve the primary purpose of being an aid, but rather that of being an *aim* in itself … posing and suggesting answers to questions about the evolution and development of mathematics, … about the inner and outer driving forces of this evolution, or the cultural and societal aspects of mathematics and its history"[9]. "The word *goal* must be understood in the sense that it is considered a goal to show the students something about the historical development of mathematics… that mathematics exists and evolves in time and space; that it is a discipline which has undergone an evolution over millennia; that [it] has developed due to human activities; that [its] evolution… is due to many different cultures throughout history, and that these cultures have had an influence on the shaping of mathematics as well as the other way round."[10]. In other words, "[It] concerns … learning something about the meta-aspects or *meta-issues* of mathematics … [that is] issues involving looking at the entire discipline of mathematics from a meta perspective level"[11].

From the above description it is clear that these two ways in which HM becomes relevant to ME are mutually exclusive, in the sense described in section 1; that the emphasis put on each case are clearly different and to a large extent incompatible with each other. Nevertheless, it should be remarked at this point that although

"…history-as-a-goal 'in itself' does not refer to teaching history of mathematics *per se*, but using history to surface meta-aspects of the discipline… in specific teaching situations, [it] may have the positive side effect of offering to students insight into

---

[9] Jankvist 2009b §1.1.

[10] Jankvist & Kjeldsen (in press), §2.

[11] Jankvist 2009c, pp239-240. E.g. "How does mathematics evolve in time and space? What forces and mechanisms cause the evolution of mathematics? How does the evolution of mathematics interact with society and culture? Can mathematics become obsolete?" (Jankvist 2009d, p.8).

254

mathematical in-issues of a specific history" (Jankvist 2009d, p.8).
Conversely, using "history-as-a-tool" to teach and learn a specific mathematical subject may stimulate reflections of a meta-perspective nature extrapolated from the particular subject considered; that is, we may have a kind of **anchoring** of meta-issues into the in-issues that constitute the study of the subject.

"By 'anchoring' we are referring to something that substantiates discussions and reflections about meta-issues on a basis of knowledge and understanding of the related in-issues, e.g. by revealing insights about the meta-issues that could not have been accessed or uncovered without knowing about the in-issues, or by providing in-issue evidence for meta-issue claims or viewpoints"[12], which may lead to "...different levels regarding anchoring of the students' metaperspective discussions (...anchored comments, arguments, and discussions)"[13];

for more details and examples of this idea, see Jankvist 2009b, §§5.3, 5.4, 6.1, 6.3, Jankvist (in press, *JRME*). These are important interrelations, stressing the indispensability of both the "history-as-a-tool" and the "history-as-a-goal" ways, which thus constitute what we called earlier a coherent *conceptual dipole*. In fact this concept of anchoring of meta-issues into the in-issues could be explored in at least two other directions: First, is there a possibility of a similar process of anchoring in the opposite direction? That is, the possibility of anchoring inner issues of mathematics during the exploration and discussion of meta-issues? More generally, the existence and depth of a two-way/reciprocal type of anchoring between issues pertaining to two different disciplines may constitute "...a possible marker for the level of

---

[12]Jankvist & Kjeldsen (in press), §3.

[13] Jankvist (in press, *JRME*), abstract.

interdisciplinarity achieved…"[14] or is possible[15].

**2.2** Quite independently, having in mind both historians and educators of mathematics, Grattan-Guinness introduced the distinction between *History* and *Heritage*. More specifically:

The *History* (*Hi*) of a particular mathematical subject $N$ refers to "… the development of $N$ during a particular period: its launch and early forms, its impact [in the immediately following years and decades], and applications in and/or outside mathematics. It addresses the question *'What happened in the past?'* by offering descriptions. Maybe some kinds of explanation will also be attempted to answer the companion question *'Why did it happen?'*"[16]. "[It] should also address the dual questions *"what did not happen in the past?"* and *"why not?"*; false starts, missed opportunities …, sleepers, and repeats are noted and maybe explained. The (near-) absence of later notions from $N$ is registered, as well as their eventual arrival; *differences* between $N$ and seemingly similar more modern notions are likely to be emphasized"[17].

On the other hand,

The *Heritage* (*He*) of a particular mathematical subject $N$ refers "…. to the impact of $N$ upon later work, both at the time and afterward, especially the **forms** which it may take, or be **embodied**, in later contexts. Some modern form of $N$ is usually the main

---

[14]Jankvist (in press, *JRME*), "Perspectives and future research".

[15]A typical example in this connection could be the exploration of this idea for (specific areas of) mathematics and (specific areas of) physics. In fact, the close interrelation between these disciplines can be further developed along these lines as a two-way process, reflected in the description of "mathematical physics" and "physical mathematics" that has been presented in Tzanakis 1999.

[16]Grattan-Guiness, 2004b, p.7.

[17]Grattan-Guiness, 2004a, p.164.

focus, with attention paid to the course of its development. Here the mathematical relationships will be noted, but historical ones ... will hold much less interest. [It] addresses the question *"how did we get here?"* and often the answer reads like "the royal road to me." The modern notions are inserted into $N$ when appropriate, and thereby $N$ is unveiled ... *similarities* between $N$ and its more modern notions are likely to be emphasized; the present is *photocopied* onto the past"[18].

Of course,

"[b]*oth kinds of activity are quite legitimate*, and indeed important in their own right; .....A philosophical difference is that inheritors [those having a heritage-like attitude] tend to focus upon knowledge alone (theorems as such, and so on), while historians also seek motivations, causes, and understanding in a more general sense. The distinction sometimes made by historians of science between "internal" and "external" history forms part of this difference"[18].

(cf. Rowe's distinction between "cultural historians" and "mathematical historians" mentioned in section 1 (Rowe 1996)).

Although, Grattan-Guinness is mainly concerned with the implications of this distinction on the way past mathematics should be approached, he clearly indicates its relevance to ME (Grattan-Guinness 2004b, §1.6). He argues that ME can profit equally well from **both** *Hi* and *He*, urging for further exploration in this context, a fact that has not passed unnoticed (Siu 2006, p.273, Rogers 2009 pp.120-121, Schubring 2008, p.5). He also gives a detailed list of the differences between these two conceptions (Grattan-Guinness 2004b, §1.3; in more elaborated form in Grattan-Guinness 2004a, §3), with emphasis on their **incompatibility**, summarized as follows:

---

[18]Grattan-Guiness, 2004a, p.165.

"The distinction between history and heritage is often sensed by people who study some mathematics of the past, and feel that there are fundamentally different ways of doing so. Hence the disagreements can arise; one man's reading is another man's anachronism, and his reading is the first one's irrelevance. The discords often exhibit the differences between the approaches to history usually adopted by historians and those often taken by mathematicians."[19] "…in particular, mathematical research often seems to be conducted in a heritage-like way …., whether the predecessors produced their work long ago or very recently. *The confusion of the two kinds of activity is not legitimate*, either taking heritage to be history (frequently the mathematicians' view—and historians' sometimes!) or taking history to be heritage (the occasional burst of excess enthusiasm by a historian); indeed, such conflations may well mess up both categories, especially the historical record. In the case of sequences of notions, a pernicious case arises when $N_1$ is a logical consequence or a generalization of $N_0$, and the claim is made that a knower of $N_0$ knew $N_1$ also"[20].

On the other hand, however, their **indispensability** in understanding the development of mathematics is clearly emphasized:

"The claim put forward here is that *both history and heritage are legitimate ways of handling the mathematics of the past; but muddling the two together, or asserting that one is subordinate to the other, is not.*"[21]

Thus, the above quotations show that the two concepts are **complementary** in the sense of section 1, constituting the two poles of a coherent *conceptual dipole*.

As far as HM in ME is concerned, the distinction between *History* and *Heritage* is related

---

[19]Grattan-Guinness 2004b, p.8.

[20]Grattan-Guiness, 2004a, p.165.

258

to the rationale underlying the distinction between pairs of methodological approaches put forward in the past, like *explicit & implicit* use of history, *direct & indirect* genetic approach, *forward & backward heuristics* (Tzanakis & Arcavi 2000, §7.3.2 and references therein):

When "...HM is *explicitly* integrated, mathematical discoveries are presented in all their aspects. Different teaching sequences can be arranged *according to the main historical events*, in an effort to show the evolution and the stages in the progress of mathematics by describing a certain *historical period*... [When]... HM enters *implicitly*, history *suggests* a teaching [approach], in which use may be made of concepts, methods and notations, that appeared later than the subject under consideration, keeping always in mind the general didactic aim, namely to *understand mathematics* in its *modern* form.... [It] does not necessarily respect the order by which the historical events appeared; rather, one looks at the historical development from the *current* stage of concept formation and logical structuring of the subject... *they have a dual character*...and *both may be used*.... in *complementary* ways... in an explicit integration of the HM, the emphasis is on a rough, but ... more or less accurate mapping of the path network that appeared historically and led to the modern form of the subject[22]; in an implicit integration [of the HM], the emphasis is on the redesigning, shortcutting and signalling this path network"[23].

Hence, the dipole (*history, heritage*) is potentially of great relevance to ME, as Grattan-

---

[21]I. Grattan-Guinness 2004b, p.8.

[22]This is close to the concept of the "history-satire" introduced by Grattan-Guinness (1973; pp.445ff), under which "...the broad features of the historical record are respected and used; but usually many detours and complications occur that, while they attract the historian, will impede teaching and so should be set aside or at most treated only in passing" (Grattan-Guinness 2004b, p.16).

[23]Tzanakis & Arcavi 2000, p.210; our emphasis. Careful reading of these quotations, reveals the co-existence of the two poles of each of the dipoles; (History, Heritage) & (History-as-a-tool, History-as-a-goal)

Guinness himself points out (see also Rogers 2009):

"Where does mathematical education lie in between history and heritage? My answer
is: exactly there, and a very nice place it is. Educators can profitably use both history
*and* heritage for their purposes"[24]. "… In particular, if notion N is to be taught, then
both its history and its heritage can be used".[25]

In fact, given that "*History* focuses on the detail, cultural context, negative influences,
anomalies, and so on, in order to provide evidence, so far as we are able to tell, of what
happened and how it happened [whereas] [*h*]*eritage*… focuses on the positive influence of
events and discoveries; on 'completion' and fairly finished forms of mathematics that can be
communicated without special concern for the dynamics of their particular production"[26], this
conceptual dipole -among other things - may contribute towards an operational answer to the
recurrent question: Why history and which history is appropriate to be used for educational
purposes? (Barbin 1997).

## 3. A LIST OF THE *WHYS & HOWS*

In this section we present the list of the *whys & hows* according to the ICMI Study volume
(Tzanakis & Arcavi 2000, §§7.2, 7.3) and Jankvist's three broad types of immplementations
(2009c, §6) that will be used in the next section, as well as, the list of possible objections
originally given in the ICMI Study volume and subsequently elaborated further in Siu 2006 and
Tzanakis 2009.

### 3.1 The "ICMI Study *whys*"

---

[24]Grattan-Guinness 2004a, p.175.

[25]Grattan-Guinness 2004b, p.16.

[26]Rogers 2009, p.2782.

The main epistemological and philosophical thesis underlying the ICMI Study *whys* is the following: The "polished" products of mathematical activity is just that aspect of mathematical knowledge, which can be communicated, criticized (in order to be finally accepted or rejected) and serve as the basis for new work. However and especially from a didactical point of view, the **process of "doing mathematics"** is an **equally important** aspect, which includes the understanding of the motivations for problems and questions, the sense-making actions and the reflective processes which are aimed at the **construction of meaning** by linking old and new knowledge, and by extending and enhancing existing conceptual frameworks.

"Here we find the obvious but vital factor in favour of historical elements... whatever mathematics we want to learn, it has historical roots and motivations and therefore important and quite possibly essential features caused by its historical background.... Put another way, the distinction between 'mathematics' and 'history of mathematics' is false in principle: *there are only mathematical problems, and they have a history...*" (Grattan-Guineess, 1973, p.446).

In this sense, any interest on the HM (both epistemologically and didactically) should not be seen as an obligation, but stems from the awareness that there are things to learn from history which otherwise would be inaccessible (cf. Grattan-Guineess, 1973, p.447).

Hence, the leitmotiv of the arguments for integrating HM in ME is that mathematics should be conceived not simply as a rigidly structured system of logically consistent and clear results, but also as a continuously evolving multifarious **human intellectual process**, tightly linked to other sciences, culture and society.

The following are the areas in which the teaching and learning of mathematics can profit from integrating the HM in the educational process (the main points of the arguments below appear in Tzanakis & Arcavi 2000, §7.2, but some of them appear in a more refined form, keeping in mind the conceptual dipoles introduced in section 2).

## A. *The learning of Mathematics*

**1.** *Historical development vs. polished mathematics*: It is a fact that "[n]o mathematical idea has ever been published in the way it was discovered…"[27] and that mathematics is usually globally and retrospectively reorganized. Although this reorganization may lead to a more concise and **logically** clear form of a mathematical subject, it often results to an ad hoc introduction of concepts, methods or proofs. However, looking at basic aspects of the historical development retrospectively and integrating (some of) them properly into teaching, can play an important role by helping to uncover/unveil the meaning and significance of the mathematics that are new to students. In this way it becomes possible to present the subject in a natural way, by keeping to a minimum logical gaps and an ad hoc introduction of concepts, methods or proofs.

**2.** *History as a re-source*: Among other things, HM constitutes a vast reservoir of **relevant** questions, problems and expositions, which - if properly introduced into ME - can motivate, raise the interest and engage the learner by linking present knowledge and learning process to knowledge and problems in the past.

**3.** *History as a bridge between mathematics and other disciplines/domains*: It is an historical fact valid for all times that the introduction and/or development of a great deal of mathematics was motivated and inspired by questions and problems coming from other (apparently unrelated) disciplines and has emerged, established and consolidated under the pressure and need to answer and solve such questions and problems. In fact, grasping the

---

[27]Freudenthal (1983, p.IX).

meaning and significance, and acquiring a good working knowledge, of a lot of *inner issues* of mathematics that is new to the learner, can be gained by properly looking in a historical perspective at such "external" issues that naturally bring in new aspects, subjects and methods. In addition, such a perspective may unfold interrelations among domains, which at first glance appear unrelated, thus providing the opportunity to appreciate that fruitful research in a scientific domain does not stand in isolation from similar activities in other domains (clearly an important *meta-issue*).

**4.** *The more general educational value of history*: A historical perspective in ME and especially the learners' and teachers' involvement in historically oriented study projects may lead to the development of personal growth and skills - not necessarily connected to mathematics: reading, writing, looking for resources, documenting, discussing, analyzing, and "talking about" (as complementary to "doing") mathematics[28].

### B. *The nature of mathematics and mathematical activity*

**1.** *Content*: HM provides a different viewpoint of *inner issues*, like (possibly familiar) mathematically relevant and historically important questions and problems that offer insights into concepts, conjectures & proofs, and help to understand why they do or do not supply satisfactory answers to the already existing questions and problems. Additionally, the answers and/or dead-end roads to such questions and problems may help to provide a more accurate view of mathematics and mathematical activity: to increase the visibility of the

---

[28]This seems to be close to aspects of some of Niss' mathematical "competencies": The mathematical thinking competency (possessing the ability to pose questions); the reasoning competency (following and assessing others' reasoning; the communication competency (understanding, examining and interpreting different kinds of written, oral, or visual expressions and texts); the tools & aids competencies (reflectively using tools and aids) (Niss 2003, pp.119-121; 2004, pp.184-186). Though this is a potentially interesting connection, which could be explored for all the *whys* in this section, it will not be pursued further in this paper.

evolutionary nature of mathematical knowledge and the time-dependent character of fundamental meta-concepts, like proof, rigor, evidence, error; to develop the awareness that mistakes, heuristics arguments, uncertainties, doubts, intuitive arguments, blind alleys, controversies and alternative approaches to problems are both legitimate and an integral part of mathematics in the making[29]; these are important *meta-issues* that could be explored in this way.

**2.** *Form*: Mathematical notations, terminology, favourite computational methods, modes of expression and representations have been evolving. Bringing the old close to the new, may help the learner to become aware of the mathematical (verbal, or symbolic) language of a given period, to re-evaluate the role of visual, intuitive and non-formal approaches that have been put forward in the past and to understand better the advantages and/or disadvantages of the modern form of mathematics, hence to motivate learning by stressing clarity, conciseness and logical completeness.

### C. *The didactical background of teachers*[30]

**1.** *Identifying motivations*: Incorporating questions and problems that historically served as prototypes to introduce new ideas, concepts and methods, may help to see the rationale lying behind the introduction of new knowledge and the substratum on which further progress was based.

**2.** *Awareness of difficulties & obstacles*: By studying historical issues related to what

---

[29]Cf. "Integrating [historical] sources in mathematics challenges the learner's perceptions through *making the familiar unfamiliar*. Getting to grips with a historical text can cause a reorientation of his views and thus deepen his mathematical understanding" (Furinghetti, Jahnke, van Maanen p.1286); the HM may act as source of "astonishment", having an "expatriating" function ("une fonction dépaysante" in Barbin's terminology; Barbin, 1997).

[30]Some items concern mathematics educators as well (C.1 & C.2), or the students (C.3 &C.5).

teachers (are intended to) teach, they are given the opportunity: (i) To identify **difficulties,** or, **obstacles**, that appeared in history and bear analogies with students' difficulties identified in the classroom that could be due to (a combination/interference) of epistemological and/or didactical reasons; this may lead to a better and more appropriate teaching design and implementation. (ii) To realize that often new knowledge has been the result of a **gradual** evolution, based on questions and problems that presuppose a **mathematical maturity** on the part of the students that may not exist yet; this may help the teacher to become aware of the pros and cons of presenting a subject at a particular level of education.

**3.** *Getting involved and/or becoming aware of the creative process of "doing mathematics":* Both the teachers and the learners can be involved into **doing** (usually known) mathematics by tackling questions and problems in historical context. This could focus on (i) solving the problems per se; (ii) connecting these questions and problems to others and profiting from this interconnection to provide a wider and deeper understanding of mathematics and its evolution, hence, to enrich mathematical literacy; (iii) pointing out meta-mathematical issues inherent in this questions and problems, thus deepening the understanding of the nature of mathematics.

**4.** *Enriching the didactical repertoire*: Because the HM constitutes a vast, stimulating and insightful resource of questions and problems, it is potentially a natural path to enrich the teachers' didactical repertoire and increase their ability to explain, approach, and understand specific issues **in** mathematics and **about** mathematics.

**5.** *Deciphering and understanding idiosyncratic and/or non-conventional approaches to mathematics*: Teachers who are got involved into a historically based and/or inspired situation in which they have to decipher/understand a known piece of "correct mathematics", expressed and treated in its original form may learn how to work on known mathematics in a different (old) context and therefore become more sensitive and tolerant towards non-conventional, idiosyncratic, or "wrong" mathematics adopted, or developed by their students.

**D.** *The affective predisposition towards mathematics*

**1.** *Understanding mathematics as a human endeavour*: History is a privileged domain in the context of which the evolutionary steps of mathematics can be shown explicitly, and consequently, mathematics can be seen as an evolving and human subject rather than as a God-given finished system of rigid truths.

**2.** *Persisting with ideas, attempting lines of inquiry, posing questions*: As already mentioned, mathematics is a human endeavour, an integral part of which consists of posing questions, putting forward heuristic arguments and following alternative (non-mainstream) approaches. Looking in detail at similar examples in the past, may indirectly have the side effect to learn how to persist with ideas, to undertake lines of inquiry, to pose questions, and feel free and legitimised to attempt to develop creative or even idiosyncratic ways of thought.

**3.** *Not getting discouraged by failures, mistakes, uncertainties, misunderstandings*: Similarly to what was mentioned in the previous paragraph, uncertainties, doubts, controversies, mistakes and blind alleys have always been present in the work of the most prominent mathematicians, or even formed its building blocks. Looking in detail at such examples in the past may help the learner not to get discouraged by failure, mistakes, uncertainties or misunderstandings, and the teacher to appreciate the fact that they constitute an integral part of learning and doing mathematics.

**E.** *The appreciation of mathematics as a cultural endeavour*

**1.** *To appreciate that mathematics evolves under the influence of factors intrinsic to it*: The detailed study of historical examples provides the opportunity to identify and appreciate the role of internal factors in the development of mathematics (like aesthetic criteria, intellectual curiosity, challenge and pleasure, recreational purposes etc), thus appreciating that mathematic is not only driven by utilitarian reasons, but also for its own sake.

**2.** *To appreciate that mathematics evolves under the influence of factors extrinsic to it*: Similarly, history can provide examples of how the internal development of mathematics, whether driven by utilitarian or "pure" reasons, has been influenced, or even determined to a large extent, by social and cultural factors.

**3.** *To appreciate that mathematics form part of local cultures*: Notwithstanding the importance of western culture in the development of mathematics, history shows that mathematics appeared within other cultures and exerted an influence on them. Studying specific examples from this perspective, gives the opportunity to become aware of the multicultural nature of mathematics. In addition, such cultural aspects may help teachers in their daily work with multiethnic classroom populations, in order to re-value local cultural heritage as a means of appropriately designing their teaching and developing tolerance and respect among fellow students.

### 3.2 The "ICMI Study *hows*"

Below is a similar account of the *hows* for integrating HM in ME according to the ICMI Study volume. The examples given below to exemplify the *hows* are indicative included in a more elaborate form in Tzanakis & Arcavi 2000 §7.4, after having taken into account Boyé et al (to appear), section 3.1[31].

#### 3.2.1. *Learning **history** by providing **direct historical information***

One broad way to integrate HM in ME is by providing direct historical information, that

---

[31]The list is not exhaustive.

is:

(a) Isolated "factual information" like, giving names, dates, famous works and events, time charts, biographies, famous problems and questions, attribution of priority, facsimiles etc. This information may appear as historical snippets, presentation, comments & interpretation of images from books, original paintings or photographs etc

(b) Courses or books on the HM, where history appears as a natural, integral and explicit part, or consisting of a simple account of historical data, or a history of conceptual developments, or something in between in the form of separate historical chapters.

(c) Suggestions for further reading possibly with historical notes and/or annotated bibliography; a guide to the literature (books, journals, websites) for further reading, research and insights etc

The **emphasis is more on learning history, than on learning mathematics**. Hence it concerns mainly the *history-as-a-goal* in the terminology of section 2, not directly intended to affect, or modify the teaching of particular mathematical contents.

*3.2.2. Learning **mathematical topics** by following a teaching **approach inspired by history***

A much more demanding way to integrate HM in ME is to follow an approach that in general is inspired by history, which appears as a natural, integral and explicit part of teaching and/or didactical material and textbooks. HM in this context may appear either explicitly, or implicitly in the sense described briefly in §2.2 above and in more detail in Tzanakis & Arcavi 2000, §7.3.2. Here, the **emphasis is more on learning mathematics, than on learning history** in the sense of the conception of *history-as-a-tool* described in section 2. Examples of this type of approach greatly vary in form, content and specific methodology, e.g.:

(a) Teaching modules and/or mathematical textbooks, permeated by history, usually in an implicit manner.

(b) Student research projects often based on original documents (possibly benefiting from e-sources and the web).

(c) Worksheets (often based on original sources), either designed as semi-structured and guided sets of questions to introduce a new topic, a set of problems, issues for discussion etc, or containing a collection of exercises, recreational problems and games in order to master a procedure, consolidate a topic etc.

(d) Possibly annotated excerpts from original documents to introduce a topic, to help the reader to grasp better the mathematical content, both in its modern form and in its original context.

(e) "Historical packages" in the form of a self-contained collection of materials, ready for use by teachers in the classroom, narrowly focused on a small mathematical topic and with strong ties to the curriculum.

(f) More "localized", history-based teaching capsules and/or didactical material that aim to take advantage of errors, alternative conceptions, change of perspective, revision of implicit assumptions, intuitive arguments etc in order to support the teaching and learning of specific pieces of mathematics.

(g) Hints to past and/or recent developments, related to outstanding old issues and problems of various kinds[32], or constituting good examples to fascinate students and reveal the evolutionary character of mathematics.

---

[32]Recreational, unsolvable, still unsolved, with clever or alternative or exemplary solution, having provoked and/or anticipated important developments etc.

*3.2.3. Developing and/or enhancing mathematical awareness*

By mathematical awareness we mean the ability to recognize and understand the general characteristics of mathematics both as a corpus of knowledge and as a human activity, appreciating their significance for mathematics per se and in its relation to other scientific disciplines, culture and society. Given that this awareness is clearly related to issues at a *meta-perspective* level, putting emphasis on its developing is more (though not exclusively) pertinent to the *history-as-a-goal* of section 2, as it will be seen from the more detailed description below. This "awareness" includes, aspects related to the **intrinsic** and the **extrinsic** nature of mathematical activity

(a) *Awareness of the intrinsic nature of mathematical activity* (***intrinsic awareness***)

Historical elements of important aspects of "doing mathematics" are integrated in ME with focus on unfolding, analysing and emphasizing:

(i) The role of general conceptual frameworks and of associated motivations, questions and problems in the development of, and in, particular mathematical domains.

(ii) The evolutionary nature of mathematical content and form: notation, terminology, computational methods, modes of expression & representations, metamathematical notions (proof, rigor, evidence etc).

(iii) The significance and basic characteristics of the mathematical activity itself in the context of specific situations: the role played by doubts, paradoxes, contradictions, heuristics, intuitions, dead ends etc in the process of generalization, abstraction an formalization.

(b) *Awareness of the extrinsic nature of mathematical activity* (***extrinsic awareness***)

Historical aspects are introduced in ME with focus on refuting the largely prevailing view of mathematics as a discipline largely disconnected from social and cultural concerns and influences. In particular, HM is introduced in order to reveal, explore and emphasize:

(i) The relations of (specific pieces of) mathematics to philosophy, arts and social sciences;

(ii) The influence of the social and cultural contexts in the development of particular mathematical issues;

(iii) The cultural character of mathematics as an integral part of different civilizations and traditions;

(iv) Aspects of the history of ME and the influence they exerted on ME.

### 3.3 A new 3-fold classification of *hows*

As already mentioned in section 1, Jankvist classified the different approaches to introduce HM in ME in three broad categories (Jankvist 2009a §§2.4, 2.6, 2009c §§6, 9):

3.3.1. *Illumination approaches*: Teaching and learning of mathematics, in the classroom or the textbooks used, is supplemented by historical information of varying size and emphasis. Both *in-issues* and *meta-issues* can be considered in this context, where, history "spice" up the presentation of the subject, so to speak (Jankvist 2009c §6.1).

3.3.2. *Modules approaches*: These consist of instructional units devoted to history, and often based on the detailed study of specific cases, tied or adjoint to the mathematics curriculum. History appears more or less directly, possibly based on original documents and the content of such modules varies greatly in size, from a sharply focused approach to a well-defined "local" issue, to a full course or textbook with a wider scope to present conceptual mathematical developments, historical facts or both (Jankvist 2009c §6.2). This is close to the *explicit* use of the HM described in Tzanakis & Arcavi 2000, §7.3.2 (cf. §2.2 above)

3.3.3. *History-based approaches*: These are approaches directly inspired by, or based on the HM. Even though, they may lean heavily upon history, they deal with studying the HM indirectly, rather than directly. The historical development is not necessarily discussed in the open, but often sets the agenda for the order and way in which mathematical topics are presented (Jankvist 2009c §6.3). This is close to the *implicit* use of the HM described in Tzanakis & Arcavi 2000, §7.3.2 (cf. §2.2 above)

Both types of *hows* correspond to possible implementations of the HM into ME, of a different character, however; the ICMI Study *hows* focus on different emphases, whereas, Jankvist's focus strictly on the adopted methodologies. We believe that both are useful in the present context and will be further considered in section 4. Their possible interrelations is an interesting issue that needs further study. An attempt is made in Jankvist 2009c, §9 and a hopefully refined perspective is offered by their classification presented in the next section. However, this remains an open issue to be further elaborated.

### 3.4 Objections against integrating HM in ME

The following is a list of objections that have been put forward in the literature against the role of HM in ME, some of which are also frequently raised by teachers at all levels of instruction. The list (and corresponding references) appeared originally in Tzanakis & Arcavi 2000, p.203 and subsequently was enlarged in Siu 2006. Below, the arguments are classified according to their character, following Tzanakis 2009, §3. We attempt to connect their eventual refutation to the ICMI *whys* of section 3.1, in the sense that the proper realization of the *whys* - seen as *tasks* that one attempts to accomplish – is potentially related to the refutation of the objections which correspond to them; this is the meaning of the references below to the items of §3.1.

### 3.4.1. *Objections of an epistemological and methodological nature*

(a) *On the nature of mathematics*[33]

---

[33]Some of these objections have been presented and criticised very early by Grattan Guinness, (Grattan-Guinness 1973, pp.446-447).

272

(1) This is not mathematics! Teach the subject first; then its history. (A.1, A.2, C.1, C.4)

(2) Progress in mathematics is to make difficult problems routine, so why bother to look back? (A.1, B.1, C.1, C.2, D.3, E.1)

(3) What really happened can be rather tortuous. Telling it as it was can confuse rather than to enlighten! (A.1, B.1, C.1, C.5, D.3)

(b) *On the difficulties inherent to this approach*

(1) Does it really help to read original texts, which is a very difficult and time-consuming task? (A.4, B.1, B.2)

(2) Is it liable to breed cultural chauvinism and parochial nationalism? (E.3)

(3) Students may have an erratic historical sense of the past, which makes historical contextualization of mathematics impossible without their having a broader education in general history. (A.4)

### 3.4.2. *Objections of a practical and didactical nature*

(a) *The background and attitude of the teachers*

(1) Lack of didactical time: no time for it in class! (A.2)

(2) Teachers should be well educated in history: "I am not a professional historian of mathematics. How can I be sure of the accuracy of the exposition?" (A.3, E.2)

(3) There is a lack of teacher training in it! (C.2, C.3, C.4, C.5)

(4) There is a lack of appropriate didactical and resource material on it! (A.2, C.4)

(b) *The background and attitude of the students*

(1) Students regard it as history and they hate history class! (A.4)

(2) Students regard it just as boring as the subject mathematics itself! (A.2, B.1)

(3) Students do not have enough general knowledge on culture to appreciate it! (D.1, E.2, E.3)

(c) *Assessment issues*

(1) How can you set question on it in a test or examination? (A.2, A.3, C.3)

(2) Is there any empirical evidence that students learn better when history of mathematics is made use of in the classroom? (A.4)

We should remark that the objections of an epistemological and methodological character (§3.4.1) are more directly related to the *whys*, than the objections of a practical and didactical character (§3.4.2), resulting mainly from a narrow or defective understanding of the nature of mathematics and/or history, hence of their intimate connection. Therefore, the classification scheme proposed in the next section is expected to be more useful for the analysis and refutation of these objections. On the other hand, some of the objections in §3.4.2 ((a1), (a3), (a4), (c2)), point to very serious obstacles in any attempt to integrate HM in ME, as already stressed in the literature (e.g. Tzanakis & Arcavi 2000, p.212).

## 4. THE 2-D CLASSIFICATION OF THE *WHYS & HOWS*

By taking into account the presentation of the conceptual dipoles in section 2 (in connection with the details provided in the corresponding references therein), a 2X2 table results, composed by the distinct elements of each dipole. Then, according to the description provided in section 3, each of the *whys* and *hows* can be placed in at least one cell, depending on how sharply and clearly it has been described there and in the corresponding references (Tzanakis & Arcavi 2000, §§7.2, 7.3; Jankvist 2009c, §6)[34]. The entries in the next table refer to the numbering in §3.1.

*Table 1*: The classification of the ICMI *whys* (cf. §3.1)

---

[34]We note however, that A.4 is not easy to be placed in the table; it is necessary to be further explored, taking into account possible connections with Niss' competencies mentioned in footnote 28.

274

|  | *History* | *Heritage* |
|---|---|---|
| *History as a goal* (emphasis on *meta-issues*)[35] | C.2(ii), C.3(iii)(?), C.4 E.1, E.2, E.3 | A.3; B.1, B.2 D.1 |
| *History as a tool* (emphasis on *inner-issues*)[35] | A.3; C.1, C.3(i), C.4, C.5 D.2, D.3 | A.1, A.2 B.1, B.2(?) C.2(i), C.3(ii) E.3 |

In this way, a classification scheme is obtained, in which the two conceptual dipoles act as a "magnifying lens", either requesting a more complete description of each *why* and *how*, or/and providing a clearer orientation of the way each *why* and *how* could be implemented. This is explained below by means of some examples and further comments.

Items appearing in more than one cell are shaded and those placed with reserve appear with an interrogation mark[36]. This suggests that

(a) the *whys* are not **irreducible** with respect to the two classification dipoles, but consist of **simpler** elements; hence, either they should be further analysed and/or be more sharply described; or,

(b) the appearance of the same *why* in two different cells exhibits a possible *anchoring* in the sense described in section 2.1.

For instance, C.3 is a good example for (a): It appears in three cells. According to section 3.1, it concerns the involvement and/or acquisition of awareness of the *creative process of "doing mathematics"*, by (i) tackling problems in historical context; (ii) enriching mathematical literacy; (iii) appreciating the nature of mathematics. Clearly, (i) is related to the *History – History as a tool* cell and (ii) to the *Heritage – History as a tool* cell (one is being involved into doing mathematics in historical context and in so doing improves his

---

[35]Relating *History-as-a-goal* and *History-as-a-tool* with *inner-issues* and *meta-issues*, respectively is done keeping in mind the possible cross-interrelations mentioned at the end of §2.1!

mathematical literacy, thus becoming more aware of what this intellectual activity has been through the ages). However, (iii) cannot be classified without ambiguity. Although it should be placed in the *History as a goal* row, this can be accomplished by looking either at a subject's *History*, or *Heritage*. This argument needs further clarification.

Similarly, A.3 appears in the *History – History as a tool* cell because working on historically driven relevant questions and problems posed and or motivated by other disciplines, a good working knowledge on mathematical issues may be acquired. This process could indirectly (i.e. not intentionally designed and implemented) lead the learner to appreciate the deep interdisciplinarity of specific mathematical subjects and that research in a certain domain profits from and affects other domains, clearly important issues at a *meta-perspective* level.

Maybe this is a general characteristic of all the *whys* that appear in more than one cells, reflected in the fact that, according to Table 1, all these *whys* appear in different columns. In this connection B.1 & C.4 potentially express the possibility of *anchoring* of *meta-issues* into corresponding *in-issues*, whereas E.3 and maybe B.2 seems to suggest an **inverse** *anchoring* of *in-issues* into corresponding *meta-issues* (cf. last paragraph of §2.1)[37]. This point however, needs to be further explored in the context of specific examples (see e.g. section 4).

These examples[38], suggest that some of the ICMI *whys* could be further sharpened, so that they are decomposed into "irreducible" arguments, in the sense that they fall into only one

---

[36]This convention is applied to all tables in this paper.

[37]This point needs more clarification, especially for B.2: In connection with the description of B.2 in §3.1, think for instance the advantages and disadvantages of the older formulation of differential geometry by using indices to describe tensor quantities, compared to its modern, often extremely compact coordinate-independent formulation. This is a very good example if one wants to contrast different symbolic languages and the role played by their procedural effectiveness and/or semiotic virtue. However, this is impossible without first acquiring a good working knowledge of both formalisms in the context of specific examples; this means, mastering much of differential geometry and tensor calculus!

cell of Table 1, or express interconnections reflected in an *anchoring* process. But this remains to be shown and further work is needed. In this way, a **finer** classification of the *whys* becomes possible, to the extent that the classification dipoles have been determined as sharply as possible, of course. Clearly, this presupposes the detailed study of the *whys* and each *conceptual dipole* in the context of specific examples (see section 5; cf. Tzanakis & Thomaidis, to appear). In addition, this and the following tables can be considered in relation to the **target population** to whom they are addressed; mathematics teachers; curriculum designers, producers of didactical material; mathematics teachers' trainers and advisors. That is, they can be useful to specify which entries are better suited to whom.

*Table 2*: The classification of the ICMI *hows* (cf. §3.2)

|  | *History* | *Heritage* |
|---|---|---|
| *History as a goal* (emphasis on *meta-issues*)[35] | Direct historical information: 3.2.1 Intrinsic awareness: 3.2.3(a)(ii) Extrinsic awareness: 3.2.3(b)(ii) | Direct historical information: 3.2.1 Extrinsic awareness: 3.2.3(b) (i) (iii) (iv) |
| *History as a tool* (emphasis on *inner-issues*)[35] | Intrinsic awareness: 3.2.3(a)(i), (iii) Learning mathematical topics (explicit use of history): 3.2.2 | Learning mathematical topics (implicit use of history): 3.2.2 |

The labels (i) to (iv) in this table refer to the corresponding items in §3.2.3 and provide another example of the "irreducibility" idea mentioned above; the development of mathematical awareness has been described more clearly in the ICMI Study volume, which in turn, allows for a clearer classification of its various aspects[39]. The same holds for learning mathematical topics by following an approach inspired by history, either explicitly or

---

[38]Others could be similarly analyzed, but lack of space does not allow for that.

[39]E.g. items 3.2.3(b) definitely concern *meta-issues*; (ii) requires to consider issues in their historical context of a particular period; (i) (iii) (iv) touch upon issues that connect the present to the past and are likely to be based on a *heritage*-like approach, though this is clearer for (ii) & (iv) than for (i). On the other hand, 3.2.3(a)(i), (iii) mainly concern specific historical examples whose mathematical content should be explored in a way that awareness of *meta-issues* may be developed (or be *anchored* there, in the sense of §2.1).

implicitly (cf. the last paragraph of section 2 on the distinction between "explicit" and "implicit" use of history). However, this is not so for learning history by providing direct historical information, which is only briefly described. Nevertheless, it would be possible to classify the different items in §§3.2.1-3.2.3 by considering their implementation in particular cases, so that the irreducible parts of the *hows* are better displayed and understood.

*Table 3*: The classification of ***hows*** proposed by Jankvist's (cf. §3.3)

| | *History* | *Heritage* |
|---|---|---|
| **History as a goal (emphasis on meta-issues)**[35] | *Modules* approaches: 3.3.2 | *Illumination* approaches: 3.3.1 *History-based* approaches(?): 3.3.3 |
| **History as a tool (emphasis on inner-issues)**[35] | *Illumination* approaches: 3.3.1 *Modules* approaches: 3.3.2 | *History-based* approaches: 3.3.3 |

The table above suggests that the three broad types of approaches introduced by Jankvist constitute an interesting and promising identification of broad categories of approaches, which can be analysed into more sharply described approaches, by means of specific examples (this is already apparent in the description given in Jankvist 2009c, §6). Due to space limitations, a detailed classification of the items in §§3.2.1-3.2.3 will not be done here.

## 5. AN EXAMPLE

The character of this paper is clearly theoretical. The basic ideas should be further elaborated by analysing specific examples. In order to illustrate the general classification presented briefly in section 4, we apply it in a specific example. However, this should be considered only as a first step towards a better understanding and sharpening of this classification and many more examples should be analysed in much more detail before the key idea of this paper could be considered sufficiently well founded and a useful tool for didactical research and applications.

## 5.1 Logarithmic notions

We consider the chain, or rather network of interrelated notions

*Power of numbers – exponent – logarithm & (logarithmic) base – exponential function – logarithmic function*

In teaching and learning these concepts, or any other mathematical subject, questions are often raised, whose answer presupposes both teacher's knowledge of the historical development and (meta)knowledge of how to deal with this historical knowledge in the classroom. The range and depth of this knowledge is closely related not only to the subject itself and the questions raised, but also to the learners' age, the level of instruction, the didactical aims and how the subject fits into the curriculum. These factors determine to a large extent the relation between HM and ME in the specific subject to be taught and learnt.

Here, we will approach such **possible questions** as specific **tasks** to be accomplished didactically, thus reflecting the *whys* of section 3[40] and corresponding answers as an outline of possible **approaches** to do that, thus reflecting the *hows* in that section[39]. In this way we will illustrate the fitting of the *whys & hows* into the 2X2 classification scheme of section 4. However, we would like to emphasize that the formulation of the questions below and the outline of their answer constitute only one **choice** among various possibilities and that different questions and/or different answers could be provided, depending on the factors mentioned in the previous paragraph. We simply aim to indicate how the suggested classification is applied in a particular case, **once** we have specified the questions raised and their possible answers.

For the reader's convenience, we indicate *questions* by • and *answers* by □ and we refer

---

[40] Cf. the first paragraph of section 1.

directly to their relation with the numbering of the *whys & hows* in section 3, respectively. We note that the questions and answers as they appear below have been formulated on the basis of the existing literature of the last 25 years or so, on a historically motivated teaching of logarithms (e.g. Katz 1986, 1995; Thomaidis 1987; Toumasis 1993; Fauvel 1995; van Maanen 1997; Clark 2006; Stein 2006; Barbin et al. (2006); Panagiotou 2011).

The concept of logarithm, which is usually introduced· for the first time to upper high school students (16-17 years old) in the context of teaching of the exponential and logarithmic functions, raises several questions, already from its definition.

- $q_1$: Why the exponent of a number's power is called the *logarithm* of that power relative to that number as a base? (A1, B2)

Any attempt to answer this question cannot avoid direct or indirect reference to the HM:

□$a_1$: When the concept of logarithm was invented in the early $17^{th}$ century, the modern exponential notation of powers did not exist yet, hence the logarithm was **not** defined as the exponent of a power relative to a base. (3.2.1, 3.3.1)

This answer immediately leads to new questions:

- $q_{2-1}$: Why was the concept of logarithm introduced at all? (C1)

- $q_{2-2}$: How was the logarithm defined originally? (A2)

- $q_{2-3}$: Why was the **term** "logarithm" used/introduced? (A2)

To answer such questions, sufficiently deep historical knowledge is required, hence the choice and use of appropriate historical sources is raised (a book on the HM[41], a treatise on

---

[41] E.g. Boyer 1968, Katz 1998.

the history of logarithms[42], relevant original sources[43] etc) and whether the answers to be used will be accompanied by historical references or not, e.g.

□a$_2$: The answer could be limited to a modern, strictly mathematical framework, if the correspondence between an arithmetical and geometrical progression is used, from which it can be easily explained the usefulness of the logarithm as a tool to simplify numerical calculation and the etymology of the word "logarithm"[44]. History is used implicitly; no traces are left in the classroom. (3.2.2-implicit use of HM, 3.3.3)

This answer raises new questions:

• q$_3$: How can this model be useful in practice? (C4)

□a$_3$: The answer to this question too can also be restricted in a modern, strictly mathematical context, explaining technically how to construct a 'dense" geometrical progression and a corresponding "dense" arithmetical one. (3.2.2-implicit use of HM, 3.3.3)

Though it may be explained that at that era numerical calculations with many-digit numbers was time-consuming, tedious job – hence there was vivid interest to exploit any idea on their simplification, like the correspondence of arithmetic and geometric progressions -, it is readily appreciated that the same holds for the construction of two practically useful "dense" progressions. Hence, new questions arise naturally:

---

[42] Naux 1966, 1971, Knott 1915.

[43] Nepair 1616/1969.

[44] *Logarithm*: From the Greek *logos* (ratio) and *arithmos* (number); every term of the arithmetic progression shows the number (multitude) of the ratios of successive terms of the geometrical progression up to the corresponding term of this progression. E.g., 6 in the arithmetic progression 0, 1, 2, 3, 4, 5, 6 (which equals $log_2 64$) indicates that to get the term 64 of the geometric progression 1, 2, 4, 8, 16, 32, 64 starting from its first term, six ratios are inserted, namely, 2:1, 4:2, 8:4, 16:8, 32:16, 64:32.

- $q_{4-1}$: What was the motivation to get involved in the construction of two practically useful "dense" progressions? (C1)

- $q_{4-2}$: Is it really true that this problem is of such great mathematical interest that it is worthwhile to get involved in it, ignoring the practical difficulties inherent to its solution?[45] (C3)

□$a_4$: The problem of simplifying tedious numerical calculations was posed in the context of dealing with economic exchanges (e.g. using tables of interests) and astronomical measurements (use of trigonometric tables). Therefore, answering these questions leads outside mathematics, and refers directly to the relation of Mathematics to other disciplines in that historical period, which evidently, cannot be considered solely in mathematical terms. (3.2.1, 3.3.2)

This leads to new questions.

- $q_5$: How did people cope with elaborated calculations in various disciplines, including Mathematics? (A3, E1, E2)

□$a_5$: The answer will definitely refer to the reciprocal relation between Mathematics and other disciplines which are based/using Mathematics, elaborating on the use of groups of calculators paid by financial institutions or observatories, the development of new methods to construct numerical tables of higher accuracy, the invention of tricky methods to simplify calculation (like the trick of "prosthaphairesis"[46]) etc. (3.2.3(bii), 3.3.2)

---

[45]Cf. "Fermat's Last Theorem" that was formulated a few years later.

- $q_6$: Who was the first to construct logarithmic tables, and how did he achieve that?

(C4)

☐$a_6$ This is a complex question the answer to which includes references to the not rare fact of "independently made similar or identical discoveries/inventions" in mathematics; the arithmetical background of Bürgi's *"red numbers"* and the kinematical-trigonometric background of Napier's *"logarithms"*. A historically complete answer should stress the essential differences between these two approaches; that is, the fact that they essentially concern different conceptions of "logarithmic" notion, though of course they have as a common starting point the correspondence between arithmetic and geometric progression and serve the same purpose. In addition, it is important from a didactical point of view to study biographical elements of the scientists involved in the invention of logarithms. (3.2.1, 3.3.2)

Though the logarithmic tables solved the crucial – at that time - problem faced by those involved in complex arithmetical calculations and their use was adopted with enthusiasm, nowadays their use has banished and what remains is the concept of logarithmic function relative to a given base. Given that at the time of the invention of logarithms both the idea of a base and the function concept were nonexistent, the following questions naturally arise:

- $q_{7-1}$: What was the reason and/or questions that led to the connection of the concept of logarithm with those of exponent and base? (B1)

- $q_{7-2}$: What was the reason and/or questions that led to the connection of the concept

---

[46]From the Greek *Prosthesis* = addition and *aphaeresis* = subtraction; a trick based on trigonometric relations to transform the product of two trigonometric numbers into sums of such numbers (e.g. Smith 1959, pp.455-472; Barbin et al 2006, ch.II; Thomaidis 1987, §3).

of logarithm with the concept of a "logarithmic function"? (B1, C1)

□$a_7$: Answers to these questions could be given in a strictly mathematical context with no, or only limited reference to the historical development. However, given that this historical development is closer to contemporary Mathematics, many of the original texts are suitable for reading and discussion in the classroom. In this way, the integration of the HM could be more direct, efficient and demanding, contributing to the development of a classroom discourse, resembling a "community of researchers" which explores mathematical questions and problems and deepens its understanding of Mathematics by studying historical texts. (3.2.2-explicit use of HM, 3.3.2)

The above questions and answers can be rearranged and seen in the context of the classification scheme of section 4 in the following table.

*Table 4*: The classification of questions and answers on logarithms

| | *History* | | *Heritage* | |
|---|---|---|---|---|
| ***History as a goal* (emphasis on *meta-issues*)** | $q_5$ (E1, E2) | $a_4$ (3.2.1, 3.3.2)<br>$a_5$ (3.2.3, 3.3.2)<br>$a_6$ (3.2.1, 3.3.2) | $q_{7-1}$ (B1)<br>$q_{7-2}$ (B1) | $a_1$ (3.2.1, 3.3.1) |
| ***History as a tool* (emphasis on *inner-issues*)** | $q_{2-1}$ (C1)<br>$q_3$ (C4)<br>$q_{4-1}$ (C1)<br>$q_{4-2}$ (C3)<br>$q_5$ (A3)<br>$q_6$ (C4)<br>$q_{7-2}$ (C1) | $a_5$ (3.2.3, 3.3.2)<br>$a_6$ (3.2.1, 3.3.2)<br>$a_7$ (3.2.2, 3.3.2) | $q_1$ (A1, B2)<br>$q_{2-2}$ (A2)<br>$q_{2-3}$ (A2)<br><br>$q_6$ (A2) | $a_2$ (3.2.2, 3.3.3)<br><br>$a_3$ (3.2.2, 3.3.3) |

*Remarks*: (a) If the questions and answers were formulated more sharply and in more detail, we expect that they would not appear in more than one cell each.

(b) The arrows link questions, which refer to meta-issues with answers connected to inner issues; this illustrates the *anchoring* process as described in section 2.

## 5.2 Comments

Caution should be taken in answering the questions presented in the previous subsection in actual implementations. The discussion of the associated issues should be properly contextualized, or at least should avoid the (tacit) introduction of elements that followed the period to which the questions refer, assigning knowledge of them to the key figures of the given period. This will certainly imply a **distorted** view of what really happened (i.e. of *History* in the sense of section 2) and on how we got here (i.e. of *Heritage* in the sense of section 2). ".... the intervening achievements are more *of a hindrance* than a help. They can easily lead to distortions... and in any case they yield only historical assessments of a theory in terms of some later standpoint *and thus illuminate that standpoint rather than the theory itself*. When analysing the work of a mathematician we should examine his *ignorance situation* as well as his knowledge situation... *we* must try to detect these ignorances in order to *reconstruct his unclarities* as well as we can. Then we can trace the limits as well as the scope of his work and hence begin to assess the achievements of his successors."[47]

The above remarks may be illustrated by some comments on the way logarithmic notions are often presented in modern treatises on the HM. For instance, in order to present Napier's definition of logarithm, avoiding lengthy explanations, some authors exploit the opportunity for a concise presentation offered by introducing retrospectively modern notions and tools of calculus, thus giving a rather misleading image of the historical process. Though, it is understandable that there is a strong motivation for that, it should be kept in mind that such an approach leads to a distorted view of history that will confuse the learner and make

---

[47]Grattan-Guinness 1973, pp.447-448. "....The reverse situation may also apply; *he* may have known theories which *we* do not know or at least have forgotten, because— either deservedly or undeservedly—they have lost status in the intervening period. But if he seems to have found them important, then we must also." (*op.cit.*).

important achievements of the past look like modern school exercises. An indicative example in this connection is given by Edwards 1979, where under the characteristic title "*Napier's Curious Definition*" writes:

"This definition involves two points moving along two different lines. The first point $P$ starts at the initial point $P_0$ of a segment $P_0Q$ of length $10^7$, with initial speed $10^7$, and moves towards $O$, with its speed decreasing in such a way that it always equals the remaining distance $PO$. The second point $L$ starts at the initial point $L_0$ of a half-line, and moves to the right with constant speed $10^7$. ... Napier then defines the segment y = $L_0L$ to be the *logarithm* of the segment $x = PO$. ..

It is **informative** to explore this somewhat obscure definition in terms of what we now call *natural* logarithms – log$x$ being the power to which e must be raised to give $x$. In calculus notation, the motion of the point $P$ is described by the differential equation

$$\frac{dx}{dt} = -x$$

with the initial condition $x(0) = 10^7$, whose solution is log$x = -t + \log 10^7$

or $t = \log \dfrac{10^7}{x}$.

The motion of the point $L$ is therefore given by $y = 10^7 x = 10^7 \log \dfrac{10^7}{x}$.

If we write $y$ = Nog$x$ for Napier's logarithm of $x$, we therefore see that the relation between Nog$x$ and the natural logarithm log$x$ is given by $\text{Nog}x = 10^7 \log \dfrac{10^7}{x}$."

(Edwards 1979, p.149, our emphasis)

Although this is a mathematically elegant presentation, for whom and in which sense is it "informative" and why and in which respect is Napier's definition "curious"? Put it another way, what type of answer does it provide to the questions raised in the previous subsection about how were logarithms originally defined and what were the means used to produce the first logarithmic tables (cf. $q_{2-2}$, $q_{4-1}$, $q_{4-2}$)? Napier faced the problem of calculating an enormous number of terms of an arithmetic and a geometric progression that were in 1-1 correspondence, a task superseding the arithmetical and algebraic tools available at the time, since any operationally useful and efficient conception of the arithmetic continuum and any tools of the calculus were absent, of course. Therefore, he needed a definition that would allow the determination of appropriate inequalities to put bounds to the unknown terms of the arithmetic progressions (the *logarithms*), in terms of the known terms of the geometric progression. His ingenious solution was based on imagining the two progressions as representations of the parallel motion of two points and deducing the sought inequalities from simple properties of motion (cf. $a_3$, $a_6$). Modern differential calculus offers a definitely concise and rigorous mathematical framework, which, being based on the arithmetical continuum as we know it today, is completely foreign to Napier's line of thought and the more general conceptual framework where it was placed, however; in particular it does not offer answers to the questions raised when attempting to deal with the didactical problems of teaching and learning logarithmic notions (cf. $q_{7-2}$) in a way that will help students to understand both the mathematics and the motivations and problems that drove their introduction (cf. $a_7$).

"This is a major matter. Later notions are *not* to be ignored... Instead, when studying the history of [a notion] N0, by all means *recognize* the place of later notions N1, N2, . . . but *avoid* feeding them back into N0 itself. For if that does happen, the novelties that attended the emergence of N1, N2, . . . will not be registered... In such situations not only is the history of N0 messed up but also that of the intruding successors, since their *absence* before introduction is not registered. [Such an approach] ....not only distorts [the pioneers' work that later led to a new theory] but also muddies the (later) emergence of that theory itself by failing to note its absence in [their work].

Sometimes such modernizations are useful to save space on notations, say, or to summarize mathematical relationships, but the ahistorical character should be stressed: "in terms of [that] theory (which [the pioneers] did not have), [their] theorem ...may be stated thus: . . . ." ...

By contrast, when studying the heritage of N0, by all means feed back N1, N2, . . . , to create new versions; it may be clarified by such procedures...... But it is only negative feedback, unhelpful for both history and heritage, to attack a historical figure for having found only naïve or limited versions of a theory that, as his innovations, helped to lead to the later versions upon which the attack is based."[48]

The above remarks and quotations clearly indicate the validity of the claims made in §2.2. that *History* and *Heritage* are complementary (in the sense of §2.2), hence, both legitimate and indispensable and that both can profitably, though cautiously, be used in ME.

## 6. CONCLUDING REMARKS

As already mentioned, this paper is mainly theoretical in nature, and further examples should be analysed to explore these classifications more thoroughly and to show how they can be useful in actual implementations. We believe this is a promising line of inquiry and expect that by analysing specific examples in this context, the arguments for and approaches of integrating HM into ME will be sharpened and the importance and possible interrelation of the conceptual dipoles described here will be better revealed and understood. In order to explore the classification schemes in this paper in a wide spectrum of situations, one possibility would be to consider in the context of this paper the many specific examples of possible implementations briefly discussed in Tzanakis & Arcavi 2000, §7.4, and Siu 2000, which cover a wide spectrum of subjects, and instructional levels.

Additionally, a number of relevant issues that are in principle interesting but need to be further elaborated were only touched upon in this paper:

- To classify the ICMI *whys* according to the target population, namely, mathematics teachers, curriculum designers, producers of didactical material, mathematics teachers' trainers and advisors. This is expected to lead to a finer classification of the ICMI *whys*, seen as tasks to be accomplished by the corresponding target population; in other words, to understand and describe better, which tasks are better suited to whom and for what purpose.

- To explore further in the context of particular cases, the phenomenon of *anchoring* of *meta-issues* in specific *inner-issues* as described in Jankvist (in press, *JRME*) and Jankvist & Kjeldsen (in press). Is there an inverse phenomenon? (see also end of §2.1).

- To explore possible interconnections of the ideas in this paper, in particular the *whys* & *hows* as well as the conceptual dipoles of section 2, with Niss' mathematical *competencies* (Niss 2003, 2004) (see §3.1, footnote 28 and Jankvist (to appear)).

- To enrich the theoretical constructs in this paper, by exploring possible links to relative constructs, in particular those considered in Kjeldsen (to appear).

### *REFERENCES*

-Barbin, E., 1997, "Histoire et Enseignement des Mathématiques. Pourquoi? Comment? », *Bulletin AMQ,* Montréal, **37** (1), 20-25.

-Barbin, E. et al. (eds), 2006, *Histoire de logarithmes*, Ellipses Edition Marketing S.A., Paris.

---

[48]Grattan-Guinness 2004a, §3.5.

-Bohr, N., 1934, *Atomic theory and the description of nature*, Cambridge University Press, ch.2.

-Bohr, N., 1958, *Atomic Physics and Human Knowledge*, New York: J. Wiley, ch.2 & 3.

-Boyé A., Demattè A., Lakoma E., Tzanakis, to appear, "The history of mathematics in school textbooks", in E. Barbin, M. Kronfellner, C. Tzanakis (eds), Proceedings of ESU 6, Vienna University of Technology, Austria.

-Boyer, C. B., 1968, *A History of Mathematics*. J. Wiley & Sons, New York.

-Clark, K., 2006, *Investigating teachers' experiences with the history of logarithms: A collection of five case studies*, Doctoral dissertation, Faculty of the Graduate School, University of Maryland, College Park, USA.

-Edwards, G. H., 1979, *The Historical Development of the Calculus*. Springer-Verlag, New York.

-Fauvel, J., 1995, "Revisiting the History of Logarithms", in F. Swetz *et al* (eds.), *Learn from the Masters*, The Mathematical Association of America, pp.39–48.

-Fauvel J. & van Maanen J. (eds), 2000, *History in Mathematics Education: The ICMI Study*, "New ICMI Study Series", vol.**6,** Kluwer Academic Publishers.

-Freudenthal, H., 1983, *Didactical Phenomenology of Mathematical Structures*, Boston/Lancaster: Dordrecht

-Furinghetti, F., Jahnke N.H., van Maanen J., 2006, *Mini-Workshop on Studying Original Sources in Mathematics Education*, Mathematisches Forschungsinstitut Oberwolfach No22, pp.1285-1318.

-Grattan-Guiness, I., 1973, "Not from nowhere. History and philosophy behind mathematical education", *Int. J. Math. Edu. in Science and Technology*, **4**, 421-453.

-Grattan-Guiness, I., 2004a, "The mathematics of the past: distinguishing its history form our heritage" *Historia Mathematica*, **31**, 163-185.

-Grattan-Guiness, I., 2004b, "History or Heritage? An important distinction in mathematics for mathematics education", in G. van Brummelen & M. Kinyon (eds), *Mathematics and the historian's craft*, Springer, pp.7-21. Also published in *Am. Math. Monthly*, **111**(1), 2004, 1-12.

-Jammer M., 1974, *The philosophy of Quantum Mechanics*, J. Wiley, New York.

-Jankvist, U.Th., 2009a, *Using history as a "goal" in Mathematics Education"*, PhD Thesis, Roskilde University, Denmark, chs.2 & 8.

-Jankvist, U.Th., 2009b, "On empirical research in the field of using history in mathematics education", *Revista Latinoamericana de Investigatión en Matemática Educativa*, **12** (1), 67-101, particularly§1.1

-Jankvist, U.Th., 2009c, "A categorization of the 'whys' and 'hows' of using history in mathematics education", *Educ. Stud. Math*, **71** (3), 235-261.

-Jankvist, U.Th., 2009d, "History of modern applied mathematics in mathematics education", *For the learning of Mathematics*, **29**(1), 8-13.

-Jankvist, U.Th, (in press), "Anchoring students' metaperspective discussion of history in mathematics", *JRME*.

-Jankvist, U.Th, (to appear), "Designing teaching modules on the history, application and philosophy of mathematics", *Proc. 7th European Conference on Research in Mathematics Education* (CERME 7), Rzeszów, Poland, 9-13/2/2011.

-Jankvist, U.Th & Kjeldsen T.H. (in press), "New Avenues for History in Mathematics Education: Mathematical competencies and anchoring", *Science & Education*.

-Katz, V., 1986, "Using History in Teaching Mathematics", *For the Learning of Mathematics* **6**(3), 13–19.

-Katz, V., 1995, Napier's Logarithms Adapted for Today's Classroom. In F. Swetz *et al* (eds), *Learn from the Masters*, The Mathematical Association of America, pp.49–55.

-Katz, V. J., 1998, *A History of Mathematics. An Introduction*, 2nd edition, Addison-Wesley, New York.

-Kjeldsen T.H. (to appear), "Uses of history in mathematics education: development of learning strategies and historical awareness", *Proc. 7th European Conference on Research in Mathematics Education* (CERME 7), Rzeszów, Poland, 9-13/2/2011.

-Knott, C.G. (ed.), 1915, *Napier Tercentenary Memorial Volume*. Longmans, Green & Co, London.

-Naux, Ch., 1966, *Histoire des Logarithmes*, Tome I : La découverte des logarithmes et le calcul des premières tables. Blanchard, Paris.

-Naux, Ch., 1971, *Histoire des Logarithmes*, Tome II : La promotion des logarithmes au rang de valeur analytique. Blanchard, Paris.

-Nepair, John1616/1969, *A Description of the Admirable Table of Logarithmes*. Printed by Nicholas Okes, London, 1616. Reprint: Da Capo Press, Amsterdam, 1969

-Niss, M. 2003, "Mathematical competencies and the learning of mathematics: The Danish KOM Project", in A. Gagatsis & S. Papastavridis (eds), *Proc. 3rd Mediterranean Conference on mathematics Education*, The Greek Mathematical Society, Athens, pp.115-124.

-Niss, M. 2004, "The Danish 'KOM' project and possible consequences for teacher education. In R. Strässer, G. Brandell, B. Grevholm & O. Helenius (eds.) *Educating for the Future*, The Royal Swedish Academy, Göteborg, pp.179-190.

-Panagiotou, E., 2011, "Using History to Teach Mathematics: The Case of Logarithms", *Science & Education* **20**, 1–35.

-Pauli, W. 1950/1994, "The philospophical significance of the idea of complementarity", in

292

C.P. Enz & K. von Meyenn (eds) *Writings on physics and philosophy*, Springer, Berlin, pp.35-42.

-Rogers, L., 2009, "History, Heritage and the UK Mathematics Classroom", in V. Durand-Guerrier, S. Soury-Lavergne & F. Arzarello (eds), *Proc. of the 6th Congress of the European Society for Research in mathematics Education (CERME 6)*, pp.2781-2790.

-Rowe, D.E., 1996, "New trends and old images in the History of Mathematics", in R. Calinger (ed.), *Vita Mathematica: Historical research and integration with teaching*, The Mathematical Association of America, Washington DC, pp.3-16.

-Schubring, G. 2008, "The debate on a "geometric algebra" and methodological implications", in R. Cantoral, F. Fasanelli, A. Garciadiego, R. Stein, C. Tzanakis (eds), *Proceedings of HPM 2008: The HPM Satellite Meeting of ICME 11*, Mexico City, Mexico (edition in cd form).

-Siu, M. K. 2000[49], "Historical support for particular subjects", in *History in Mathematics Education: The ICMI study*, J. Fauvel & J. van Maanen (eds.), Dordrecht: Kluwer, pp.241-290.

-Siu, M. K. 2006, "No, I don't use history of mathematics in my class. Why?", in F. Furinghetti, S. Kaijer & C. Tzanakis [eds.], Proceedings of HPM 2004 & ESU 4 (revised edition), University of Crete, Greece, pp.268-277.

-Smith, D.E., 1959, *A Source Book in Mathematics*, Dover, New York.

-Stein, B., 2006, "The fascinating history of logarithms", in F. Furinghetti, S. Kaijser & C. Tzanakis (eds), *Proceedings of HPM 2004 & ESU 4* revised edition, University of Crete, Greece, pp.134–151.

-Thomaidis, Y., 1987, *Genesis and development of logarithmic concepts*. Study Group on the

---

[49]This is a collective work of many authors in the context of the ICMI Study directed and edited by Fauvel & van Maanen (2000); for brevity, only the convenor(s) are mentioned.

History of Mathematics, no4, Thessaloniki, Greece (in Greek).

-Toumasis, Ch., 1993, "Teaching Logarithms via Their History", *School Science and Mathematics* **93**(8), 428–434.

-Tzanakis, C. 1999, "Mathematical Physics and Physical Mathematics: A historical approach to didactical aspects of their relation", in P. Radelet-de Grave & D. Janssens (eds), *Proceedings of the 3rd European Summer University on the History and Epistemology in Mathematics Education*, Leuven, Belgium, vol.I, pp.65-80

-Tzanakis, C. 2009, "Benefits from the interrelations between the history of mathematics and mathematics education: the pros and cons on the basis of the international experience" in Greek Society of the Didactics of Mathematics (eds), *The value of the history of mathematics in mathematics education*, (in Greek) Ziti Publications, Thessaloniki, Greece, pp.7-39

-Tzanakis, C. & Arcavi, A., 2000[48], "Integrating history of mathematics in the classroom: an analytic survey", in *History in Mathematics Education: The ICMI study*, J. Fauvel & J. van Maanen (eds.), Dordrecht: Kluwer, pp.201-240.

-Tzanakis C. & Thomaidis Y., to appear, "Complementary routes to integrate history in mathematics education: In search of an appropriate theoretical framework", in E. Barbin, M. Kronfellner, C. Tzanakis (eds), Proceedings of ESU 6, Vienna University of Technology, Austria.

-van Maanen, J., 1997, "New Maths may profit from Old Methods", *For the Learning of Mathematics* **17**(2), 39–46.

# A first attempt to identify and classify empirical studies on history in mathematics education[1]

Uffe Thomas Jankvist
University of Southern Denmark[2]

## Abstract

The field of history in mathematics education is beginning to pay more attention to empirical studies than has been the case earlier. Through a general survey of relevant literature this paper identifies the shifts in focus of this field as having gone from more advocating studies over describing studies to actual research studies, lately with an increase in the empirical research studies. Unlike more traditional surveys, the idea of this one is not to discuss the identified empirical studies in detail, nor to provide long arguments for classifying studies in one way or another. The idea is to provide a list of the empirical studies which has been identified in the survey, a comprehensive but of course not an exhausting list, and then a rough participation of these studies according to an interpretation of what appears to be their overall objective. The motives behind this approach is first and foremost to obtain an idea of where the emphasis of the empirical studies is located, both presently and in the past, and thereby also point out future domains and research questions. Yet a motive is to provide researchers in the field, newcomers as well as more established ones, with a reference list of studies that they may resort to upon engaging in similar and related empirical studies themselves. In this sense, I believe that such a list can provide both background and inspiration for future studies.

**Keywords:** History in mathematics education; empirical studies; history as a goal; history as a tool;
cognitive tool; motivating tool; historical parallelism.

## Introduction

In this paper I address the question of using history of mathematics in mathematics education from a research perspective. I first describe the subfield of mathematics education that considers the use of history in mathematics teaching and learning, both in terms of its academic fora and the field's present state, the latter done amongst other by quoting from interviews with experienced researchers in the field. Having done that I provide a general survey of samples of the field's literature in order to illustrate the different kinds of objectives of the field's research as well as to argue that the field's object of study is slowly moving towards a more empirically oriented type of research. This leads to the paper's main purpose, namely, as indicated by the title, to try to identify and classify empirical studies on history in mathematics education. The approach taken to classify the identified studies rests on a previously published

---

[1] The paper is an extension and update of parts of a larger study (Jankvist; 2009e).

[2] Department of Mathematics and Computer Science (IMADA), Campusvej 55, 5230 Odense M, Denmark, e-mail: utj@imada.sdu.dk

categorization of the arguments and purposes usually given for using history in mathematics education (the so-called 'whys') (Jankvist; 2009a). I end the paper by discussing possible perspectives for future empirical research in the field of history in mathematics education, therein also drawing on some of the 'trends' spotted in some of the empirical studies found.

# The field of history in mathematics education

When describing the field of using history in mathematics education, including its present state, the sociological setting with conferences, meetings, etc. (as well as the field's object of study) must be described. I begin with the sociological setting.

## The academic fora

History of mathematics in the teaching and learning of mathematics is an area which has attracted an increasing amount of interest within didactics of mathematics. This has given rise to several publications, newsletters, conferences more or less dedicated to this topic, and working groups at more general conferences on mathematics education. *The International Study Group on the Relations between the History and Pedagogy of Mathematics* (HPM) can be traced back to the second *International Congress on Mathematical Education* (ICME) in 1972, when it began as a working group. At the third ICME of 1976, HPM was set up as an affiliated study group to the *International Commission on Mathematical Instruction* (ICMI) together with the presently more widely known *International Group for the Psychology of Mathematics Education* (PME) (see Fasanelli & Fauvel, 2007). Every fourth year HPM holds a satellite conference to the ICME, and three times a year HPM publishes a newsletter. Also, HPM has an Americas section which as of now have two smaller meetings a year, one in Washington DC in March and another in Pasadena in October. At the ICME itself, there is a *Topic Study Group* (TSG) on the role of history of mathematics in mathematics education. Another conference dedicated to the history of mathematics and mathematics education is the *European Summer University on the History and Epistemology in Mathematics Education* (ESU), which is a more recent initiative taken by the French Mathematics Education community (IREM) in 1993 (see Barbin et al., 2007). ESU has until now been held every third year. Also the *International Colloquium on the Didactics of Mathematics* (ICDM), which is usually held every second year, has history in mathematics education as one of its themes. A new initiative is the working group on *History in Mathematics Education*, a group set up for the 6th *Congress of the European Society for Research in Mathematics Education* (CERME) in Lyon 2009.[3]

Needless to say, all these conferences, working groups, etc. all produce proceedings in one form or another, adding substantially to the amount of papers on history in mathematics education. The proceedings from the combined conference of HPM2004 and ESU4 held in Uppsala as a satellite meeting to ICME10, for instance, includes a total of 78 papers. But also in more general journals on mathematics education it seems as if the involvement of history is becoming increasingly prominent, especially in journals

---

[3] At CERME-6 this group had the following name: *The Role of History of Mathematics in Mathematics Education:Theory and Research*, but the name was changed for CERME-7.

such as *Educational Studies in Mathematics* (ESM) and *For the Learning of Mathematics* (FLM) of which special issues on history in mathematics education have appeared in 2007 and 1991, respectively. A journal such as *Zentralblatt für Didaktik der Mathematik* (ZDM)[4] has had several papers on history in past times (Gulikers & Blom; 2001), although the number has decreased during the last decade (Jankvist; 2009a). In 2004 the *Mediterranean Journal for Research in Mathematics Education* (MJRME) published a double special issue on the role of history in mathematics education consisting of the papers from the TSG17 at ICME10. But also entire books on the subject are available. Examples are: Fauvel (1990); Swetz et al. (1995); Jahnke et al. (1996); Calinger (1996); Katz (2000);[5] Shell-Gellasch & Jardine (2005); and most importantly Fauvel & van Maanen (2000), the ICMI Study on history in mathematics education. To the best of my knowledge, this ICMI Study is to date the most comprehensive sample on the topic.

## Present state of the field

But what then is the present state of the field? One way of trying to answer this question is to ask some of the more experienced researchers who have been in the field for some time. As a part of my postgraduate studies, I had the opportunity to interview some of the authors of the ICMI Study, who are also deeply involved in the HPM society, and ask them about their professional as well as personal view of things. Their insightful opinions on where and how to direct further research on history in mathematics education draw a perspective picture of this field today. (All quotations below are brought with their acceptance and have partly been displayed in Jankvist (2009e) also.)

In June, 2007 I met Abraham Arcavi, the co-author of chapter 7 in the ICMI Study (Tzanakis & Arcavi; 2000), at a summer school in Iceland.[6] Arcavi revealed:

> The community of HPM has been successful in at least two fronts: it called the attention to the potential of history of mathematics in mathematics education and it also provided a lively 'home' to learn from each other for all the professions (teachers, mathematics educators, mathematicians, and historians) who work with history. However, HPM still needs much more empirical research on teaching and learning related to history than it is the case now, and there is no lack of research questions to pursue. This avenue is important in order to strengthen HPM both internally and externally. Internally, research, as I envision it, would provide insights which confirm, extend or challenge some of our assumptions and proposals, it may reveal directions not yet pursued and it would certainly sharpen our own views and future plans. Externally, research can be a way to reach out and communicate with other communities within mathematics education like

---

[4] Now, *ZDM - The International Journal on Mathematics Education*.
[5] Calinger (1996) and Katz (2000) are, in fact, proceedings from HPM1992/ICME7 and HPM1996/ICME8, respectively, which have been published by the Mathematical Association of America as separate books in the series of MAA notes. Also, yet a new MAA volume entitled *Recent Developments on Introducing a Historical Dimension in Mathematics Education*, edited by Victor Katz and Constantinos Tzanakis, is currently in press.
[6] *Nordic Graduate School of Mathematics Education* (NoGSME).

PME, CERME, and others and would open opportunities for its themes to appear more in journals like ESM, JRME [*Journal for Research in Mathematics Education*], JMB [*Journal for Mathematical Behavior*] and many others. Pursuing such opening of the current 'borders' will give history a wider stage and will be instrumental in attracting more people. Probably, HPM should aim at working in a similar way than other 'thematic' communities already do (such as technology in mathematics education, modeling, and the like) – they nurture inner meetings and discussions, but at the same time they pursue a strong presence in general conferences (plenaries, working sessions, discussion groups) and publish in general journals. In my opinion, research is the main way to pursue a wider and visible presence which would make HPM stronger and ever growing. (Arcavi; 2007)

Arcavi's call for more empirical research studies in the field of using history in mathematics education may be seen as a consequence of some of the critiques of the available literature. In 2001, Iris van Gulik-Gulikers and Klaske Blom provided a large systematic survey listing "the recent literature on the use and value of history in mathematics education" with a special emphasis on geometry (Gulikers & Blom; 2001). Based on this survey Gulikers and Blom noted:

Most publications are anecdotic and tell the story of one specific teacher, whereas it is unclear whether and how the (generally positive) experiences can be transferred to other teachers, classes and types of schools. [...] The amount of general articles that contribute to the debate outnumbers the practical essays which contain suggestions for resources or lessons. [...] [A] gap exists between historians, writing 'general' articles, and teachers, writing 'practical' articles. Most of the essays lack a legitimation of the ideas and suggestions. For example, the following questions have hardly been answered: What makes one think that the use of history deepens the mathematical understanding? Is it really motivating to stress the human aspect of mathematics or is it the enthusiastic teacher who motivates his class? Has any research been done to confirm these previous thoughts? Is there any psychological theory to confirm it? And how do people justify their choice of resources? (Gulikers & Blom; 2001, p. 223, 241-242)

Gulikers & Blom (2001) end by stating that they intend "to design a method for evaluating the *effectiveness* of mathematics teaching which uses classroom material in which history is an integral part" (italics added). In the case of using history of geometry, van Gulik-Gulikers' did this in her own dissertation (van Gulik-Gulikers; 2005) by carrying out a large quantitative, empirical study in upper secondary and pre-university education. She primarily investigated the conceptual and the motivational arguments for using history – some of the "assumptions and proposals" which Arcavi also refers to above. van Gulik-Gulikers' supervisor was Jan van Maanen, co-editor of the ICMI Study, former chair of HPM (1996-2000), and since the director of the Freudenthal Institute in the Netherlands. When visiting him in March 2007, I asked him about the results of van Gulik-Gulikers' research, the present state of the field, and in what direction it should be heading now:

We need more studies like the one Iris van Gulik-Gulikers did. Maybe that should be published in a more international source, because the fundamental article about the teaching of geometry was in *Educational Studies in Mathematics*, but all the observational, quantitative material is not published internationally. It would be important to have that. It would be good to have some studies about the *effectiveness* of integrating historical elements in maths teaching, and to study the influence of the teacher in that, for example. We don't know about that. We have no information about the teacher conduct and how classes react – there is no clear information about that at the moment. [...] Maybe an important thing is about creating better facilities for teachers, publishing a source book or something like that. A good source book with texts which pupils in school can use to read Euler, to read Cantor, to read Descartes, maybe, and there are other accessible authors, after some editorial work, that is. That would be a good thing and useful for teachers. (van Maanen; 2007, italics added)

Man-Keung Siu, author of chapter 8 of the ICMI Study (Siu; 2000), and Constantinos Tzanakis, co-author of chapter 7 of the ICMI Study (Tzanakis & Arcavi; 2000) and former chair of HPM (2004-2008), concluded in the evaluation of the TSG17 on history at the ICME10 that "it became clear that enough has been said on a 'propagandistic' level, that rhetoric has served its purpose", and hence argue that what is needed now are empirical investigations on the *effectiveness* of using history in the learning and teaching of mathematics (Siu & Tzanakis; 2004, p. 3). In his paper in the revised proceedings from the HPM2004&ESU4, Siu (2007, p. 269) mentions that he, at the time of the conference, was only aware of a total of five such empirical studies evaluating the effectiveness of the use of history in mathematics education within the English literature.[7]

Now, Siu's list is of course not meant to be a comprehensive one and, in fact, more empirical studies on the effectiveness of using history may be found in the selfsame proceedings. Out of the total of 78 papers in these proceedings, about ten percent are either clear-cut or somewhat empirical studies, though not all concerning the effectiveness (Jankvist; 2007, p. 84).[8] This relatively high percentage may, perhaps, indicate that the field of using history in mathematics education is beginning to slowly direct itself towards conducting more empirical research studies. When scanning through the four major mathematics education journals of ESM, FLM, ZDM, and JRME in the years after the ICMI Study (1998-2010)[9] I was able to identify 38 papers on the use of history in mathematics education. Out of these 38 papers, 14 may be regarded as

---

[7] McBride & Rollins (1977); Fraser & Koop (1978); Philippou & Christou (1998); Gulikers & Blom (2001); Lit
et al. (2001).
[8] For the exact papers, see table 2.
[9] Although the ICMI Study was published in year 2000, it was completed earlier than this and therefore does not include much literature from the years 1998 and 1999.

empirical research studies in some sense: 12 of these in the ESM (four in the 2007 special issue); 1 in FLM; 1 in ZDM; and 0 in JRME.[10] (For the exact papers, see table 2).

Based on my knowledge of the available literature on using history of mathematics in mathematics education, I'd say it is rich on advocating arguments as to why to include the history, thoughts on how to do it, ideas on what elements of the history to include, and on what levels of education to do this. But when it comes to testing these arguments, thoughts, and ideas on an empirically founded basis, then the 'richness' is not as overwhelming, even though "the times they are a-changin'". The shift in focus, for instance, is indicated by the earlier mentioned CERME working group "dedicated *primarily* to theory and research on all aspects of the role, effect and efficacy of history in mathematics education."[11] In the call for papers for this working group, nine topics were listed, which, except for the two first, should receive special empirical attention:

1. Theoretical, conceptual and/or methodological frameworks for including history in mathematics education;
2. Relationships between (frameworks for and empirical studies on) history in mathematics education and theories and frameworks in other parts of mathematics education;
3. The role of history of mathematics at primary, secondary, and tertiary level, both from the cognitive and affective points of view;
4. The role of history of mathematics in pre- and in-service teacher education, from cognitive, pedagogical, and/or affective points of view;
5. Possible parallelism between the historical development and the cognitive development of mathematical ideas;
6. Ways of integrating original sources in classrooms, and their educational effects, preferably with conclusions based on classroom experiments;
7. Surveys on the existing uses of history in curricula, textbooks, and/or classrooms in primary, secondary, and tertiary levels;
8. Design and/or assessment of teaching/learning materials on the history of mathematics;
9. Relevance of the history of mathematical practices in the research of mathematics education.

When I asked Man-Keung Siu about the present state of the field while in Hong Kong in October 2006, he revealed the following with regard to number 8 and his own experiences in holding workshops for teachers:

---

[10] One empirical paper on history in mathematics education is, however, due to appear later this year in JRME: Jankvist (2011a).

[11] Quoted from the working group's latest 'call for papers' to be found at: http://www.cerme7.univ.rzeszow. pl/WG/CERME7-WG12.pdf At CERME-6 in Lyon, France 2009 this group was chaired by Fulvia Furinghetti, former chair of HPM (2000-2004), and co-chaired by Jan van Maanen, Constantinos Tzanakis, Jean-Luc Dorier, and Uffe Thomas Jankvist. (Abraham Arcavi assisted in writing the group's first 'call for papers' for CERME-6.) At CERME-7 in Rzeszów, Poland 2011 the group is chaired by Uffe Thomas Jankvist and co-chaired by Jan van Maanen, Constantinos Tzanakis, and Snezana Lawrence.

> You have to have something in between, not just the research results in history of mathematics, not just the primary texts, not just the storytelling popular accounts. You have to have something in between, and those are materials that would be useful for teachers in the classroom. (Siu; 2006)

In other words, the topics to research empirically within the field of using history in mathematics education are plenty, and there seems to be a general acknowledgement within the community of such empirical research studies being both relevant and highly needed. But before we enter more deeply into the issue of empirical studies, let us have a look at the different kinds of publications that the field of history in mathematics education has offered in the past decades.

## The general literature survey
Although the publications on history in mathematics education are of various kinds, I shall claim that they generally are of three fundamentally different types:

1. publications *advocating* in one way or another for history in mathematics education;
2. publications *describing* either concrete uses by teachers or developments of teaching material etc.;
3. actual *research* on history in mathematics education.

In this section I shall make some 'downstrokes' in the literature in order to exemplify the three different types, and I shall discuss a few publications within each. But first an overview of the literature samples surveyed.

### Overview of the literature samples surveyed
The literature which make up this survey are collective samples on the use of history in mathematics education. These consist of special books or collections of papers, as mentioned previously, special issues of journals devoted to history in mathematics education and of proceedings from the more recent HPM and ESU conferences. The samples mainly used are displayed in table 1.

| — | FLM Special Issue on History in Mathematics Education: Vol. 11(1) 1991 |
|---|---|
| Swetz et al. (1995) | Learn from the Masters |
| Jahnke et al. (1996) | History of Mathematics and Education: Ideas and Experiences |
| Calinger (1996) | Vita Mathematica: Historical Research and Integration with Teaching (proceedings from HPM1992) |
| — | FLM Issue addressing History: Vol. 17(1) 1997 |
| Katz (2000) | Using History to Teach Mathematics: An International Perspective (proceedings from HPM1996) |
| Fauvel & van Maanen (2000) | History in Mathematics Education: The ICMI Study |
| Horng & Lin (2000) | Proceedings of the HPM 2000 Conference: History in Mathematics Education: Challenges for a New Millennium |
| Bekken & Mosvold (2003) | Study the Masters: The Abel-Fauvel Conference (proceedings) |

| — | MJRME Double Special Issue on The Role of the History of Mathematics in Mathematics Education (proceedings TSG17, ICME10): Vol. 3(1-2) 2004 |
|---|---|
| Shell-Gellasch & Jardine (2005) | From Calculus to Computers – Using the last 200 years of mathematics history in the classroom |
| Furinghetti et al. (2007) | Proceedings HPM2004 & ESU4 |
| — | ESM Special Issue on The History of Mathematics Education: Theory and Practice: Vol. 66 2007 |
| Barbin et al. (2008) | History and Epistemology in Mathematics Education: Proceedings of the 5th European Summer University (ESU5) |
| — | Papers presented at TSG23 on The Role of History of Mathematics in Mathematics Education at ICME11 |
| Cantoral et al. (2008) | Proceedings HPM2008 (only on CD-ROM) |
| — | Proceedings CERME-6, WG15: The Role of History of Mathematics in Mathematics Education: Theory and Research |
| — | Accepted papers for CERME-7, WG12: History in Mathematics Education |

**Table 1** A table of the collective samples used when surveying the literature on history in mathematics education.

## Comments on and examples of the advocating samples

The papers of the FLM 1991 special issue on history in mathematics are mostly of an advocating nature, the paper by Russ et al. (1991) on the experience of history in mathematics education being the most 'propagandistic' one. The papers do, however, provide various examples from the history of mathematics serving as inspiration for teachers, and some of the authors also describe their experiences from teaching situations (see the examples of descriptive papers below). Something similar may be said about the collection of 23 papers edited by Swetz et al. (1995). The collection is organized according to papers discussing the integration of history at secondary and tertiary level, respectively, and mainly focusses on selected mathematical topics, although a few papers also focus on design (e.g. Helfgott, 1995). The collection edited by Calinger (1996) is equally concerned with the history of mathematics as with the role of it in mathematics education. Concerning the latter, the papers are not very different in nature from those in Swetz et al. (1995), except, perhaps, for the fact that some papers address specifically the role of history for pre-service and in-service teachers (e.g. Heiede, 1996; Kleiner, 1996). Also, the collection edited by Jahnke et al. (1996) includes mainly advocating studies, but in contrast to the above mentioned samples it uses these to propose some interesting visions for future research in the field. For instance, when the editors write in their introduction:

> We need more sound knowledge about what is going on when students of a certain age are confronted with history of mathematics. We urgently need conceptual ideas about how history could be originally embedded into normal teaching. And, above all, we need a continuous process of exchange between interested mathematics educators, historians of mathematics, and research mathematicians. (Jahnke et al.; 1996, pp. viii)

The authors then go on to discuss the relations between historians of mathematics and mathematics educators, arguing that these fields have much more to offer each other than is usually considered the case by their practitioners:

> History of mathematics is considered by many as fundamental research, and integrating history into teaching seems to be a mere application of some more or less trivial by-products of the fundamental historical work. This idea is misleading. The significance of history lies in its contribution to the general culture. Even more than for general history, it is true for history of science that the fundamental relation to culture is bounded by what is termed 'Bildung' in German. If this is accepted, the immediate consequence is that we cannot live any longer with a situation in which mathematics educators have to fumble for subject matter which just might be adequate for teaching uses. (Jahnke et al.; 1996, pp. viii-ix)

'Bildung' (or 'Algemeinbildung') in relation to the role of history of mathematics in mathematics education is embedded in the notion of using 'history as a goal' (Jankvist, 2009a), which will be explained in a subsequent section.

As examples of advocating papers I shall discuss two from the *International Journal of Mathematical Education in Science and Technology*, i.e. not included in table 1, one of older date and one which is contemporary with the ICMI Study. The reason for choosing exactly these is that the first one is one of the better examples of advocating papers in that it lists a series of arguments for integrating history, the second is typical because of another reason, namely due to its discussion of the possible benefits for teachers of knowing something about the history of mathematics. The one of older date is by Siu & Siu (1979). They argue for six different 'profits' of taking history into consideration, three for curriculum planning and three for classroom learning. First, a look at the history may help structure the development of content in a curriculum in a more natural way (here they refer to the so-called recapitulation argument – ontogenesis recapitulates philogenesis). Second, mathematics is a discipline focusing on rigor in the sense that a minimum number of basic essential facts are selected and propositions are derived in a logical way from these. However, from an educational viewpoint, if focusing too much on rigor this may restrict the students' original thinking and provide them with an incorrect image of mathematics. Siu & Siu (1979, p. 563) say: "Ironically, the most important as well as the most difficult task in mathematical education is to make students realize that 'mathematics-as-an-end-product' as presented in textbooks can be very different from 'mathematics-in-the-making'." In this respect, looking to the history may illustrate that even the concept of rigor within mathematics continually evolves. Third, the history of mathematics may also serve as a useful guide to pointing out and illustrating interrelations between various branches of mathematics. Fourth, entering the three 'profits' for classroom learning, history is embedded in present mathematics (ideas, notation, etc.) and may therefore also help us understand it. Besides assisting in the understanding of specific concepts, history may also help students understand the global picture of mathematics and make them able to place the more local and fragmentary parts they have to study in a broader context. Fifth, history may show that

mathematics is a human endeavor and part of the culture of mankind. Sixth, and last, history may also give the students confidence in the sense that it
may show them how mathematics has not come into being exclusively by the hand of geniuses, that it is human to err, and that co-operation is often a key to success.

The other advocating paper, from the same journal, is one by Furinghetti (2000).[12] In the context of prospective teachers, she discusses the history of mathematics as a coupling link between secondary and university teaching. Although the paper mainly concerns the use of history as a means for the learning and understanding of mathematics, it also addresses the topics of students' beliefs and images of mathematics, and it is in this respect I shall discuss it here. In Italy, the author states, students arrive at the university with a good disposition towards mathematics, but after having been exposed to the teaching at university their feeling changes "'very quickly towards a strictly formalist view of the discipline' (these are a student's words). Their conception of mathematics becomes so poor that one student says 'mathematics exists because it is taught'" (Furinghetti; 2000, p. 44). According to Furinghetti, history has a part to play in remedying this, since "history is a good vehicle for reflecting on cognitive and educational problems, for working on students' conceptions of mathematics and its teaching, and for promoting flexibility and open-mindedness in mathematics" (Furinghetti; 2000, p. 51). Concerning the idea of promoting flexibility and open-mindedness, Furinghetti has addressed this in other papers, also in relation to her viewpoint that "all kinds of mathematics students (prospective teachers and others) should know the history of mathematics for its own cultural value" (Furinghetti; 2000, p. 51). In a previous FLM-paper she argues that teachers should "approach mathematics as a set of human activities [...] and not as a body of rigidly defined knowledge" (Furinghetti; 1993, p. 38), which is very much in line with the advocating viewpoints of Siu & Siu (1979). Furinghetti states her personal belief that it is possible to provide students with the opportunity to develop what she calls an *'ecological' image of mathematics*, meaning "an image respectful of the peculiarities of this protean discipline" (Furinghetti; 1993, p. 38).

## Comments on and examples of descriptive samples

The collection edited by Katz (2000) is quite similar in nature to Calinger (1996), due to the fact that it also includes selected proceedings from an HPM meeting, but an interesting difference is that there is an increase in descriptive papers (and research papers). The ICMI Study (Fauvel & van Maanen; 2000) in particular reflects this increase in descriptive approach, although it also provides its share of advocating arguments for the use of history. Part of the descriptive nature of the ICMI Study is, of course, explained by the fact that it is itself a very extensive survey of the literature on using history in mathematics education. Still, it reflects very well the increase in descriptive papers from the earlier samples mentioned in table 1. Examples of descriptions of using history in the ICMI Study are those of current practice in teacher training as described in chapter 4 (Schubring; 2000); the section on ideas and examples of classroom implementation in chapter 7 (Tzanakis & Arcavi; 2000); descriptions of

---

[12] Other examples of papers addressing the topic of history in mathematics education in the same journal are those of Grattan-Guinness (1973) and Grattan-Guinness (1978).

teaching projects inspired by history in chapter 8 (Siu; 2000); the integration of original sources in pre-service teacher education and classrooms in general described in chapter 9 (Jahnke; 2000); and examples of using other media such as role plays, ancient instruments, computer software, and the Internet for integrating the history in class described in chapter 10 (Nagaoka; 2000). This increase in descriptive studies somehow seems to follow through in the remaining samples displayed in table 1 along with an increase in research studies, and I shall discuss the remaining samples in this context later. For now I shall exemplify two different types of descriptive samples from the literature: a teacher's concrete use of history in class and an extensive development of teaching materials relying on original sources.

On several occasions, Jan van Maanen has described his use of the history of mathematics in upper secondary classrooms etc., uses which are interesting since van Maanen at the time was a postgraduate student in the history of mathematics (van Maanen; 2007). One of his first descriptions appears as a rarity of its kind in the FLM 1991 special issue (van Maanen; 1991) – a frequently cited paper in the literature, in which he outlines a couple of lessons where his upper secondary students were to study a problem from L'Hôpital's 1696-textbook on differential calculus, *Analyse des infiniment petits*, the so-called weight problem. In a paper appearing in the collection edited by Swetz et al. (1995), van Maanen (1995) continues his descriptions based on personal experiences with three historical cases: seventeenth century instruments for drawing conic sections, improper integrals, and one on the division of alluvial deposits in medieval times. The latter, which I shall describe in more detail, is different from the other two in that it describes a project in three first-year classes of the Dutch grammar school, pupils about age 11, and because it was an interdisciplinary project with Latin.

In a medieval setting of a case of three landowners, who all had land on the bank of a river, and fought over an alluvial deposit bordering their land, the students were to investigate the problem by means of a method proposed by the Italian professor Bartolus of Saxoferrato in 1355 (the example with the landowners was that used by Bartolus himself). The ideas of the project were, among others, to demonstrate the importance of mathematics in society, to let pupils 'invent' a number of constructions by ruler and compass, to have them apply these 'inventions' to solve the legal problem of the medieval example, and to have them read excerpts from Bartolus' treatise in the original Latin language, illustrating "that it is impossible to interpret the sources of Western culture without knowledge of classical languages" (van Maanen; 1997, p. 79). Van Maanen's evaluation of the implementations of the project is: "Making contact with Bartolus was only possible via deciphering and translating, but that was simply an extra attraction to most of the pupils. They learned to work with point-sets in plane geometry, and simultaneously their knowledge of general history increased. Last but not least, they were greatly stimulated to learn Latin." The project on Bartolus is just one of several of van Maanen's which all illustrate that "new maths may profit from old methods" (van Maanen; 1997).

The other example of descriptive papers is one in a long line by David Pengelley and collaborators[13] describing their work on developing materials for classroom projects using original sources (Pengelley; 2008). In particular, the group is interested in "learning discrete mathematics and computer science via primary historical sources", a project which has resulted in 18 sets of material for students projects in classrooms since 2008.[14] At HPM2008 in Mexico City, I attended the workshop by Pengelley and Barnett, in which two of their materials were discussed: *Treatise on the Arithmetical Triangle* relying on the original work of Blaise Pascal; and *Early Writings on Graph Theory: Euler Circuits and the Königsberg Bridge Problem* using Euler's solution to the Königsberg bridge problem[15] from 1736, today considered the starting point of modern graph theory (Barnett et al.; 2008). The idea of these materials, as well as the others displayed on their web site, is to guide students through a reading of the original sources, providing them with clarifying and elaborating questions on the mathematics of the source along the way. Barnett et al. (2008) describe their design and aim as:

> The projects are designed to introduce or provide supplementary material for topics in the curriculum, such as induction in a discrete mathematics course, or compilers and computability for a computer science course. Each project provides a discussion of the historical exigency of the piece, a few biographical comments about the author, excerpts from the original work, and a sequence of questions to help the student appreciate the source and learn how to do the relevant mathematics. The main pedagogical idea is to teach and learn certain course topics directly from the primary historical source, thus recovering motivation for studying the material. (Barnett et al.; 2008, pp. 2-3)

Alongside the excerpts from original sources, the students are often provided with the terminology of modern mathematics, and in the case of the material on the graph theory, the students were asked to 'fill in the gaps' in a modern proof of Euler's main theorem for solving the bridge problem, thereby also drawing their attention to current standards regarding formal proof.[16]

## Comments on and examples of research samples

As should be clear from the above, the earliest sources mentioned in table 1, i.e. those from the 1990s,[17] offer almost no examples of research papers on the use of history in

---

[13] Other papers in the samples from table 1 are: Pengelley (2003a), Pengelley (2003b), Laubenbacher & Pengelley (1996).

[14] These projects may be found at: http://www.cs.nmsu.edu/historical-projects/projects.html. 16 other projects, developed before 2008, may be found at http://www.math.nmsu.edu/hist_projects.

[15] The Königsberg bridge problem: is it possible to plan a stroll through the town of Königsberg which crosses each of the town's seven bridges once and only once? See e.g. Biggs et al. (1976).

[16] In March 2010 I visited David Pengelley and Jerry Lodder in Las Cruces and Janet Barnett in Pueblo to discuss with them my own adoption of their design for developing teaching materials based on readings of original sources. For a description of this adoption, see Jankvist (2011c).

[17] Swetz et al. (1995); Jahnke et al. (1996); Calinger (1996); and Katz (2000) since this includes papers from HPM1996.

mathematics education.[18] However, around the end of the 1990s and at the beginning of the new decade, a minor shift in focus seems to occur. The ICMI Study, although this mainly is of a descriptive (and advocating) nature, also occasionally refers to actual research on the use of history in mathematics education, e.g. in chapter 3 on research perspectives (Barbin; 2000). The HPM2000 proceedings has as one of its themes "the effectiveness of history in teaching mathematics: empirical studies", and a total of 25 papers are listed under that topic (although some of them are to be considered descriptive rather than empirical). As mentioned previously, the HPM2004&ESU4 proceedings, the MJRME 2004 double special issue, and the ESM 2007 special issue also include a number of empirical research studies. The same goes for the ESU5 proceedings, the HPM2008 proceedings, and, naturally, the papers from the CERME-6 and CERME-7 working group on history in mathematics education.[19]

Now, research studies on the use of history in mathematics education need not only be empirical. They can also be analytical studies and discussions. One such example is found in the proceedings from the Abel-Fauvel conference (Bekken & Mosvold; 2003). Mosvold (2003) discusses the so-called genetic principle and its history, mainly with reference to Schubring (1978), but also relates it to the, of course much younger, Norwegian tradition.[20] Another example of such an analytical study is found in a FLM 1997 issue, also partly dealing with history in mathematics education. Here Radford (1997) discusses (and criticizes) the historicoepistemological obstacles in mathematics, as discussed by Brousseau (1997), as a means for understanding and overcoming students' learning difficulties. Analytical research also includes, for instance, studies in the use of history in existing curricula and textbooks. An example of this is the paper by Smestad (2003), also in the proceedings of the Abel-Fauvel conference. Smestad reviewed the textbooks which came out after the 1997 reform in Norway, in which history of mathematics was included in the curriculum for elementary school (ages 6-16). Smestad concludes that the treatment of history is problematic in this first generation of textbooks. The authors seem to have had problems including the history in a meaningful way and lots of factual errors occur. Smestad also describes his ongoing study at the time, to review the use of history in the textbooks for the Norwegian upper secondary school, in which the history became part of the curriculum in 1994 (inspired by the Danish upper secondary mathematics program of 1987). The situation for these seem more promising, according to Smestad. Fewer factual errors occur, but still the authors have problems actually integrating the history. Often it is 'pasted on' in either the beginning or the end of chapters.[21]

A special kind of analytical research studies are those which discuss and/or categorize the different arguments for using history and the different approaches to doing so, i.e.

---

[18] For the exceptions, see table 2.

[19] Also, once the ESU6 proceedings are ready they too will offer further empirical samples.

[20] Mosvold's discussion is based on his master's thesis (Mosvold; 2001) and other publications on the same topic (Mosvold, 2002b; Mosvold, 2002a).

[21] Something similar is, in fact, the case in many of the newer Danish textbook systems for upper secondary mathematics, where the historical information often is added on in special colored boxes (Jankvist; 2008a).

the 'whys' and the 'hows' of using history. An early example of such, although not in any way a categorization, is the one by Fauvel (1991) in the FLM special issue. Proposed categorizations, however, are found in the papers by Gulikers & Blom (2001), Fried (2001), Furinghettti (1997) and Furinghetti (2004), and as part of the ICMI Study in the chapter by Tzanakis & Arcavi (2000), a categorization closely related to that of Tzanakis & Thomaidis (2000), and later put into a somewhat different perspective by Tang (2007).

## Two different purposes for using history

Generally speaking, it may be argued that there are mainly two different purposes ('whys') for using history of mathematics in mathematics education: *history as a goal* and *history as a tool* (Jankvist; 2009a). Regarding the first, it is *not* a matter of teaching history *per se*. The word *goal* must be understood in the sense that it is considered a goal to show the students something about the historical development of mathematics. For example to show them that mathematics exists and evolves in time and space; that it is a discipline which has undergone an evolution over millennia; that human beings have taken part in the evolution; that the evolution of mathematics is due to many different cultures throughout history, and that these cultures have had an influence on the shaping of mathematics as well as the other way round. Of course, a use of history as a goal may have a positive side effect of assisting the learning of mathematics, but the important thing is that it is not the primary purpose for using history here.

That is, however, the case when using history as a *tool*. In this sense, history is used as an assisting means, or an aid, in the teaching and learning of mathematics – it becomes a vehicle for teaching and learning mathematical ideas, concepts, theories, methods, algorithms, ways of argumentation and proof, etc. When restricting ourselves to the use of history as a tool we may (sub)categorize such uses into at least three different types:

- history as a motivational and/or affective tool,
- history as a cognitive tool (e.g. the idea of epistemological obstacles), and
- the role of history in what may be referred to as the evolutionary arguments (the recapitulation argument or historical parallelism).

Another way of distinguishing the use of history as a goal from that of a tool is to say that, where 'history as a goal' concerns the learning of some meta-perspective issues – or *meta-issues* – of mathematics and its history, 'history as a tool' focuses primarily on the internal or inner issues – or *in-issues* – of mathematics itself (Jankvist; 2009a). In the following section, I shall use the above outlined categorization to classify the empirical studies identified in the literature.

## Identification of empirical studies

As mentioned earlier, the search for empirical studies in this survey bases itself on the literature of collective samples, special journal issues, proceedings, etc. as listed in table 1 as well as the past 13 years (1998-2010) of four major journals in mathematics education: ESM, JRME, FLM, and ZDM.

| Jahnke et al. (1996) | Bartolini Bussi & Pergola (1996) |
|---|---|
| Calinger (1996) | Bero (1996) |
| Katz (2000) | Bruckheimer & Arcavi (2000); Dorier (2000); Isaacs et al. (2000); Winicki (2000) |
| Horng & Lin (2000) | Hsiao&Chang (2000); Hsieh (2000); Hsieh&Hsieh (2000); Lakoma (2000); Lin (2000); Liu (2000); Ming (2000); Ohara (2000); Prabhu & Czarnocha (2000); Su (2000); Troy (2000); Tsukahara (2000);Winicki-Landman (2000) |
| Shell-Gellasch & Jardine (2005) | McGuire (2005); Greenwald (2005) |
| MJRME Special Issue | Barabash & Guberman-Glebov (2004); Fung (2004);Waldegg (2004); Zormbala & Tzanakis (2004) |
| Furinghetti et al. (2007) | Demattè (2007); Horng (2007); Isoda (2007); Liu (2007); Smestad (2007); Su (2007); Tzanakis & Kourkoulos (2007); Winicki-Landman (2007) |
| ESM (1998-2010) | Philippou & Christou (1998); Radford (2000b); van Amerom (2003); Durand-Guerrier & Arsac (2005); Bakker & Gravemeijer (2006); Farmaki & Paschos (2007); Arcavi & Isoda (2007); Furinghetti (2007); Radford & Puig (2007); Thomaidis & Tzanakis (2007); Charalambous et al. (2009); Jankvist (2010) |
| FLM (1998-2010) | Jankvist (2009b) |
| ZDM (1998-2010) | Kjeldsen & Blomhøj (2009) |
| Barbin et al. (2008) | Dimitriadou (2008); Glaubitz (2008); González-Martín & Correia de Sá (2008); Liu (2008); Morey (2008); Paschos & Farmaki (2008); Thomaidis & Tzanakis (2008) |
| TSG23, ICME11 | Lawrence (2008) |
| Cantoral et al. (2008) | Ceylan Alibeyoglu (2008); Gonulates (2008); Jankvist (2008c); Kourkoulos & Tzanakis (2008); Nataraj & Thomas (2008); Peard (2008); Reed (2008); Smestad (2008) |
| WG15, CERME6 | Blanco & Giovart (2009); Jankvist (2009d); Kjeldsen (2009); Lawrence (2009); Tardy & Durand-Guerrier (2009); Thomaidis & Tzanakis (2009) |
| WG12, CERME7 | Alpaslan et al. (2011); Clark (2011); Kjeldsen (2011); Kotarinou et al. (2011); Lawrence & Ransom (2011); OReilly (2011); Ransom (2011) |
| Other samples | McBride & Rollins (1977); Fraser & Koop (1978); Arcavi et al. (1982); Arcavi et al. (1987); Harper (1987); Sfard (1995); Demattè & Furinghetti (1999); Lit et al. (2001); Jankvist (2008d); Jankvist (2008b); Jankvist (2009c); Jankvist & Kjeldsen (2010); Haverhals & Roscoe (2010); Kjeldsen & Blomhøj (2011); Jankvist (2011a); Jankvist (2011b) |
| Ph.D. dissertations | van Amerom (2002); Bakker (2004); van Gulik-Gulikers (2005); Clark (2006); Goodwin (2007); Jankvist (2009e) |

**Table 2** A list of the 98 empirical studies found in the samples in table 1 as well as others, which I have come across in my survey. Remarks: Samples which were not found to include empirical studies are not listed in the table. The surveys of proceedings edited by Horng & Lin (2000), Barbin et al. (2008), and Cantoral et al. (2008) do not include the papers written in Taiwanese, French, and Spanish. Nor does the survey of contributions in Barbin et al. (2008) take into account abstracts, since these do not display any empirical data. Also, the survey does not include the forthcoming proceedings of ESU6 in Vienna, since these were not ready at the time of submission of this paper.

As a result of surveying the mentioned literature almost one hundred more or less empirical studies were identified. Of course, empirical studies may be many different things: from large scale quantitative studies to small scale qualitative studies, from experimental investigations to a teacher testing out a course using methods of questionnaires and interviews. When scanning the literature and deciding which studies to regard as empirical, focus was put on whether or not a paper had any display of empirical data to support findings and conclusions, and if there seemed to be some kind of (implicit or explicit) underlying discussions or descriptions of research design and methodology. These criteria have resulted in the list of 98 somewhat or clear-cut empirical studies depicted in table 2.

Surely, the list of table 2 is not a complete list of every empirical study ever made on the use of history in mathematics education. Nevertheless, it is a fairly comprehensive one, and as such it may provide us with a sound indication of where the emphasis of the empirical studies within the field lies. And furthermore, it may provide a sound starting point for researchers who wish to engage with empirical studies on history in mathematics education.

## Classification of the identified empirical studies

Of the (relatively few available) empirical studies the vast majority of these concern a use of history as a tool. I shall list these according to their main purpose regarding the three subcategories of using history as a tool.

The studies concerning mostly the use of history a motivational and/or affective tool are:

| Table 3. Studies focusing on history as a motivational and/or affective tool |
|---|
| Charalambous et al. (2009); Haverhals & Roscoe (2010); Hsiao & Chang (2000); Hsieh (2000); Hsieh & Hsieh (2000); Lawrence (2008); Lin (2000); Liu (2007); McBride & Rollins (1977); McGuire (2005); Philippou & Christou (1998); Troy (2000); Ransom (2011). |

Studies focusing on history as a cognitive tool (including epistemological obstacles, guided reinvention, etc.) make up the largest group of empirical studies:

| Table 4. Studies focusing on history as a cognitive tool |
|---|
| Arcavi et al. (1982); Arcavi et al. (1987); Arcavi & Isoda (2007); Bakker (2004); Bakker & Gravemeijer (2006); Barabash & Guberman-Glebov (2004); Blanco & Giovart (2009); Bruckheimer & Arcavi (2000); Bartolini Bussi & Pergola (1996); Ceylan Alibeyoglu (2008); Dimitriadou (2008); Dorier (2000); Fung (2004); Furinghetti (2007); Glaubitz (2008); González-Martín & Correia de Sá (2008); Goodwin (2007); van Gulik-Gulikers (2005); Horng (2007); Isoda (2007); Jankvist & Kjeldsen (2010); Kjeldsen (2009); Kjeldsen & Blomhøj (2011); Kourkoulos & Tzanakis (2008); Lakoma (2000); Lawrence (2009); Lawrence & Ransom (2011); Lit et al. (2001); Liu (2000); Liu (2008); Ming (2000); Morey (2008); Nataraj & Thomas (2008); Ohara (2000); Paschos & Farmaki (2008); Peard (2008); Prabhu&Czarnocha (2000); Radford (2000b); Reed (2008); Su (2000); Su (2007); Tardy & Durand-Guerrier (2009); Thomaidis & Tzanakis (2008); Thomaidis & Tzanakis (2009); Tsukahara (2000); van Amerom (2002); van Amerom (2003);Winicki (2000);Winicki-Landman (2000);Winicki-Landman (2007). |

Of studies somehow referring to the evolutionary arguments for using history, historical parallelism, etc., we find:

| Table 5. Studies focusing on historical parallelism etc. |
| --- |
| Bero (1996); Durand-Guerrier & Arsac (2005); Farmaki & Paschos (2007); Harper (1987); Radford & Puig (2007); Sfard (1995); Thomaidis & Tzanakis (2007); Tzanakis & Kourkoulos (2007);Waldegg (2004); Zormbala & Tzanakis (2004). |

Of course, some studies recognize history as a cognitive tool and at the same time as a motivational or affective tool (or some other combination of the roles), but I have tried to classify them according to what appears to be their main objective. Also, it should be mentioned that some of the studies concerning history as a tool also acknowledge the role of history as a goal, although this is not their main objective (e.g. Hsieh & Hsieh, 2000; Liu, 2000; Su, 2007). Other studies focus on history as a goal, but mention the possible side effects in terms of history as a tool (e.g. Kjeldsen & Blomhøj, 2009; Jankvist, 2009b).[22] Of studies mainly concerned with using history as a goal we find:

| Table 6. Studies focusing on history as a goal |
| --- |
| Demattè (2007); Demattè & Furinghetti (1999); Greenwald (2005); Isaacs et al. (2000); Jankvist (2008b); Jankvist (2008d); Jankvist (2008c); Jankvist (2009b); Jankvist (2009c); Jankvist (2009d); Jankvist (2009e); Jankvist (2010); Jankvist (2011a); Jankvist (2011b); Jankvist & Kjeldsen (2010); Kjeldsen & Blomhøj (2009); Kotarinou et al. (2011); OReilly (2011). |

Categorizing the above empirical studies was not always an easy task. Some of the classifications are results of interpreting what seemed to be the authors' underlying motivation. In particular, this was the case when having to separate some of the tool studies into those using history as a motivational tool and those using it as a cognitive tool. This being said, it was fairly easy to separate the studies focusing mainly on history as a tool from those focusing mainly on history as a goal.

In addition to the above, there is yet a category, namely empirical studies that somehow are of a more overarching nature. As an example take the studies that address in-service teachers' actual use of history in their classrooms or their general views and attitudes towards why and how to use history.

| Table 7. Overarching studies |
| --- |
| Alpaslan et al. (2011); Clark (2006); Clark (2011); Fraser & Koop (1978); Gonulates (2008); Kjeldsen (2011); Smestad (2007); Smestad (2008). |

## Current observations and future perspectives

A few observations should be made based the list of empirical studies as well as the distribution of the studies according to the chosen classification. First of all, the vast majority of the studies are from the last 10-15 years, if one had tried to write up a similar list 20 years ago it would hardly have been existing. To some degree this of course has to do with the work of the ICMI Study on history in mathematics education (Fauvel & van Maanen; 2000), which clearly showed a lack of empirical studies in this particular subfield of mathematics education. But at the same time, it also shows an

---

[22] The article by Jankvist & Kjeldsen (2010) exploits equally the avenue of using history as a cognitive tool and that of using it as a goal, and for that reason it appears in both tables 4 and 6.

increasing interest in actually taking on this task of conducting empirical research on history in mathematics education. In this light, also taking into consideration how relatively small this subfield is, a number of almost one hundred studies makes up a reasonable point for departure for future empirical research projects. And, as illustrated through the quotes from interviews with experienced researchers, "there is no lack of research questions to pursue", as Arcavi put it. In particular, he points out that empirical studies may provide insights that could confirm, extend, or challenge some of the assumptions, conjectures, hypotheses, and proposals of the more advocating literature. Also, it is a way to test some of the, often positive, results reported in the describing literature, to see in what ways these work in different or more general settings. That is to say to investigate the effectiveness and/or efficacy of actual uses of historical elements in the teaching and learning of mathematics. And as van Maanen pointed out, to attempt to clarify the role and influence of the teacher's discourse when integrating historical elements. Certainly one conjecture is that a 'successful' integration of history requires the enthusiasm of the teacher, which I do not necessarily disagree with. But from a research point of view, it would be interesting to know what happens if the voice (and discourse) of the teacher is kept to a minimum, for example by having students work with materials based on guided readings of original sources (the before mentioned approach of Pengelley and collaborators). What is the effect of the material's discourse, and therefore to some extent also the history's discourse itself? Such a question cannot be answered based on 'armchair research' alone, good as this may also be, it will require empirical investigations.[23]

To linger a little with the teachers' 'enthusiasm', as mentioned above, this also came up in my interview with Man-Keung Siu when discussing the preparation of materials and design of historical teaching modules to be used in classrooms:

> Some teachers hope that one day someone will write all these materials and distribute them, or they can get it in the bookstore and use it directly in the classroom. But I don't think that would work, because you need the enthusiasm of the teacher himself or herself in order to use this kind of material well. Just having the material there is not enough. (Siu; 2006)

This constitutes another line of research questions to be pursued empirically, namely questions on how such 'enthusiasm' can be mobilized? One suggestion is to have the teachers construct their own materials, which of course requires arousing some kind of awareness that this would be helpful in order to get teachers to cooperate. Siu explained:

> I hold workshops for large groups of teachers now and then. If I can arouse the interest of a few teachers to begin this kind of collaborative activities then it will be a good start. You have to have this self-motivated initiative to do the thing. And usually their excuses are: 'I do not know history, I'm not an historian of mathematics, I don't know how to do it', but I always give my own example, that I didn't know it at the beginning either. But it's not very

---

[23] For a paper addressing the design aspects of these questions, see Jankvist (2011c).

persuading or convincing, because they say, well, you have been doing it for thirty years. My answer is that you have to start somewhere! I don't like to paint a rosy picture for them, letting them think that they don't have to do anything, just sit there and then they have the material and then it works wonders. I don't think it ever works that way. But one has to start somewhere. It is hard work and it is going to be hard work. In my case, or in my days, I did it alone. But one of the things we can achieve is, I think, to have a group collaborating in a collective effort, which will make the task easier. (Siu; 2006)

Similar initiatives have been taken in other parts of the world. In Taiwan, for instance, researchers have followed and coached teachers who met for three hours a week and discussed texts on the history of mathematics, and based on these discussions prepared materials to use in their teaching (Su, 2007; Horng, 2007). At Roskilde University in Denmark, courses have been offered where in-service upper secondary teachers could receive assistance from professionals to develop their own small teaching modules on topics from the history of mathematics (and modeling of mathematics), as well as in evaluating their implementations of these (Kjeldsen; 2011). The produced materials for the teaching modules and the evaluations were then posted on the web page of the course, so other teachers could benefit from them as well. Such activities are of course much more demanding for the teachers than just grapping prepared material from the bookshelf. However, it seems likely that a teacher should feel more at ease with a material he/she had taken part in preparing himself/herself, and the 'enthusiasm' might then come easier too. At any rate, much more empirical research on such matters is required before any parts of this territory can be reclaimed.

Regarding the distribution of the empirical studies, it is very clear that the emphasis to some degree has been – and still is – on the use of history as a cognitive tool in one form or another (tables 3, 4, and 5 vs. table 6). And when weighing the total number of history-as-a-tool studies (the affective and motivational, the cognitive, and those on historical parallelism) against the number of history-as-a-goal studies, then there is no question which one comes out with the upper hand: only around 1/5 of the identified empirical studies may be said to mainly address matters of history as a goal. One reason for this may of course be that it is difficult to argue why we need studies on the use of history as a goal, in particular when many mathematics educators (and policy makers) only see mathematics education as having as its sole purpose to make students better at *mathematics*. Of course, this is *the* purpose, but in my mind there is more to it than just that. Mathematics education includes also the elements of 'Algemeinbuildung' as mentioned by Jahnke et al. (1996), or what Niss & Jensen (2002, p. 66) refer to as 'overview and judgment' which, as a dimension supplementing actual mathematical competencies, consists of "a set of viewpoints, which provide overview and judgment about the connection of mathematics to circumstances and allotting in nature, society, and culture",[24] or mathematics' and its history's contribution to "students' growing into whole human beings" as argued by Fried (2007, p. 203). I shall return to these aspects

---

[24] For an English description of Niss' and Jensens' three types of overview and judgment, see Jankvist & Kjeldsen (2010).

shortly, but first let me point out that the first time I did the review of empirical studies presented in this paper, the fraction of history-as-a-goal papers were closer to 1/10 (Jankvist; 2009e). Of course, I am aware of the fact that my own publications have contributed in increasing the number of history-as-a-goal studies, but there are also recent entries from other authors into this category, not least in the context of the CERME working group. And from my line of interviews with experienced researchers in the field, I clearly sensed an acknowledgement of the role of 'history as a goal' in mathematics education. My interview with Luis Radford, author of chapter 5 of the ICMI Study (Radford; 2000a), illustrates this. When visiting him in Sudbury, Canada in June 2008 he provided the following observations:

> Usually, the history of mathematics has been considered an *instrument* to reach an end. This end is usually formulated in terms of improving the students' learning of a target mathematical content. I think that this way of considering the history of mathematics in math education was the one I was defending when many of us met in Marseille in 1998 for the ICMI Study organized by John Fauvel and Jan van Maanen. Now, I would characterize my conception of the use of the history of mathematics in math education not as an instrument (or tool), but as an end. Perhaps it was in the panel organized by Evelyne Barbin two years ago in Prague [ESU5] that it was possible for me to see the value of this conception for the first time. It was during the preparation of my panel intervention that I realized that the history of mathematics is not just a tool, but something more complex. It is a kind of goal – but not a goal in itself. The idea is not to resort to the history of mathematics *per se*, but, through the history of mathematics, to become a more critical citizen. And I think that we need that in the 21st century. So, the idea is not to use the history of mathematics to make the students better problem solvers only, but to make them better citizens.
>
> And how can the history of mathematics make them better citizens? Well, by making them aware that the reality that we have in front of us is a reality that can't really be understood if it is not comprehended through an analysis of the history that lies behind the processes of contemporary mathematics. So, this means that the history of mathematics is something that will allow us to become critical practitioners, critical individuals in a society where we have to face multiple choices and different voices, people coming from different cultures. In this very complex place that has become our contemporary world I think that the history of mathematics has an empowering role. But again, not because it makes us better problem solvers, but rather better people in the sense that we will be able to better understand other rationalities and other ways of thinking about the world. (Radford; 2008)

In a sense, Radford's argument is in line with one of Skovsmose (1990), who argues that knowledge about mathematical modeling will provide students with a democratic competence and allow them to practice critical citizenship. Surely knowledge of the history of mathematics will also add to such a democratic competence, but unlike mathematical modeling it will do so from a more cultural point of view, contributing to students' understanding of "other rationalities and other ways of thinking", as Radford says. But to actually see if and how this will work in practice when students are exposed to elements of history in their mathematics teaching is a different matter, and it will have to be considered from an empirical point of view – a viewpoint Radford also agreed with when discussing the need for and role of empirical studies on history in mathematics education:

> We need to refine the theoretical principles of a conception of mathematics that takes seriously into account the role of history and culture. Such a conception, if clearly defined, should allow us to envision new ways in which to use and integrate the history of mathematics in the classroom. The problem, however, is not merely a theoretical one. We also need to have empirical studies to inform us of the effectiveness of didactic designs – like teaching modules, working with primary sources, etc. First of all, we need to design modules and teaching sequences that will not only integrate the history of mathematics but also open spaces for classroom reflection and discussion. Resorting to history to improve the teaching and learning of mathematics is not just a question of providing information that cubic equations were solved in a certain year or that a certain event happened in the 18th century, or that somebody called X did this and that. In resorting to history in the teaching and learning of mathematics, we have to go beyond historical facts. We live in a historical world – a world populated by historically formed concrete and abstract objects, modes of thinking and values. The main point is that the reality in front of us is intrinsically historical and its understanding and disclosure can only be achieved through a critical stance. As far as I can see, a critical, historical approach is the only key to penetrate the deepness of reality. In this context, one of the fundamental questions for us mathematics educators is how to offer the students contexts for critical action, discussion, and reflection. And history, for the reasons already mentioned, must play a fundamental mediating role here. But we need empirical studies to see how it works, how critical and empowered the students become as a consequence of being exposed to a space of discussion that instead of stopping at the ephemeral present reveals its historical nature. (Radford; 2008)

Of course, history as a goal is also related to students' beliefs or images of mathematics as an academic discipline, and the argument that history has a role to play in making students images of such more profound, balanced, multifaceted, and reflected (Demattè & Furinghetti, 1999; Jankvist, 2009e; Jankvist, 2011b). But also this is an area that needs much more empirical research.

One final issue which I shall draw attention to is one related to the use of history as a tool. The vast majority of the empirical studies on history as a cognitive tool (see table 4) concern the learning of more or less specific mathematical concepts, theorems, methods, etc. For people not already familiar with the history of mathematics, the use of history to teach students matters of mathematical content knowledge may seem a

316

roundabout way to take, in particular when so many other methods within mathematics education offer the same, e.g. mathematical modeling. However, as recent studies point out, the use of history in mathematics education may come about much more naturally in the context of developing students' mathematical competencies (Kjeldsen, 2009; Jankvist & Kjeldsen, 2010) or in the context of learning what Sfard (2008) refers to as meta-discursive rules in mathematics (Kjeldsen & Blomhøj; 2011). Furthermore, these different ways of using history as a tool also show that a use of history as a tool not necessarily leads to a 'Whig' interpretation of history, as previously argued by Fried (2001).22²⁵ Perhaps, the integration of history in mathematics education from the perspective of developing students' mathematical competencies could be an eye opener for mathematics educators regarding the effectiveness and efficacy of using history as a tool in mathematics education.

In any case, whether promoting history as a tool through mathematical competencies or history as a goal through empowerment, critical citizenship, and more reflective images of mathematics, the propagation of history in mathematics education must be made through a significant increase in the body of empirical research studies. In this respect, the HPM has a pivotal role to play in the nurturing and promotion of such studies. So far, the HPM has done some, e.g. with the working group at CERME, but it is my sincere hope that over time the organization will increase its focus on the empirical aspects of using history in mathematics education. This may be done by providing researchers with other dedicated fora within already established mathematics education contexts, e.g. PME, ICME, etc., where their work may be linked to and weighted against the general mathematics education research. Needless to say, general mathematics education research has a lot to offer future empirical studies on history in mathematics education with is theoretical constructs, methodologies, etc., but surely, once having gained an empirical foothold, history in mathematics education will have something to offer the general mathematics education research in return.

# Bibliography

Alpaslan, M., Isiksal, M. & Haser, C. (2011). The development of attitudes and beliefs questionnaire towards using history of mathematics in mathematics education, *Proceedings from CERME-7 (Working Group 12)*, ERME, pp. 1–10. (Preprint).

Arcavi, A. (2007). Interview with Professor Abraham Arcavi, June 8th 2007. Conducted by Uffe Thomas Jankvist at the NoGSME summer school in Laugarvatn, Iceland.

---

[25] When discussing this, Fried (2001, p. 395) says "That mathematics teachers are committed to teaching modern mathematics and modern mathematical techniques naturally makes their relationship to the history of mathematics quite different from that of an historian of mathematics." By modern mathematics Fried is referring to modern presentations of curriculum mathematics, and on this basis he further argues that "the history of mathematics becomes something not studied but something *used*", and continues to say that "when *history* is being *used* to justify, enhance, explain, or encourage distinctly modern subjects and practices, it inevitably becomes what is 'anachronical' [...] or 'Whig' history."

Arcavi, A., Bruckheimer, M. & Ben-Zvi, R. (1982). Maybe a mathematics teacher can profit from the study of the history of mathematics, *For the Learning of Mathematics* **3**(1): 30–37.

Arcavi, A., Bruckheimer, M. & Ben-Zvi, R. (1987). History of mathematics for teachers: the
case of irrational numbers, *For the Learning of Mathematics* **7**(2): 18–23.

Arcavi, A. & Isoda, M. (2007). Learning to listen: from historical sources to classroom practice, *Educational Studies in Mathematics* **66**: 111–129. Special issue on the history of mathematics in mathematics education.

Bakker, A. (2004). *Design research in statistics education – On symbolizing and computer tools*, PhD thesis, The Freudenthal Inistitute, Utrecht.

Bakker, A. & Gravemeijer, K. P. E. (2006). An historical phenomenology of mean and median, *Educational Studies in Mathematics* **62**: 149–168.

Barabash, M. & Guberman-Glebov, R. (2004). Seminar and graduate project in the history
of mathematics as a source of cultural and intercultural enrichment of the academic teacher education program, *Mediterranean Journal for Research in Mathematics Education* **3**(1-2): 73–88. Special double issue on the role of the history of mathematics in mathematics education (proceedings from TSG 17 at ICME 10).

Barbin, E. (2000). Integrating history: research perspectives, *in* J. Fauvel & J. van Maanen (eds), *History in Mathematics Education*, The ICMI Study, Kluwer Academic Publishers, Dordrecht, pp. 63–90. Chapter 3.

Barbin, E., Stehlíková, N. & Tzanakis, C. (2007). Europan summer universities on the history and epistemology in mathematics education, *in* F. Furinghetti, S. Kaijser & C. Tzanakis (eds), *Proceedings HPM2004 & ESU4*, revised edn, Uppsala Universitet, pp. xxix–xxxi.

Barbin, E., Stehlíková, N. & Tzanakis, C. (eds) (2008). *History and Epistemology in Mathematics Education: Proceedings of the 5th European Summer University (ESU5)*, Vydavatelský servis, Plzen.

Barnett, J., Bezhanishvili, G., Leung, H., Lodder, J., Pengelley, D. & Ranjan, D. (2008). Learning discrete mathematics and computer science via primary historical sources: Student projects for the classroom, *in* R. Cantoral, F. Fasanelli, A. Garciadiego, B. Stein & C. Tzanakis (eds), *Proceedings of HPM2008, The satellite meeting of ICME 11*, HPM, Mexico City, pp. 1–15. CD-ROM.

Bartolini Bussi, M. G. & Pergola, M. (1996). History in the mathematics classroom: Linkages and kinematic geometry, *in* H. N. Jahnke, N. Knoche & M. Otte (eds), *History of Mathematics and Education: Ideas and Experiences*, number 11 in *Studien zur*

*Wissenschafts-, Sozial- und Bildungsgeschichte der Mathematik*, Vandenhoeck & Ruprecht, Göttingen, pp. 39–67.

Bekken, O. B. & Mosvold, R. (eds) (2003). *Study the Masters: The Abel-Fauvel Conference*, Nationellt Centrum för Matematikutbildning, NCM, Göteborg.

Bero, P. (1996). Pupils' perception of the continuum, *in* R. Calinger (ed.), *Vita Mathematica*
*– Historical Research and Integration with Teaching*, number 40 in *MAA Notes*, The Mathematical Association of America, Washington, pp. 303–308.

Biggs, N., Lloyd, E. & Wilson, R. (1976). *Graph Theory: 1736-1936*, Clarendon Press, Oxford.

Blanco, M. & Giovart, M. (2009). Introducing the normal distribution by following a teaching
approach inspired by history: An example for classroom implementation in engineering education, *Proceedings from CERME-6 (Working Group 15)*, ERME, pp. 2702–2711.
**URL:** *http://www.inrp.fr/editions/editions-electroniques/cerme6/*

Brousseau, G. (1997). *Theory of Didactical Situations in Mathematics*, Kluwer Academic Publishers, Dordrecht. Edited and translated by Nicolas Balacheff, Martin Cooper, Rosamund Sutherland, and Virgina Warfield.

Bruckheimer, M. & Arcavi, A. (2000). Mathematics and its history: An educational partnership, *In* V. Katz (ed.), *Using History to Teach Mathematics – An International Perspective*, number 51 in *MAA Notes*, The Mathematical Association of America, Washington, pp. 135–146.

Calinger, R. (ed.) (1996). *Vita Mathematica – Historical Research and Integration with Teaching*, The Mathematical Association of America, Washington.

Cantoral, R., Fasanelli, F., Garciadiego, A., Stein, B. & Tzanakis, C. (eds) (2008). *Proceedings of HPM2008, The satellite meeting of ICME 11*, HPM, Mexico City. CD-ROM.

Ceylan Alibeyoglu, M. (2008). Mathematics in zeugma, *in* R. Cantoral, F. Fasanelli, A. Garciadiego, B. Stein & C. Tzanakis (eds), *Proceedings of HPM2008, The satellite meeting of ICME 11*, HPM, Mexico City, pp. 1–3. CD-ROM.

Charalambous, C. Y., Panaoura, A. & Philippou, G. (2009). Using the history of mathematics to induce changes in preservice teachers' beliefs and attitudes: insights from evaluating a teacher education program, *Educational Studies in Mathematics* **71**(2): 161–180.

Clark, K. M. (2006). *Investigating teachers' experiences with the history of logarithms: a collection of five case studies*, PhD thesis, University of Maryland, College Park.

Clark, K. M. (2011). Voices from the field: Incorporating history of mathematics in secondary and post-secondary classrooms, *Proceedings from CERME-7 (Working Group 12)*, ERME, pp. 1–10. (Preprint).

Demattè, A. (2007). A questionnaire for discussing the 'strong' role of the history of mathematics in the classroom, *in* F. Furinghetti, S. Kaijser & C. Tzanakis (eds), *Proceedings HPM2004 & ESU4*, revised edn, Uppsala Universitet, pp. 218–228. Demattè, A. & Furinghetti, F. (1999). An exploratory study on students' beliefs about mathematics as a socio-cultural process, *in* G. Philippou (ed.), *Eighth European Workshop: Research on Mathematical Beliefs – MAVI-8 Proceedings*, University of Cyprus, Nicosia, pp. 38–47.

Dimitriadou, E. (2008). Didactical and epistemological issues related to the concept of proof: Some mathematics teachers' ideas about the role of proof in Greek secondary curriculum, *in* E. Barbin, N. Stehlíková & C. Tzanakis (eds), *History and Epistemology in Mathematics Education: Proceedings of the 5th European Summer University (ESU5)*, Vydavatelský servis, Plzen, pp. 363–372.

Dorier, J.-L. (2000). *On the Teaching of Linear Algebra*, Kluwer Academic Publishers, Dordrecth, The Netherlands.

Durand-Guerrier, V. & Arsac, G. (2005). An epistemological and didactic study of a specific calculus reasoning rule, *Educational Studies in Mathematics* **60**: 149–172.

Farmaki, V. & Paschos, T. (2007). Employing genetic 'moments' in the history of mathematics in classroom activites, *Educational Studies in Mathematics* **66**: 83–106.

Fasanelli, F. & Fauvel, J. G. (2007). The international study group on the relations between the history and pedagogy of mathematics: The first twenty-five years, 1967-2000, *in* F. Furinghetti, S. Kaijser & C. Tzanakis (eds), *Proceedings HPM2004 & ESU4*, revised edn, Uppsala Universitet, pp. x–xxviii.

Fauvel, J. (1991). Using history in mathematics education, *For the Learning of Mathematics* **11**(2): 3–6. Special Issue on History in Mathematics Education.

Fauvel, J. (ed.) (1990). *History in the Mathematics Classroom*, The IREM Papers, The Mathematical Association, Leicester.

Fauvel, J. & van Maanen, J. (eds) (2000). *History in Mathematics Education – The ICMI Study*, Kluwer Academic Publishers, Dordrecht.

Fraser, B. J. & Koop, A. J. (1978). Teachers' opinions about some teaching material involving history of mathematics, *International Journal of Mathematical Education in Science and Technology* **9**(2): 147–151.

Fried, M. N. (2001). Can mathematics education and history of mathematics coexist?, *Science & Education* **10**: 391–408.

Fried, M. N. (2007). Didactics and hisotry of mathematics: knowledge and self-knowledge, *Educational Studies in Mathematics* **66**: 203–223. Special issue on the history of mathematics in mathematics education.

Fung, C.-I. (2004). How history fuels teaching for mathematising: Some personal reflections, *Mediterranean Journal for Research in Mathematics Education* **3**(1-2): 125–146. Special double issue on the role of the history of mathematics in mathematics education (proceedings from TSG 17 at ICME 10).

Furinghetti, F. (1993). Images of mathematics outside the community of mathematicians: Evidence and explaniations, *For the Learning of Mathematics* **13**(2): 33–38.

Furinghetti, F. (2000). The history of mathematics as a coupling link between secondary and university teaching, *International Journal of Mathematical Education in Science and Technology* **31**(1): 43–51.

Furinghetti, F. (2004). History and mathematics education: A look around the world with
particular reference to Italy, *Mediterranean Journal for Research in Mathematics Education*
**3**(1-2): 1–20. Special double issue on the role of the history of mathematics in mathematics education (proceedings from TSG 17 at ICME 10).

Furinghetti, F. (2007). Teacher education through the history of mathematics, *Educational Studies in Mathematics* **66**: 131–143. Special issue on the history of mathematics in mathematics education.

Furinghetti, F., Kaijser, S. & Tzanakis, C. (eds) (2007). *Proceedings HPM2004 &ESU4*, Uppsala Universitet, Uppsala.

Furinghettti, F. (1997). History of mathematics, mathematics education, school practice: Case studies in linking different domains, *For the Learning of Mathematics* **17**(1): 55–61.

Glaubitz, M. R. (2008). The use of original sources in the classroom: Theoretical perspectives and empirical evidence, *in* E. Barbin, N. Stehlíková & C. Tzanakis (eds), *History and Epistemology in Mathematics Education: Proceedings of the 5th European Summer University (ESU5)*, Vydavatelský servis, Plzen, pp. 373–382.

Gonulates, F. (2008). Prospective teachers' views on the integration of history of mathematics in mathematics courses, *in* R. Cantoral, F. Fasanelli, A. Garciadiego, B. Stein & C. Tzanakis (eds), *Proceedings of HPM2008, The satellite meeting of ICME 11*, HPM, Mexico City, pp. 1–11. CD-ROM.

González-Martín, A. S. & Correia de Sá, C. (2008). Historical-epistemological dimension of the improper integral as a guide for new teaching practices, *in* E. Barbin, N. Stehlíková

& C. Tzanakis (eds), *History and Epistemology in Mathematics Education: Proceedings of the 5th European Summer University (ESU5)*, Vydavatelský servis, Plzen, pp. 211–224.

Goodwin, D. M. (2007). *Exploring the relationship between high school teachers' mathematics history knowledge and their images of mathematics*, PhD thesis, University of Massachusetts, Lowell.

Grattan-Guinness, I. (1973). Not from nowhere – history and philosophy behind mathematical education, *International Journal of Mathematical Education in Science and Technology* **4**: 421–453.

Grattan-Guinness, I. (1978). On the relevance of the history of mathematics to mathematics education, *International Journal of Mathematical Education in Science and Technology* **8**: 275–285.

Greenwald, S. J. (2005). Incorporating the mathematics achievements of women and minority mathematicians into classrooms, *in* A. Shell-Gellasch & D. Jardine (eds), *From Calculus to Computers – Using the last 200 years of mathematics history in the classroom*, number 68 in *MAA Notes*, The Mathematical Association of America, Washington, pp. 183–200.

Gulikers, I. & Blom, K. (2001). 'A historical angle', a survey of recent literature on the use and value of the history in geometrical education, *Educational Studies in Mathematics* **47**: 223–258.

Harper, E. (1987). Ghost of Diophanthus, *Educational Studies of Mathematics* **18**: 75–90.

Haverhals, N. & Roscoe, M. (2010). The history of mathematics as a pedagogical tool: Teaching the integral of the secant via Mercator's projection, *The Montana Mathematics Enthusiast* **7**(2-3). 339-368.

Heiede, T. (1996). History of mathematics and the teacher, *in* R. Calinger (ed.), *Vita Mathematica – Historical Research and Integration with Teaching*, number 40 in *MAA Notes*, The Mathematical Association of America, Washington, pp. 231–244.

Helfgott, M. (1995). Improved teaching of the calculus through the use of historical materials, *in* F. Swetz, J. Fauvel, O. Bekken, B. Johansson & V. Katz (eds), *Learn From The Masters*, The Mathematical Association of America, pp. 135–144.

Horng,W.-S. (2007). Teachers' professional development in terms of the HPM: A story of Yu, *in* F. Furinghetti, S. Kaijser & C. Tzanakis (eds), *Proceedings HPM2004 & ESU4*, revised edn, Uppsala Universitet, pp. 346–358. Note that due to an error in the printing of the proceedings it is not the correct version of this paper that appears in these. I obtained a correct copy from the editors.

Horng, W.-S. & Lin, F.-L. (eds) (2000). *Proceedings of the HPM 2000 Conference: History*

*in Mathematics Education: Challenges for a New Millennium*, Department of Mathematics,
National Taiwan Normal University, Taipei, Taiwan.

Hsiao, Y.-H. & Chang, C.-K. (2000). Using mathematics history and pcdc instruction model to activate underachievement students' mathematics learning, *in* W.-S. Horng & F.-L. Lin (eds), *Proceedings of the HPM 2000 Conference: History in Mathematics Education: Challenges for a New Millennium*, Department of Mathematics, National Taiwan Normal University, Taipei, Taiwan, pp. 162–170. Volume I.

Hsieh, C.-J. & Hsieh, F.-J. (2000). What are teachers' views of mathematics? – an investigation of how they evaluate formulas in mathematics, *in* W.-S. Horng & F.-L. Lin (eds), *Proceedings of the HPM 2000 Conference: History in Mathematics Education: Challenges for a New Millennium*, Department of Mathematics, National Taiwan Normal University, Taipei, Taiwan, pp. 98–111. Volume I.

Hsieh, F.-J. (2000). Teachers' teaching beliefs and their knowledge about the history of negative numbers, *in* W.-S. Horng & F.-L. Lin (eds), *Proceedings of the HPM 2000 Conference: History in Mathematics Education: Challenges for a New Millennium*, Department of Mathematics, National Taiwan Normal University, Taipei, Taiwan, pp. 88–97. Volume I.

Isaacs, I., Ram, V. M. & Richards, A. (2000). A historical approach to developing the cultural significance of mathematics among first year preservice primary school teachers, *in* V. Katz (ed.), *Using History to Teach Mathematics – An International Perspective*, number 51 in *MAA Notes*, The Mathematical Association of America, Washington, pp. 123–128.

Isoda, M. (2007). Why we use historical tools and computer software in mathematics education: Mathematics activity as a human endeavor project for secondary school, *in* F. Furinghetti, S. Kaijser & C. Tzanakis (eds), *Proceedings HPM2004 & ESU4*, revised edn, Uppsala Universitet, pp. 229–236.

Jahnke, H. N. (2000). The use of original sources in the mathematics classroom, *in* J. Fauvel & J. van Maanen (eds), *History in Mathematics Education*, The ICMI Study, Kluwer Academic Publishers, Dordrecht, pp. 291–328. Chapter 9.

Jahnke, H. N., Knoche, N. & Otte, M. (eds) (1996). *History of Mathematics and Education: Ideas and Experiences*, number 11 in *Studien zur Wissenschafts-, Sozial- und Bildungsgeschichte der Mathematik*, Vandenhoeck & Ruprecht, Göttingen.

Jankvist, U. T. (2007). Empirical research in the field of using history in mathematics education: Review of empirical studies in HPM2004 & ESU4, *Nomad* **12**(3): 83–105.

Jankvist, U. T. (2008a). Den matematikhistoriske dimension i undervisning – gymnasialt set, *MONA* **4**(1): 24–45. English translation of title: The Dimension of the History of Mathematics in Teaching – The Case of Upper Secondary Level.

Jankvist, U. T. (2008b). Evaluating a teaching module on the early history of error correcting codes, *in* M. Kourkoulos & C. Tzanakis (eds), *Proceedings 5th International Colloquium on the Didactics of Mathematics*, The University of Crete, Rethymnon, pp. 447–460.

Jankvist, U. T. (2008c). History of modern mathematics and/or modern applications of mathematics in mathematics education, *in* R. Cantoral, F. Fasanelli, A. Garciadiego, B. Stein & C. Tzanakis (eds), *Proceedings of HPM2008, The satellite meeting of ICME 11*, HPM, Mexico City, pp. 1–10. CD-ROM.

Jankvist, U. T. (2008d). A teaching module on the history of public-key cryptography and RSA, *BSHM Bulletin* **23**(3): 157–168.

Jankvist, U. T. (2009a). A categorization of the 'whys' and 'hows' of using history in mathematics education, *Educational Studies in Mathematics* **71**(3): 235–261.

Jankvist, U. T. (2009b). History of modern applied mathematics in mathematics education, *For the Learning of Mathematics* **29**(1): 8–13.

Jankvist, U. T. (2009c). On empirical research in the field of using history in mathematics education, *ReLIME* **12**(1): 67–101.

Jankvist, U. T. (2009d). Students' beliefs about the evolution and development of mathematics, *Proceedings from CERME-6 (Working Group 15)*, ERME, pp. 2732–2741. **URL:** *http://www.inrp.fr/editions/editions-electroniques/cerme6/*

Jankvist, U. T. (2009e). *Using History as a 'Goal' in Mathematics Education*, PhD thesis, IMFUFA, Roskilde University, Roskilde. Number 464 in Tekster fra IMFUFA. **URL:** *http://milne.ruc.dk/imfufatekster/pdf/464.pdf*

Jankvist, U. T. (2010). An empirical study of using history as a 'goal', *Educational Studies in Mathematics* **74**(1): 53–74.

Jankvist, U. T. (2011a). Anchoring students' meta-perspective discussions of history in mathematics, *Journal for Research in Mathematics Education*. In Press.

Jankvist, U. T. (2011b). Changing students' images of mathematics as a discipline. In review.

Jankvist, U. T. (2011c). Designing teaching modules on the history, application, and philosophy of mathematics, *Proceedings from CERME-7 (Working Group 12)*, ERME, pp. 1–10. (Preprint).

Jankvist, U. T. & Kjeldsen, T. H. (2010). New avenues for history in mathematics education: Mathematical competencies and anchoring, *Science & Education* . In Press: Available Online First.

Katz, V. (ed.) (2000). *Using History to Teach Mathematics – An International Perspective*, number 51 in *MAA Notes*, The Mathematical Association of America, Washington.

Kjeldsen, T. H. (2009). Using history as a means for the learning of mathematics without losing sight of history: The case of differential equations, *Proceedings from CERME-6 (Working Group 15)*, ERME, pp. 2742–2751.
**URL:** *http://www.inrp.fr/editions/editions-electroniques/cerme6/*

Kjeldsen, T. H. (2011). Uses of history in mathematics education: Development of learning
strategies and historical awareness, *Proceedings from CERME-7 (Working Group 12)*, ERME, pp. 1–10. (Preprint).

Kjeldsen, T. H. & Blomhøj, M. (2009). Integrating history and philosophy in mathematics education at university level through problem-oriented project work, *ZDM – The International Journal on Mathematics Education* **41**: 87–103.

Kjeldsen, T. H. & Blomhøj, M. (2011). Beyond motivation – history as a method for the learning of meta-discursive rules in mathematics, *Educational Studies in Mathematics* . In Press.

Kleiner, I. (1996). A history-of-mathematics course for teachers, based on great quotations, *in* R. Calinger (ed.), *Vita Mathematica – Historical Research and Integration with Teaching*, number 40 in *MAA Notes*, The Mathematical Association of America, Washington, pp. 261–268.

Kotarinou, P., Stathopoulou, C. & Chronaki, A. (2011). Establishing the 'meter' as citizens of French national assembly during the French revolution, *Proceedings from CERME-7 (Working Group 12)*, ERME, pp. 1–10. (Preprint).

Kourkoulos, M. & Tzanakis, C. (2008). Contributions from the study of the history of statistics in understanding students' difficulties for the comprehension of the variance, *in* R. Cantoral, F. Fasanelli, A. Garciadiego, B. Stein & C. Tzanakis (eds), *Proceedings of HPM2008, The satellite meeting of ICME 11*, HPM, Mexico City, pp. 1–25. CD-ROM.

Lakoma, E. (2000). On the effectiveness of history in teaching probability and statistics, *in* W.-S. Horng & F.-L. Lin (eds), *Proceedings of the HPM 2000 Conference: History in Mathematics Education: Challenges for a New Millennium*, Department of Mathematics, National Taiwan Normal University, Taipei, Taiwan, pp. 171–177. Volume I.

Laubenbacher, R. C. & Pengelley, D. J. (1996). Mathematical masterpieces: Teaching with original sources, *in* R. Calinger (ed.), *Vita Mathematica – Historical Research and Integration with Teaching*, number 40 in *MAA Notes*, The Mathematical Association of America, Washington, pp. 257–260.

Lawrence, S. (2008). History of mathematics making its way through the teacher networks:
Professional learning environment and the history of mathematics in mathematics curriculum. Contribution to Topic Study Group 23 at ICME11.
**URL:** *http://tsg.icme11.org/tsg/show/24*

Lawrence, S. (2009). What works in the classroom – project on the history of mathematics and the collaborative teaching practice, *Proceedings from CERME-6 (Working Group 15)*, ERME, pp. 2752–2761.
**URL:** http://www.inrp.fr/editions/editions-electroniques/cerme6/

Lawrence, S. & Ransom, P. (2011). How much meaning can we construct around geometric constructions?, *Proceedings from CERME-7 (Working Group 12)*, ERME, pp. 1–10. (Preprint).

Lin, Y.-Y. (2000). F&B mathematics teaching in vocational school: A team work with HPM perspectives, *in* W.-S. Horng & F.-L. Lin (eds), *Proceedings of the HPM 2000 Conference: History in Mathematics Education: Challenges for a New Millennium*, Department of Mathematics, National Taiwan Normal University, Taipei, Taiwan, pp. 61–66. Volume II.

Lit, C.-K., Siu, M.-K. & Wong, N.-Y. (2001). The use of history in the teaching of mathematics: Theory, practice, and evaluation of effectiveness, *Educational Journal* **29**(1): 17–31.

Liu, P.-H. (2000). The use of history of mathematics to increase students' understanding of
the nature of mathematics, *in* W.-S. Horng & F.-L. Lin (eds), *Proceedings of the HPM 2000 Conference: History in Mathematics Education: Challenges for a New Millennium*, Department of Mathematics, National Taiwan Normal University, Taipei, Taiwan, pp. 178–188. Volume I.

Liu, P.-H. (2007). The historical development of the fundamental theorem of calculus and its implications in teaching, *in* F. Furinghetti, S. Kaijser & C. Tzanakis (eds), *Proceedings HPM2004 & ESU4*, revised edn, Uppsala Universitet, pp. 237–246.

Liu, P.-H. (2008). Investigation of students' perceptions of the infinite: A historical dimension, *in* E. Barbin, N. Stehlíková & C. Tzanakis (eds), *History and Epistemology in Mathematics Education: Proceedings of the 5th European Summer University (ESU5)*, Vydavatelský servis, Plzen, pp. 385–394.

McBride, C. C. & Rollins, J. H. (1977). The effects of history of mathematics on attitudes toward mathematics of college algebra students, *Journal for Research in Mathematics Education* **8**(1): 57–61.

McGuire, L. E. (2005). Using 20th century history in a combinatorics and graph theory class, *in* A. Shell-Gellasch & D. Jardine (eds), *From Calculus to Computers – Using the last*

*200 years of mathematics history in the classroom*, number 68 in *MAA Notes*, The Mathematical Association of America, Washington, pp. 101–108.

Ming, Y. F. (2000). Using mathematics texts in classroom: The case of Pythagorean theorem, *in* W.-S. Horng & F.-L. Lin (eds), *Proceedings of the HPM 2000 Conference: History in Mathematics Education: Challenges for a New Millennium*, Department of Mathematics, National Taiwan Normal University, Taipei, Taiwan, pp. 67–82. Volume II.

Morey, B. (2008). Navigation instruments and teacher training, *in* E. Barbin, N. Stehlíková & C. Tzanakis (eds), *History and Epistemology in Mathematics Education: Proceedings of the 5th European Summer University (ESU5)*, Vydavatelský servis, Plzen, pp. 587–594.

Mosvold, R. (2001). *Det genetiske prinsipp i matematikkdidaktikk*, Master's thesis, Høgskolen i Agder, Kristiansand.

Mosvold, R. (2002a). Genesis principles in mathematics education, *Telemarksforskning Notodden* **02**(09): 1–21.

Mosvold, R. (2002b). "Genetisk" – begrepsforvirring eller begrepsavklaring?, *Telemarksforskning Notodden* **02**(10): 1–16.

Mosvold, R. (2003). Genesis principles in mathematics education, *in* O. Bekken & R. Mosvold (eds), *Study the Masters*, Nationellt Centrum för Matematikutbildning, NCM, Göteborgs Universitet, pp. 85–96. Proceedings from the Abel-Fauvel Conference in Kristiansand.

Nagaoka, R. (2000). Non-standard media and other resources, *in* J. Fauvel & J. van Maanen (eds), *History in Mathematics Education*, The ICMI Study, Kluwer Academic Publishers, Dordrecht, pp. 329–370. Chapter 10.

Nataraj, M. S. & Thomas, M. O. J. (2008). Using history of mathematics to develop student understanding of number system structure, *in* R. Cantoral, F. Fasanelli, A. Garciadiego, B. Stein & C. Tzanakis (eds), *Proceedings of HPM2008, The satellite meeting of ICME 11*, HPM, Mexico City, pp. 1–12. CD-ROM.

Niss, M. & Jensen, T. H. (eds) (2002). *Kompetencer og matematiklæring – Ideer og inspiration til udvikling af matematikundervisning i Danmark*, Undervisningsministeriet. Uddannelsesstyrelsens temahæfteserie nr. 18. English translation of title: Competencies and Learning of Mathematics – Ideas and Inspiration for the Development of Mathematics Education in Denmark.

Ohara, Y. (2000). Epistemological complexity of multiplication and division from the view of dimensional analysis, *in* W.-S. Horng & F.-L. Lin (eds), *Proceedings of the HPM 2000 Conference: History in Mathematics Education: Challenges for a New Millennium*,

Department of Mathematics, National Taiwan Normal University, Taipei, Taiwan, pp. 189–195. Volume I.

OReilly, M. (2011). Using students' journals to explore their affective engagement in a module on the history of mathematics, *Proceedings from CERME-7 (Working Group 12)*, ERME, pp. 1–10. (Preprint).

Paschos, T. & Farmaki, V. (2008). The integration of genetic moments in the history of mathematics and physics in the designing of didactic activities aiming to introduce first-year undergraduates to concepts of calculus, *in* E. Barbin, N. Stehlíková & C. Tzanakis (eds), *History and Epistemology in Mathematics Education: Proceedings of the 5th European Summer University (ESU5)*, Vydavatelský servis, Plzen, pp. 297–310.

Peard, R. (2008). Quantitative literacy for pre-service elementary teachers within social and historical contexts, *in* R. Cantoral, F. Fasanelli, A. Garciadiego, B. Stein & C. Tzanakis (eds), *Proceedings of HPM2008, The satellite meeting of ICME 11*, HPM, Mexico City, pp. 1–9. CD-ROM.

Pengelley, D. J. (2003a). The bridge between the continous and the discrete via original sources, *in* O. Bekken & R. Mosvold (eds), *Study the Masters*, Nationellt Centrum för Matematikutbildning, NCM, Göteborg, pp. 63–74.

Pengelley, D. J. (2003b). A graduate course on the role of history in teaching mathematics, *in* O. Bekken & R. Mosvold (eds), *Study the Masters*, Nationellt Centrum för Matematikutbildning, NCM, Göteborg, pp. 53–62.

Pengelley, D. J. (2008). Teaching mathematics with primary historical sources: Should it go mainstream? Can it?, *in* R. Cantoral, F. Fasanelli, A. Garciadiego, B. Stein & C. Tzanakis (eds), *Proceedings of HPM2008, The satellite meeting of ICME 11*, HPM, Mexico City, pp. 1–17. CD-ROM.

Philippou, G. N. & Christou, C. (1998). The effects of a preparatory mathematics program in changing prospective teachers' attitudes towards mathematics, *Educational Studies in Mathematics* **35**: 189–206.

Prabhu, V. & Czarnocha, B. (2000). Method of indivisibles in calculus instruction, *in* W.-S. Horng & F.-L. Lin (eds), *Proceedings of the HPM 2000 Conference: History in Mathematics Education: Challenges for a New Millennium*, Department of Mathematics, National Taiwan Normal University, Taipei, Taiwan, pp. 196–203. Volume I.

Radford, L. (1997). On psychology, historical epistemology, and the teaching of mathematics: towards a socio-cultural history of mathematics, *For the Learning of Mathematics* **17**(1): 26–33.

Radford, L. (2000a). Historical formation and student understanding of mathematics, *in* J. Fauvel & J. van Maanen (eds), *History in Mathematics Education*, The ICMI Study, Kluwer Academic Publishers, Dordrecht, pp. 143–170. Chapter 5.

Radford, L. (2000b). Signs and measuring in students' emergent algebraic thinking: A semiotic analysis, *Educational Studies in Mathematics* **42**: 237–268.

Radford, L. (2008). Interview with Professor Luis Radford, June 5th 2008. Conducted by Uffe Thomas Jankvist at the Laurentian University, Sudbury, Ontario, Canada.

Radford, L. & Puig, L. (2007). Syntax and meaning as sensuous, visual, historical forms of algebraic thinking, *Educational Studies in Mathematics* **66**: 145–164. Special issue on the history of mathematics in mathematics education.

Ransom, P. (2011). A cross-curricular approach using history in the mathematics classroom with students aged 11-16, *Proceedings from CERME-7 (Working Group 12)*, ERME, pp. 1–10. (Preprint).

Reed, B. M. (2008). The effects of studying the history of the concept of function on student understanding of the concept, *in* R. Cantoral, F. Fasanelli, A. Garciadiego, B. Stein & C. Tzanakis (eds), *Proceedings of HPM2008, The satellite meeting of ICME 11*, HPM, Mexico City, pp. 1–14. CD-ROM.

Russ, S., Ransom, P., Perkins, P., Barbin, E., Arcavi, A., Brown, G. & Fowler, D. (1991). The experience of history in mathematics education, *For the Learning of Mathematics* **11**(2): 7–16. Special Issue on History in Mathematics Education.

Schubring, G. (1978). *Das genetische Prinzip in der Mathematik-Didaktik*, Klett-Cotta, Stuttgart.

Schubring, G. (2000). History of mathemtics for trainee teachers, *in* J. Fauvel & J. van Maanen (eds), *History in Mathematics Education*, The ICMI Study, Kluwer Academic Publishers, Dordrecht, pp. 91–142. Chapter 4.

Sfard, A. (1995). The development of algebra: Confronting historical and psychological perspectives, *Journal of Mathematical Behaviour* **14**: 15–39.

Sfard, A. (2008). *Thinking as Communicating: Human development, the growth of discourses, and mathematizing*, Cambridge University Press, New York.

Shell-Gellasch, A. & Jardine, D. (eds) (2005). *From Calculus to Computers – Using the last 200 years of mathematics history in the classroom*, number 68 in *MAA Notes*, The Mathematical Association of America, Washington.

Siu, F.-K.&Siu, M.-K. (1979). History of mathematics and its relation to mathematical education, *International Journal of Mathematical Education in Science and Technology* **10**(4): 561–567.

Siu, M.-K. (2000). Historical support for particular subjects, *in* J. Fauvel & J. van Maanen (eds), *History in Mathematics Education*, The ICMI Study, Kluwer Academic Publishers, Dordrecht, pp. 241–290. Chapter 8.

Siu, M.-K. (2006). Interview with Professor Man-Keung Siu October 5th 2006. Conducted by Uffe Thomas Jankvist at Hong Kong University, Pokfulam, Hong Kong.

Siu, M.-K. (2007). 'No, I don't use history of mathematics in my class. Why?', *in* F. Furinghetti, S. Kaijser & C. Tzanakis (eds), *Proceedings HPM2004 & ESU4*, revised edn, Uppsala Universitet, pp. 268–277.
**URL:** *http://hkumath.hku.hk/ mks/*

Siu, M.-K. & Tzanakis, C. (2004). History of mathematics in classroom teaching – appetizer?, main course? or dessert?, *Mediterranean Journal for Research in Mathematics Education* 3(1-2): v–x. Special double issue on the role of the history of mathematics in mathematics education (proceedings from TSG 17 at ICME 10).

Skovsmose, O. (1990). Mathematical education and democracy, *Educational Studies in Mathematics* **21**: 109–128.

Smestad, B. (2003). Historical topics in Norwegian textbooks, *in* O. Bekken & R. Mosvold (eds), *Study the Masters*, Nationellt Centrum för Matematikutbildning, NCM, Göteborg, pp. 163–168.

Smestad, B. (2007). History of mathematics in the TIMSS 1999 video study, *in* F. Furinghetti, S. Kaijser & C. Tzanakis (eds), *Proceedings HPM2004 & ESU4*, revised edn, Uppsala Universitet, pp. 278–283.

Smestad, B. (2008). Teachers' conceptions of history of mathematics, *in* R. Cantoral, F. Fasanelli, A. Garciadiego, B. Stein & C. Tzanakis (eds), *Proceedings of HPM2008, The satellite meeting of ICME 11*, HPM, Mexico City, pp. 1–10. CD-ROM.

Su, Y.-W. (2000). Using mathematical text in classroom: The case of probability theory, *in* W.-S. Horng & F.-L. Lin (eds), *Proceedings of the HPM 2000 Conference: History in Mathematics Education: Challenges for a New Millennium*, Department of Mathematics, National Taiwan Normal University, Taipei, Taiwan, pp. 215–221. Volume I.

Su, Y.-W. (2007). Mathematics teachers' professional development: Integrating history of mathematics into teaching, *in* F. Furinghetti, S. Kaijser & C. Tzanakis (eds), *Proceedings HPM2004 & ESU4*, revised edn, Uppsala Universitet, pp. 368–382.

Swetz, F., Fauvel, J., Bekken, O., Johansson, B. & Katz, V. (eds) (1995). *Learn from the Masters*, The Mathematical Association of America, Washington.

Tang, K.-C. (2007). History of mathematics for the young educated minds: A Hong Kong reflection, *in* F. Furinghetti, S. Kaijser & C. Tzanakis (eds), *Proceedings HPM2004 & ESU4*, revised edn, Uppsala Universitet, pp. 630–638.

Tardy, C. & Durand-Guerrier, V. (2009). Introduction of an historical and anthropological perspective in mathematics: An example in secondary school in France, *Proceedings from*
*CERME-6 (Working Group 15)*, ERME, pp. 2791–2800.
**URL:** *http://www.inrp.fr/editions/editions-electroniques/cerme6/*

Thomaidis, Y. & Tzanakis, C. (2007). The notion of historical 'parallelism' revisited: historical evolution and students' conception of the order relation on the number line, *Educational Studies in Mathematics* **66**: 165–183. Special issue on the history of mathematics in mathematics education.

Thomaidis, Y. & Tzanakis, C. (2008). Original texts in the classroom, *in* E. Barbin, N. Stehlíková & C. Tzanakis (eds), *History and Epistemology in Mathematics Education: Proceedings of the 5th European Summer University (ESU5)*, Vydavatelský servis, Plzen, pp. 49–62.

Thomaidis, Y. & Tzanakis, C. (2009). The implementation of the history of mathematics in the new curriculum and textbooks in Greek secondary education, *Proceedings from CERME-6 (Working Group 15)*, ERME, pp. 2801–2810.
**URL:** *http://www.inrp.fr/editions/editions-electroniques/cerme6/*

Troy, W. S. (2000). The royal observatory in greenwich; ethnomathematics in teacher training, *in* W.-S. Horng & F.-L. Lin (eds), *Proceedings of the HPM 2000 Conference: History in Mathematics Education: Challenges for a New Millennium*, Department of Mathematics, National Taiwan Normal University, Taipei, Taiwan, pp. 132–139. Volume I.

Tsukahara, K. (2000). The practical use of the history of mathematics and its usefulness in
teaching and learning mathematics at high school: The development of materials in teaching calculus, *in* W.-S. Horng & F.-L. Lin (eds), *Proceedings of the HPM 2000 Conference: History in Mathematics Education: Challenges for a New Millennium*, Department of Mathematics, National Taiwan Normal University, Taipei, Taiwan, pp. 1–16. Volume II.

Tzanakis, C. & Arcavi, A. (2000). Integrating history of mathematics in the classroom: an analytic survey, *in* J. Fauvel & J. van Maanen (eds), *History in Mathematics Education*, The ICMI Study, Kluwer Academic Publishers, Dordrecht, pp. 201–240. Chapter 7.

Tzanakis, C. & Kourkoulos, M. (2007). May history and physics provide a useful aid for introducing basic statistical concepts?, *in* F. Furinghetti, S. Kaijser & C. Tzanakis (eds), *Proceedings HPM2004 & ESU4*, revised edn, Uppsala Universitet, pp. 284–295.

Tzanakis, C. & Thomaidis, Y. (2000). Integrating the close historical development of mathematics and physics in mathematics education: Some methodological and epistemological remarks, *For the Learning of Mathematics* **20**(1): 44–55.

van Amerom, B. A. (2002). *Reinvention of early algebra – Developmental research on the transition from arithmetic to algebra*, PhD thesis, The Freudenthal Inistitute, Utrecht.

van Amerom, B. A. (2003). Focusing on informal stragies when linking arithmetic to early
algebra, *Educational Studies in Mathematics* **54**: 63–75.

van Gulik-Gulikers, I. (2005). *Meetkunde opnieuw uitgevonden – Een studie naar de waarde en de toepassing van de geschiedenis van de meetkunde in het wiskundeonderwijs*, PhD thesis, Rijksuniversiteit Groningen, Groningen.

van Maanen, J. (1991). L'Hôpital's weight problem, *For the Learning of Mathematics* **11**(2): 44–47. Special Issue on History in Mathematics Education.

van Maanen, J. (1995). Alluvial deposits, conic sections, and improper glasses, *in* F. Swetz, J. Fauvel, O. Bekken, B. Johansson & V. Katz (eds), *Learn From The Masters*, The Mathematical Association of America, pp. 73–92.

van Maanen, J. (1997). New maths may profit from old methods, *For the Learning of Mathematics* **17**(2): 39–46.

van Maanen, J. (2007). Interview with Professor Jan van Maanen March 6th 2007. Conducted by Uffe Thomas Jankvist at the Freudenthal Institute, Utrecht Overvecht, the Netherlands.

Waldegg, G. (2004). Problem solving, collaborative learning and history of mathematics: Experiences in training in-service teachers, *Mediterranean Journal for Research in Mathematics Education* **3**(1-2): 63–72. Special double issue on the role of the history of mathematics in mathematics education (proceedings from TSG 17 at ICME 10).

Winicki, G. (2000). The analysis of regula falsi as an instance for professional development of elementary school teachers, *in* V. Katz (ed.), *Using History to Teach Mathematics – An International Perspective*, number 51 in *MAA Notes*, The Mathematical Association of America, Washington, pp. 129–134.

Winicki-Landman, G. (2000). An episode in the development of mathematics teachers' knowledge? The case of quadratic equations, *in* W.-S. Horng & F.-L. Lin (eds), *Proceedings of the HPM 2000 Conference: History in Mathematics Education: Challenges for a New Millennium*, Department of Mathematics, National Taiwan Normal University, Taipei, Taiwan, pp. 82–87. Volume I.

Winicki-Landman, G. (2007). Another episode in the professional development of mathematics teachers: The case of definitions, *in* F. Furinghetti, S. Kaijser & C. Tzanakis (eds), *Proceedings HPM2004 & ESU4*, revised edn, Uppsala Universitet, pp. 383–388.

Zormbala, K. & Tzanakis, C. (2004). The concept of the plane in geometry: Elements of the historical evolution inherent in modern views, *Mediterranean Journal for Research in Mathematics Education* **3**(1-2): 37–62. Special double issue on the role of the history of mathematics in mathematics education (proceedings from TSG 17 at ICME 10).

# Reflections on and benefits of uses of history in mathematics education
## exemplified by two types of student work in upper secondary school

Tinne Hoff Kjeldsen

IMFUFA, Department of Science, Systems and Models, Roskilde University

## Introduction[1]

It is not at all clear what the role of history in mathematics education is or should be. At conferences where designs of lessons, results, and experiences with using history in teaching situations, or claiming to do so, are presented two camps can often be distinguished during the discussions according to whether the emphasis is placed on a genuine approach to history or to teach a certain subject matter within mathematics. One camp often accuses activities aiming at using history as an effective tool for teaching a certain concept or method in mathematics of distorting history. The other camp often finds that activities aiming at introducing students to historical issues of the development of mathematics have no relevance for the mathematics that needs to be taught.

Several recent papers have discussed this seemingly dilemma between genuine history and relevant mathematics that teachers are faced with if they want to use or integrate history in their classrooms.[2] While these discussions have focused on transforming views of mathematics and mathematics education, the conception of history has been taken to be more or less synonymous with a traditional professional historians' approach to history – at least in the methodological approaches and the criteria for what counts as a genuine approach to history. However, perhaps we also need to broaden our view of history as well to move beyond the two-split divide if we want history to play a more significant role for teaching and learning mathematics. The past can serve many purposes and we need to develop a more nuanced view of history in mathematics education in order to capture the complex ways in which history can benefit students learning of and about mathematics.

In this chapter such a broadened view of history is outlined, and its implications for history in mathematics education are discussed. It should be seen as the first step towards the aim of developing an adequate theoretical framework for integrating history of mathematics in mathematics education that can be used to analyze specific implementations and to provide a tool for orienting the design of future implementations of history in mathematics education. The main focus of the chapter is theoretical, but it also contains an empirical part that illustrates the theory in two types of student work in upper secondary school: (1) a carefully

---

[1] The ideas of this chapter were first presented at the European conference for research in mathematics education *CERME 7* in Polen in February 2011. Parts of the chapter constitute the conference paper which will be published in the proceedings for *CERME 7* (Kjeldsen, forthcoming b). However, the paper has been extended considerably for this chapter, especially regarding the second part where the presentation and discussion of the second type of student work is new. I would like to thank Constantinos Tzanakis for helpful comments on a preliminary version of the paper for the proceedings.

[2] (Freid, 2001; Jankvist and Kjeldsen, forthcoming; Kjeldsen, forthcoming a; Tzanakis and Thomaidis, 2010).

designed and implemented case study of experimental teaching in problem oriented group work in history of Egyptian mathematics developed in an in-service course; and (2) an inter-disciplinary student project in history and mathematics with the title *Transfer of knowledge in the Middle Ages*.

## History of mathematics from a professional point of view

Mathematical knowledge is produced and used by humans so we can think of mathematical activities as integrated elements of historical-social reality and of human life. We can perceive mathematical activities as creations of history as well as acts that create a history of mathematics. The development of mathematics and changes within our perceptions, views, and treatments of mathematics can to a certain extent be understood as realisations (intended as well as unintended) of goals set by people. If we want to understand historical-social processes of change in the development of mathematics as products of human activities, we must pay attention to intentions and thoughts of the actors, as well as their understanding of the subject matter and the context in which they performed and made their choices. Hence, we are dealing with an action oriented multiple perspective approach to history.

At a first sight it might seem that while such an approach can be used to study the history of sociological aspects of mathematics, such as the development of its profession in different countries and/or places or the history of mathematical journals, it cannot be used to study the history of the subject-matter of mathematics due to the universal character of mathematics. But if the development of mathematics is studied from its practice, where the historian focuses on concrete practices of mathematics, acknowledging that, despite its universal character, mathematical knowledge is produced by mathematicians, who live, interact and communicate in concrete social settings, the history of mathematical ideas, concepts and theories can also be pursued within such a framework.

Such a position is in accordance with recent trends in the history of mathematics that have emerged as reactions towards the well-known critic of the widely used anachronistic (whiggish) approach to history of mathematics and the methodological debate of internalism versus externalism (Epple, 2000), (Kjeldsen et al. 2004), and Science in Context, 2004, 17(1/2). Within the last decades many studies in the history of mathematics focus on the practice of mathematics within social, intellectual, and cultural contexts of mathematical activities. Here professional historians of mathematics have a critical approach to source material they analyze in order to understand its significance in its proper historical context.

However, when we deal with integrating history in teaching and learning of mathematics we must keep in mind that the education and training of professional historians of mathematics is not the main purpose of general mathematics education. In some countries development of students' historical awareness is part of the curriculum[3], but that is not always the case, and if it is it only plays a minor part. With this in mind it seems too restrictive to require that the history of mathematics taught within mathematics education should be presented as traditional academic history. A didactical transposition is needed, just as is the case with

---

[3] See e.g. (Jankvist and Kjeldsen, forthcoming; Jankvist 2010; Kjeldsen and Blomhøj, 2009).

school mathematics, which is also not identical with the discipline of (academic) mathematics. In the following, Jensen's (2010) broader view of history will be introduced along with several pairs of concepts that can be useful for a nuanced analysis and discussion of the role of past mathematical episodes for the learning and teaching of mathematics.

## Theoretical constructs for analysing different uses of the past

Jensen (2010) sees the academic research subject *history*, as professional historians think and work with it, as just one of many approaches to history. According to him, history is employed every time a person or a group of people is interested in something from the past, and uses their knowledge about it for some purpose. People use history for many different purposes and in many different connections, and consequently there are major differences between a lay person's and a professional historian's use of history. Recent investigations (Rosenzweig and Thelen, 1998) have shown that lay persons' and professional historians' conceptions of history differ in various respects and on several levels. Lay-history has a reputation of being naïve viewed from the academic discipline of history, while on the other hand lay historians view academic history as lifeless and remote from the real world. For professional historians it is important to place past episodes and artefacts in their historical contexts. Their historical awareness is conceived of as an interpretation of the past whereas lay persons view history more as a source of memoirs.

Jensen distinguishes between pragmatic and scholarly approaches to history. In a pragmatic approach the study of the past is guided by the idea that we can learn from history. The "usefulness" of history is an underlying perspective or principle in a pragmatic approach to history. The idea is that through history we can gain knowledge about our world of today, that history can teach us better ways to live our lives. In a pragmatic approach to history, past events are studied from a utility perspective. Jensen (2010, p. 51) contrasts a pragmatic approach to history with a scholarly approach, where historians retain a critical distance to past events and emphasize differences between past and present. In the professional, academic discipline of history both traditions can be found, but since the mid 19th century the scholarly approach to history has been more and more dominant.

Observer history and actor history is a third pair of concepts through which we can discuss and understand uses of past events and sources. Jensen (2010, p. 41) talks about uses of the past from an actor perspective, if we use history to orient ourselves and act in a present context. He calls this an intervening use of history. If the past is viewed retrospectively with a purpose to enlighten instead of a purpose to act or intervene he talks about uses of past from an observer perspective.

A fourth pair of concepts, introduced by Jensen (2010, p. 52) is the distinction between identity neutral and identity concrete history writing. These concepts are distilled from discussions of how history can be used in an intervening sense not only because people use their knowledge of the past in their day-to-day life, but because episodes from the past can be used to form people's identity. What counts as an identity concrete writing of history changes over time. The characteristics of an identity neutral and an identity concrete history depend on culture and time. A history that one group of people will consider identity neutral might not be considered as such by another group of people, and what was considered concrete at one point in time might not be considered as such at another point in time.

Finally, the so-called "living history" use of history is a way of using the past to help participants develop historical awareness and learning strategies. In Denmark living history takes place at some museum centres and at some yearly events. One such centre is The Medieval Centre. On their homepage (http://www.middelaldercentret.dk/Engelsk/welcome.html) they state that the centre: "is an experimental museum where you can experience life in a reconstructed late 14th century market town: Daily life, knights tournaments, trebuchets, canons, ships, markets, … and a lot more...". According to Jensen (2010, p. 145) living history appeals to so many not only because the participants actively take part in the events, but also because they use other types of learning strategies where the focus can be, for example, to develop the skills of past craftsmen.

## Bridging between history and mathematics education

These concepts of, approaches to, and thinking about history and uses of past episodes and artefacts present a framework for a refined discussion and systematic analysis of how past episodes and sources can be/are used in the integration of history for the teaching and learning of mathematics. They open up a variety of approaches to history and uses of the past for teachers who want history to play a role for teaching and learning mathematics. Which approach to choose depends on the intended learning. For example, Kjeldsen and Blomhøj (forthcoming) argue, based on Sfard's (2008) theory of thinking as communicating, that history presents itself as the obvious tool for developing students' proper meta-discursive rules, because meta-discursive rules are contingent and as such can be studied at the object level of history discourse. This presupposes a scholarly approach to history. The idea is to use past mathematical activities and sources with the intention of creating learning and teaching situations where students can experience what Sfard calls commognitive conflicts. Hence, the past is used with the purpose of intervening, and therefore the scholarly approach to history is from an actor perspective.

Kjeldsen (forthcoming a) discusses the role of history for the teaching and learning of mathematics with reference to a competence based understanding of mathematics education (Niss, 2004). Here the development of students' mathematical competence is the main purpose of mathematics education along with the development of some second order competencies, including historical overview and awareness. For the development of historical overview and awareness, a scholarly approach from an observer perspective can be chosen. For development of specific mathematical competencies, a pragmatic approach from an actor perspective might be considered.

## An experimental teaching course in history of Egyptian mathematics

In the following the theoretical framework will be used to analyze the implementation of a specific experimental teaching course in problem oriented group work in history of Egyptian mathematics which was developed in an in-service course for upper secondary teachers in Denmark. Especially the "living history" approach will be examined to see how it might be adapted as a way for mathematics teachers to use past episodes and sources to develop students' learning strategies and historical awareness.

**An in-service course in problem oriented group work in history of mathematics**
The in-service course for teachers was developed in response to a reform of the Danish upper secondary school system that was implemented in 2005. The reform challenged the teachers in several ways. First of all, history of mathematics has a more prominent position in the new curriculum where teachers have to include episodes from the history of mathematics in their teaching. Secondly, the reform requires that teachers bring mathematics into play in interdisciplinary projects in cooperation with other subjects, from science, from the humanities, and from the social sciences. Thirdly, to fulfil the new curriculum teachers have to design, organize and carry out group and project organized work in their mathematics teaching. On this basis the objective of the in-service course was to support teachers in (1) their development of an experimental teaching course in problem oriented group work in history of mathematics, (2) their implementation of the experimental course in their own classes, (3) their evaluation of their course, and (4) their documentation of their course through a written report.

The core element of the in-service course was the development of the teachers' experimental practice with history of mathematics and problem-oriented group work.[4] In problem-oriented group work students work together on a specific problem over a time period long enough to allow them to work independently of the teacher, to find information themselves, and to decide among themselves how they want to solve the problem. The problem should be challenging for the students. It should be a more or less open task for which the students do not posses a "ready-made" solution method. The problem should function as the "guiding star" for the students' work. In the ideal case every decision made during the group work should be justified by its contribution to the solution of the problem. This is crucial, since engaging in decisions provides opportunities for students to work independently, to gain control, and to direct the group work. In order for this to happen, though, the teacher needs to set a scene for the work, that is to formulate the task[5] for the work, the conditions for the working process, the time constraints, and the requirements for the end product, for example a written report or a power point presentation fulfilling some specific requirements. In this way it is possible for the teacher to have some control while at the same time to leave room for the students to work independently of the teacher, to take responsibility and to make decisions.

**Objectives and design of the experimental teaching course on Egyptian mathematics**
The experimental teaching course on Egyptian mathematics was implemented in a classroom of 1. year students (10th grade, age 16) in a Danish upper secondary school. The course was meant to be interdisciplinary, with history about Ancient Egypt in combination with their mathematics. The mathematics teacher had no experience neither with group-organized nor

---

[4] The development of experimental teaching courses in problem oriented group work in mathematical modelling was also part of the in-service course, see (Blomhøj and Kjeldsen, 2006).
[5] In this particular case the teacher formulated the problem for the group work, but the formulation of the problem can be part of the task for the students.

problem-oriented teaching in mathematics, which was his focus for his own professional development. His objectives for the students' learning were to:

a) enhance the students' competence to work in teams

b) enhance the students' independent learning

c) enhance the students' oral presentation skills

d) have the students gain experiences with power point

e) have the students appreciate that mathematics has been different from what it is today

f) develop the students' awareness that mathematical results have evolved, that mathematics is not static, which is contrary to the way it is often presented

g) develop the students' awareness that mathematics develops in an interplay with culture and society. (Wulff, 2004, p. 2-3; my translation)

The objectives fall into two parts: the first four address independent study skills, the development of which problem-oriented project work is an excellent pedagogical tool, whereas the last three concern the history of mathematics requirements of the new mathematics curriculum in Danish upper secondary school.[6]

The teacher orchestrated the students' project work in three stages:

(1)  The first stage was a two-lesson introduction to Egyptian mathematics using a text from the students' textbook (Carstensen and Frandsen, 2002), where the teacher introduced the Egyptians' method of multiplication by repeated doubling, their number symbols, and their way of formulating problems.

(2)  The introduction was followed by eight lessons during which the students worked in groups of four, guided by the teacher's description of

i)   the problem, which was given by the teacher (see below)
ii)  the learning objectives
iii) the product of the group work
iv)  the topic for each group

The groups worked independently. The teacher took the role of a consultant who could be called in for advice. When that happened he focused on posing questions and challenging the groups instead of providing answers. All groups worked on the same problem: *How and why did the Egyptians calculate?* The seven groups worked with different topics of Egyptian mathematics corresponding to different chapters of a textbook on Egyptian mathematics (Frandsen, 1996): fractions; Bread and beer (Pesu) exercises; first degree equations; two equations with two unknowns and second degree equations; the circle and approximations of $\pi$; the volume of a truncated pyramid; and computations of areas.[7]

---

[6] Note that a)-c) and e)-g) are elaborated versions of some of the ICMI Study *whys*, see Fauvel and van Maanen (2000, pp. 205, 207, 211-212).

[7] To have a whole textbook on an episode from the history of mathematics in Danish is a rare circumstance, and one of the reasons why Egypt was chosen for this group work.

(3)  Each group had to present its results for the rest of the class in an oral presentation supported by a power point presentation. This took up four lessons.

**Aspects of Egyptian mathematics**

To give a flavour of what the students worked with I will present examples from the topics of some of the group's work. The Egyptians had special notation for a few fractions e.g. 2/3 but except for those they only worked with unit fractions, i.e. fractions with 1 in the numerator, or sums of unit fractions. Writing in hieroglyphs, the Egyptians would represent a fraction by a number (written in hieroglyph) with an oval super-imposed over the number. Instead of the oval we will use a horizontal stroke, e.g. we denote 1/6 with $\bar{6}$. Also, the Egyptians did not write a symbol for addition, so $\bar{6}\ \bar{11}\ \bar{16}\ \bar{27}$ in hieroglyphs means $1/6 + 1/11 + 1/176 + 1/27$. For students of the twenty-first Century the Egyptians' way of working with fractions seemed extremely difficult and complicated.

The so-called *pesu* or bread and beer problems are significant for Egyptian mathematics. In the *Rhind Mathematical Papyrus* it concerns the problems 69-78, which have been classified as problems teaching administrative mathematics. Whereas the scribes were supposed to learn mathematical techniques from the mathematical texts they were supposed to use their mathematical skills to deal with the administrative texts (Imhausen, 2003, p. 6).

Pesu is a measure for the utilization of grain when baking bread or making beer. The pesu represents the number of bread loaves that is made from one hekat of grain, where hekat is an Egyptian measure of volume. Hence, it can be calculated as the number of loafs divided by the number of hekats of grain that is used to produce the loaves. Problem 69 of the *Rhind Mathematical Papyrus* is reproduced in the students' text book in the following form in a direct translation from the Danish text book into English (Frandsen, 1996, p. 39; the translation is mine):

| | | |
|---|---|---|
| 1 | 3 $\bar{2}$ hekat grain bakes 80 loafs | |
| 2 | Tell me, how much grain is there in each loaf, and what is its pesu. | |
| 3 | 1 | 3 $\bar{2}$ |
| 4 | 10 | 35 |
| 5 | \ 20 | 70 |
| 6 | \ 2 | 7 |
| 7 | \ $\bar{\bar{3}}$ | 2 $\bar{3}$ |
| 8 | \ $\overline{21}$ | $\bar{6}$ |
| 9 | \ $\bar{7}$ | $\bar{2}$ |
| 10 | The pesu is 22 $\bar{\bar{3}}$ $\bar{7}$ $\overline{21}$ | |
| 11 | Control: | |
| 12 | \ 1 | 22 $\bar{\bar{3}}$ $\bar{7}$ $\overline{21}$ |
| 13 | \ 2 | 45 $\bar{3}$ $\bar{4}$ $\overline{14}$ $\overline{28}$ $\overline{42}$ |
| 14 | \ $\bar{2}$ | 11 $\bar{3}$ $\overline{14}$ $\overline{42}$ |
| 15 | Result | 80 |

| | | | |
|---|---|---|---|
| 16 | 3 $\overline{2}$ hekat is 1120 ro, since | | |
| 17 | | \1 | 320 |
| 18 | | \2 | 640 |
| 19 | | \$\overline{2}$ | 160 |
| 20 | | Result | 1120 |
| 21 | Hence, calculate with 80 to get 1120. | | |
| 22 | Do as follows: | | |
| 23 | | 1 | 80 |
| 24 | | \10 | 800 |
| 25 | | 2 | 160 |
| 26 | | \4 | 320 |
| 27 | | Result | 14 |
| 28 | Each bread contains 14 ro, or $\overline{32}$ hekat 4 ro, grain | | |
| 29 | Control: | | |
| 30 | | 1 | $\overline{32}$ hekat 4 ro |
| 31 | | 2 | $\overline{16}\ \overline{64}$ hekat 3 ro |
| 32 | | 4 | $\overline{8}\ \overline{32}\ \overline{64}$ hekat 1 ro |
| 33 | | 8 | $\overline{4}\ \overline{16}\ \overline{32}$ hekat 2 ro |
| 34 | | \16 | $\overline{2}\ \overline{8}\ \overline{16}$ hekat 4 ro |
| 35 | | 32 | 1 $\overline{4}\ \overline{8}\ \overline{64}$ hekat 3 ro |
| 36 | | \64 | 2 $\overline{2}\ \overline{4}\ \overline{32}\ \overline{64}$ hekat 1 ro |
| 37 | It becomes 3 $\overline{2}$ hekat grain to 80 loaves | | |

The students are encouraged to try to perform the exercise formulated in line 1-2 before they read the remaining part – according to the author this will make it easier for the students to understand the thoughts (the procedure) of the scriber.

The problem is the following: 3 $\overline{2}$ hekats of grain produces 80 loaves of bread find its pesu and the amount of grain in each loaves of bread. As explained in the Danish textbook the pesu is found by dividing 80 with 3 $\overline{2}$. This division is performed in lines 3-9, since the sum of the numbers in the marked lines of the right column is 80. In the textbook the columns are interpreted as follows:

| Egyptians: | | explanation: |
|---|---|---|
| 1 | 3 $\overline{2}$ | $1 \times 3\ \overline{2} = 3\ \overline{2}$ |
| 10 | 35 | $10 \times 3\ \overline{2} = 35$ |
| \20 | 70 | $20 \times 3\ \overline{2} = 70$ |
| \2 | 7 | $2 \times 3\ \overline{2} = 7$ |
| \$\overline{\overline{3}}$ | 2 $\overline{3}$ | $2/3 \times 3\ \overline{2} = 2\ 1/3$ |
| \$\overline{21}$ | $\overline{6}$ | $1/21 \times 3\ \overline{2} = 1/6$ |
| \$\overline{7}$ | $\overline{2}$ | $1/7 \times 3\ \overline{2} = 1/2$ |

By adding the marked lines we get:

$$3 \ 1/2(20 + 2 + 2/3 + 1/21 + 1/7) = (70 + 7 + 2 \ 1/3 + 1/6 + ½)$$

Hence,

$$3 \ ½ \times 22 \ 18/21 = 80 \qquad \text{and} \qquad \text{the pesu} = 22 \ \bar{\bar{3}} \ \bar{7} \ \overline{21} \quad [=22 \ 18/21]$$

In line 11-15 this result is checked by multiplying the pesu $22 \ \bar{\bar{3}} \ \bar{7} \ \overline{21}$ with $3\bar{2}$ to see that it gives 80.

The next part of the problem (lines 16-20) is to calculate the amount of grain in one loaf. The $3\bar{2}$ hekat is converted into the unit ro. One hekat equals 320 ro. This is multiplied by $3\bar{2}$ to give 1120 ro. In the next part (lines 21-27) the division 1120/80 is performed. The result is 14. Hence, each loaf contains 14 ro, or, which is the same, $\overline{32}$ hekat 4 ro grain. This is then controlled by multiplying $\overline{32}$ 4 with 80 in lines 29-37.

In the chapter of first degree equations the method of *false position* is introduced with problem 24 of the *Rhind Mathematical Papyrus* as an example. The problem is the following: A quantity, and $\bar{7}$ of it added to it, becomes 19. The computation looks as follows

| | | | | |
|---|---|---|---|---|
| /1 | 7 | | | |
| $/\bar{7}$ | 1 | | | |
| | | | | |
| 1 | 8 | | | |
| /2 | 16 | | | |
| $\bar{2}$ | 4 | | | |
| $/\bar{4}$ | 2 | | | |
| $/\bar{8}$ | 1 | Do as follows: | | |
| | | | | |
| /1 | $2 \ \bar{4} \ \bar{8}$ | The quantity is | | $16 \ \bar{2} \ \bar{8}$ |
| | | | | |
| /2 | $4 \ \bar{2} \ \bar{4}$ | | $\bar{7}$ | $2 \ \bar{4} \ \bar{8}$ |
| /4 | $9 \ \bar{2}$ | | Total | 19 |

The idea underneath the method of false position is to guess an answer and propose this as a solution after which the deviation is found and the proposed solution is then corrected. The above problem can then be understood as follows: The guess of a solution is 7. If we add $\bar{7}$ to it, i.e. 1/7 of 7 added to 7, it becomes 8 (line 1-2). This is wrong, it should have been 19. Hence, the real solution is 19/8 times bigger, i.e. $\dfrac{19}{8} \cdot 7$. In line 3-7 the fraction 19/8 is calculated. It equals $2 \ \bar{4} \ \bar{8}$. The true value of the unknown is $\dfrac{19}{8} \cdot 7$ which can be calculated as $7 \cdot 2\ \overline{4}\ \overline{8}$. This calculation is performed in the last three lines, with $16\ \overline{2}\ \overline{8}$ (i.e. $16 + \frac{1}{2} + 1/8$) as the result.

## Analysis of the implementation of the experimental teaching course

As mentioned above the first set of the teacher's learning objectives for the students, deals with issues of enhancing students' independent study skills. In his evaluation the teacher emphasized in particular that the students acquired the mathematical knowledge of the Egyptians by themselves (in contrast to ordinary teaching where he explained everything), that they "cracked the code" themselves, and that they were conscious about it. Regarding item e) and f) of the second set of learning objectives, the teacher wrote: "they were all about gaining insights into current mathematics precisely by studying the mathematics of another time" (Wulff, 2004, p. 3), from which we can infer that the teacher used a pragmatic approach to history. He used past episodes of mathematics from a utility perspective. This also becomes clear from his description of a discussion that took place between him and the students during the introduction: "Already during the first module [the first two lessons] came the classical question, why are we going to learn this? And we had a good talk about the intended learning issues e), f), and g), during which the class apparently accepted that historical mathematics, besides being interesting as such, could contribute to a more nuanced view on current mathematics." (Wulff, 2004, p. 5). Regarding the learning objective of realizing that mathematics has evolved over time, the teacher was rather critical, explaining that this aspect was not really complied with, since a comparison of Egyptian and modern mathematics only shows that mathematics has changed; it does not give insights into the actual process of change. Regarding the last item g) of the second part of the learning objectives, the teacher wrote in his evaluation: "here is where the subject of history can be involved. From a general knowledge about Ancient Egypt and its society, students can discuss how society and culture have been driving forces for the mathematics of that time. At the same time the historians' method of source criticism is an essential tool for interpreting ambiguous and defective papyri" (Wulff, 2004, p. 4). In contrast to items e) and f) the teacher here takes a scholarly approach to history. The teacher used the past from an observer perspective in both approaches.

The students' work with the sources and exercises in the textbook on Egyptian mathematics to answer the "How" part of the problem formulation can be considered a "living history" approach. They put themselves in the place of Ancient Egyptians, trying to understand and learn how they calculated, how they dealt with geometry, how they proposed mathematical problems, and so forth. The teacher reported the following situation he observed in the classroom: "Many students wondered about how "stupid" the Egyptians were. Why did they only use unit fractions? Why should a number be expressed as a sum of different unit fractions? On the other hand their methods were very difficult to understand; that is rather advanced, so in that respect they weren't stupid at all. I think that many of the students realized that current mathematics is not "just" like today, but is a result of a long development, during which many things have been simplified." (Wulff, 2004, p. 7). This shows a development of historical awareness among the students. That the students' learning strategies were developed through this kind of "living history" approach can be inferred from the following observation made by the teacher: "This [that mathematics had made progress] became especially obvious when the students constantly rewrote the Egyptian notation to current notation with $x$'s, formulas, etc. After they had finished an Egyptian calculation they would say: 'but that just corresponds to …' followed by a solution of an equation in our way. It was very inspiring to see how students, who normally were a bit alienated towards $x$'s and equations now had taken those to themselves as their own, and all of a sudden perceived equations as an easy way to solve problems. The students became aware that modern notation makes the calculations much easier than they would have been otherwise" (Wulff, 2004, p. 7).

As mentioned above the teacher found that item g) in the list of learning objectives, which was supposed to link the development of mathematics with a scholarly approach to history, was not realized. The "why" part of the problem formulation was designed especially towards this goal. The mathematics teacher had hoped that the students would have been able to experience concrete examples of how needs of society sometimes act as driving forces for the development of mathematical ideas. This is a very ambitious goal, and since the history teacher focused more on religion and dynasties, the mathematics teacher felt that the students did not get opportunities to gain real insights into why mathematics was developed in interaction with the needs of society and culture. A less ambitious teacher would probably evaluate this part differently, pointing towards the fact that was explained above, that the students gained genuine historical knowledge about Egyptian mathematics situated in the proper historical context. Finally, the teacher concluded that the students afterwards showed signs of possessing a more mature and reflective approach to mathematics than they had before. Unfortunately, the teacher did not document this with observations from the classroom.

## An inter-disciplinary student project: *Transfer of knowledge in the Middle Ages*

The second example is also taken from upper secondary school in the Danish school system, but in all other respects it is completely different. Whereas the first example is an experimental course that was designed and taught by a mathematics teacher in an ordinary mathematics classroom the second example was defined and carried out by a single student who worked on his/her own on an inter-disciplinary project (the SRP-project) with history and mathematics and with limited time (a few hours) with a teacher who functioned as a consultant. In this construct both subjects enter on an equal footing, so to speak, and hence, from a "history in mathematics education" point of view, history enters in a very different way than in the first example. In the following the student project and its content will be presented and analysed with respect to the theoretical framework of different uses of the past and its potentials for benefiting students' learning of mathematics.

### The educational structure of SRP-projects
The so-called SRP-projects[8] entered the Danish gymnasium (upper secondary school) with a reform of 2003 which was implemented in 2005. The SRP-project is an independent and inter-disciplinary unit of study with its own organization which runs across the other school subjects. Every student writes a SRP-project in their final year of the gymnasium. Each

---

[8] SRP is short for "Studieretningsprojekt" and in daily talk it goes by the name *SRP-project*. A student's choice of subjects that can be combined in this project depends on the line of study (studieretning) the student has chosen – hence the name.

student chooses two subjects that have to be combined in the project. They will be assigned a teacher from one of the subjects who will function as a supervisor or a consultant.

The student is supposed to suggest (in discussion with the supervising teacher) a problem formulation, and ideas to how the two chosen subjects in combination will fulfil the problem formulation. This also includes suggestions for what kind of experiments, theoretical frameworks, texts and/or artefacts, the arguments of the projects should be based on. These suggestions of course depend on whether the subjects come from science, the humanities, or the social sciences (or from two of them). There are no subject specific requirements for the content of the SRP-project work, but it is required that the problem formulation contains both disciplinary and inter-disciplinary elements, that its fulfilment requires that the student study some aspects of the disciplinary subjects that goes beyond the curriculum of that subject; and that it include subject elements that the teacher have not taught the student in advance. It is also required that the problem formulation is concrete and delineated which is often achieved through the formulation of subtasks; that it indicates a clear direction that can guide the student's work with the task and help him/her with structuring the report. The problem formulation has to fulfil these requirements in such a way that the student still has to make some decisions and to prioritize.

The supervising teacher writes the problem formulation and hand it over to the student at a given date decided by the school. The student has two weeks to write the project. At some schools these are two consecutive weeks while at other schools there will be e.g. a two day writing period followed by a three day period and ending with a week, with ordinary school days in between. The student does the research and writes the project report on his/her own and hand in a written report of approximately 20 pages with references. The project reports are distributed among a group of external evaluators who are upper secondary school teachers, university professors or similar. The project report is graded by the external examinator and a teacher from the school who represents the other subject with respect to the expertise for the examinator.

The overall goal of the project is to prepare the students for further studies beyond upper secondary school. The idea is to give the student opportunity to conduct an independent, in depth, study within an area of his/her own choice that combines several subjects, and disseminate his/hers findings and analyses systematically in a written report of approximately 20 pages. It is a very ambitious study unit and the quality of the project reports vary a lot from a kind of research projects in miniature to reports that have been produced by a kind of cut-and-paste procedure or in some, though rare, cases by copying published materials. This last strategy is plagiarising and students can be expelled completely from upper secondary school if they are caught in this act.

Just to give an impression of the amount of project reports written every year, 3414 project reports with mathematics as one of the two subjects was handed in the first year of the implementation of the reform of these 38 % was in combination with history making this combination the most frequent followed by physics (32 %).

**Presentation of the SRP-project *Transfer of knowledge in the Middle Ages***
The report of the project *Transfer of knowledge in the Middle Ages* was written in the school year of 2009-2010. Hence, it belongs to the third generation of these reports and as such the worst "child diseases" had been remedied.

The problem formulation for the project work was the following:

> Describe Leonardo Fibonacci's treatment of 2. degree equations and compare with the Arabic sources he drew on, especially al-Khwarismi. Examine the contemporary Arabic treatment of third degree equations and compare the mathematical level of the two. Evaluate if reconstructions of al-Tusi's treatment of 3. degree equations use differential calculus as is claimed by Rashed. Describe the European view of science in medieval and perform an analysis of the Arabs' influence on European science. Perform a historiographical analysis of the changed view on Arabic medieval science. (Student report, 2009; my translation)

The problem formulation is clearly divided into a mathematics part and a history part with specific subtasks that have to be performed. The reading of historical source material from the history of mathematics, the performing of historiographical analyses, and the treatment of the 3. degree equation are all requirements that are beyond the curriculum for history and mathematics respectively. The evaluation of whether al-Tusi used differential calculus requires a historiographical analysis and the analysis of the Arabic influence on European science requires scientific (mathematical) insights, so the requirement of inter-disciplinary aspects is present at least in these two subtasks. The problem formulation does not specify any text or sources that the student has to study. The choice of literature is left to the student.

In accordance with the intentions of the SRP-project, the student wrote a coherent presentation and analysis fulfilling the requirements of the problem formulation instead of answering the subtasks separately in the report. The student interpreted the problem formulation in the introduction of the report (Student report, 2009, p. 2; my translation):

> One of the big issues in history is how cultures meet, and what influence they have on each other. In the Middle Ages the Christian European culture met the Arabic culture and the two cultures exchanged knowledge. The knowledge transfer was very one-sided. European culture had little to offer in comparison with the Arabic culture. The Arabs had in the early Middle Ages made many important contributions to, among others, science and mathematics, and these contributions are the focus of the report. The report starts with a description of the European view of science in medieval and an analysis of the Arab influence on European science. Then the Italian mathematician Leonardo Pisano's (1170 - 1250), better known as Fibonacci, work with the quadratic equation is explained and compared with the Arab sources he draws on specifically al-Khwarizmi (ca. 790-850). The report also discusses the difference of the mathematical level between Leonardo and the contemporary Arab mathematician al-Tusi (1135-1213), who worked on the solution of the third degree equation. To do this, I review some of al-Tusi's work. In the problem formulation it is stated that an evaluation of whether reconstructions of al-Tusi's treatment of 3. degree equations use differential calculus, as claimed by Rashed, must be performed. It is clear that reconstructions,

claiming that al-Tusi used differential calculus, use differential calculus themselves. Therefore I have interpreted this part of the assignment as if it says that I, from reconstructions of al-Tusi's work, must assess whether al-Tusi has taken advantage of differential calculus as claimed by the historian of mathematics Roshdi Rashed. This is followed by a historiographical analysis of the changed view on Arab medieval science with a focus on the mathematical contributions of the Arabs during this period.

The student structured the report following what could be called an *hourglass model* starting with a broad description of Europe in the Middle Ages with focus on the view of science and scientific progress from the 11th century and the Arabic influence, followed by a short description of Arabic science. Then the report narrows and zooms in on Leonardo Pisano's treatment of quadratic equations and al-Tusi's treatment of third degree equations. In these sections the student presents a detailed in depth study of the mathematics involved both regarding the type of problems dealt with, the algorithms uses to solve the equations, and the different proofs that were given for the algorithms. The student discussed differences between the original texts and how we would interpret the mathematics today in our modern notation, and analyzed and criticized different interpretations of al-Tusi's texts given by historians of mathematics. After that the student zooms out again and applies a broader perspective comparing Leonardo's and al-Tusi's work in order to perform a historiographical analysis of our changing views of Arabic contributions to mathematics in the Middle Ages. Finally, the student tied it all together in a conclusion.

**Aspects of the SRP-project *Transfer of knowledge in the Middle Ages***
I the flowing I will present some aspects of the actual content of the student's report in order to be able to analyse (1) how the past is used in this SRP-project within the framework outlined above and (2) the potentials of such uses for how history can benefit students' learning of mathematics – exemplified with this particular SRP-project report.

The problem formulation indicates that we are dealing with a scholarly approach to history from an observer perspective. It asks for a critical distance to the past and to our interpretations of the past both in the history part and the mathematics part. In the mathematics part the formulation explicit asks for a treatment of al-Khwarizmi's and al-Tusi's way of dealing with second and third degree equations respectively, not our modern way. The requirements to compare Leonardo's treatment of second degree equations with that of al-Khwarizmi and to asses if al-Tusi's work on the third degree equation uses differential calculus points explicitly towards developing the student's historical insights and awareness of mathematics as a subject that has been developed in different cultures that have influences one another. From the point of view of mathematics, history is not studied form a utility perspective in the SRP-project, and it is not used in an intervening sense in order for the student to orient him- or herself in a present context of e.g. the mathematics of their curriculum.[9]

However, this does not mean that this inter-disciplinary SRP-project with history and mathematics did not benefit the student's learning of mathematics. On the contrary, these kinds of projects have huge potentials for students' learning of mathematics and for the

---

[9] A similar analysis can be performed from the point of view of history. However, this will not be performed here where we only focus on mathematics and the use of the past in relation to mathematics.

development of students' mathematical competencies. To illustrate this I will discuss pieces of the student's treatment of Leonardo's and al-Tusi's work.

*Leonardo and the second degree equation*
The student's treatment of Leonardo's work with second degree equations is based on Seigler's (2002) commented translation of Leonardo's *Liber Abaci* presumably from 1202. Leonardo did neither use negative numbers nor zero, which meant that he dealt with 6 different versions or forms of second degree equations. Since he had no algebraic notation he wrote mathematics in usual language (Student project, 2009, p. 8):

> *"the census plus roots are equal to a number."* (Quoted from (Siegler, 2002, p. 555))

Translated into modern notation using the usual symbols the above becomes:

$$ax^2 + bx = c$$

Census refers to $x^2$, root to the root of census i.e. $x$, and since the word "roots" is in plural there are more than one root i.e. $bx$. The sum of these is equal to a number, $c$. Hence, we can see that Lenoardo had a word that represented the unknown, but nothing to represent general coefficients. As a consequence he used examples with given numbers for the coefficients but, as pointed out by the student, his method of solution, which he gave as an algorithm, was general.

His first example is the following:

> *"When two census are equal to $X^{10}$ roots, then you divide the number of roots by the number of census, namely the 10 by 2; the quotient will be 5 roots that are equal to one census, that is the root of the census is 5, and the census is 25."* (Student report, 2009, p. 9; quoted from (Siegler, 2002, p. 554)

In the student's translation into our notation, this corresponds to the following:

If $2x^2 = 10x$ then $x^2 = (10/2)x$, and $x^2 = 5x$, hence $x = 5$ and $x^2 = 25$

Leonardo gave an algorithm for solving the three versions of the second degree equation that contains all three elements: census, root and number. The student presented Leonardo's algorithm for the equation we would write as $ax^2 + bx = c$:

> You take the square of half the number of roots, and you add it to the given number, and of that which will result you take the root; from this you subtract half the number of roots, and that which will remain will be the root of the sought census. (Student report, 2009, p. 9; quoted from (Siegler, 2002, p. 555))

[10] In this example Leonardo used the symbol $X$ from the Roman number system. Everywhere else in the original Latin text, the numbers are spelled out in words. In Siegler's translation they are denoted with the Arabic number symbols.

The student translated Leonardo's description of "what to do" into the formula:

$$\sqrt{(b/2)^2 + c} - b/2 = x$$

and then showed how Leonardo's algorithm can be derived from our modern formula in the following way:

First we rewrite the case so it looks like our modern way of writing the second degree equation:

$$ax^2 + bx = c \Leftrightarrow ax^2 + bx - c = 0$$

We can then find $x$ from our modern solution formula,

$$x = \frac{-b \pm \sqrt{b^2 - 4a(-c)}}{2a}$$

Since $a = 1$, because we have reduced the equations with $a$ before the rewriting, we get:

$$x = \frac{-b \pm \sqrt{b^2 - 4(-c)}}{2} = -b/2 \pm \sqrt{\frac{b^2 - 4(-c)}{2^2}} = -b/2 \pm \sqrt{(b/2)^2 + c}$$

Since

$$-b/2 + \sqrt{(b/2)^2 + c}$$

is positive it is a solution according Leonardo, and this is exactly the solution above. The other solution is negative, and was discarded by Leonardo. (Student report, 2009, p.10-11; my translation)

Leonardo did not only present the algorithm for solving the second degree equation, he also provided geometrical proofs. Actually, he gave two proofs for the equation above; one where he drew on al-Khwarizmis proof and one where he used Euclid. The student presented both these proofs in the report, but here I will only give the student's treatment of the first of Leonardo's proofs:

Leonardo uses the specific second degree equation $x^2 + 10x = 39$ to perform the proof. It is the same one that al-Khwarizmi used in his proofs. I [i.e. the student] will give the proof for the more general equation $x^2 + bx = c$.

Leonardo began by drawing a square $abcd$ (see figure), in which all the sides are greater than 5, that is $b/2$.

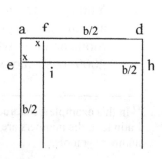

He then marks a point $f$ such that $df = 5$, i.e. $b/2$. In the same way he marks a point $h$ on $dc$ such that $ch = 5$, i.e. $b/2$. He does that also for the two remaining sides, and draws the lines between $f$ and $g$, and $h$ and $e$. Leonardo describes very carefully which sides are equal and why; for example:

> *"Because the rectangle .ac. is a square the line segment .da. will be equal to the line segment .ba.; and when equals are subtracted from equals, then those which remain are equal"* (quoted from (Seigler, 2002, p. 556))

Here Leonardo uses an axiom from the first book of Euclid. From the figure one can see that a new square $gh$ is constructed with side length 5, i.e. $b/2$, and a smaller square, $ef$, the area of which represents $x^2$, hence the length of the line segments is $x$. Besides the two squares we also get two rectangles, $bi$ and $id$. These have the sides 5, i.e. $b/2$, and $x$. The area of those two together is $bi + id = 5x + 5x = 10x$, i.e. $bx$. Now we have two of the elements of the second degree equation $ax^2 + bx = c$, hence

$$ef + bi + id = x^2 + 10x = 39$$

From this deduction we can see that 39, i.e. $c$, corresponds to the gnomon *abgihd*. Leonardo then adds the square $gh$ with area 25, i.e. $(b/2)^2$, to the equation:

$$x^2 + 10x + 25 = 39 + 25 = 64 \qquad (x^2 + bx + (b/2)^2 = c + (b/2)^2)$$

64, or $(b/2)^2 + c$, corresponds to the area of the larger square *abcd*, hence the length of the sides of the square *ad*, is

$$\sqrt{(b/2)^2 + c} = \sqrt{64} = 8$$

Leonardo then calculated $x$ by subtracting the side length $eb = b/2$ from the side length $ab$:

$$x = \sqrt{(b/2)^2 + c} - b/2 = 8 - 5 = 3$$

Hence, Leonardo has proved that his algorithm for solving $ax^2 + bx = c$ is correct. (Student report, 2009, p. 10-11).

*Al-Tusi and the third degree equation*
One of the subtasks within the student's problem formulation was to describe and explain the Arabic treatment of third degree equations which took place at the same time period that Leonardo was working in – and to compare the mathematical level. To accomplish this, the student treated al-Tusi's work with the third degree equation using the historian of mathematics Roshdi Rashed's translations of al-Tusi's work (Rashed, 1994).

As explained by the student, al-Tusi grouped the general cubic equation into 24 cases. To explain al-Tusi's way of dealing with cubic equations the student chose the case

$$x^3 + c = ax^2 + bx$$

Al-Tusi divided this case into three sub-cases: $a > \sqrt{b}$, $a = \sqrt{b}$ and $a < \sqrt{b}$.

The student used the last one of the sub-cases to illustrate how al-Tusi dealt with cubic equations:

First he determines a line segment $x_0$ such that

$$b/3 = x_0(x_0 - 2a/3) \quad \text{and } a < \sqrt{b}$$

al-Tusi then presents his argument for how $x$ can be found. He rewrote the case into the following:

$$c = bx + ax^2 - x^3$$

and gave the right hand side of the equation a particular name; I [i.e. the student] am going to name it $f(x)$. He then shows that for $f(x) = bx + ax^2 - x^3$ the following is true:

For $c = f(x_0)$ the equation $x^3 + c = ax^2 + bx$ has one solution, namely $x_0$
For $c > f(x_0)$ the equation have no solutions
For $c < f(x_0)$ the equation have two solutions.

al-Tusi then deduced that $f(x) < f(x_0)$ for all $x > 0$ and $x \neq x_0$, hence $x_0$ is a maximum point. (Student report, 2009, p. 14-15).

The student used Rashed's shortened version of al-Tusi's proof to illustrate how al-Tusi's argued. I will skip that here and instead move directly to the student's discussion of the different interpretations of al-Tusi's method for solving cubic equations that have been given by historians of mathematics:

al-Tusi was, according to Rashed, inspired by the graph for the function [ $f(x) = bx + ax^2 - x^3$, in determining the condition for the maximum point, i.e. $b/3 = x_0(x_0 - 2a/3)$ ]. Rashed illustrates where the different values are placed on the graph[11], and claims that al-Tusi did the same. From the graph is can be seen that al-Tusi's conclusion is true: the

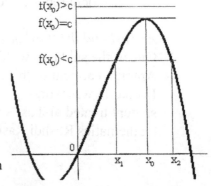

_____

[11] Notice that the student made a mistake in the drawing – the signs for less than an interchanged.

number of solutions depends on how big $c$ is compared to $f(x_0)$. But not all historians of mathematics agree that al-Tusi himself used graphs to illustrate or for inspiration. Jan hogendijk, for example, comments on Rashed's interpretation in a paper on al-Tusi:

> *"As far as is known, cubic curves were never drawn by medieval mathematicians."* (quoted from (Hogendijk, 1989, p. 16)

This is not the only point of disagreement regarding the work of al-Tusi. Rashed is also of the opinion that al-Tusi used a form for differential calculus to derive his condition for the maximum point $x_0$. In our modern time we know that if $x_0$ is a maximum point for $f(x)$ then $f'(x_0) = 0$. We also know that: $f'(x) = 2ax + b - 3x^2$. Hence, if $x_0$ is a maximum point then $2ax_0 + b - 3x_0{}^2 = 0$ which is equivalent to the condition on $x_0$ given by al-Tusi.

But differential calculus was not invented until the 17th century, so many historians of mathematics, including Hogendijk, think that al-Tusi could not have used that method. Rashed defends his position by stating that al-Tusi's work show that al-Tusi was ahead of his time and thereby anticipated some of the ideas of differential calculus especially the methods used by Fermat (1601-1665). Jan Hogendijk believes that al-Tusi used a derivation in the Greek style to find the condition on the maximum point $x_0$. Hogendijk's reconstruction is geometrical like all al-Tusi's arguments.

Hence, there is another explanation for how al-Tusi found the condition on $x_0$, than that he used third degree curves and a form for differential calculus, as claimed by Rashed. Since Hogendijk's explanation is much closer to al-Tusi's other arguments, his explanation is a possible reconstruction of al-Tusi's original analysis, and there is no reason for ascribing methods such as third degree curves and differential calculus to al-Tusi for which there are no evidence in the sources. (Student report, 2009, p. 15-16).

*The student's historiographical analysis of the changed view on Arabic medieval mathematics*
The student took evaluations of Arabic mathematics from Morris Kline, Seyyed Hossein Nasr, H. G. Zeuthen and Victor Katz as point of departure for the historiographical analysis. Morris Kline wrote the following about Arabic contributions to mathematics in his book *Mathematical thought from Ancient to Modern Times*:

> *"The Arabs made no significant advance in mathematics. What they did was absorb Greek and Hindu mathematics, preserve it and ultimately, ... transmit it to Europe."* (Student report, 2009, p. 19; quoted from (Kline, 1972, p. 197))

This is then contrasted to the evaluation given by the Muslim scholar Seyyed Hossein Nasr:

*"As in trigonometry so in algebra Muslims must be considered as the founders of this science whose name (from the Arabic al-jabr) reflects its origin ...but it was the early Muslim mathematicians of the 3rd /9th century leading to Muhammad ibn Musa al-Khwarizmi who firmly established this branch of mathematics."* (Student report, 2009, p. 19; quoted from (Nasr, 1976, p. 84))

The student explained the huge difference between these two evaluations of the Arabs' contribution to mathematics by considering the background of the authors. Kline was trained as a mathematician. He also worked in physics, and he held the opinion that the mathematics taught in classrooms should be usable for solving problems outside of mathematics, e.g. in physics. The student reflected on the differences between the evaluations by linking the evaluations to the background of the authors: Kline was a mathematician who thought mathematics should be useful in problem solving outside of mathematics. The student doubted that he had read the Arabic sources himself, and the student thinks that since he values mathematics usefulness in physics, the Arab contributions in mathematics was not held in high regard by Kline. Considering Nasr's background, on the other hand, the student is convinced (and with good reasons) that he has read the Muslim sources. The fact that he is a Muslim himself, the student questioned whether this has influenced his view of the Arabs contribution to mathematics causing him to over estimate their contribution.

This was a comparison of two authors who wrote in the same time period and nonetheless held very different opinions of the Arabs' contribution to mathematics – a difference the student explained with regard to their background and their motives.

As an example of evaluations from different time periods the student contrasted H.G. Zeuthen's work from 1893 with Victor J. Katz's writing of the twenty-first century:

*"I [Zeuthen] wished – among other things by comparison with the Romans – to emphazise the value and the extent of the Arabs' mathematics work as to avoid the drawing of negative conclusions based on the few positive results, they achieved that went beyond what the Greeks already knew."* (Student report, 2009, p. 20; quoted from (Zeuthen, 1893, p. 261))

And Victor Katz:

*"We will consider some of the highlights of the Islamic mathematics. In particular, we will see that Islamic mathematicians fully developed the decimal place value system to include fractions; systematized the study of algebra and began to consider the relationship between algebra and geometry."* (Student report, 2009, p. 20; quoted from (Katz, 2004, p. 163))

The student had the following reflections regarding the differences between Zeuthen and Katz:

> From these two quotations we can distinguish two different attitudes towards the Arab mathematicians. Zeuthen was of the opinion that they contributed a lot if one compared their contributions with those of the Romans, but compared to the Greeks their originality was minimal. Katz on the other hand has the opinion that they made significant contributions especially within algebra and geometry. The changed view on Arabic medieval contributions to mathematics is due to, among other things, new findings of many Arabic texts, which show the scientific originality of the Arabs. But it might also be because of our changed views on non-Western cultures. At the end of the 19th century the Arabic and other non-Western cultures were considered to be underdeveloped, but less so to day. Since our views on the Arabs have changed we are

more willing to credit them the honour for the scientific progress they made in the
medieval. (Student report, 2009, p. 20).

**Analyses of the SPR-project with respect to uses of the past and benefits of history in
math**

As already mentioned during the discussion of the problem formulation we are dealing with a
scholarly approach to history from an observer perspective. Differences between the past and
the present is emphasized in the student's reflections both of the mathematics that is
presented from the different cultures and of the reconstructions and historical narratives
produced by historians at different times and from different backgrounds.

In the historiographical analysis the student uses, though without really commenting on it,
view points from both lay-historians and professional historians. Here Kline and Zeuthen,
even though they both are held in high regards by historians of mathematics nevertheless can
be considered to be lay-historians in Jensens' terminology, where as Rashed and Hogendijk
are professional historians of mathematics. The student's discussion of the influence of the
background of the authors for their evaluation of the Arabs' contribution to mathematics can
be analyzed with respect to the concepts of identity concrete and identity neutral history
writing. Within a Western group, Kline's way of writing history could be thought of as being
identity neutral, while Rashed's could be accused of being identity concrete – and Arabs
might consider the two ways of writing history in a vice verse perspective.

Finally, I will discuss the benefits of using history in this way in mathematics education using
the framework of mathematical competence, as developed by Niss (2004). Niss distinguishes
between eight mathematical competencies that together span mathematical competence. Here
we will only evaluate the SRP-project report *Transfer of knowledge in the Middle Ages* for its
potential to develop four of the competencies, namely *mathematical thinking competency*;
*mathematical reasoning competency*; *symbols and formalism competency*; and
*communicating competency* in, with, and about mathematics.

The ability to translate between natural language and formal/symbolic language as well as the
ability to handle and manipulate statements and expressions that contain symbols and
formulae are parts of what constitute symbols and formalism competency. It is clearly
documented in the student's report that he/she on several occasions had translated between
the historical sources use of natural language and our modern way of writing the mathematics
as equations using symbols. The student also evoked his/her competency of thinking
mathematically at several places. A part of this competency is to be able to understand what it
means to generalize mathematical results – a competency the student possessed in his/her
generalization of Leonardo's proof for his algorithm for the specific quadratic equation
$x^2 + 10x = 39$ where the student clearly understood that on the one hand Leonardo only
proved his algorithm for this specific quadratic equation, but on the other hand, realized that
Leonardo's proof could be generalised for the $x^2 + bx = c$, which the student did in the
report. Reasoning competency involves being able to follow and understand a chain of
mathematical arguments, a competency the student evoked at several places in the report
where he/she explained Leonardo's and al-Tusi's proofs of their solution methods for second
and third degree equations, respectively. Finally, the communicating competency include
being able to understand other's mathematical texts and to express oneself about

mathematics. The student used and trained this competency in the reading and interpretation of the sources and the secondary literature, and least but not last, during the writing of his/her own report.

Included in mathematical competence is, besides the eight main competencies mentioned above, also some second order competencies of possessing overview and judgement regarding mathematics as a discipline, where one of them regards the historical development of mathematics. The student gained insight into the historical development of mathematics, how mathematical knowledge has travelled from one culture to another; and from one time period to another.

## Concluding remarks

The purpose of the chapter was to present a theoretical framework for a systematic analysis of uses of history for teaching and learning mathematics in order to propose a didactical transposition of history from the academic research subject to history in mathematics education. The analysis of the teacher's report on the experimental course on Ancient Egyptian mathematics with respect to the described framework of different uses of past episodes shows that in this course, history was used in different ways to provide a very rich teaching and learning environment. The teacher used different approaches to history and used past episodes from various perspectives for different purposes, thereby creating learning situations that developed students' historical awareness and mathematical learning strategies at the same time. History was used in ways in which students gained genuine historical insights, developed learning strategies, and enhanced their mathematical problem solving skills even though they worked on mathematics that might not be part of the core curriculum. The analysis of the SRP-project *Transfer of knowledge in the Middle Ages* within the framework clarified a use of a scholarly approach to history from an observer perspective, where the student analysed and compared lay-persons' and professional historians' interpretations of Arabic mathematics, which could be distinguished using the concepts of identity neutral and identity concrete writings of history. Besides developing the student's historical awareness the project work also benefited his/her learning of mathematics, in the sense of its potential to develop several of the student's mathematical competencies.

## References

Blomhøj, M. and Kjeldsen, T.H. (2006). Teaching mathematical modelling through project work – Experiences from an in-service course for upper secondary teachers. *ZDM*, 38, 2, 163-177.

Carstensen, J. and Frandsen, J. (2002). *Mat 1*. Viborg: Systime.

Epple, M. (2000). Genies, Ideen, Institutionen, mathematische Werkstätten: Formen der Mathematikgeschichte. *Mathematische Semesterberichte*, 47, 131-163.

Fauvel, J., and van Maanen, J. (eds.), (2000). History in mathematics education – The ICMI study. Dordrecht: Kluwer.

Frandsen, J. (1996). *Ægyptisk matematik*. Viborg: Systime.

Fried, M.N. (2001). Can Mathematics Education and History of Mathematics Coexist? *Science & Education*, 10, 391-408.

Hogendijk, J.P. (1989). Sharaf al-Din al-Tusi on the Number of Positive Roots of Cubic Equations. *Historia Mathematica*, 16, 69-85.

Imhausen, A. (2003). Calculating the daily bread: Rations in theory and practice. *Historia Mathematica*, 30, 3-16.

Jankvist, U.T. (2010). An empirical study of using history as a 'goal'. *Educational Studies in Mathematics*, 74(1), 53-74.

Jankvist, U.T. and Kjeldsen, T.H. (forthcoming). New Avenues for History in Mathematics Education: Mathematical Competencies and Anchoring. Published first on line December 2010. DOI: 10.1007/s11191-010-9315-2

Jensen, B.E. (2010). *Hvad er historie*. Copenhagen: Akademisk Forlag.

Katz, V. (2004). *A History of Mathematics, Brief Edition*, Boston: Pearson Education, Inc.

Kjeldsen, T.H. (forthcoming a). A multiple perspective approach to the history of the practice of mathematics in a competency based mathematics education: history as a means for the learning of differential equations. To appear in Katz, V., & Tzanakis, C. (eds.) *Recent developments on introducing a historical dimension in mathematics education*, to be published in 2010.

Kjeldsen, T.H. (forthcoming b). Uses of History in Mathematics Education: Development of Learning Strategies and Historical Awareness. Accepted for publication. To be published in *CERME 7, Proceedings of the seventh Congress of the European Society for Research in Mathematics Education*. 2011.

Kjeldsen, T.H., Pedersen, S.A. and Sonne-Hansen, L.M. (2004). *New Trends in the History and Philosophy of Mathematics*, Odense: SDU University Press, 11-27.

Kjeldsen, T.H. and Blomhøj, M. (2009). Integrating history and philosophy in mathematics education at university level through problem-oriented project work. *ZDM Mathematics Education, Zentralblatt für Didaktik der Mathematik*, 41(1-2), 2009, 87-103.

Kjeldsen, T.H. and Blomhøj, M. (forthcoming). Beyond Motivation – History as a method for the learning of meta-discursive rules in mathematics. Accepted for publication in *Educational Studies in Mathematics*.

Kline, M. (1972). *Mathematical Thought from Ancient to Modern Times*, New Yok: Oxford University Press.

Nasr, S.H. (1976). *Islamic Science, An Illustrated Study*. World of Islam Festival Publishing Company Ltd.

Niss, M. (2004). The Danish "KOM" project and possible consequences for teacher education. In R. Strässer, G. Brandell, B. Grevholm and O. Helenius (eds.) *Educating for the Future*. (pp. 179-190). Göteborg: The Royal Swedish Academy.

Rashed, R. (1994). *The Development of Arabic Mathematics, between arithmetic and algebra*, Kulmer Acud Press.

Rosenzweig, R. and Thelen, D. (1998). *The Presence of the Past. Popular Uses of History in American Life*. New York: Colombia University Press.

Siegler, L.E. (1979). *Fibonacci's Liber Abaci*, New York: Springer Verlag, Inc.

Sfard, A. (2008). *Thinking as Communicating*. Cambridge: Cambridge University Press.

Student report (2009). *Transfer of knowledge in the Middle Ages.* SRP-report with history and mathematics, 25 pp., Nærum Gymnasium, 2009/2010.

Tzanakis, C. and Thomaidis, Y. (2010). The Implementation of the history of mathematics in the new curriculum and textbooks in Greek Secondary Education. *6, Proceedings of the sixth Congress of the European Society for Research in Mathematics Education.* Viviane Durand-Guerrier; Sophie Soury-Lavergne; and Ferdinando Arzarello (eds.). 2742-2751.

Wulff, P. (2004). *Ægyptisk matematik.* Report from the in-servise course: Problem oriented project work in, with and about mathematics, Roskilde University. (http://dirac.ruc.dk/mat/efteruddannelse/rapporter/2004/Peter.pdf)

Zeuthen, H.G. (1893). *Mathematikkens Historie*, Kjøbenhavn, Høst og Søns Forlag.

# Adversarial and friendly interactions: Progress in 17th century mathematics

Shirley B. Gray
California State University, Los Angeles
sgray@calstatela.edu

Libby Knott
Washington State University
lknott@wsu.edu

Many agencies, including the National Council of Teachers of Mathematics (NCTM, 2000) and mathematics standards boards (e.g., the Common Core State Standards (CCSS, 2010) and state teacher certification agencies, for example, suggest that it is very important that future teachers of mathematics be familiar with the history of mathematics. Krussel (2000), Sfard (1991), and others have noted that students benefit from an understanding of how mathematicians over the centuries have grappled with and solved significant mathematical problems. Ball and Bass (2000) have recognized that mathematics teachers not only need to know mathematics content and mathematics pedagogy but they also need to know how these ideas are integrated.

Further, college students often learn about mathematics history in a piecemeal fashion. They learn fragments of calculus history in their calculus class, fragments of statistics history in their probability and statistics class, and a bit about the history of geometry in their geometry class. Lost in this approach is any appreciation for the connections between and among different contemporary mathematicians. Krussel (2000) found that students who understood the struggles suffered by mathematicians trying over centuries to understand a new idea had a much better tolerance for and appreciation of their own struggles with difficult mathematical ideas. Students who have wrestled with the idea of limits in calculus, for example, are heartened and encouraged to learn that mathematicians struggled with the same idea for at least two hundred years before they fully understood and explained it fully. In addition, students will gain a sense of the non-linearity of the advancement of mathematical ideas and work, as well as an appreciation for the symbiosis of ideas in both adversarial and collegial relationships.

The mathematical accomplishments of these featured mathematicians are inextricably intertwined. Too often students view the pre-internet world as one of individual struggle and isolated, solitary work, with little communication and connection among mathematicians, particularly across international borders. But this is perhaps surprisingly not the case. European mathematicians in the 17th and 18th centuries did not work in isolation; rather they worked in collaboration and sometimes in competition with their counterparts across the continent.

Just as today, it was a competitive world; everyone wanted to garner the recognition that came with being the first with a particular result. But this only fueled the need for collaboration and the sharing of ideas, through written communication, both personal and in professional journals, personal visits and travel, and presentations at professional society meetings. But this underscores the fact that mathematics is a social endeavor. It progresses by a hodgepodge of encounters and collaborations in non-linear fashion. As Newton wrote in a letter to his rival Robert Hooke, in 1676: "If I have seen a little further it is by standing on the shoulders of giants."

In this chapter, an octahedral model is offered as an interpretation of the connections in the lives and accomplishments of six well-known 17th – 18th century contemporary mathematicians. Readers will be familiar with many of these names; however, they may not realize that each offered the other intellectual competition, support, encouragement and stimulation.

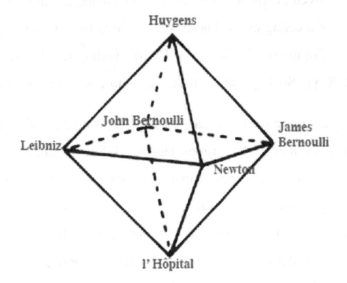

Figure 1: The connections among six famous 17<sup>th</sup> century mathematicians

While our six selected protagonists – Huygens, two Bernoullis, Newton, Leibniz, and l'Hôpital were geographically spread across Europe, in Holland, Switzerland, England, Germany and France, they were fortunate to be contemporaries and were united through publishing in Latin, a written language common to them all. Moreover, they were devoted throughout their lives to understanding the nature of advanced mathematics. We demonstrate how this community of scholars interacted to advance the frontiers of mathematics. Students will understand how we borrow the idea of using a paradigm or model, so common in the world of education, to create an understanding of the interaction of ideas and lives to advance mathematics.

**Christiaan Huygens (1629-1695)**                                          **Netherlands**

Huygens is at the apex of our model in part because he was born first, but also due to his personality and background. Huygens was a descendant of several generations of aristocrats. As such, he was privileged to travel extensively and meet an amazing array of scholars. He was "home schooled" until the age of sixteen, yet his education was clearly cosmopolitan. Descartes was an occasional guest in his home. His extensive correspondence with the mathematician Father Mersenne started at this time. As a twenty year-old he went on a diplomatic tour to Denmark but was disappointed to be unable to continue to Stockholm to see Descartes. In his

30s in Paris, he often enjoyed the meetings of various social and scientific societies. He wrote to his brother, "... there is a meeting every Tuesday . . . where twenty or thirty illustrious men are found together. I never fail to go ...." (Huygens, 1660). Having traveled to London, he became a member of the British Royal Society and later, the Académie Royale des Sciences in Paris.

Today, we find evidence of an exchange of ideas on mathematics and science with van Schooten, Pascal, Fermat, Roberval, Desargues, Hooke, Halley, Wren, Boyle, Newton and many others. In science, we recognize Huygens primarily for has original contributions to astronomy. His name is virtually synonymous with the pendulum clock so necessary in his day for the navigational problems in measuring longitude. In mathematics, however, his greatest accolade may arguably be none other than that of "teacher" for he mentored his now famous student, the great German mathematician Gottfried Wilhelm Leibniz.

## Gottfried Wilhelm Leibniz (1646-1716)        Germany

Like Huygens, Leibniz went to Paris as a young man, then only 26. Yet Leibniz arrived trained as a lawyer and was "not even a novice in mathematics" (Hoffman, 1964). Still, Paris provided the milieu for the two to meet and exchange ideas. Surely Huygens recognized that the young German was indeed mathematically talented if not well versed in the mathematical ideas and problems of the day. We are certain the "teacher" and the "student" successfully exchanged a challenge problem of finding the sum of the series for inverse triangular numbers (Burton, 2010). But Leibniz, while preoccupied with his new mathematical adventures, was obliged to make a diplomatic side trip to London. Many young adults have traveled to Paris and London but few have had a more memorable and hotly debated journey.

While in London, Leibniz made the acquaintance of another German mathematician, Henry Oldenburg. Oldenburg was none the permanent secretary of the British Royal Society, the gathering place of all notable scientists, amateur and otherwise, of the day. Historians will never know for certain what was said between the two, what was their exchange of ideas in 1673. But certainly, Leibniz and Oldenburg would speak to one another freely in their common language, German. Further, Oldenburg knew of Newton's work on sums of infinite series. Historians are certain that later on, Newton's two letters to Oldenburg in 1676 were translated into Latin and forwarded to Leibniz as a member of the Royal Society. We are certain, also, that in 1684 Leibniz published the first ever article on differential calculus in a new scholarly periodical, *Acta Eruditorum* while Newton's earlier ideas remained in his "wastebooks", i.e., his notebooks, in Cambridge.

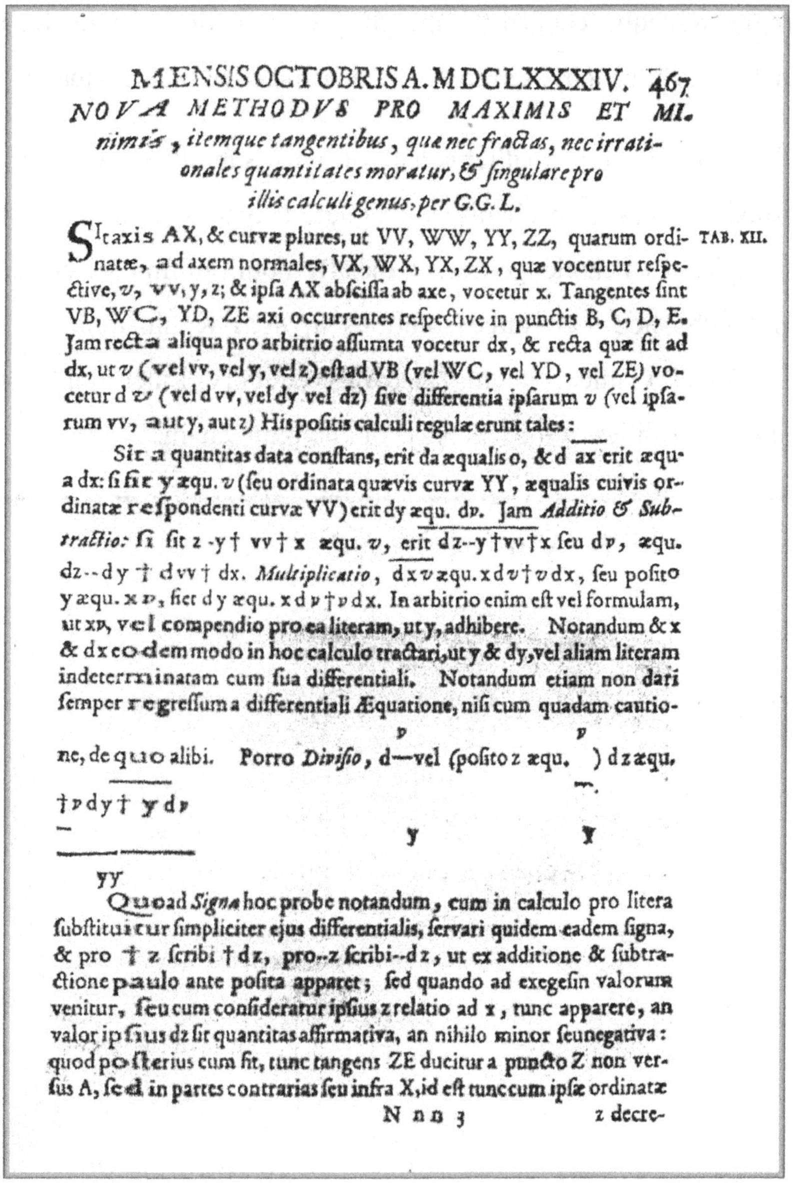

MENSIS OCTOBRIS A. MDCLXXXIV. 467
*NOVA METHODVS PRO MAXIMIS ET MI-nimis, itemque tangentibus, quæ nec fractas, nec irrationales quantitates moratur, & singulare pro illis calculi genus, per G.G.L.*

SIt axis AX, & curvæ plures, ut VV, WW, YY, ZZ, quarum ordi-TAB. XII. natæ, ad axem normales, VX, WX, YX, ZX, quæ vocentur respective, v, vv, y, z; & ipsa AX abscissa ab axe, vocetur x. Tangentes sint VB, WC, YD, ZE axi occurrentes respective in punctis B, C, D, E. Jam recta aliqua pro arbitrio assumta vocetur dx, & recta quæ sit ad dx, ut v (vel vv, vel y, vel z) est ad VB (vel WC, vel YD, vel ZE) vocetur d v (vel d vv, vel dy vel dz) sive differentia ipsarum v (vel ipsarum vv, aut y, aut z) His positis calculi regulæ erunt tales:

Sit a quantitas data constans, erit da æqualis o, & d ax erit æqu- a dx: si sit y æqu. v (seu ordinata quævis curvæ YY, æqualis cuivis ordinatæ respondenti curvæ VV) erit dy æqu. dv. Jam *Additio & Subtractio*: si sit z -y + vv + x æqu. v, erit dz--y + vv + x seu dv, æqu. dz--dy + dvv + dx. *Multiplicatio*, dxv æqu. xdv + vdx, seu posito y æqu. xv, fiet dy æqu. xdv + vdx. In arbitrio enim est vel formulam, ut xv, vel compendio pro ea literam, ut y, adhibere. Notandum & x & dx eodem modo in hoc calculo tractari, ut y & dy, vel aliam literam indeterminatam cum sua differentiali. Notandum etiam non dari semper regressum a differentiali Æquatione, nisi cum quadam cautione, de quo alibi. Porro *Divisio*, d--vel (posito z æqu. ) dz æqu.

+ v dy + y dv
yy

Quoad *Signa* hoc probe notandum, cum in calculo pro litera substituitur simpliciter ejus differentialis, servari quidem eadem signa, & pro + z scribi + dz, pro-z scribi--dz, ut ex additione & subtractione paulo ante posita apparet; sed quando ad exegesin valorum venitur, seu cum consideratur ipsius z relatio ad x, tunc apparere, an valor ipsius dz sit quantitas affirmativa, an nihilo minor seu negativa: quod posterius cum sit, tunc tangens ZE ducitur a puncto Z non versus A, sed in partes contrarias seu infra X, id est tunc cum ipsæ ordinatæ
N nn 3     z decre-

Figure 2. First publication on the Derivative, *Acta,* 1684. 'G.G.L.' are the initials used by Leibniz when publishing.

**James Bernoulli (1654-1705)**                                    **Switzerland**

Almost all of what today we call introductory calculus along with differential equations was published in the late 1600s in the *Acta Eruditorum*. The mathematics articles were mostly written by Leibniz, John Bernoulli and an elderly Huygens. Newton contributed anonymously from across the English Channel but was recognized as the "lion known by his claws". The Bernoulli brothers undoubtedly scanned the monthly mathematical articles for accuracy, yet the eclectic content included philosophy, sermons, science and even magic squares.

Perhaps the most famous mathematical articles submitted were initiated by James Bernoulli who issued challenges to all his colleagues. Huygens, Leibniz, and James' brother, John Bernoulli, all replied to a challenge from James Bernoulli to find the catenary's actual shape (1690-1691). James published their three solutions in *Acta* within months (June, 1691). See below. Leonardo da Vinci, Galileo, Stevin, Descartes, Gregory, Euler and Thomas Jefferson all remarked on the special qualities of the catenary, or "hanging chain."

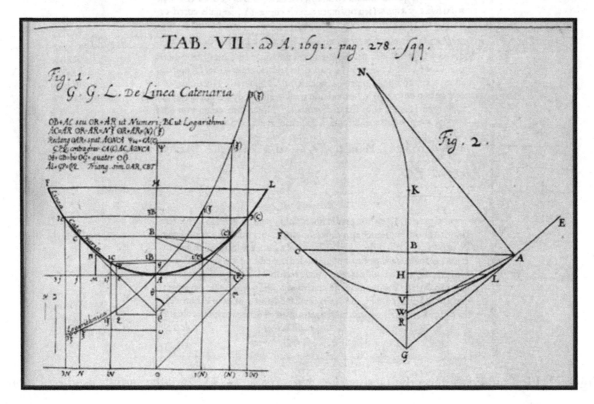

Figure 3. The Catenary, in *Acta*, June 1691.

To underscore the historical importance of the catenary, we have selected the figures submitted by Leibniz and Huygens for publication in the widely acclaimed *Acta Eruditorum*, 1691. Huygens' illustration is on the right (see Figure 3.). Huygens even proved to Father Mersenne that the catenary could not be a parabola. ("The Catenary," 2011).

The name "catenary" is an Anglicized version of the Latin word *cateneria* first used by Huygens in a letter sent to Leibniz in 1690. Both words are derived from the Latin noun *catena* meaning "chain."

**John (Johann) Bernoulli (1667-1748)**                              **Switzerland**

John Bernoulli was not about to let his older brother abscond with his mathematical reputation. Indeed, their rivalry is legendary. John posed the "linea brachystochrona" problem in a leaflet (1696) sent to "the shrewdest mathematicians of all the world." (Beckmann, 1971). According to a husband of Newton's niece, Newton's leaflet containing his solution to this problem arrived at the Royal Society in London in the afternoon. Newton had solved the problem before retiring that evening and posted his solution the following morning. The solution arrived in Basel, Switzerland unsigned. John Bernoulli immediately recognized the notation and terminology to exclaim, "tanquam ex ungue leonem" or "the lion is known by its claw." (Beckmann, 1971). Shortly thereafter, solutions from John Bernoulli, James Bernoulli, Leibniz and Newton all appeared in the 1697 *Acta*. See Figure 4.

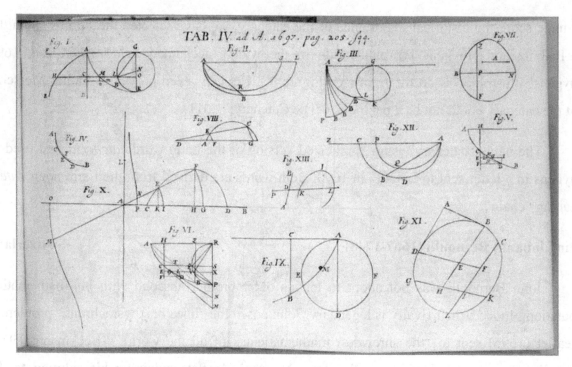

Figure 4. Illustrations for the solutions to the brachistochrone challenge issued by Joh. Bernoulli. (*Acta* 1697)

## Marquis de l'Hôpital (1661-1704)                                            **France**

Like Huygens, l'Hôpital was "of the manor born." As a wealthy citizen of Paris, he very much wanted to participate in its intellectual life. Much in the tradition of Huygens tutoring Leibniz, l'Hôpital sought guidance in understanding mathematics, in this instance turning to John Bernoulli, six years his younger, as his mentor. The wealthy l'Hôpital engaged John to spend several months in Paris between the publication of the catenary (1691) and the publication of the brachistochrone (1697) problems. At some point John returned to Switzerland, but continued as a paid tutor by sending letters to l'Hôpital, by an arrangement that had a special stipulation. For John to be paid, he would be required to send his written correspondence exclusively to l'Hôpital. (Those of us familiar with book publishing, writing for foundations, media scripts, etc., will recognize modern contract similarities.) These letters resulted in l'Hôpital's publication of *Analyse des Infiniment Petits* (1696) as a single author text. While the Marquis (de l'Hôpital ) died eight short years later, John Bernoulli would spend the next 52 years of his life seeing the

*Analyse* full of his ideas - but not his name – in what was to become the most successful calculus text in Europe.

Still, this is not an altogether clear illustration of 'good' and 'evil' in credit being awarded for originality in mathematics. Across the Channel, English mathematicians were certainly acceding to Newton's vocabulary and notation, e.g., 'fluxions' and 'fluents' for what we now use from Leibniz, $\dfrac{dy}{dx} = 0$. We turn now to an Englishman, who translated l'Hôpital's *Analyse,* including the following, often ignored, tribute to the others:

From the Preface of l'Hôpital's 1696 volume of Analyse des infiniment petits Pour l'intelligence des lignes courbes and its 1730 English translation, prepared by E. Stone and printed by William Innys:

"*I must own my self very much obliged to the Labours of Messiers Bernoulli, but particularly to those of the present Professor at Groeningen, as having made free with their Discoveries as well as those of Mr. Leibnitz(sic): So that whatever they please to claim as their own, I frankly return to them.*"

And elsewhere:

"*I must here in justice own, (as Mr. Leibnitz (sic) himself has done in Journal des Scavans for August, 1694) that the learned Sir Isaac Newton likewise discovered something like the Calculus Differentialis, as appears by his excellent 'Principia', published first in the year 1687 which almost wholly depends upon the use of the said Calculus. But the method of Mr. Leibnitz's (sic) is much more easy and expeditious, on account of the notation he uses, not to mention th wonderful assistance it affords on many occasions.*" Guillaume Francois-Antoine, Marquis de l'Hôpital. (l'Hôpital, 1696).

So what was l'Hospital's written explanation of what today we call his "Rule"? He writes: "*...the infinitely small part by which a quantity is continually increased or diminished, called the Difference be considered as the assemblage of an infinity of straight lines, each* *infinitely small, or as a polygon having an infinite number of sides"* (Stone, 1730).

Today we denote the "Difference" of a variable $x$ quantity by $dx$. Clearly, l'Hôpital considers a segment of a line as part of the curve. Thus, we encounter the concept of a curve, the concept of tangent, the concept of slope and the problem of what method to use when handling indeterminate forms, i.e., 0/0 and $\infty/\infty$.

To our young, but careful readers, some observations are suggested. First, all of these men were clearly talented in mathematics, yet they were not young when they first published. The cliché of being a "young genius" in mathematics is clearly overrated. It is a mistake to think that an "Aha" moment is the exclusive domain of youth. Moreover, judging by the hesitation of others, these men were probably not completely convinced that finding the sums of infinite series would lead to anything other than a purely intellectual pursuit. Thus, patience and judgment became crucial factors in mathematical research.

All six of these men were keenly interested in the literature and historical development of mathematics. Moreover, all participated in a community of scholars that today is analogous to our professional mathematics organizations. While poster sessions, conferences, internet and email did not exist in their time, these men certainly knew others in their field and followed the flow of ideas with careful scrutiny.

Modern historians can only speculate on how deeply they felt the relationship, or competition among themselves. The Bernoulli brothers were known to support Leibniz and exchanged letters with him over several years (The Open University (2011). The Newton versus Leibniz rivalry over priority in developing calculus is still debated after 300 years. Several countries have honored Huygens, Newton, Leibniz and James Bernoulli by issuing postage stamps, but not so for John or l'Hôpital. The impressive country estates of Huygens and Newton are popular tourist destinations well worth any traveler's time. The Newton memorabilia at Trinity College, Cambridge draws thousands annually.

One clue to the reward and respect they earned from their contemporaries is evidenced in their burials. Newton became a celebrity in his own time. Voltaire wrote that he was buried like a king in Westminster Abbey ("National Curve Bank – Newton Gallery," 2011). His sarcophagus is one of the most outstanding in the Abbey. Leibniz died in Hanover, Germany. Historians record that the only person at his funeral was his servant. Still, the university is named for him and every single word he ever wrote has been collected in the Gottfried Wilhelm Leibniz Bibliothek ("National Curve Bank – Newton Gallery," 2011). His collection of letters on mathematical notation is especially impressive (Cajori, 1928). James Bernoulli has a magnificent tomb in the cathedral of Basel (4) while John supposedly spent many years in bitterness feeling both his brother and l'Hopital received credit for work that was really his. In Paris, 72 scientists and mathematicians have their names carved on the four sides of the pedestal of the Eifel Tour. l'Hôpital is not among them. Yet every calculus text in the world has l'Hôpital's Rule.

Note: The National Curve Bank website (http://curvebank.calstatela.edu/home/home.htm) is an invaluable source of information, both text and in animated graphical representations associated with these (and many other) mathematicians. A list of entries related to this chapter is provided in the references.

## References

Ball, D. & Bass, H. (2000). Interweaving content and pedagogy in teaching and learning to teach: Knowing and using mathematics. In J. Boaler, (Ed.), *Multiple perspectives on the teaching and learning of mathematics*, 83-104. Westport, CT: Ablex.

Beckmann, Petr (1971). *A history of $\pi$ (PI)*, 139-140. New York: St. Martin's Press.

Burton, David (2011). *A history of mathematics*, 412. New York: McGraw Hill.

Cajori, Florian (1928). *A history of mathematical notations*, vol. 1. Chicago: Open Court.

Common Core State Standards [CCSS]. (2010). Retrieved April 24, 2011, from http://www.corestandards.org/.

Hoffman, Joseph E. (1964). *Leibniz in Paris* 1672-1676. Cambridge University Press; letter, Leibniz to Gabrield Wagner, 1696. (in *Journal of Philosophy*, VII, 522)

Katz, Victor (2004). Calculus in the seventeenth century, in *A history of mathematics: Brief edition*, 282-328. Boston, MA: Addison Wesley.

Krussel, L. (2000). Using history in the teaching of mathematics *PRIMUS*, X (3), 273 – 276.

Leibniz, G. (1684). New methods for maxima and minima…, *Acta Eruditorum*, 467.

L'Hôpital, Marquis de (1696). *Analyse des infiniment pour l'intelligence des lignes courbes.* English translation by E. Stone, printed by William Innys.

National Council of Teachers of Mathematics [NCTM]. (2000). *Principles and standards for school mathematics.* Reston, VA: NCTM.

Sfard, A. (1991). On the dual nature of mathematical conceptions: Reflections on processes and objects as different sides of the same coin. *Educational Studies in Mathematics, 22* (1), 1-36.

Stone, E. (1730). Translation of l'Hôpital's (1696) *Analyse des infiniment pour l'intelligence des lignes courbes.*

The Open University. (2011) DVD, *The story of maths,* Athena, Disc 2. Previewed April 24 at http://www.ouworldwide.com/video_maths.asp

**On-line sources by subject:**

James Bernoulli

National Curve Bank – The catenary. (2011). Retrieved April 24, 2011, from http://curvebank.calstatela.edu/catenary/catenary.htm
National Curve Bank – The lemniscate. (2011). Retrieved April 24, 2011, from http://curvebank.calstatela.edu/lemniscate/lemniscate.htm
National Curve Bank – Bernoulli and the history of probability (2011). Retrieved April 24, 2011, from http://curvebank.calstatela.edu/jabernoulli/jabernoulli.htm

John Bernoulli

National Curve Bank – The lemniscate. (2011). Retrieved April 24, 2011, from http://curvebank.calstatela.edu/lemniscate/lemniscate.htm
National Curve Bank – The brachistochrone, part III. (2011). Retrieved April 24, 2011, from http://curvebank.calstatela.edu/brach3/brach3.htm
National Curve Bank – The brachistochrone part II (2011). Retrieved April 24, 2011, from http://curvebank.calstatela.edu/brach2/brach2.htm

l'Hôpital

National Curve Bank – l'Hopital (2011). Retrieved April 24, 201, from http://curvebank.calstatela.edu/lhospital/lhospital.htm

Huygens
Biography of Christian Huygens. (2011). Retrieved April 24, 2011, from

http://www-history.mcs.st-and.ac.uk/history/Biographies/Huygens.html
(Note: 1660 letter from Huygens in Paris to his brother.)
The Catenary. (2011). Retrieved April 24, 2011, from
http://www-history.mcs.st-and.ac.uk/history/Curves/Catenary.html
(Note: Catenary could not be a parabola.)

Leibniz

The Logic of Leibniz, Chapter 3. (2011). Retrieved April 24, 2011, from
http://philosophyfaculty.ucsd.edu/faculty/rutherford/Leibniz/ch3.htm
National Curve Bank – A calculus connection – Leibniz. (2011). Retrieved April 24,
2011, from http://curvebank.calstatela.edu/derrules/derrules.htm
The Open University. (2011) DVD, *The Story of Maths*, Athena, Disc 2. Previewed at
http://www.ouworldwide.com/video_maths.asp

Newton

National Curve Bank – Newton Gallery. (2011). Retrieved April 24, 2011, from
http://curvebank.calstatela.edu/newtongallery/newtongallery.html
National Curve Bank – Cambridge mathematics. (2011). Retrieved April 24, 2011, from
http://curvebank.calstatela.edu/cambridge/cambridge.htm

# The Mathematics Enthusiast

Volume 13
Number 1 *Numbers 1 & 2*

Article 1

2-2016

# TME Volume 13, Numbers 1 and 2

**The Mathematics Enthusiast**
**Volume 13, Numbers 1-2**
**Special issue: Mathematical Knowledge for Teaching: Developing Measures and Measuring Development**
Guest Edited by Reidar Mosvold and Mark Hoover

**Guest Editorial:**

**Mathematical Knowledge for Teaching: Developing Measures and Measuring Development**

Reidar Mosvold (reidar.mosvold@uis.no)
University of Stavanger, Norway


Mark Hoover (mhoover@umich.edu)
University of Michigan, USA

In recent years, researchers' interest in mathematical knowledge that is specific to teaching has escalated. Its increasing importance has also influenced the content of this journal. Less than two years ago, a special issue that focused on mathematical knowledge for teaching was published in *The Mathematics Enthusiast* (volume 11, no. 2). The title of this issue was: "The Mathematical Content Knowledge of Elementary Prospective Teachers". All of the articles in that special issue were written in connection with a review of 112 studies from 1978 to 2012 on the knowledge of prospective elementary teachers in the following five content areas: numbers and operations, fractions, decimals, geometry and measurement, and algebra. In their final article of that special issue, Thanheiser, Browning, Edson, Lo, Whitacre, Olanoff and Morton (2014) concluded that the number of articles on prospective teachers' mathematical content knowledge had increased over the years. They also found that most of the studies focused on "static studies of knowledge" rather than on how this knowledge developed. Fraction content knowledge was the area with most of the studies, and — across all five content areas — the studies showed that prospective teachers seemed to rely on procedural understanding, and most of the literature focused on identifying deficits in their understanding. In their conclusions, Thanheiser et al. suggested that more research should focus on characterizing prospective teachers' content knowledge and investigate how this knowledge develops.

Our aim is that the articles in the present special issue, at least to a certain extent, contribute to moving the field forward in the directions pointed out by Thanheiser et al. (2014). The focus of the articles in this special issue is on mathematical knowledge for teaching more broadly, and not only on prospective elementary teachers' knowledge. As a whole, the articles in this special issue build upon an assumption that strong instruments are of vital importance to measure — and thus to characterize the nature and development of — mathematical knowledge for teaching. Although all of the studies in this special issue draw upon the practice-based theory of mathematical knowledge for teaching that has been developed by researchers at the University of Michigan (e.g., Ball, Thames, & Phelps, 2008), the lessons learned from these studies may be relevant to the work of researchers who adhere to other views about the mathematical knowledge that is distinctly related to the teaching of mathematics.

The articles in this special issue differ in type as well as in scope, ranging across theoretical and conceptual studies, studies that focus on instruments and instrument development, and more standard empirical studies. All of them contribute either to broader efforts to develop measures of mathematical knowledge for teaching or to reflections on research that has been done and new directions important for advancing. Thus, the title of this special issue, where the first use of the word "measure" refers to the literal assigning of numbers for the purpose of comparison and the second use, "measuring development," refers to a figurative perusing and appraising of headway in the field. Although authors of most of the articles have some connection with the research group at the University of Michigan, the authors represent different universities and institutions from five different countries: USA, Norway, Finland, South Korea and Malawi. Many are by junior faculty and may suggest a shifting interest and focus of study taking place in the field. It is our hope that this special issue will be of interest to all researchers in the field of mathematics education, and especially to those who are involved in research on mathematical knowledge for teaching. We believe that the articles in the special issue provide a relevant snapshot of where this field of research is today, and we also hope that they provide relevant suggestions for the further development of the field.

## References

Ball, D. L., Thames, M. H. & Phelps, G. (2008). Content knowledge for teaching: What makes it special? *Journal of Teacher Education, 59*(5), 389–407.

Thanheiser, E., Browning, C., Edson, A. J., Lo, J.-J., Whitacre, I., Olanoff, D., & Morton, C. (2014). Prospective Elementary Mathematics Teacher Content Knowledge: What Do We Know, What Do We Not Know, and Where Do We Go? *The Mathematics Enthusiast, 11*(2), 433–448.

# Making Progress on Mathematical Knowledge for Teaching

Mark Hoover

University of Michigan, USA


Reidar Mosvold

University of Stavanger, Norway


Deborah Loewenberg Ball

University of Michigan, USA


Yvonne Lai

University of Nebraska-Lincoln, USA

**Abstract:** Although the field lacks a theoretically grounded, well-defined, and shared conception of mathematical knowledge required for teaching, there appears to be broad agreement that a specialized body of knowledge is vital to improvement. Further, such a construct serves as the foundation for different kinds of studies with different agendas. This article reviews what is known and needs to be known to advance research on mathematical knowledge for teaching. It argues for three priorities: (i) finding common ground for engaging in complementary studies that together advance the field; (ii) innovating and reflecting on method; and (iii) addressing the relationship of such knowledge to mathematical fluency in teaching and to issues of equity and diversity in teaching. It concludes by situating the articles in this special issue within this emerging picture.

*Keywords:* mathematical knowledge for teaching, MKT, specialized knowledge, pedagogical content knowledge, PCK, mathematics teacher education, method, mathematical fluency, equity, diversity.

## Introduction

A century ago, a central focus of teacher education in the United States was on developing a thorough understanding of subject matter, but the mid-twentieth century witnessed a steady shift to an emphasis on pedagogy generalized to be largely independent of subject matter. By the 1980s, an absence of content focus was so prevalent that Shulman (1986) referred to this as a "missing paradigm" in teacher education. A similar tendency can be seen in other countries. For example, a few decades ago, it was possible to become qualified for teaching mathematics in grades 1–9 in Norway with no more mathematics than a short course in didactics. A widespread assumption seemed to be that prospective teachers already knew the content they needed, from their experiences as students, and they only required directions in how to teach this content. Shulman's call for increased attention to subject matter reoriented research and practice. However, the connection between the formal education of mathematics teachers and the content understanding important for their work is not straightforward. Teachers' formal mathematics education is not highly correlated with their students' achievement (Begle, 1979) or with the depth of understanding they seem to have of the mathematical issues that arise in teaching (Ma, 1999).

One of Shulman's (1986) most important contributions was the suggestion that the work of teaching requires professional knowledge that is distinctive for the teaching profession. He proposed different categories of professional knowledge for teaching. One of these categories was distinctive content knowledge, which Shulman described as including a deep knowledge of the structures of the subject (e.g., Schwab, 1978), beyond procedural and factual knowledge. Another category of knowledge was what Shulman termed "pedagogical content knowledge," which is aspects of the content most germane to its teaching (1986, p. 9). The idea about an amalgam of subject matter knowledge and pedagogical knowledge has continued to appeal to researchers working in different subject areas, and Shulman's foundational publications are among the most cited references in the field of education. (Google Scholar identifies over 13000 publications that cite his 1986 article.)

In the last two decades, researchers and mathematics educators have increasingly emphasized the significance of mathematical knowledge that is teaching-specific. Such knowledge is seen as different from the mathematics typically taught in most collegiate mathematics courses and from the mathematics needed by professionals other than teachers. Although it includes knowing the mathematics taught to students, the kind of understanding of the material needed by teachers is different than that needed by the students. Even though the literature suggests a general consensus that mathematics teaching requires special kinds of mathematical knowledge, agreement is lacking about definitions, language, and basic concepts. Many scholars draw on Shulman's notion of pedagogical content knowledge (or PCK) and view this knowledge as being either a kind of "combined" knowledge or a kind of "transformed" knowledge. Grounded in Shulman's proposals, the phrases "for teaching" and "practice-based" have been emphasized to indicate the relationship of the knowledge to specific work of teaching (e.g., Ball, Thames, & Phelps, 2008). For this article, we adopt these phrases but maintain an ecumenical view of a more extended literature.

With growing interest in ideas about specialized professional content knowledge, the early 2000s saw a spate of large-scale efforts to develop measures of such knowledge and the use of such a construct as the basis for a wide range of research studies, such as evaluating professional development (e.g., Bell, Wilson, Higgins, & McCoach, 2010), examining the impact of structural differences on the mathematical education of teachers (e.g., Kleickmann et al., 2013), arguing for policies and programs (e.g., Hill, 2011), and investigating the role of professional content knowledge on mathematics teaching practice (e.g., Speer & Wagner, 2009). Instruments for measuring such knowledge represent a crucial tool for making meaningful progress in a field. They operationalize emerging thinking, invite scrutiny, and support the investigation of underlying models.

This special issue on developing measures and measuring development of mathematical knowledge for teaching continues this focus on instruments, along with a concomitant regard for broader purposes and potential ways to advance the field. In an effort to situate this special issue, this introductory article provides some selected highlights from the field — focusing on what is being studied, how, and to what ends. To accomplish this, we draw on both a detailed review of articles sampled from 2006 to 2013 and our wider reading of the literature. We then nominate some key areas for making progress on research and development of the specialized mathematical knowledge teachers need and we use this framing to characterize the agendas and contributions of the collection of articles assembled in this special issue. The article consists of three major sections.

1. Lessons from Empirical Research
2. Next Steps for the Development of Mathematical Knowledge for Teaching
3. Articles that Develop Measures and Measure Development

The first describes a review we conducted and discusses three broad arenas of work suggested by this review. The second discusses three proposals for future research. The last briefly situates the articles in this issue within the lessons and directions discussed.

**Lessons from Empirical Research**

In our reading of empirical literature concerned with the distinctive mathematical knowledge requirements for teaching, several broad strands of research stand out. We begin by describing a formal review we conducted of empirical research that began appearing in about 2006, in the wake of a number of conceptual proposals (beyond PCK), and that began using these proposals as a conceptual basis for empirical study. This review informs our overall reading of the field. Combining this review with our wider reading in the field, we then identify and discuss three major arenas of work.

**Reviewing the literature.** In the course of other research we were conducting, we reviewed international empirical literature published in peer-reviewed journals in English between 2006 and 2013.[1] Wanting to survey the topic across theoretical perspectives, we developed and tested inclusive search terms:

- Mathematics
  - math* (the asterisk is a placeholder for derived terms)
- Content knowledge
  - know* AND (content OR special* OR pedagog* OR didact* OR math* OR teach* OR professional OR disciplin* OR domain) OR "math for teaching" OR "mathematics for teaching" OR "math-for-teaching" OR "mathematics-for-teaching

---

[1] This more formal review, which we use to inform our wider reading of the literature, was funded by the National Science Foundation under grant DRL-1008317 and conducted in collaboration with Arne Jakobsen, Yeon Kim, Minsung Kwon, Lindsey Mann, and Rohen Shah, who we wish to thank for their assistance with searching, conceptualizing codes, coding, and analysis. The opinions reported here are those of the authors and do not necessarily reflect the views of the National Science Foundation or our colleagues.

- Teaching
    - teaching OR pedagog\* OR didact\* OR instruction\*

These search terms initially yielded over 3000 articles from the following six databases:

- PsycInfo
- Eric
- Francis
- ZentralBlatt
- Web of Science
- Dissertation Abstracts

Broadened search terms, additional databases, and inclusion of earlier publication years yielded none to negligible additional articles.

Based on a reading of abstracts, 349 articles were identified as potential empirical articles (as characterized by the American Educational Research Association, 2006) in which some concept of distinctive mathematics needed for teaching was used as a conceptual tool to formulate research questions or structure analysis. Our goal was not to reach high standards of reliability, but rather to use a systematic process to collect a corpus of relevant studies representing the literature from this period. In coding the articles, we sought to be descriptive rather than evaluative and iteratively worked between an inductive examination of a sample of articles and initial conceptualizations of empirical research combined with a basic model of educational change. After reading full articles, 190 of the 349 remained in the final set. A set of core codes were developed for the following categories:

1. Genre of the study
2. Research problem used to motivate the study
3. Variables used
4. Whether or not and how causality was addressed
5. Findings

Additional codes included sample size, instruments used for measuring mathematical knowledge for teaching, school level or setting, professional experience of the teachers, geographic region, and mathematical area addressed. Each article was read and coded by two team members, with a decision as to whether it satisfied our inclusion criteria, and if so, codes were reconciled. (For a more detailed description of the methods used, see Kim, Mosvold, and Hoover (2015).)

In table 1, we present some patterns that emerged from some of the additional, descriptive codes.

Table 1. *Selected descriptive codes for sample size, instrument used, level of schooling, and geographic context.*

| Categories and codes | Number of papers |
|---|---|
| **Sample size** | |
| Small scale (<10) | 60 |
| Medium 1 (10–29) | 51 |
| Medium 2 (30–70) | 34 |
| Large scale (>70) | 43 |
| None | 2 |
| **Instrument** | |
| COACTIV | 4 |
| CVA | 3 |
| DTAMS | 3 |
| LMT (including adaptations) | 31 |
| TEDS-M | 2 |
| Non-standardized | 56 |
| None | 91 |
| **Level of teachers** | |
| Primary (K–8) | 81 |
| Middle (5–9) | 45 |
| Secondary (7–13) | 41 |
| Tertiary | 3 |
| Across levels | 20 |
| **Regions** | |
| Africa | 7 |
| Asia | 27 |
| Europe | 22 |
| Latin America | 3 |
| North America | 112 |
| Oceania | 15 |
| Across regions | 4 |

We observe that many studies are small-scale, and a large number of the studies apply non-standardized instruments or no instruments. In the studies where standardized instruments were used to measure teachers' knowledge, the instruments developed to measure mathematical knowledge for teaching in the Learning Mathematics for Teaching (LMT) project were most common. An abundance of studies focuses on primary teachers, and most studies were carried out in North America.

Table 2 provides the fourteen categories developed for coding the research problem. We have grouped these into three domains and use these groups to discuss the literature in the following sections.

Table 2. *Research problems addressed.*

| Problems | Number of papers | % |
|---|---|---|
| Nature and composition of SM | 55 | 28.9 |
|     What is SM? | 34 | |
|     What relationships exist among aspects of SM or with other variables? | 21 | |
| Improvement of SM | 81 | 42.6 |
|     What professional development improves teachers' SM? | 28 | |
|     What teacher education improves teachers' SM? | 28 | |
|     What curriculum/tasks improve teachers' SM? | 10 | |
|     What teaching practice improves teachers' SM? | 0 | |
|     How SM develops? | 15 | |
|     How to scale up the teaching and learning of SM? | 0 | |
| Contribution of SM | 33 | 17.4 |
|     Does SM contribute to teaching practice? | 6 | |
|     What does SM contribute to teaching practice? | 12 | |
|     Does SM contribute to student learning? | 15 | |
|     What does SM contribute to student learning? | 0 | |
| Other | | |
|     What SM do teachers know? | 21 | 11.1 |
|     How policy influences teachers' SM? | 0 | 0 |
| *Total* | *190* | *100* |

In order to make table 2 more readable, we use the abbreviation SM to signify any of the variety of ways in which mathematical knowledge for teaching might be conceptualized and named. The intention is not to introduce yet another term or acronym for such knowledge. In this article, we have adopted more generic language to express an inclusive notion of such knowledge and we avoid the use of any specific acronym label.

For the purpose of this introductory article, we used patterns evident in the review described above to inform our extended reading of the field. Together, these efforts led us to identify three broad themes. First, a number of studies investigate the nature and composition of teacher content knowledge. Given that foundational research into teaching-specific mathematical knowledge pointed to its elusiveness and complexity, it is not surprising that scholars continue to investigate what it is — its components, measurement, features, and related constructs. A second group of studies, which constitutes the majority of published articles, investigates approaches to increasing teacher knowledge, in both the context of pre-service teacher education and the professional education of practicing teachers. A third group of studies, fewer in number, investigates effects of teachers' knowledge on both teaching and student learning. In the following sections, we use these three broad themes to organize our comments on selected highlights from the literature. Following these, we provide suggestions about possible next steps for further research in this field.

**Nature and composition of mathematical knowledge for teaching.** Current studies continue to probe ideas about the nature and composition of teaching-specific knowledge of mathematics. Some studies consider the construct in broad terms. They may identify or elaborate aspects or frameworks, characterize or critique the construct, compare different representations or sub-domains, or compare such knowledge with other kinds of mathematical knowledge. Others examine a constrained area of knowledge: some in relation to specific mathematical topics; some in relation to specific practices of teaching, or at specific levels (such as interpreting and responding to student thinking, curriculum use, or proving in high school geometry); and some in relation to specific qualities (such as connectedness). However, these studies do not build on each other in obvious ways and clear lessons are hard to identify. The one avenue of work that represents progress for the field is the development of instruments, and we focus our discussion there.

Instruments provide a crucial tool for investigating the nature and composition of mathematical knowledge needed for teaching. They serve to operationalize ideas about mathematical knowledge for teaching and test assumed models of the role it plays. They are used to investigate the teaching and learning of such knowledge, relationships with other variables, and other questions important for practice and policy. On the one hand, rigorous instrument development is expensive relative to budgets available for most studies and many instruments are used in a single study and limited in the extent to which they meet psychometric standards and establish validity. On the other hand, several larger efforts have invested in building instruments for large-scale studies and wider use in the field. The Learning Mathematics for Teaching (LMT) instruments for practicing elementary and middle school teachers (Hill, Schilling, & Ball, 2004) include nearly 1000 items on over a dozen different instruments and have been used in numerous program evaluations and studies of relationships and effects. They have been extensively validated (Schilling & Hill, 2007) and adapted internationally (Blömeke & Delaney, 2012). The Diagnostic Teacher Assessment in Mathematics and Science (DTAMS) instruments for practicing middle school teachers (Saderholm, Ronau, Brown, & Collins, 2010) include 24 forms in four content areas, have been administered and rigorously analyzed with a sample of several thousand teachers, and are currently being expanded. The Teacher Education and Development Study in Mathematics (TEDS-M) instruments for pre-service primary and lower secondary teachers (Tatto et al., 2008; Senk et al. 2012) include over 100 items and were originally administered to 23,000 pre-service teachers in 17 countries.

These instruments represent an important contribution to the field. Extensive cross-professional-community review and the building of agreed-on formulations of important content knowledge have played a major role in the development of these measures. The synthesis of ideas and the integration of expertise from multiple professional communities have helped to clarify and improve ideas about mathematical knowledge for teaching. In addition, the availability of common instruments has enabled meaningful comparison and interpretation across programs, countries, and studies in ways that contribute to the maturity of research on mathematical knowledge for teaching.

Several other efforts have developed instruments with less focus on broad consensus or widespread use. The COACTIV instrument for practicing secondary teachers (Kunter, Klusmann, Baumert, Voss, & Hacfeld, 2013) produced items of a genre similar to those described above and used these to investigate relationships to other variables and to understand issues of practice and policy related to the mathematical education of teachers. Some instruments have been developed to focus on mathematical knowledge related to a specific topic, such as fractions (Izsak, Jacobson, de Araujo, & Orrill, 2012), geometry (Herbst & Kosko, 2012), algebra (McCrory, Floden, Ferrini-Mundy, Reckase, & Senk, 2012), and continuous variation and covariation (Thompson, 2015). Others have focused on specific aspects of teaching, and the mathematical knowledge required in these specific teaching practices, such as choosing examples (Chick, 2009; Zodik & Zaslavsky, 2008) and scaffolding whole-class discussions to address mathematical goals (Speer & Wagner, 2009). Many instruments have been developed in relation to specific lines of research and often in response to perceived issues with more established instruments. A number of researchers are concerned about a potentially narrow interpretation of knowledge as declarative or about a possible discrepancy between knowledge and knowledge use. These concerns have led some scholars to explore different conceptualizations of the mathematics teachers need and to look for alternative formats for measuring it (e.g., Kersting, Givvin, Thompson, Santagata, & Stigler, 2012; McCray & Chen, 2012; Thompson, 2015).

Although the development of instruments is an important step toward building a robust conception of teaching-specific knowledge of mathematics, these efforts also reveal a lack of shared language and meaning of foundational concepts. Differences in meaning for the construct PCK have been noted in the past (Ball, Thames, & Phelps, 2008; Depaepe, Verschaffel, & Kelchtermans, 2013; Graeber & Tirosh, 2008). These differences persist, yet they are often overlooked with regard to instruments.

For example, Kaarstein (2014) examined whether the LMT, TEDS-M, and COACTIV instruments, each referencing Shulman and stating that the respective instrument measures PCK, measure the same thing. To study this issue, she constructed a taxonomy of the different levels of categories in Shulman's initial framework as well as the frameworks that were used to develop the three instruments. She then selected three items — one from each instrument — and categorized them according to each of the three frameworks. Her main argument is that content knowledge and pedagogical content knowledge are supposed to be distinct categories, and therefore three projects that use the same basic categories should categorize items in the same way. However, from her analysis the items would be placed in different basic categories using the criteria reported by the projects. As an example, an item that was categorized as a specialized content knowledge item (measuring content knowledge) in the LMT project would probably have been categorized as a PCK item in TEDS-M and COACTIV. Kaarstein's argument does not necessarily threaten the validity of the measures from each of the three projects, but her observation deserves further attention.

Similarly, a study by Copur-Gencturk and Lubienski (2013) echoes this concern. In order to investigate growth in pre-service teacher knowledge, they used two different instruments: LMT and DTAMS. When comparing groups of teachers who had participated in different kinds of courses, they concluded that the LMT and DTAMS instruments measure aspects of mathematical knowledge for teaching that are substantially different. Teachers who participated in a hybrid mathematics content/methods course had the most significant increase in their LMT score, and this score remained stable although they took an additional content course. Teachers' DTAMS score also increased during the hybrid course; during the content knowledge course, only the content knowledge part of their DTAMS increased. This study thus supports the idea that there is specialized mathematical content knowledge not influenced by general mathematics content courses. That different instruments measure different aspects of knowledge is not necessarily surprising, but it is worrying if instruments ostensibly designed to capture the same construct in fact measure significantly different facets of that knowledge, with little clarity about these differences.

The concerns raised by Kaarstein (2014) and Copur-Gencturk and Lubienski (2013) suggests that the limited specification of the construct and the different ways of operationalizing it makes it difficult to interpret results. This limits the extent to which results from these instruments, taken individually or together, can inform the conceptualization of mathematical knowledge for teaching or practical decisions needed to design learning opportunities.

**Developing teachers' mathematical knowledge for teaching.** With a growing sense of the mathematics important for improving teaching and learning, practitioners have turned their attention to increasing teachers' knowledge of professionally relevant mathematics and scholarly work has followed suit. A large number of studies make it clear that the design and evaluation of teacher education and professional development programs in developing teachers' mathematical knowledge for teaching are top priorities. From several decades of research, we propose what we see as a few related emerging lessons:

- Teaching teachers additional standard disciplinary mathematics beyond a basic threshold does not increase their knowledge in ways that impact teaching and learning.

- Providing teachers with opportunities to learn mathematics that is intertwined with teaching increases their mathematical knowledge for teaching.
- The focus of the content, tasks, and pedagogy for teaching such knowledge requires thoughtful attention to ways of maintaining a coordination of content and teaching without slipping exclusively into one domain or the other.

These lessons are rooted in early efforts to document effects of teachers' mathematical knowledge on student learning and are reinforced by current research on the design and implementation of teacher education and professional development. We begin by briefly reflecting on that early work and then tracing these lessons into current research.

Much of the impetus for the surge in research on teaching-specific knowledge began with reviews of several decades of large-scale research that found surprisingly little to no effect of teachers' mathematical knowledge on their students' learning (Ball, Lubienski, & Mewborn, 2001). The studies reviewed were often conducted with large datasets but very coarse measures. Taking Shulman's (1986) suggestion that the content knowledge needed by teachers was characteristically different from that needed by other professionals, researchers began to look more closely at the measures used in those studies and at the findings. The clearest finding that emerged was that methods courses consistently showed positive effects while content courses did not (e.g., Begle, 1979; Ferguson & Womack, 1993; Guyton & Farokhi, 1987; Monk, 1994). The second was that positive effects were more likely when the content taught to teachers was more closely related to the content they subsequently taught. For instance, several scholars found effects when using student exams to measure teachers' knowledge (Harbison & Hanushek, 1992; Mullens, Murnane, & Willett, 1996). Reinforcing these results, Monk (1994) found that coursework in calculus influenced the achievement of secondary teachers' students in algebra classes, but not in their geometry classes. In general, when the mathematics taught or measured is meaningfully connected to classroom materials or interactions, it is modestly associated with improved teaching and learning.

For some practitioners and policy-makers, the implication of these empirical studies, combined with logical arguments for teaching-specific professional knowledge, has been enough to lead to prioritizing mathematical knowledge for teaching in the mathematical education of teachers. Nevertheless, many policies continue to press for increases in the number of mathematics courses required of teachers, regardless of their connection to teaching, despite abundant evidence that such policies are unlikely to improve teaching and learning (e.g., Youngs & Qian, 2013). Such policies have probably been less the result of lingering doubt about empirical results and more the result of overextending the notion that knowing content well is key to good teaching, even in the face of disconfirming evidence. Of course, a certain threshold level of knowledge of the subject is essential, but preparing teachers by requiring mathematics courses that are not directly connected to the content being taught or to the work involved in teaching that content is misguided.

More recent studies continue to reinforce these established lessons. One recent line of inquiry is the investigation of features of innovative, well-received professional development programs. To us, the most compelling result emerging from these studies is that professional development requires designing pedagogically relevant movement between mathematical and pedagogical concern both to motivate teachers' investment in mathematical issues and to keep the mathematical attention on mathematics that matters for the work of teaching. To elaborate, we offer several examples that contribute to this claim.

With deep regard for the limited effects of decades of substantial national investment in professional development, several research groups have organically developed approaches informed by thoughtful reflection and attention to disciplined observation of teachers' engagement with and actual uptake of ideas and practices. One important insight emerging from these decades-long investments is that cycling through mathematical considerations, pedagogical considerations, and reflective enactment is vital to the design of professional development. For instance, Silver, Clark, Ghousseini, Charalambous, and Sealy (2007) set out to provide evidence for whether and how teachers might enhance their mathematical knowledge for teaching through monthly practice-based professional development workshops designed to cycle from activities of doing mathematics, to examining case-based pedagogical and student-related issues, to planning, teaching and debriefing lessons collaboratively (all related to a common mathematics task or set of tasks). Examining the interactions of one teacher, they document ways these activities provided opportunities for teachers to build connections among mathematical ideas and to consider these ideas in relation to student thinking and teaching. They do not measure teacher learning. Nor do they disentangle effects of what they refer to as a professional-learning-task cycle from a number of other important features of their professional development program. However, they document dynamics in which the teacher, from an initial experience solving a nontrivial mathematics problem, supported by mathematically sensitive facilitation, successively engages in mathematical issues and pedagogical issues in ways that visibly build connections among mathematical ideas, pedagogical practice, and growing mathematical knowledge for teaching. In addition, they argue that their cyclic design increased teachers' motivation for learning mathematics, both in the workshops and in their daily practice.

> Through successive opportunities to consider mathematical ideas in relation to the activities of classroom practice, our participants came to see their pedagogical work as permeated by mathematical considerations. (p. 276)

Similarly, in working to close the gap between a reform vision and the actual practice of mathematics teaching and learning, Koellner et al. (2007) implemented a model of professional development designed to help teachers deepen their mathematical knowledge for teaching through a cycle of solving a mathematics problem, teaching the problem, and analyzing first teacher questioning and then student thinking in videos of their teaching. In order to understand the learning opportunities afforded by what they refer to as a problem-solving-cycle design, they analyzed artifacts from two years of a series of monthly, full-day workshops with ten middle school mathematics teachers, including workshop videos and interviews with facilitators. The researchers used the knowledge domains identified in Ball, Thames, and Phelps (2008) to analyze several teacher interactions. They found that different learning opportunities were afforded by different activities: specialized content knowledge was developed by comparing, reasoning, and making connections between the various solution strategies; knowledge of content and teaching was developed by analyzing teacher questioning in the video clips from the teachers' lessons; and knowledge of content and students was developed by analyzing students' solution methods (interpreting them and considering their implications for instruction). More importantly, the researchers found that reflecting on and discussing the nature of student thinking and teacher questioning of students evident in videos of their own teaching led teachers to extend their mathematical knowledge for teaching as they re-engaged with the mathematics problem and reconsidered how they might teach the problem in light of their new regard for how students might approach the problem. Throughout the analysis, the authors found that specialized content knowledge interacts with pedagogical content knowledge in interpreting student thinking and planning lessons. The authors argue that the workshops developed teachers' mathematical knowledge for teaching by supporting teachers' current knowledge, while gradually challenging them to gain new understanding for the purpose of their work as teachers.

The lessons from studies such as these are subtle. The movement between mathematical study and pedagogical practice is central, but attention needs to be given to dynamics regarding teachers' motivation, the timing of different activities, and specific mathematical opportunities arising from specific pedagogical activities. In reading these reports, one gets the sense that really smart enactment of the professional development was key to success and that replicating effects might be challenging. From this work, it would seem important to discern the essential design features and elaborate the necessary character of facilitation.

One effort along these lines is a study by Elliott, Kazemi, Lesseig, Mumme, and Kelley-Petersen (2009). In the context of supporting facilitators' enactment of mathematically focused professional education, they analyzed facilitators' learning and the use of two frameworks provided as conceptual tools: (i) sociomathematical norms for cultivating mathematically productive discussion in professional development, adapted from Yackel and Cobb (1996) and (ii) practices for orchestrating productive mathematical discussions, adapted from Stein, Engle, Smith, and Hughes (2008). In their study, Elliot and colleagues collected extensive documentation and analyzed the learning of 5 of the 36 facilitators trained at two sites in 6 two-day seminars across an academic year. They found that although facilitators responded positively to the frameworks, they experienced tensions in using the frameworks to ask questions about colleagues' mathematical thinking and they struggled with the fact that teachers positioned themselves and others as better or worse in mathematics. These dynamics got in the way of productive mathematical discussions and frustrated facilitators. The analysis revealed that one way to mitigate these tensions was by helping facilitators to identify mathematical ideas that teachers would readily see as worth developing. This led the researchers to see a need for developing more nuanced and detailed purposes for doing mathematics in professional development in ways that teachers would see as relevant to their work.

This then led the researchers to realize that they needed a way to focus the purpose and work of professional development on connections between mathematics and the work of teaching. To accomplish this, they added a third framework to their design. The authors argue that the mathematical-knowledge-for-teaching framework engaged facilitators in understanding the ways in which specialized content knowledge (SCK) connects mathematics to teaching and that the framework provided a meaningful articulation of the purpose of the professional development and a helpful focus for the mathematical tasks and discussions that took place.

> By understanding how a SCK-oriented purpose for PD is tied to classroom teaching and being able to articulate that understanding to teachers in accessible ways, leaders will be able to begin to address the pressure they felt to assure relevance in their PD. (Elliott et al., 2009, p. 376)

Again, the dynamics between mathematics and the motivation and use of that mathematics is key to effective teacher learning of professionally relevant mathematics.

The field is also beginning to see evidence that these insights have measurable yield. For instance, Bell, Wilson, Higgins, and McCoach (2010) argue that it is the practice-based character of the nationally disseminated Developing Mathematical Ideas (DMI) mathematics professional development program that best explains participating teachers' learning of mathematical knowledge for teaching. The researchers examined pre and post teacher content knowledge for 308 treatment and comparison teachers across 10 well-established sites. They found significantly larger gains for treatment teachers' scores and that these gains were related to breadth of opportunity to learn provided by facilitators. Methodically considering a number of alternative explanations for treatment teachers' improvement, the researchers emphasize the classroom-practice feature of the professional development, where teachers move back and forth between seminars and their own classrooms, receiving written feedback from regularly observing facilitators. Referring to Ball and Cohen's (1999) argument that teacher learning needs to be embedded in practice, they point out that connecting to practice can leverage teacher learning in and from their daily work, greatly expanding overall capacity for teacher learning and improvement. They argue that the practice-based nature of their design contrasts with professional development that takes place apart from teachers' practice.

> DMI is quite different in this regard, for it encourages teachers to take their nascent SCK, KCS, and KCT into their classrooms and try things out. Repeatedly, teachers told us of their revelations — both in seminars and in their own schools — as they drew on their growing knowledge of and enthusiasm for mathematics and teaching mathematics in their classrooms. This anecdotal evidence aligns with results from S. Cohen's (2004) yearlong study of changes in teachers' thinking and practices over the course of their participation in DMI seminars. (Bell et al., 2010, p. 505)

These different studies compellingly add to the arguments that teachers need mathematical knowledge that is connected to the work they do and that situating the learning of mathematical knowledge in teachers' practice supports the learning of mathematical knowledge for teaching. Bell et al.'s (2010) large-scale study of the effect of professional development on teacher learning corroborates the qualitative, small-scale findings of the other studies. The professional development models highlighted set teachers up to learn in and from their practice. Together, the studies discussed above point to the coordinated nature of mathematical knowledge for teaching and the ways in which the coordination between mathematics and pedagogy is essential to teaching and learning mathematical knowledge for teaching.

**Impact of mathematical knowledge for teaching.** Whereas more studies have investigated the nature and composition of mathematical knowledge for teaching and developing teachers' knowledge, fewer studies have investigated the impact such knowledge has on teaching and learning. As mentioned earlier, several studies report positive effects of mathematical knowledge for teaching on student learning. Crucial to this research has been the development of robust instruments assessing mathematical knowledge for teaching. The field has found evidence linking mathematical knowledge for teaching to student achievement using the LMT instrument (e.g., Hill, Rowan, & Ball, 2005; Rockoff, Jacob, Kane, & Staiger, 2011), the COACTIV instrument (e.g., Baumert et al., 2010; Kunter et al., 2013), and the Classroom Video Analysis (CVA) instrument (e.g., Kersting et al., 2010, Kersting et al., 2012). A fewer number of studies have investigated links between teaching practice and mathematical knowledge for teaching and/or student achievement (e.g., Hill, Kapitula, & Umland, 2011). In these studies, student learning is mostly measured by standardized test scores, and the studies vary in how they measure teaching quality. These studies indicate that, generally speaking, mathematical knowledge for teaching impacts teaching and learning.

We acknowledge the importance of studies that identify an influence of teachers' mathematical knowledge on teaching and learning, but are particularly excited about studies that unpack the dynamics of how mathematical knowledge for teaching impacts teaching and learning. In their study of 34 teachers, Hill, Umland, Litke, and Kapitula (2012) demonstrated that the connection between mathematical knowledge for teaching (measured with the LMT instrument) and the quality of instruction is complex. While weaker mathematical knowledge for teaching seemed to predict poorer quality of instruction, and stronger mathematical knowledge for teaching seemed to predict higher quality of instruction, teachers who performed in the midrange on the LMT measure varied widely in the quality of their instruction. Student achievement also varied widely for teachers with mid-range mathematical knowledge for teaching. Furthermore, Hill et al.'s (2008) study of 10 teachers found that although use of supplemental curriculum materials, teacher beliefs, and professional development are factors of potential influence, these factors might all cut both ways depending on the teachers' mathematical knowledge for teaching. These two studies underscore that simply establishing impact of knowledge on teaching is not enough to make decisions about teacher education or policy.

To frame a fuller consideration of impact, we reflect briefly on the nature of teaching and learning. Teaching mathematics involves managing instructional interactions, including everything teachers say and do together with students focused on content, where teacher knowledge is a resource for the work (Cohen, Raudenbush, & Ball, 2003). This observation suggests that in addition to general effect studies on teaching and learning, it would be helpful to know more about which specific aspects of teaching and learning are influenced by teacher content knowledge, which specific aspects of teacher content knowledge are influential, and how the influences impact interactions among teacher and students around content. In other words, we propose that Cohen et al.'s conceptualization of teacher content knowledge as a resource impacting instructional interactions is important for framing an investigation of mathematical knowledge for teaching suited to informing the improvement of teaching and learning.

A promising direction in recent work has been initial investigation of the specific influence that mathematical knowledge for teaching has on teaching. One example of this kind is Speer and Wagner's (2009) case study of one undergraduate instructor's scaffolding of classroom discussions. Using Williams and Baxter's (1996) constructs of social and analytic scaffolding as a frame, Speer and Wagner argue that aspects of pedagogical content knowledge are important for helping students find productive ways of solving particular problems and for understanding which student contributions — correct or incorrect — are important to emphasize in a discussion. They trace ways in which particular knowledge of students' understanding aids teachers in assuring that the lesson reaches intended mathematical goals and in understanding the role of particular mathematical ideas in students' development.

In a similar vein, an exploratory study by Charalambous (2010) investigated teachers' knowledge in relation to selection and use of mathematical tasks. He investigated the teaching of two primary mathematics teachers with different levels of mathematical knowledge for teaching and found notable differences in the quality of their teaching. He used Stein and colleagues' mathematical tasks framework to examine the cognitive level of enacted tasks, and he formulated three tentative hypotheses about mechanisms of how mathematical knowledge for teaching impacts teachers' selection and use of mathematical tasks. First, he hypothesizes that strong mathematical knowledge for teaching may contribute to a use of representations that supports students in solving problems, whereas weaker mathematical knowledge for teaching may limit instruction to memorizing rules. Second, he proposes that mathematical knowledge for teaching appears to support teachers' ability to provide explanations that give meaning to mathematical procedures. Third, he proposes that teachers' mathematical knowledge for teaching may be related to their ability to follow students' thinking and responsively support development of understanding.

These two studies exemplify potential analyses of mathematical knowledge for teaching in relation to frameworks of teaching and learning. They leverage findings about teaching to probe the contributions of mathematical knowledge for teaching in ways that begin to unpack the specific role such knowledge plays. They are not the only studies to do so, but to date studies in this realm are rare. Building on these ideas, further conceptualization of distinctly mathematical tasks of teaching might provide even more focused contexts for studying mathematical knowledge for teaching as a resource for teaching. Establishing agreed-upon conceptualizations of mathematical knowledge for teaching related to well-studied components of the work of teaching and using these as a common ground for instrument development would provide a solid foundation for advancing the field.

From this brief review of recent progress on identifying, developing, and understanding the impact of mathematical knowledge for teaching, we now turn our attention to proposing directions for future work.

**Next Steps for the Development of Mathematical Knowledge for Teaching**

As described above, compelling examples of mathematical knowledge for teaching and evidence associating it with improved teaching and learning have sparked interest in making it a central goal in the mathematical education of teachers. However, various impediments exist. The lack of rigorous, shared definitions and the incomplete elaboration of a robust body of knowledge create problems for meaningful measures and curricula development. Underlying these challenges are competing ideas about how to conceptualize the knowledge, questions about the relationship among knowledge, knowledge use, and outcome, and the need for ways to decide claims about whether or not something constitutes professional knowledge.

We suggest three priorities for research and development of mathematical knowledge for teaching: (1) focused studies that together begin to compose a more coherent, comprehensive, and shared understanding of what it is, how it is learned, and what it does; (2) innovation and reflection on method for investigating it; and (3) studies of mathematical fluency in teaching and the nature of mathematical knowledge for equitable teaching. Below, we argue that each of these is vital to long-term progress in improving the mathematical education of teachers and the mathematics teaching and learning that depends on it.

**Investigating focused issues while contributing to a larger research program.** Scores of articles in the previous decade have argued for particular ways of distinguishing and conceptualizing important knowledge, and many others have sought to establish its presence and overall impact. With a sense of the importance of mathematical knowledge for teaching, additional studies explored the teaching of such knowledge. However, on the whole, conceptual work has been exploratory, measures have been general, and studies of the mathematical education of teachers have been limited by under-specification of the body of knowledge. We suggest that the field would benefit from focused studies that build on each other in ways that begin to put in place the machinery needed to develop an overall system for educating teachers mathematically. Such a system would include clear content-knowledge standards for professional competence, comprehensive content-knowledge course and program curricula, robust exit or professional content-knowledge exams, and rationale for what is to be taught in pre-service programs and what is better addressed in early career professional development or later on. To get there, we propose collectively pursuing several focal areas of study.

First, mathematical knowledge for teaching needs to be elaborated — for specific mathematical topics and tasks of teaching, across educational levels. Some of this work is underway, but we suggest that more needs to be done in ways that research studies, taken together, define a body of professional knowledge and provide a basis for curricula, standards, and assessments. One area of need that stands out is the investigation of the mathematical knowledge demands associated with particular domains of the work of teaching, such as leading a discussion, launching students to do mathematical work, or deciding the instructional implications of particular student work. This is a particularly challenging area of study because the field lacks comprehensive, robust specifications of the work of teaching. It is also a potentially promising area of study. Where initial decompositions of teaching are available, such as for orchestrating discussions, awareness of the mathematical knowledge entailed in the teaching can position teachers to learn both the domain of teaching and the mathematical knowledge more productively (Boerst, Sleep, Ball, & Bass, 2011; Elliott et al., 2009). Nonetheless, domains of teaching need additional parsing before they can be fully leveraged.

A second proposed area of study is determining meaningful "chunking" of mathematical knowledge for teaching and practical progressions for teaching and learning it. In considering the mathematics that students need to learn, topics are typically decomposed into a sequence of small-sized learning goals. In contrast, teachers' mathematical knowledge for teaching is not simply a mirror image of student curriculum. Teachers need knowledge that is different in important ways from the knowledge students need to learn. Mathematical knowledge for teaching is related to student curriculum, but it is not clear what this relationship implies for how it is best organized. In contrast to the mathematics that students need to learn, the specialized mathematics that teachers need to learn appears to be constituted in ways that span blocks of the student curriculum.

For instance, a teacher who learns how to model the steps of the standard addition algorithm using base ten blocks might still need to think through modeling subtraction, but as a minor extension of what is already learned, not as a new topic, requiring a new program of instruction. The question deserves more careful examination, but our experience is that teachers who participate in professional development related to a particular strand of work on place value exhibit significantly increased mathematical knowledge for teaching more generally across whole number computation, but with little to no impact on their mathematical knowledge for teaching topics related to geometry, data analysis, or even rational number computation. This is just a conjecture, but we offer it as a way to indicate an area of study that would contribute to improved approaches to the mathematical education of teachers. How big are these chunks? What are possibilities for structuring the chunks? Which have the greatest impact for beginning teachers? Some of these questions could be investigated as part of the elaboration research described above. Our point is that beyond the important goal of identifying knowledge for specific mathematical topics and tasks of teaching, across educational levels, research on how best to organize that knowledge might usefully inform the mathematical education of teachers.

This discussion leads to a third proposed line of investigation, one that explores mathematical knowledge for teaching along a professional trajectory from before teachers enter teacher preparation, through their training and novice practice, and into their maturation as professionals. This would require navigation among questions about what teachers know, what might be learned when, what is essential to responsible practice, and what can be sensibly coordinated with growing professional expertise. For this, the field would need to know more about the mathematical knowledge for teaching that prospective teachers bring to teacher education and whether there are things that might more readily be learned in the program and others that might be more productively required before admission. The field would need to know more about mathematical knowledge for teaching that is readily acquired from experience, as well as the supports needed to do so. Researchers would need to investigate how to distinguish between the mathematical knowledge for teaching that is essential to know before assuming sole responsibility for classroom instruction and the knowledge that can be safely left to later professional development. We suggest that such studies would contribute to developing coherence, efficiency, and responsibility in an overarching picture of the mathematical education of teachers.

Another proposed area of study would extend work that examines effects of specific mathematical knowledge on specific teaching and learning in ways that identify underlying mechanisms and informs views of when and how mathematical knowledge is used in teaching. We noted above a need for more studies that unpack relationships among mathematical knowledge for teaching, teaching practice, and student learning. Such studies might examine the nature of student learning gains resulting from specific teacher knowledge or they might investigate the mechanisms by which teachers' mathematical knowledge for teaching has an impact. They would provide a better understanding of the nature and role of mathematical knowledge in teaching, informing both its conceptualization and validating underlying assumptions about its significance.

Finally, we suggest that the field would benefit from more studies of effects at a mid-range level, above that of idiosyncratic, individual programs and courses and below that of large-scale, international studies. In their 2004 International Congress on Mathematics Education plenary, Adler, Ball, Krainer, Lin, and Novotna (2008) observed that the majority of studies in teacher education are small-scale qualitative studies conducted by educators studying the teachers with whom they are working within individual programs or courses. The TEDS-M study and the development of some of the instruments described above have supported an increase in large-scale and cross-case studies, but as Adler and her colleagues point out, the study of courses, programs, and teachers by researchers who are also the designers and educators of those programs and teachers creates both opportunities and risks. From our review, our sense is that many small studies are driven by convenience and reduced cost, but at the expense of rigorous design and skeptical stance. Mid-sized studies would be enhanced by efforts such as developing collaborative investigations across remote sites with either similar or contrasting interventions. This is consistent with arguments about research on professional development made by Borko (2004).

Next, we argue that the agenda sketched above will require explicit development of methods for conducting such research efficiently and effectively.

**Innovating and reflecting on method.**[2] We propose that a central problem for progress in the field is a lack of clearly understood and practicable methodology for the study and development of mathematical knowledge for teaching. First, many researchers, including graduate students, seem eager to conduct studies in this arena, but choices about research design and approaches to analysis are uncertain. In our review of the literature, we found that methods vary widely, are relatively idiosyncratic, and are in general weak — in some cases attempting to make causal claims from research designs poorly suited for such claims and in others providing thoughtful claims but from unclear

[2] Material in this section is based on work supported by the National Science Foundation under grant number 1502778. Opinions are those of the authors and do not necessarily reflect the views of the National Science Foundation.

processes and underdeveloped logical rationale. We suspect that a lack of clarity and rigor of methods, including in our own work, are a result of several factors: unresolved and underdeveloped conceptualization of the terrain; competing purposes of research (often within a single study); and uncertain grounds for making claims about whether something does or does not constitute professional knowledge. Struggles to design robust studies and to articulate methods used suggest a need for increased attention to method.

This should not be surprising. The vitality of research in areas still in early stages of theory development requires a concomitant consideration of method. Importing method from other arenas is appropriate, but regard for the theoretical foundations of the object of study and their implications for all aspects of method is also important. We propose that reflective innovation of method, grounded in emerging theory of teaching, can better account for confounding variables that are relevant to teaching and can inform the alignment among research questions, design, analysis of data, claims, and interpretations. To ground this proposal, we reflect on two approaches that have been evident in efforts to study the nature and composition of mathematical knowledge for teaching (interview studies and observational studies) and then suggest directions for potential innovations.

Early investigations of teacher content knowledge were mostly limited to correlational studies (e.g., Begle, 1979). Correlational studies remain prominent in the field (e.g., Baumert et al., 2010 Hill, Rowan, & Ball, 2005; Kersting et al., 2010), but in the 1980s and 1990s studies began using teacher interviews to investigate teacher knowledge (see Ball, Lubienski, & Mewborn, 2001). This early work was limited in two ways. First, it tended to focus on identifying deficits in teachers' mathematical knowledge instead of clarifying the mathematical knowledge requirements of teaching. Second, although some good interview prompts emerged and supported a surfeit of studies, generating additional high-quality prompts has not been easy. The strength of these early interview prompts was that they were focused, specific, and offered compelling examples of specialized mathematical knowledge that would be important for teachers to know. The weaknesses were that they focused the conversation on teachers' lack of knowledge, while providing little insight into how to rectify these lacks, and they left the difficult work of generating good prompts invisible.

Similarly, methods for observational studies have often been weakly specified and hard to use by other scholars as the basis for complementary study. For example, because of the shortcomings of teacher interviews, the research group at the University of Michigan developed a practice-based approach to the study of video records of instruction (Ball & Bass, 2003; Thames, 2009). This approach requires simultaneously conceptualizing the work of teaching together with the mathematical demands of that work. It is empirical, interdisciplinary, analytical-conceptual research that involves developing concepts and conceptual framing by parsing the phenomenon and systematically testing proposals for consistency with data and with relevant theoretical and practice-based perspectives. The approach is time-intensive and expensive, requires skillful use of distributed expertise, and is sufficiently underspecified to make broader use challenging. For instance, early on, these researchers wrote about ways in which inter-disciplinary perspectives were central to their analyses (e.g., Ball, 1999; Ball &

Bass, 2003), but this characterization, although it captures an important feature of the approach, is inadequate as a characterization of their approach and as a method for others to use. It is underspecified, relies more on experienced judgment than on independently usable criteria or techniques, and leaves key foundational issues in doubt (Thames, 2009).

Reflecting on our own use of these approaches, we offer several, somewhat ad hoc, observations.

- Teaching is purposeful work and, as such, imposes logical demands on the activity, and these logical demands play a role in warranting claims about the work of teaching and mathematical knowledge needed for teaching.
- Mathematical knowledge for teaching is professional knowledge, and central to its development and articulation is professional vetting or consensus building based on cross-community professional judgment.
- The pedagogical context provided in crafted items and prompts entails engagement in the work of teaching and in the use of mathematical knowledge and, as such, provides crafted instances for the study of specialized knowledge for teaching.
- There is an iterative process among the development of instructional tasks, assessment items, and interview prompts and our increasing capacity for eliciting and engaging mathematical knowledge for teaching.
- Analysis of mathematical knowledge as professional knowledge for teaching, whether in situ or in constrained instances, is fundamentally an empirical, conceptual-analytic, normatively informed process, not a strictly descriptive one.

We believe that the first two observations have important methodological implications that are as of yet unrealized. Key to understanding teaching and its knowledge demands is understanding its contextual rationality. In other words, meaningful study of teaching must account for the directed and contextual nature of the work. We suspect that such study will require the development and use of methods fit to the work of teaching and that this means greater reliance of underlying theory of teaching in designing studies and choosing methods of analysis. As Gherardi (2012, p. 209) succinctly summarizes in her writing about conducting practice-based studies, "Hence, empirical study of organizing as knowing-in-practice requires analysis of how, in working practices, resources are collectively activated and aligned with competence." She argues for thoughtful consideration of how methodological approaches are positioned in relation to the nature of practice and its constant reconstitution in the context of professional work. We agree with this position and suggest that it is exactly these issues that need to be taken up in an investigation of method for studying mathematical knowledge for teaching.

The second observation in our list above raises additional concerns for the development of new methods. Mathematical knowledge for teaching is professional knowledge, in the sense that it is shared, technical knowledge determined by professional judgment (Lortie, 1975; Abbott, 1988), but it is distinctive as a body of knowledge in that it requires the coordination of mathematics with teaching, which are different areas of expertise resident in different professional communities (Thames, 2009). Thus, the study of such knowledge requires coordination across different professional communities with different disciplinary foundations. In other words, the study of mathematical knowledge for teaching requires, or is at least enhanced by, collective work across distinct professional communities with different expertise and different professional norms and practices, and such work requires special consideration and support (Star & Griesemer, 1989).

With the call for cross-professional coordination, the study of mathematical knowledge for teaching involves much more than an assembly line model where different professional constituents inject their specific expertise into a product handed down the line. It calls for specification of processes for collectively considering whether a proposed claim of professional knowledge is warranted. It is about establishing protocols for merging and melding different expertise in the midst of improvement work that attends to overall coherence and practical merit. It requires specific ways of working together, tools for organizing the scholarly work, and boundary objects that provide meaningfully bridges among communities (Akkermann & Bakker, 2011). Each of these adds to the need for new methodology.

Innovation and reflection on method can be carried out in numerous ways. Researchers can simply attend more closely to decisions of method and explicit reporting of method. Alternatively, they can deliberatively develop, implement, and study methods. In order to provide a sense of the kind of innovation and reflection that might be done, we discuss some of the ways we have begun to explore methodological approaches for the study of mathematical knowledge for teaching.

An emerging approach we find promising is to use sites where professional deliberation about teaching are taking place as sites where we might productively research the work of teaching and its mathematical demands. In recent studies, we have designed interview protocols as a tool for generating data useful for studying the mathematical work of teaching. For instance, to investigate the work involved in providing students with written feedback, Kim (this issue) provides a strategic piece of student work and asks interviewees to provide written feedback and to explain the rationale for the feedback. Here, instead of videotaping classroom instruction and analyzing the mathematical demands of teaching, Kim analyzes those demands as they play out in a constrained slice of the mathematical work of teaching as evidenced in responding to a teaching scenario provided.

We see this approach as an instance of a more general phenomenon, one of using sites of professional deliberation about teaching as research sites for studying teaching. For example, a group of mathematics teachers and mathematics educators in a professional development setting might discuss responses to a particular pedagogical situation in ways akin to the pedagogical deliberations of a teacher engaged in teaching. Thus, this professional development event can be useful for studying professional practice. It may even have the advantage that professional action and reasoning are more explicitly expressed, yet of course, with certain caveats in place as well, such as recognizing that real-time demands of teaching are suspended. Similar opportunities can arise in other settings where pedagogical deliberations take place, such as teacher education or the development of curriculum or assessment. For instance, recent investigation of the design process for producing tasks to measure mathematical knowledge for teaching suggests that writing and reviewing such tasks can provide insight into teaching and its mathematical demands, even to the point of serving as a site for investigating mathematical knowledge for teaching (Jacobson, Remillard, Hoover, & Aaron, in press; Herbst & Kosko, 2012).

We propose that such an approach is distinctively different from general interview techniques that have teachers reflect on their teaching. Crucial to this difference is that the prompts are designed to provide authentic pedagogical contexts with essential, yet minimal, constraints for directing targeted pedagogical work (such as a crucial instructional goal, a key excerpt from a textbook, or strategically selected student work). Good pedagogical context needs to be based on initial conceptions of key aspects of the work, and constraints need to be designed to engage initial ideas about the nature and demands of the work. Otherwise, the pedagogical context of the tasks is unlikely to engage people in authentic pedagogical work.

Our recent experience with interview prompts of this kind has convinced us of their potential for studying teaching and teacher knowledge. Several advantages are evident: constraints provided can be manipulated; different professional communities can be engaged; and bounded instances of work examined. The development of this approach would support new lines of research that specify teaching and its professional knowledge demands in ways that can better inform professional education and evaluation. They are also easy to use and require only modest time and expense.

Such innovations begin to suggest the development of a "laboratory science" approach for studying mathematical knowledge for teaching that takes advantage of the tools of constrained prompts, the generative analytic techniques of instructional analysis, and the multiple sites available for such study. By a "lab science" approach we mean direct interaction with the world of instruction or slices of instruction using tools, data collection techniques, and models and theories of teaching. Analogous to the ways in which experimental psychologists isolate phenomena under controlled conditions in a laboratory setting or biochemists manipulate protein processes at the bench, we propose that the study of specialized teacher content knowledge can isolate activities of teaching and the use of resources, examine those activities and resources in detail, and systematically manipulate constraints to better understand phenomena. This work can be done deductively, to test specific hypotheses, inductively, to discern functional relationships, or abductively, to refine current understanding. Such an investigation of method should be intimately tied to underlying foundational issues, both shaped by theoretical commitments and giving precise definition and form to underlying theory.

In conclusion, we suggest that the development of usable, practical, and defensible method, whether along the lines we have sketched here or along other lines, will be critical to carrying out the extensive agenda described earlier for building a understanding of mathematical knowledge for teaching adequate for sustainable improvement of the mathematical education of teachers. We now sketch two areas of study largely missing from the literature on knowledge distinctive for teaching mathematics and argue that both of these are essential to viable progress on building a theory and practice of mathematical knowledge for teaching.

**Addressing Two Key Issues: Mathematical Knowledge for Fluent and Equitable Teaching.** Although there has been substantial progress in conceptualizing and understanding the mathematical understanding needed for the practice of teaching, significant issues remain. We focus here on two aspects that seem to us to be particularly critical to progress on mathematical knowledge for teaching. One centers on the communicative demands of teaching, the other on what is involved in teaching to disrupt the historical privileging of particular forms of mathematical competence and engagement, resulting in persistent inequity in access and opportunity. We argue that both of these are key to the long-term viability of efforts to improve the mathematical education of teachers.

Teaching is inherently a communication-intensive practice. Teachers listen to their students, explain ideas, and pose questions. They read their students' written work and drawings, and provide written feedback. Throughout these communications, they use mathematics in a range of specialized ways. They must hear what their students say, even though students talk and use mathematical and everyday language in ways that reflect their emergent understanding. Similarly, they must interpret students' writing and drawings. When they talk, teachers must attune their language to students' current understanding, and yet do so in ways that are intellectually honest and do not distort mathematical ideas to which they are responsible for giving their students access.

What is involved in this sort of mathematical communication in the context of teaching? Because teaching is fast-paced and interactive, the demands are intense. Talk and listening cannot be fully scripted or anticipated. A special kind of mathematical fluency is required, tuned to the work of teaching. Asking a question in the moment; explaining in response to a student's puzzlement; listening to, interpreting, and responding to a child's explanation — each of these involves hearing and making sense of others' mathematical ideas in the moment, speaking on one's feet while seeking to connect with others. Although much of the work on mathematical knowledge needed by teachers is situated in relation to what teachers do, including using representations and interpreting students' thinking, as yet little of it has focused on the mathematical fluency needed for the work teachers do in classrooms, live, in communicating with students. As compellingly argued by Sfard (2008) and others (e.g., Resnick, Asterhan, & Clarke, 2015), it is this communicative work that is central to the practice of education. Failing to investigate and squarely address communicative mathematical demands of teaching may result in an impoverished theory of mathematical knowledge for teaching in ways that sorely limit its utility and impact.

Another major area of work centers on the need to address the persistent inequities in mathematics learning both produced and reproduced in school. Goffney and her colleagues have begun to identify a set of practices of equitable mathematics teaching (Goffney, 2010; Goffney & Gonzalez, 2015; Goffney, 2015), and several of the articles in this volume explore the measurement of mathematical knowledge for equitable teaching. The driving question is what do teachers need to appreciate and understand about mathematics in order to be able to create access for groups that have been historically marginalized? Part of this has to do with a flexible understanding of the mathematics that enables teachers to build bridges between mathematics and their students. One aspect of this is to represent mathematics in ways that connect with their students' experience. Another is to be able to recognize mathematical capability and insight in their students' out-of-school practices. Each of these entails a flexibility of mathematical understanding, particularly of mathematical structure and practice. But it also involves the ability to recognize as mathematical a range of specific activities, reasoning processes, and ways of representing. Being able to do this can enable teachers to broaden both what it means to be "good at math" as well as what can be legitimated as "mathematics."

Equity is not a new focus in mathematics education (e.g., Schoenfeld, 2002), and there have been studies on the effect of gender and language on mathematics teachers' knowledge (Blömeke, Suhl, & Kaiser, 2011) as well as the distribution of teacher knowledge in different populations of teachers (Hill, 2007). In our review of the literature, we observed that most studies on equity were focused on aspirations and imperatives (i.e., arguments for teaching for equity). Few studies focused directly on specific practices of equitable mathematics teaching or knowledge for equitable mathematics teaching. We argue that increased focus in this area is crucial for three reasons. First is the underlying principle that extant inequity in mathematics teaching and learning is morally reprehensible in a civilized society (Perry, Moses, Cortez, Delpit, & Wynne, 2010). Second is our contention that, while certainly not in itself a solution, teacher content knowledge is both an indispensible and an untapped resource for disrupting the historical privileging of particular forms of mathematical competence and engagement. Third, as with nearly all achievement measures in early stages of development, current instruments are significantly biased because of the contextual features of where, as well as for and by whom, they are developed. The field needs good instrumentation, for research and for practice. Overly delaying the development of unbiased instruments may well undermine the political viability of well-meaning efforts to improve the mathematical education of teachers. Such development will require solid research in this difficult yet important arena.

In proposing these two areas of study, we acknowledge the conceptual and methodological challenges each presents. We suspect that research in these areas has been underdeveloped in large part because these foci involve subtle social dynamics less readily captured in print and in more conventional measures. These challenges simply add to our concern that concerted attention be given them. Our argument here is that these two areas of study are not merely our favored topics, but that that they are essential to long-term success.

## Articles that Develop Measures and Measure Development

The agenda sketched above is both a reflection of emerging work in the field and a proposal for future work. In many ways, the articles in this special issue, though specifically addressing measurement, resonate with themes above. For instance, the discussion about focused studies that contribute to a larger research program suggests some benefits of creating a common framework for describing mathematics teaching. In their article, Selling, Garcia and Ball (this issue) present a framework for unpacking the mathematical work of teaching that is promising in this respect. Whereas other frameworks often start with what teachers do, they focus first on the mathematical objects involved in the work of teaching and then follow up by describing actions that teachers do on these objects. This idea builds upon and extends the notion of mathematical tasks of teaching that has been highlighted in previous publications on the practice-based theory of mathematical knowledge for teaching (e.g., Ball et al., 2008; Hoover, Mosvold, & Fauskanger, 2014), as well as in previous efforts to conceptualize the work of teaching (e.g., Ball & Forzani, 2009). A main aim with this framework is to inform and assist future development of items and instruments for measuring mathematical knowledge for teaching.

Phelps and Howell (this issue) discuss the role of teaching contexts in items developed to measure mathematical knowledge for teaching. Given that mathematical knowledge for teaching is understood as knowledge applied in the work of teaching, a teaching context that illustrates a certain component of this work is typically included in items. Phelps and Howell discuss different ways in which context can be critical to assessing mathematical knowledge for teaching. They argue that attention to the role of context might provide better understanding of the knowledge assessed in particular items and might also inform further development of a theory in which teaching context is used to define knowledge.

Whereas both of these first articles point to core issues regarding the conceptualization of mathematical knowledge for teaching — in the context of item and instrument development — the next two articles deal more directly with measurement. Kim (this issue) focuses on designing interview prompts for assessing mathematical knowledge for teaching. In particular, her discussion focuses on the task of providing written feedback to students. To model this task, she combines a decomposition of the task with aspects of the pedagogical context involved and sub-domains of mathematical knowledge for teaching.

Where Kim's study is more qualitative and conceptual in nature, Orrill and Cohen (this issue) draw on psychometric models in their work. Their study hinges on the issue of defining the construct measured, and they use a mixture Rasch model to analyze different subsets of items to support an argument that the domain definition has strong implications on the claims one tries to make about teachers' performance. In light of our observations about the lack of consensus about how to define the constructs that are being measured and discussed across studies, a focus on careful construct definition and implications is particularly relevant.

In the international literature on teaching and learning, a focus on equity is prevalent. In research on mathematical knowledge for teaching, the discussion of knowledge for teaching equitable mathematics also receives some attention — although issues of equity and diversity have not been emphasized in frameworks of mathematical knowledge for teaching. In this connection, Wilson's (this issue) and Turkan's studies of mathematical knowledge for teaching English language learners draw attention to this missing area of research. Both involve design and application of measures. Wilson proposes a new aspect of pedagogical content knowledge that is connected specifically to the work of teaching mathematics to English language learners. Turkan addresses practicing teachers' reasoning about teaching mathematics to ELLs. Based on analysis of data from cognitive interviews, she argues that there is a unique domain of knowledge necessary for teaching ELLs — thus supporting Wilson's argument — and she calls for further investigations to identify and assess this knowledge.

Finally, this special issue includes two articles that investigate teachers' views. Koponen, Asikainen, Viholainen and Hirvonen investigate the views of teachers as well as teacher educators about the content of mathematics teacher education. Results from their survey indicate that teachers as well as teacher educators in the Finnish context emphasize the need for courses in content knowledge that is distinctive for teaching — not just more advanced. They argue that the mathematical content of teacher education needs to be tightly connected to the mathematics being taught, and even pedagogical courses need to include knowledge connected with mathematics, in particular focusing on knowledge of teaching and learning of mathematics. In the last article of the special issue, Kazima, Jakobsen and Kasoka investigate Malawian teachers' views about mathematical tasks of teaching and the potential usefulness of adapted measures of mathematical knowledge for teaching among Malawian pre-service mathematics teachers. The measures as well as the applied framework of mathematical knowledge for teaching were developed in the United States. Despite the significant cultural differences between Malawi and the United States, the authors argue that the framework as well as most of the items function well in the Malawian context.

Together, this collection of articles on the development and use of measures lies at a transition from the lessons of past studies of mathematical knowledge for teaching into vital arenas of research needed for systemic improvement on the mathematical education of teachers.

# References

Abbott, A. (1988). *The system of professions*. Chicago: University of Chicago Press.

Adler, J., Ball, D., Krainer, K., Lin, F., & Novotna, J. (2008). Mirror images of an emerging field: Researching mathematics teacher education. In M. A. Niss (Ed.), *ICME-10 Proceedings: Proceedings of the 10th International Congress on Mathematical Education, 4-11 July 2004* (pp. 123–139). Roskilde: Roskilde Universitet.

Akkerman, S. F., & Bakker, A. (2011). Boundary crossing and boundary objects. *Review of Educational Research, 81*(2), 132–169.

American Educational Research Association. (2006). Standards for reporting on empirical social science research in AERA publications. *Educational Researcher, 35*(6), 33–40.

Ball, D. L. (1999). Crossing boundaries to examine the mathematics entailed in elementary teaching. In T. Lam (Ed.), *Contemporary mathematics* (pp. 15–36). Providence, RI: American Mathematical Society.

Ball, D. L., & Bass, H. (2003). Toward a practice-based theory of mathematical knowledge for teaching. In B. Davis & E. Simmt (Eds.), *Proceedings of the 2002 annual meeting of the Canadian Mathematics Education Study Group* (pp. 3–14). Edmonton, Alberta, Canada: Canadian Mathematics Education Study Group (Groupe Canadien d'étude en didactique des mathématiques).

Ball, D. L., & Cohen, D. K. (1999). Developing practice, developing practitioners: toward a practice-based theory of professional education. In G. Sykes, & L. Darling-Hammond (Eds.), *Teaching as the learning profession: Handbook of policy and practice* (pp. 3–32). San Francisco: Jossey Bass.

Ball, D. L., & Forzani, F. M. (2009). The work of teaching and the challenge for teacher education. *Journal of Teacher Education, 60*(5), 497–511.

Ball, D. L., Lubienski, S. T., & Mewborn, D. S. (2001). Research on teaching mathematics: the unsolved problem of teachers' mathematical knowledge. In V. Richardson (Ed.), *Handbook of research on teaching* (4th ed., pp. 433–456). New York, NY: Macmillan.

Ball, D. L., Thames, M. H., & Phelps, G. (2008). Content knowledge for teaching: What makes it special? *Journal of Teacher Education, 59*(5), 389–407.

Baumert, J., Kunter, M., Blum, W., Brunner, M., Voss, T., Jordan, A., …, Tsai, Y. (2010). Teachers' mathematical knowledge, cognitive activation in the classroom, and student progress. *American Educational Research Journal, 47*(1), 133–180.

Begle, E. (1979). *Critical variables in mathematics education: Findings from a survey of empirical research*. Mathematical Association of America/National Council of Teachers of Mathematics, Washington, DC.

Bell, C. A., Wilson, S. M., Higgins, T., & McCoach, D. B. (2010). Measuring the effects of professional development on teacher knowledge: The case of developing mathematical ideas. *Journal for Research in Mathematics Education, 41*(5), 479–512.

Blömeke, S., & Delaney, S. (2012). Assessment of teacher knowledge across countries: a review of the state of research. *ZDM, 44*(3), 223–247.

Blömeke, S., Suhl, U., & Kaiser, G. (2011). Teacher education effectiveness: Quality and equity of future primary teachers' mathematics and mathematics pedagogical content knowledge. *Journal of Teacher Education, 62*(2), 154–171.

Boerst, T. A., Sleep, L., Ball, D. L., & Bass, H. (2011). Preparing teachers to lead mathematics discussions. *Teachers College Record, 113*(12), 2844–2877.

Borko, H. (2004). Professional development and teacher learning: Mapping the terrain. *Educational researcher, 33*(8), 3–15.

Charalambous, C. (2010). Mathematical Knowledge for Teaching and Task Unfolding: An Exploratory Study. *The Elementary School Journal, 110*(3), 247–278.

Chick, H. (2009). Choice and Use of Examples as a Window on Mathematical Knowledge for Teaching. *For the Learning of Mathematics, 29*(3), 26–30.

Cohen, D., Raudenbush, S., & Ball, D. L. (2003). Resources, instruction, and research. *Educational Evaluation and Policy Analysis, 25*(2), 1–24.

Copur-Gencturk, Y., & Lubienski, S. T. (2013). Measuring mathematical knowledge for teaching: a longitudinal study using two measures. *Journal of Mathematics Teacher Education, 16*(3), 211–236.

Depaepe, F., Verschaffel, L., & Kelchtermans, G. (2013). Pedagogical content knowledge: a systematic review of the way in which the concept has pervaded mathematics educational research. *Teaching and Teacher Education, 34*, 12–25.

Elliott, R., Kazemi, E., Lesseig, K., Mumme, J., & Kelley-Petersen, M. (2009). Conceptualizing the Work of Leading Mathematical Tasks in Professional Development. *Journal of Teacher Education, 60*(4), 364–379.

Ferguson, P., & Womack, S. T. (1993). The impact of subject matter and education coursework on teaching performance. *Journal of Teacher Education, 44*, 155–163.

Gherardi, S. (2012). *How to conduct a practice-based study: Problems and methods.* Northhampton, MA: Edward Elgar Publishing.

Goffney, I. M. (2010). *Identifying, defining, and measuring equitable mathematics instruction.* (Unpublished doctoral dissertation). University of Michigan, Ann Arbor, MI.

Goffney, I. M., & Gonzalez, M. (2015). Mathematics methods as a site for developing ambitious and equitable teaching practices for prospective teachers. Manuscript submitted for publication.

Goffney, I. M. (2015). Developing a theory of mathematical knowledge for equitable teaching. Manuscript in preparation.

Graeber, A., & Tirosh, D. (2008). Pedagogical content knowledge: Useful concept or elusive notion. In P. Sullivan & T. Woods (Eds.), *Knowledge and beliefs in mathematics teaching and teaching development* (pp. 117–132). Rotterdam: Sense Publishers.

Guyton, E., & Farokhi, E. (1987). Relationships among academic performance, basic skills, subject matter knowledge, and teaching skills of teacher education graduates. *Journal of Teacher Education, 38*, 37–42.

Harbison, R.W., & Hanushek, E. A. (1992). *Educational performance for the poor: Lessons from rural northeast Brazil.* Oxford, UK: Oxford University Press.

Herbst, P., & Kosko, K. (2012). Mathematical knowledge for teaching high school geometry. In L. R. Van Zoest, J.-J. Lo, & J. L. Kratky (Eds.), *Proceedings of the 34th annual meeting of the North American Chapter of the International Group for the Psychology of Mathematics Education* (pp. 438–444). Kalamazoo, MI: Western Michigan.

Hill, H. (2007). Mathematical knowledge of middle school teachers: implications for the no child left behind policy initiative. *Educational Evaluation and Policy Analysis, 29*(2), 95–114.

Hill, H. C. (2011). The nature and effects of middle school mathematics teacher learning experiences. *Teachers College Record, 113*(1), 205–234.

Hill, H. C., Blunk, M. L., Charalambous, C. Y., Lewis, J. M., Phelps, G. C., Sleep, L., & Ball, D. L. (2008). Mathematical knowledge for teaching and the mathematical quality of instruction: An exploratory study. *Cognition and Instruction, 26*(4), 430–511.

Hill, H. C., Kapitula, L., & Umland, K. (2011). A validity argument approach to evaluating teacher value-added scores. *American Educational Research Journal, 48*(3), 794–831.

Hill, H. C., Rowan, B., & Ball, D. L. (2005). Effects of teachers' mathematical knowledge for teaching on student achievement. *American Education Research Journal, 42*(2), 371–406.

Hill, H. C., Schilling, S. G., & Ball, D. L. (2004). Developing measures of teachers' mathematics knowledge for teaching. *The Elementary School Journal, 105*(1), 11–30.

Hill, H. C., Umland, K., Litke, E., & Kapitula, L. (2012). Teacher quality and quality teaching: Examining the relationship of a teacher assessment to practice. *American Journal of Education, 118*(4), 489–519.

Hoover, M., Mosvold, R., & Fauskanger, J. (2014). Common tasks of teaching as a resource for measuring professional content knowledge internationally. *Nordic Studies in Mathematics Education, 19*(3–4), 7–20.

Izsak, A., Jacobson, E., de Araujo, Z., & Orrill, C. H. (2012). Measuring mathematical knowledge for teaching fractions with drawn quantities. *Journal for Research in Mathematics Education, 43*(4), 391–427.

Jacobson, E., Remillard, J., Hoover, M., & Aaron, W. (in press). Chapter 8: The interaction between measure design and construct development: Building validity arguments. *Journal for Research in Mathematics Education. Monograph.*

Kaarstein, H. (2014). A comparison of three frameworks for measuring knowledge for teaching mathematics. *Nordic Studies in Mathematics Education, 19*(1), 23–52.

Kersting, N. B., Givvin, K. B., Sotelo, F. L., & Stigler, J. W. (2010). Teachers' analyses of classroom video predict student learning of mathematics: further explorations of a novel measure of teacher knowledge. *Journal of Teacher Education, 61*, 172–181.

Kersting, N. B., Givvin, K. B., Thompson, B. J., Santagata, R., & Stigler, J. W. (2012). Measuring usable knowledge teachers' analyses of mathematics classroom videos predict teaching quality and student learning. *American Educational Research Journal, 49*(3), 568–589.

Kim, Y., Mosvold, R., & Hoover, M. (2015). A review of research on the development of teachers' mathematical knowledge for teaching. Manuscript in preparation.

Kleickmann, T., Richter, D., Kunter, M., Elsner, J., Besser, M., Krauss, S., & Baumert, J. (2013). Teachers' content knowledge and pedagogical content knowledge: The role of structural differences in teacher education. *Journal of Teacher Education, 64*(1), 90–106.

Koellner, K., Jacobs, J., Borko, H., Schneider, C., Pittman, M. E., Eiteljorg, E., Bunning, K., & Frykholm, J. (2007). The Problem-Solving Cycle: A model to support the development of teachers' professional knowledge. *Mathematical Thinking and Learning, 9*(3), 271–300.

Kunter, M., Klusmann, U., Baumert, J., Voss, T., & Hacfeld, A. (2013). Professional competence of teachers: Effects on instructional quality and student development. *Journal of Educational Psychology, 105*(3), 805–820.

Lortie, D. C. (1975). *School teacher: A sociological inquiry*. Chicago, IL: University of Chicago Press.

Ma, L. (1999). *Knowing and teaching elementary mathematics: Teachers' understanding of fundamental mathematics in China and the United States*. Mahwah, NJ: Lawrence Erlbaum Associates.

McCray, J. S., & Chen, J.-Q. (2012). Pedagogical content knowledge for preschool mathematics: Construct validity of a new teacher interview. *Journal of Research in Childhood Education, 26*, 291–307.

McCrory, R., Floden, R., Ferrini-Mundy, J., Reckase, M. D., & Senk, S. L. (2012). Knowledge of algebra for teaching: A framework of knowledge and practices. *Journal for Research in Mathematics Education, 43*(5), 584–615.

Monk, D. H. (1994). Subject area preparation of secondary mathematics and science teachers and student achievement. *Economics of Education Review, 13*, 125–145.

Morris, A. K., Hiebert, J., & Spitzer, S. (2009). Mathematical knowledge for teaching in planning and evaluating instruction: What can preservice teachers learn? *Journal for Research in Mathematics Education, 40*(5), 491–529.

Mullens, J. E., Murnane, R. J., & Willett, J. B. (1996). The contribution of training and subject matter knowledge to teaching effectiveness: A multilevel analysis of longitudinal evidence from Belize. *Comparative Education Review, 40*, 139–157.

Perry, T., Moses, R. P., Cortes Jr, E., Delpit, L., & Wynne, J. T. (2010). *Quality education as a constitutional right: Creating a grassroots movement to transform public schools.* Boston, MA: Beacon Press.

Resnick, L. B., Asterhan, C., & Clarke, S. N. (Eds.). (2015). *Socializing intelligence through academic talk and dialogue.* Washington, D.C.: American Educational Research Association.

Rockoff, J. E., Jacob, B. A., Kane, T. J., & Staiger, D. O. (2011). Can you recognize an effective teacher when you recruit one? *Education Finance and Policy, 6*(1), 43–74.

Saderholm, J., Ronau, R., Brown, E. T., & Collins, G. (2010). Validation of the Diagnostic Teacher Assessment of Mathematics and Science (DTAMS) instrument. *School Science and Mathematics, 110*(4), 180–192.

Schilling, S. G., & Hill, H. C. (2007). Assessing measures of mathematical knowledge for teaching: a validity argument approach. *Measurement, 5*(2–3), 70–80.

Schoenfeld, A. H. (2002). Making mathematics work for all children: Issues of standards, testing, and equity. *Educational researcher, 31*(1), 13–25.

Schwab, J. J. (1978). Education and the structure of the disciplines. In I. Westbury & N. Wilkof (Eds.), *Science, curriculum, and liberal education: Selected essays* (pp. 229–272). Chicago, IL: University of Chicago Press. (Original work published 1961.)

Senk, S. L., Tatto, M. T., Reckase, M., Rowley, G., Peck, R., & Bankov, K. (2012). Knowledge of future primary teachers for teaching mathematics: An international comparative study. *ZDM, 44*(3), 307–324.

Sfard, A. (2008). *Thinking as communicating: Human development, development of discourses, and mathematizing.* Cambridge: Cambridge University Press

Shulman, L. S. (1986). Those who understand: knowledge growth in teaching. *Educational Researcher, 15*(2), 4–14.

Silver, E. A., Clark, L. M., Ghousseini, H. N., Charalambous, C. Y., & Sealy, J. T. (2007). Where is the mathematics? Examining teachers' mathematical learning opportunities in practice-based professional learning tasks. *Journal of Mathematics Teacher Education, 10*(4–6), 261–277.

Speer, N. M., & Wagner, J. F. (2009). Knowledge needed by a teacher to provide analytic scaffolding during undergraduate mathematics classroom discussions. *Journal for Research in Mathematics Education, 40*(5), 530–562.

Star, S. L., & Griesemer, J. R. (1989). Institutional ecology, "translations" and boundary objects: Amateurs and professionals in Berkeley's Museum of Vertebrate Zoology, 1907–39. *Social Studies of Science, 19*, 387–420.

Stein, M. K., Engle, R. A., Smith, M. S., & Hughes, E. K. (2008). Orchestrating productive mathematical discussions: Five practices for helping teachers move beyond show and tell. *Mathematical Thinking and Learning, 10*(4), 313–340.

Tatto, M. T., Schwille, J., Senk, S., Ingvarson, L., Peck, R., & Rowley, G. (2008). *Teacher Education and Development Study in Mathematics (TEDS-M): Conceptual framework*. East Lansing, MI: Teacher Education and Development International Study Center, College of Education, Michigan State University.

Thames, M. H. (2009). *Coordinating mathematical and pedagogical perspectives in practice-based and discipline-grounded approaches to studying mathematical knowledge for teaching (K-8)*. (Unpublished doctoral dissertation). University of Michigan, Ann Arbor, MI.

Thompson, P. W. (2015). Researching mathematical meanings for teaching. In L. D. English & D. Kirshner (Eds.), *Third handbook of international research in mathematics education* (pp. 435–461). New York: Taylor & Francis.

Williams, S. R., & Baxter, J. A. (1996). Dilemmas of discourse-oriented teaching in one middle school mathematics classroom. *The Elementary School Journal, 97*, 21–38.

Yackel, E., & Cobb, P. (1996). Sociomathematical norms, argumentation and autonomy in mathematics. *Journal for Research in Mathematics Education, 27*, 458–476.

Youngs, P., & Qian, H. (2013). The influence of university courses and field experiences on Chinese elementary candidates' mathematical knowledge for teaching. *Journal of Teacher Education, 64*(3), 244–261.

Zodik, I., & Zaslavsky, O. (2008). Characteristics of teachers' choice of examples in and for the mathematics classroom. *Educational Studies in Mathematics, 69*(2), 165–182.

# What Does it Take to Develop Assessments of Mathematical Knowledge for Teaching?:

# Unpacking the Mathematical Work of Teaching

Sarah Kate Selling
University of Michigan


Nicole Garcia
University of Michigan


Deborah Loewenberg Ball
University of Michigan

**Abstract:** In the context of the increased mathematical demands of the Common Core State Standards and data showing that many elementary school teachers lack strong mathematical knowledge for teaching, there is an urgent need to grow teachers' MKT. With this goal in mind, it is crucial to have research and assessment tools that are able to measure and track aspects of teachers' MKT at scale. Building on the concept of "mathematical tasks of teaching" (Ball et al., 2008), we report on a new framework that unpacks the mathematical work of teaching that could serve as a scaffold for item writers who are developing assessments of MKT. We argue that this framework supports a focus on the *mathematical* work of teaching that moves beyond common content knowledge but without moving into a space of pedagogical choice. We also illustrate how the framework was constructed to highlight connections within and across the mathematical content of elementary school. The mathematical work of teaching framework has implications for assessment development at scale, and could be useful as an organizing tool in mathematics teacher education efforts to grow teachers' MKT.

*Keywords:* mathematical knowledge for teaching, teacher knowledge, assessment development

## Introduction

Broad consensus exists about the importance of teachers' mathematical knowledge (Adler & Venkat, 2014; Ball, Lubienski, & Mewborn, 2001; Baumert et al. 2010; Döhrmann, Kaiser, & Blömeke, 2014). Studies have linked mathematical knowledge for teaching to the quality of teachers' mathematics instruction (Eisenhart, Borko, Underhill, Brown, Jones, & Agard, 1993; Hill et al., 2008). Mathematical knowledge for teaching has also been linked to

*The Mathematics Enthusiast,* **ISSN 1551-3440, vol. 13, no. 1&2**, pp. 35–51
2016© The Author(s) & Dept. of Mathematical Sciences-The University of Montana

student achievement gains in the elementary grades (Hill, Rowan, and Ball, 2005). However, many U.S. teachers lack the deep, nuanced, and specialized mathematical knowledge needed for responsible teaching. This finding is persistent over time, grade levels, and both national and international contexts (e.g., Hill & Ball, 2004; Ma, 1999; Tatto et al., 2008). Simultaneously, the Common Core State Standards for Mathematics (National Governors Association Center for Best Practices, 2010), which have been adopted by 47 states and territories, have set out rigorous standards for K-12 mathematics learning that consequently increase the mathematical demands of teaching. To ensure that teachers are well-positioned to help students meet these more challenging learning goals, it is now — more than ever — critically important to focus on developing their mathematical knowledge for teaching.

To investigate and grow teachers' mathematical knowledge for teaching (MKT) it is crucial to be able to measure and track the development and uses of MKT. Most work to develop measures of MKT has typically been done by groups of experts in relevant fields, such as mathematicians, mathematics educators, and teachers who have worked together to draft and revise assessment items (Hill, Schilling, & Ball, 2004). The early work in this area was focused on developing and refining the construct of mathematical knowledge for teaching while simultaneously and iteratively developing measures of the construct. The process of item development was therefore often time consuming and challenging. Because of the promising results of these earlier efforts, there is now a broad need for assessments of MKT. Building tests at scale means, however, that people who are not deeply immersed in research on MKT will have to be able to write valid MKT items. This will require detailed supports to help test developers understand the nuances of the construct of MKT and ways to assess it. In this paper, we present a framework that identifies the different ways that teachers make use of mathematical knowledge as they go about the work of teaching and provides support to assessment developers. We begin by articulating and specifying what we mean by mathematical knowledge for teaching and its relationship with the mathematical work of teaching that arises in everyday practice.

## Theoretical Framing

### Conceptualizing Mathematical Knowledge for Teaching

Building supports for assessment development of mathematical knowledge for teaching (MKT) rests on a clear conceptualization of what we mean by MKT, how MKT is drawn upon in practice, and the specific areas of the work of teaching that we seek to assess. Scholars of mathematical knowledge have examined such knowledge in action as it is used in the practice of teaching (Ball & Bass, 2002; Rowland, Huckstep, & Thwaites, 2005).

Our work builds on a particular practice-based perspective on mathematical knowledge for teaching that begins with the premise that, to understand the specific knowledge of mathematics needed in teaching, one must first examine the mathematical work that arises in the context of teachers' instruction in classrooms, a form of job analysis (Ball & Bass, 2002). Through detailed analysis of instruction in a 3rd grade classroom over an entire year, Ball and her colleagues identified mathematical problems that teachers regularly encounter and must solve while teaching, such as "interpreting and evaluating students' non-standard mathematical ideas" (Ball & Bass, 2002, p. 9). These analyses reveal that teaching entails significant mathematical work on the part of the teacher. To highlight the complexity and variety of ways that teachers engage in mathematical work, Ball and her colleagues (2008) present a list of 16 "mathematical tasks of teaching" that may occur within every day teaching practice that involve mathematical work on the part of the teacher. This list includes tasks such as "responding to students' 'why' questions", "finding an example to make a specific mathematical point", "evaluating the plausibility of students' claims (often quickly)", "choosing and developing useable definitions", or "recognizing what is involved in using a particular representation" (p. 10). These mathematical tasks of teaching provide the contexts in which teachers must draw on mathematical knowledge for teaching, and therefore offer a window into the mathematical knowledge entailed by teaching.

Based on their analyses of these ubiquitous tasks of teaching mathematics, Ball and her colleagues (2008) identified a provisional map of domains of mathematical understanding and skill. They argued that teaching requires both "pure" subject matter knowledge and pedagogical content knowledge (Shulman, 1986, 1987; Wilson, Shulman, and Richert, 1987). Pedagogical content knowledge comprises blends of mathematical knowledge together with other kinds of knowledge, such as knowledge of students' thinking in a particular content domain, or knowledge of likely effective approaches to or materials for teaching specific content ideas. For example, in teaching integers, teachers need to appreciate that notions of "debt," "assets" and "net worth" are unfamiliar to elementary age learners and that therefore financial contexts are not likely to be useful as a representation of integer arithmetic. Knowing ways to use number line models as a context for integer arithmetic is another example of pedagogical content knowledge — knowledge of teaching approaches and models combined with a particular topic. But knowing integers for teaching also involves content knowledge. "Common" content knowledge is the term Ball and her colleagues use to describe the knowledge that 0 is neither negative or positive or that $(-3) - (-7) = 4$. By this they denote knowledge that is also relevant to people who do not teach — that is, known *in common* with others. They argue that teaching also requires "specialized" content knowledge — for example, being able to explain the meaning of subtraction of a negative number and connect it to moves on the number line in ways that make conceptual sense, or being able to represent the difference – even though they might produce the same result — between subtracting -4 from 10 and adding 4 to 10. Horizon knowledge is the perspective needed to understand connections among topics or to see where ideas are headed, or to notice when students are onto a sophisticated mathematical point (Ball & Bass, 2009). In our assessment development work, we focus on specialized content knowledge, as a form of subject matter knowledge that is particularly needed in the work of teaching.

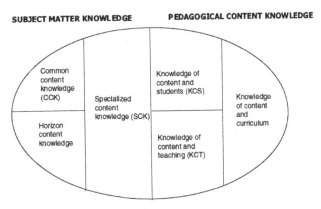

*Figure 1.* Domains of Content Knowledge for Teaching (Ball, Thames, & Phelps, 2008)

Research on specialized content knowledge has acknowledged that the line between specialized and common content knowledge might not be well-defined, and that particular mathematical tasks of teaching may elicit different types of knowledge by teachers or others asked to engage in these tasks (Delaney et al., 2005; Hill, Dean, & Goffney, 2007). In our work, we are less concerned with classifying assessment items as eliciting only specialized or common content knowledge; instead, we have chosen to focus on the mathematical work of teaching demanded by teaching practice and the knowledge that teachers would need to do that work, acknowledging that some mathematical work of teaching may elicit different domains of subject matter knowledge or even knowledge from multiple domains.

## Building Assessments of Content Knowledge for Teaching: Challenges and Supports

Existing assessments of teacher knowledge at scale, often licensure tests, tend to focus on common content knowledge (i.e., the mathematics content that teachers teach) or horizon knowledge (i.e., perspective on how what the students are working on now connects with other mathematics). Few assessments have attempted to assess specialized content knowledge at scale. The Learning Mathematics for Teaching project (Hill & Ball, 2004; Hill, Schilling & Ball, 2004) has developed elementary and middle school level measures for research purposes that have been widely adopted and implemented. However, these measures are not intended as assessments of individual teachers. This leaves unaddressed how to develop assessments of SCK at scale and how to support item writing by test designers. Our investigation of this question has been situated in a project in which we collaborated with others to build items to measure teachers' specialized content knowledge at scale. Our goal was to develop tools that could be used to guide the development of assessments of SCK with item writers who have different expertise than the groups who have in the past worked to develop items like those in the Learning Mathematics for Teaching project.

To understand what tools and supports might be needed to accomplish this, we considered what might be challenging for assessment developers when constructing measures of specialized content knowledge. First, we hypothesized that item writers might have difficulty developing measures of more than just common content knowledge, as this is the typical focus for assessments of teacher knowledge. In particular, we anticipated that there would be challenges in understanding the differences between CCK and SCK. A second related challenge concerns the possibility that in attempting to shift from writing items focused on common content knowledge, item writers might end up going too far and focusing items on pedagogical tasks of teaching that involve more than mathematical work, such as making instructional decisions about the best ways to teach a topic. In other words, we were concerned that writers might develop items focused on pedagogical content knowledge or even pedagogical choices, both which were beyond the scope of a subject matter knowledge for teaching assessment. A third challenge might arise if item writers are not familiar with the work of teaching that draws on teachers' specialized mathematical knowledge, such as the tasks of teaching set out by Ball and colleagues (2008). Finally, we hypothesized that it might also be difficult for item writers to understand how specialized content knowledge might be used across the K-6 curriculum, and how those uses might vary. Based on these four hypothesized areas of difficulty, we developed a framework that identifies the mathematical work of teaching and is strategically designed to address each of these challenges. We highlight below how the framework supports a focus on the *mathematical* work of teaching that moves beyond common content knowledge but without moving into a space of pedagogical choice. We also illustrate how the framework was constructed to highlight connections within and across the mathematical content of elementary school.

### Unpacking the Mathematical Work of Teaching Framework

The mathematical work of teaching framework expands on the mathematical tasks of teaching (Ball et al., 2008) to produce a tool that can support development of assessments of mathematical knowledge for teaching at the elementary level. The framework addresses three main goals. First, the framework supports a focus on the *mathematical* work of teaching, the mathematics that a teacher engages with while teaching content to students, as opposed to the pedagogical task of making choices about instructional strategies. Second, the framework highlights connections between the mathematical work of teaching and the mathematics content at the elementary level. Finally, the framework is usable by item writers to construct written measures of MKT, specifically measures of subject matter knowledge with a focus on specialized content knowledge. In the following sections, we unpack the mathematical work of teaching framework with respect these three goals, referencing an excerpt from the framework shown below in Table 1.

Table 1: *Mathematical work of teaching framework organized by (1) mathematical objects, (2) actions with and on those objects in teaching, and (3) specific examples.*

| | MWT: Actions with and on objects | Examples |
|---|---|---|
| **Explanations (includes justifications & reasoning)** | **Comparing explanations** to determine which is more/most valid, generalizable, or complete explanation | Given two explanations, choose which is more complete<br>Given multiple student explanations, determine which is most valid<br>Given several explanations, choose the best explanation.<br>Given conflicting explanations, determine which is valid and why.<br>Select an explanation that best captures an underlying idea. |
| | **Critiquing explanations to improve them** with respect to completeness, validity, or generalizability. | Given an incomplete but valid explanation, determine what, if anything, is missing or needs to be added to be more complete. |
| | **Critiquing explanations** with respect to validity, generalizability, or explanatory power. | Given an explanation, determine if it is mathematically valid.<br>Given several explanations, determine which ones are valid.<br>Given a text, determine what may be misleading about an explanation. |
| | **Writing mathematically valid explanations** for a process, conjecture, relationship, etc. | Write a mathematically valid explanation for a process or concept.<br>Write a mathematically valid explanation for a conjecture.<br>Given student strategies, determine properties that could be used to justify the strategy's validity. |
| **Mathematical structure** | **Determining, analyzing, or posing** problems with the same (or different) mathematical **structure** | Given a set of problems, determine which have the same structure.<br>Given a set of problems, choose the description of the structure type.<br>Given a set of problems, determine which does NOT have the same structure.<br>Write a problem that has the same structure as given problems.<br>Given a description of a structure, determine which problems fit that structure. |
| | **Analyzing structure in student work** by determining which strategies or ideas are most closely connected with respect to mathematical structure | Given a set of student strategies, determine which have similar mathematical structure.<br>Given a set of student strategies most of which use the same core idea but slightly differently, determine which one does not fit.<br>Given a set of strategies and a structure, determine which strategies fit the structure. |
| | **Matching word problems and structure** | Given a structure, choose a word problem with that structure.<br>Given a word problem, choose another problem with the same structure. |

| | | |
|---|---|---|
| Representations | **Connecting or matching representations** | Match a representation to a given interpretation of an operation. |
| | | Determine how different representations are connected. |
| | | Given two claims about a representation, determine which is correct and why. |
| | **Analyzing representations** by identifying correct or misleading representations in a text, talk, or written work. | Given a written representation (e.g., number line, table, diagram), determine what may be misleading. |
| | | Given a set of representations, choose which does or does not show a particular idea (table?) |
| | **Selecting, creating, or evaluating representations** for a mathematical purpose | Create a representation for a given number or operation. |
| | | Select a representation that highlights a particular mathematical idea. |
| | **Talking a representation** (i.e., using words to talk through the meaning of a representation and connecting it to the key ideas) | Given a suggested way to talk about a representation in a text, evaluate whether the talk clearly connects the representation and the ideas. |
| | | Given a colleague's request for feedback, determine how their talking about a representation could be improved to highlight mathematical meaning |

## Organization around "Mathematical Objects"

Our first task in developing the mathematical work of teaching framework was to organize the mathematical work that arises in the context of teaching in a way that would maintain a focus on the mathematics. The mathematical tasks of teaching, as set out by Ball and colleagues (2008), include a list of 16 illustrative tasks that arise in everyday practice and that entail mathematical work for the teacher. This list includes tasks such as "responding to students' 'why' questions", "finding an example to make a specific mathematical point" or "recognizing what is involved in using a particular representation" (p. 10). These provide useful examples of the mathematical work of teaching; however, they were not intended to be an exhaustive list. Therefore, we built on and expanded this list. As the list of teachers' mathematical work grew, we needed to create an organizational structure that would make the list more orderly, systematic, and useful for item writers.

The mathematical work of teaching framework is organized around a set of what we call "mathematical objects" that teachers encounter and with which they work while teaching. Examples include explanations, representations, mathematical errors, and definitions. We called these "mathematical objects" because they are the mathematical instructional objects that teachers encounter and with which they interact while teaching. For example, teachers regularly give, use, and encounter mathematical explanations; in this case, the mathematical explanation is the "object". Teachers give mathematical explanations themselves, but they also make sense of student explanations, compare different explanations in textbooks, determine if a student's explanation is valid, or critique written explanations for the purpose of improving them. We recognize that we define "mathematical objects" here in a way that is different from the way "objects" is typically used in mathematics to refer to objects such as numbers, functions, and polygons. Table 2 provides the set of mathematical objects around which we built the framework. Although this list is by no means exhaustive, we hoped to describe the diverse sets of mathematical objects that teachers typically interact with in teaching.

Table 2. *Organizing mathematical objects for the mathematical work of teaching.*

| Explanations (including justification and reasoning) | Errors and incorrect thinking |
|---|---|
| Conjectures | Representations |
| Mathematical structure | Manipulatives |
| Examples, non-examples, and counter-examples | Language and definitions |
| Mathematical problems | Mathematical goals and topics |
| Strategies | |

Establishing the framework around a set of mathematical objects focuses attention on the *mathematical* work of teaching. By basing the framework in these mathematical objects, the mathematics in teachers' work is foregrounded. Another way to organize such a framework would be to organize the mathematical work of teaching into pedagogical tasks or domains (e.g., the mathematical work that arises when leading a discussion). However, there are two main limitations in organizing the framework in this way. First, many of the mathematical tasks of teaching arise in the context of enacting multiple different instructional practices, but require the same mathematical work on the part of the teacher, regardless of context. For example, interpreting student mathematical errors, a key mathematical task of teaching, could arise in the context of a class discussion, but it might also arise while teachers are interpreting written work on assessments or during teachers' interactions with individuals or small groups of students. Representing the mathematical work of teaching in each of these instructional practices would result in a lengthy list of work with much repetition. A second reason to avoid organizing the framework around pedagogical domains or tasks is to keep the focus on the *mathematical* work of teaching to help item writers avoid developing items that simply assessed teachers' instructional choices. Organization around instructional practices emphasizes the teaching practice rather than the mathematics necessary to engage in that practice. Consider the instructional practice of giving oral or written feedback to students. This practice requires teachers to engage in mathematical work such as determining how a student's explanation could be improved to be more complete. Organizing the framework around mathematical objects, as opposed to instructional practices or other pedagogically focused categories, supports a focus on the mathematics in the work of teaching.

**Organization around Mathematical Work of Teaching with Respect to these Objects**

The framework is organized around a diverse set of mathematical objects to illuminate the varied mathematical terrain of teachers' work, ranging from the mathematics that arises in interacting with explanations and strategies, to the mathematics involved in using language and definitions carefully, to the mathematical work of choosing or constructing mathematical examples. Explicitly naming and building the framework around this set of objects highlights the diversity of mathematical work of teaching.

Each domain of mathematical objects is further defined by a set of mathematical work of teaching, or actions on that particular object. For example, the mathematical object of "representations" includes five categories of work, i.e., "connecting or matching representations", "analyzing representations", "choosing or creating representations", and "talking a representation". The categories of work for each object answer the question, "what mathematical work do teachers <u>do</u> with these objects while teaching"? The grain-size of these categories manages the tension between (1) defining a useful set of categories that adequately captures the nuance and variability of the ways in which teachers interact with these objects, and (2) constructing a list that is manageable to use rather than a long list of specific verbs and scenarios. These categories comprise a larger domain of mathematical work that could be defined by introducing different mathematical criteria. For example, explanations, as objects, could be critiqued by teachers with respect to the validity, generalizability, or completeness of the explanations. Rather than define a separate category for each type of critique, the framework groups them as one category of mathematical work of teaching around the mathematical object of explanations. Each category might also refer to a range of contexts in which teachers might engage in this particular mathematical work. For example, as shown in Table 1 the category of analyzing representations includes reference to a set of contexts in which teachers might encounter representations, such as written work, talk, or texts. This organization of the framework keeps the set of categories concise while also mapping the dimensions of variation in teachers' mathematical work. This could support item writers in sampling across the varied terrain of teacher's mathematical work. Although the framework is detailed, it is not fully intended to represent all of the work of teaching, but to focus on common ways that teachers interact with particular objects.

It is important to note that the high-level categories built around mathematical objects are not necessarily disjoint. We utilize the focus of the mathematical work to determine where particular tasks of teaching fit in the framework. For example, consider the mathematical work of teaching that involves analyzing student strategies in written work for evidence of use of a particular mathematical structure (e.g., examining whether strategies reveal evidence of a comparison or take-away interpretation of subtraction). This work involves both student strategies and mathematical structure. However, the framework classifies the work of teaching by the main mathematical focus or goal, so looking for structure in student strategies would be classified as belonging to mathematical structure because looking for mathematical structure is the primary mathematical work, while the student strategy was the context in which it arose.

The third level of the framework further illustrates each category of work with examples that could serve as "shells" for items or "item starters." These examples do not include all of the necessarily details that might exist in a finished item, but could serve as a starter for beginning item developer to write items in this category. The examples are specific enough to provide help in beginning to write an item in that category. Consider the category of critiquing explanations. One example in that category is "Given several explanations, determine which ones are valid." This includes information about key elements that would need to be specified in the item (i.e. several explanations) and provides the desired action on the part of the test taker (i.e. determining validity). Each category includes several examples to help item writers attend to the different contexts and ways the work might play out, but the framework makes clear that the set of examples is not exhaustive and there are other ways to construct items and scenarios in each category.

## Interactions with Mathematical Content

The framework also provides support in understanding the mapping between particular K-6 mathematical content and the mathematical work of teaching. For example, the framework helps answer questions such as "in which mathematical content areas do teachers most likely interact with mathematical definitions"? Although the mathematical work of teaching can be mapped to all K-6 mathematics content, the framework focuses the supports on the most critical, or high-leverage K-6 topics (Ball & Foranzi, 2011). The mathematical work of teaching interacts with this content in a number of ways. First, there are some categories of work with respect to particular mathematical objects that are likely to emerge in instruction across all content areas. For example, interpreting students' mathematical errors is part of the mathematical work of teaching all mathematical topics (e.g., number and operations, measurement, fractions). In contrast, there are other parts of the mathematical work of teaching that are more likely to arise when teaching particular mathematical content. Consider "critiquing strategies", a category of teacher's work. While certainly possible that teachers might engage in this work across all content areas, there are some content areas in which teacher might need to do this work more frequently and with a set of common strategies. For example, when teaching multi-digit subtraction, teachers are likely to have to analyze and critique students' non-standard strategies. Similarly, making sense of student strategies is also likely to be part of the work when students are learning to compare fractions when there are many common strategies for doing so (e.g., common numerators, benchmarking). In contrast, this type of work is less likely to emerge when teaching aspects of geometry. The framework serves as a scaffold for item writers to think about in which mathematical content particular work is most likely to happen.

Annotations are included in the fourth column of the framework to foreground these connections and to highlight other considerations for writing items. These annotations address multiple areas of concern for item writers, often suggesting mathematical topics that are a good fit (or less good fit) with that category of work. In the case of the critiquing strategies example described above, the framework includes a note that "ordering numbers, operations with numbers" are fruitful areas for writing items in this category. Other times, the framework indicates that all content areas are a good fit. This column also includes annotations about the challenges of writing items in certain categories and with particular content. For example, in the topic of comparing fractions, there are a number of strategies that will result in the correct answer in some but not all cases, which makes this a productive terrain for writing items that assess candidates' ability to critique the validity and generalizability of strategies. In contrast, with whole number operations, it is much more difficult to find examples of strategies that either only work for a subset of whole numbers or strategies that result in a correct answer but are not valid. Developing items in this space is therefore quite challenging and requires very careful and strategic section of numbers. To help item writers understand this interaction between the MWT category and this content, the fourth column includes a note to indicate this difficulty.

This fourth column provides additional varied supports for understanding the MWT framework, including identifying potential item types and interactions with content. For example, some categories of the framework are areas in which others have developed items that serve as model items noted in the annotations, whereas others are novel in the sense that very few (or even no) examples of items in that space exist. The framework includes these annotations to describe the range of work teachers do and to inspire the development of new types of items that assess this range. However, items in this space are likely to be more difficult to write without examples from which to build. Therefore, notes in the fourth column alert item writers to the fact that the category was new and potentially challenging to write to. An example of this is the category of "Critiquing the use of a representation", meant to capture the work that teachers might need to do when making sense of the use of representations in particular ways by students, other teachers, or curricular materials. The final type of annotation in the framework consists of notes to support item writers in maintaining the focus on the mathematical work and refraining from building items that focus on pedagogical choices. Early observations of item writing indicate that certain categories of work (e.g., manipulatives, errors) are more likely to lead item writers into pedagogical terrain, such as presenting a student error and asking the candidate to decide the next step that would best help that student or choosing the best manipulative to help a child see his or her mistake. Annotations are included to alert item writers when a particular category of work provides challenge for retaining the focus on the mathematics, along with common mistakes made in writing items too focused on pedagogy.

## Utilizing the Framework for Item Development

To further specify the use of the mathematical work of teaching framework, we will examine how the table can be used to develop items, with a focus on explanations and mathematical structure. We begin by selecting the mathematical object of explanations and one key piece of the mathematical work of teaching with that object, "critiquing explanations with respect to validity, generalizability, or explanatory power." For this example, we will focus on the criteria of generalizability. To develop an item focused on the mathematical work a teacher does when critiquing explanations for generalizability, we must consider what mathematics content makes available a variety of explanations that may or may not be generalizable. One area of mathematics that is ripe with both explanations and methods that may or may not be generalizable is numbers and operations. For this example, we focus on operations with decimals, specifically decimal multiplication. The item shown in Table 3 requires a teacher to determine for each given explanation, whether or not the explanation represents generalizable methods for multiplying any two decimals. This is mathematical work that teachers do on a regular basis in a variety of contexts.

The second sample item is focused on the mathematical work of teaching involved in creating problems with a particular mathematical structure. In this case, the selected object for the item is "mathematical structure" and is combined with the work of "determining, analyzing, or posing problems with the same (or different) mathematical structure." Again, we must consider the mathematics content that teachers are most likely to encounter the need to determining or highlighting the mathematical structure of the work. Division is one area of elementary mathematics where particular interpretations of the operation require teachers to attend to mathematical structure. The problem shown in Table 3 requires one to apply a measurement (or quotitive) interpretation of division to develop a word problem. This involves careful attention to the structure of measurement division problems with attention to the meaning of each of the parts of the problems and then transferring this meaning to a particular context.

Table 3. *Examples of items written at the intersection of the mathematical work of teaching and content.*

| Explanations Sample Item |
|---|
| Mr. Reinke is working with his students on decimal multiplication. He asked them to solve the problem 1.2 x 0.3 and explain their process.<br><br>Which of the following student explanations for multiplying 1.2 x 0.3 represent methods that are generalizable for multiplying any two decimal numbers? Select <u>all</u> that apply.<br><br>(A) "I just multiplied 3x12 and got 36. I counted the total numbers behind the decimal points. That was two, so I need to have two numbers behind the decimal point in my answer."<br><br>(B) " I split the problem into two problems to make it easier. So I did .2 x .3 and that got me 0.06. Then I added 0.3 to that and got 0.36."<br><br>(C) "I like to change the problem so that I can use a whole number. I changed this problem to 3 x 0.12 because I can just multiply the 0.3 by ten, but I have to divide the other number by ten so I don't change the answer."<br><br>(D) "I just multiply like they are fractions. So it's like multiplying 12/10 and 3/10. I multiply the 12x3 and the 10x10 and get 36/100. That's 0.36 when I write it as a decimal.<br><br>(E) " I need to make the length of the numbers the same, so I can line them up. My new problem is 1.2x.30. I multiply them like regular numbers, then I just bring down the decimal point from the .30, so there are two numbers behind the decimal point." |
| Mathematical Structure Sample Item |
| Ms. Fischer is working with her students on fraction division using a measurement (or quotitive) interpretation, meaning that the quotient specifies the number of equal groups. She wants to give them word problems that use this interpretation of division so that students can practice giving explanations of fraction division.<br><br>Write a word problem that uses a measurement interpretation of division and could be solved using the problem 13 ÷ ½. |

## Reflecting on the Framework: Affordances and Constraints

In constructing this framework for the mathematical work of teaching, we sought to develop a tool that could support a focus on the mathematical work that teachers do in the context of every day practice. In reflecting on the framework in its current version, we believe that it does provide a mapping of a practice-based view of contexts in which teachers need to draw on specialized content knowledge. Furthermore, by focusing on a wide array of mathematical objects with which teachers interact, the framework provides insight into the diverse terrain of teachers' knowledge use in teaching. By focusing on the *mathematical* work of teaching, this framework attempts to push the envelope on the types of items that could be written in ways that may not emerge when approaching item writing by starting with the knowledge to be assessed. This framework also provides key insights into the interactions between the mathematical work of teaching and the mathematical content of elementary school in ways that could support item writers to develop assessment tasks within and across different mathematical topics. Despite the potential affordances of the mathematical work of teaching framework, we also recognize that the framework, as a tool for item writing, does not necessarily provide all of the support that assessment developers might need to write items to measure teachers' specialized content knowledge. In the following section, we describe a set of additional supports that we hypothesize might be needed for item development.

The focus of the framework on the mathematical work of teaching with the additional supports of highlighting interactions with mathematical content may not provide sufficient support for item development if the item writers do not have well-developed mathematical knowledge for teaching across mathematical topics and with respect to actions on different mathematical objects. For example, to write productive items about mathematical structure, item writers would need to know common mathematical structures relevant to K-6 mathematics, such as different interpretations of subtraction or division (i.e., take-away vs. comparison, partitive vs. measurement) or common different problem structures in early addition and subtraction tasks (i.e. result unknown, change unknown). Similarly, in order to write items about the validity and generalizability of student strategies, item developers would need to know common strategies used by children in that content area, such as knowing different valid (or invalid) strategies for comparing fractions (e.g., McNamara & Shaughnessy, 2010). For items about representations as objects, item writers would need to know relevant representations for a particular content area (e.g., area models, number lines, sets, and fraction bars for fractions concepts) and how the key ideas of that content are highlighted or not in different representations, such as knowing how different mathematical ideas of decimal multiplication and place value play out in an area model (Ball, Lubienski, & Mewborn, 2001). To develop items about interpreting student errors, item writers would need to know what are likely errors within particular content areas and what the reasoning behind the errors might be, such as knowing what key mathematical ideas of place value are violated when students incorrectly regroup across zero in multi-digit subtraction (Ball et al., 2008; Fuson, 1990). All of these examples highlight the need for item writers themselves to have well-developed mathematical knowledge for teaching or the tools to access and learn this knowledge themselves in order to develop assessment tasks.

Support for item writers with respect to mathematical knowledge for teaching may be particularly needed for developing items at the lower elementary grades. Much of the key mathematics of those grades is so tacit for adults that they are likely to struggle to determine what ideas could be addressed in items, such as knowing the different mathematical ideas that must be coordinated by children when counting an ill-structured set of objects to determine "how many", such as one-to-one correspondence, verbal counting, cardinality principle, and strategies for keeping track of what has been counted (Clements & Sarama, 2014; Richardson, 2012). Similarly, to develop items about early place value, item writers would need to know the key, but often tacit, ideas of place value that are not related to operations, such as the role of zero as a place holder or that quantities are represented symbolically left to right. Item writers might also need support in knowing how to construct item scenarios with reasonable approximations of student work at the relevant grade. For example, what would a student's drawing look like when trying to represent fractions with area models? What are reasonable student explanations of their thinking around particular content? Adults with less experience in K-6 classrooms struggle to construct student talk and written work that is reasonably authentic, as adult's own ways of thinking, talking, and representing mathematics are likely to be much more sophisticated than those of children, especially at the lower elementary grades.

Another area in which the mathematical work of teaching, as written, might not be sufficient to support item development around specialized content knowledge is related to the distinction between what counts as common content knowledge and specialized content knowledge for teaching. In our initial work with this framework, we have found that some sets of actions on mathematical objects seem to sit more clearly sit in the space of specialized content knowledge, such as analyzing the validity and generalizability of student non-standard strategies, which is in alignment with what was found by Hill, Schilling, and Ball (2004). Other sets of actions emerge at times closer to the line between common and specialized content knowledge, such as making a conjecture. One could argue that this is work that students also often do in mathematics classrooms, but it is important work that is not done in other fields.    The line between SCK and CCK maybe particularly challenging to distinguish in the context of the Common Core, since the Standards for Mathematical Practice now ask students (and teachers) to engage with content through mathematical practices such as constructing arguments and critiquing the thinking of others, actions which in some ways align with some of the mathematical work of teaching, such as analyzing the strategies used by others. The intent and nuance of the work may be different when students critique the explanations of peers in a classroom and when teachers are making sense of those strategies but there are some interesting similarities. This points to the possibility that item developers might need further support in how to write items that focus more squarely on specialized content knowledge.

These hypothesized supports serve as an initial set that address some key areas of concern when supporting the development of assessments. There may be other additional challenges that would arise when using the framework for this and other purposes that could require different types of support.

## Discussion and Conclusion

In this paper, we have presented a framework to support the development of assessments of mathematical knowledge for teaching (Ball et al., 2008). To consider the theoretical and practical implications of this framework, we first acknowledge that this framework offers one decomposition of the mathematical work of teaching; there may be other useful ways to parse teachers' mathematical work that would foreground different aspects of practice and knowledge use. Furthermore, this framework was developed based on the concept of mathematical tasks, or the mathematical work of teaching (Ball & Bass, 2002; Ball et al., 2008) which were conceptualized based on work in elementary mathematics and with the intended goal of supporting assessment development around elementary MKT. This raises the question of whether the mathematical work of teaching framework proposed here would apply equally well for the work of teaching secondary mathematics or whether there may need to be revisions or additions. For example, at the elementary level, we chose to group explanations and justifications together as an object given the nature of mathematical arguments typically constructed at the elementary level. At the secondary level, it might be more appropriate to include "justification and proof" as a separate mathematical object with which secondary mathematics teachers interact. As part of our future work, we will be pursuing this line of inquiry as we work to support assessment development around secondary MKT.

Despite these potential limitations, the mathematical work of teaching framework offers a contribution that has both theoretical and practical implications. First, this framework builds on and expands upon the mathematical tasks of teaching (Ball & Bass, 2002; Ball et al., 2008) to provide a comprehensive and nuanced identification of the mathematical work that teachers do in the context of teaching. This provides a detailed and practice-based lens (Ball & Bass, 2002) for the contexts when teachers must draw on mathematical knowledge for teaching in their practice.

A novel contribution of this framework is the idea of organizing the work around "mathematical objects" that teachers encounter and interact with in practice. This organization highlights the central role of mathematics in the framework and also affords seeing the diverse ways that teachers interact with different types of mathematical objects (e.g., representations, explanations, mathematical structure). Drawing on a practice-based perspective on mathematical knowledge for teaching, the framework offers a systematic way to identify and examine MKT by focusing first on teaching practice and the nature of teachers' mathematical work and then including the knowledge needed to manage that work. This perspective is different from starting with teachers' mathematical knowledge. Our approach explicitly highlights the use of knowledge in practice.

The MWT framework could also serve a number of practical purposes in both assessment development and teacher education. First, the framework was designed with the purpose of supporting the development of assessments of mathematical knowledge for teaching at the elementary level. This tool, along with additional supports described in the previous section, can provide item writers ways to develop assessments of MKT at a larger scale than has previously been possible, when items such as those developed by the Learning Mathematics for Teaching project (Hill, Schilling, & Ball, 2004) have been crafted by groups of experts involving mathematicians, mathematics educators, and teachers. Specifically, this framework could serve as a tool for developing items that appraise mathematical knowledge for teaching in the context of how the knowledge might be used in practice. Furthermore, the MWT framework could provide support in maintaining the focus on the *mathematics* in the work of teaching in ways that help item writers avoid developing items that assess teachers' pedagogical choices and decisions. Similarly, the framework could help illuminate the specialized knowledge that teachers draw on in their work to support item writers in writing items that assess more than common content knowledge of particular topics. Finally, the MWT framework offers systematic ways for assessment developers to manage the connections between the mathematical work of teaching and mathematical content in ways that would allow for building assessments that tap into the diverse specialized content knowledge needed to teach the K-6 curriculum. As we work with item writers from various backgrounds to develop a content knowledge for teaching assessment for elementary mathematics, we are able to examine the utility and limitations of the MWT framework for supporting assessment development at scale.

Although the mathematical work of teaching framework was developed for the purposes of building assessments, we believe that the framework has the potential to be used for other purposes related to teacher education and professional development, because it offers a systematic identification of the mathematical work of teaching. For example, this framework could be used as an organizing principle for designing mathematics content courses for pre-service teachers to support a focus on the ways that mathematical knowledge is used in practice. Similarly, the framework could be used for as a tool for curricular mapping in mathematics teacher education so that programs could systematically design learning experiences for complementary parts of the framework in different courses (e.g., content vs. methods courses, content courses in different topics such as number or algebra). This framework could also be a useful tool for increasing the MKT of teachers and faculty, including supporting professors and instructors of mathematics content for teachers courses (who likely did not teach elementary school themselves) in better understanding the ways elementary teachers need to use mathematics in practice and the nature of this knowledge.

In the context of the increased mathematical demands of the Common Core State Standards for Mathematics and data showing that many U.S. elementary school teachers lack strong MKT (e.g., Hill & Ball, 2004; Ma, 1999; Tatto et al., 2008), there is an urgent need to develop elementary teachers' mathematical knowledge for teaching. These mathematical demands on teachers are not new (Ball, Hill, & Bass, 2005) and are likely to continue with goal of preparing skillful and responsive practitioners (Ball & Forzani, 2011) Along with ways to support teacher knowledge development, the field needs assessment tools that will allow us to measure and track teachers' growth. The mathematical work of teaching framework contributes a tool to aid in these efforts, especially in working at scale.

**Author Note**

This research was supported in part by a grant from Bill & Melinda Gates Foundation, entitled "Building Practical Infrastructure for Learning to Teach the Common Core." The content of this paper is solely the responsibility of the authors and does not necessarily represent the official views of the Bill & Melinda Gates Foundation.

Correspondence concerning this article should be addressed to Sarah Kate Selling, 1005D, School of Education Building, University of Michigan, Ann Arbor, MI 48109. Contact: sselling@umich.edu.

**References**

Adler, J., & Venkat, H. (2014). Mathematical knowledge for teaching. In S. Lerman (Ed.), *Encyclopedia of Mathematics Education* (pp. 385–388). Springer Netherlands.

Ball, D. L., & Bass, H. (2002). Toward a practice-based theory of mathematical knowledge for teaching. In B. Davis & E. Simmt (Eds.), *Proceedings of the 2002 annual meeting of the Canadian Mathematics Education Study Group* (pp. 3–14). Edmonton, AB: CMESG/GDEDM.

Ball, D. L. & Bass, H. (2009). With an eye on the mathematical horizon: Knowing mathematics for teaching to learners' mathematical futures. Paper prepared based on keynote address at the 43rd Jahrestagung für Didaktik der Mathematik held in Oldenburg, Germany, March 1 – 4, 2009.

Ball, D. L., Hill, H.C, & Bass, H. (2005). Knowing mathematics for teaching: Who knows mathematics well enough to teach third grade, and how can we decide? *American Educator*, 29(1), p. 14–17, 20–22, 43–46.

Ball, D. L., Lubienski, S. T., & Mewborn, D. S. (2001). Research on teaching mathematics: The unsolved problem of teachers' mathematical knowledge. In V. Richardson (Ed.), *Handbook of research on teaching (4ᵗʰ edition)* (pp. 433–456). New York: Macmillan.

Ball, D. L., & Forzani, F. M. (2011). Building a common core for learning to teach: And connecting professional learning to practice. *American Educator*, 35(2), 17–21, 38–39.

Ball, D. L., Thames, M. H., & Phelps, G. (2008). Content knowledge for teaching: What makes it special? *Journal of Teacher Education*, 59(5), 389–407.

Baumert, J., Kunter, M., Blum, W., Brunner, M., Voss, T., Jordan, A., ... & Tsai, Y. M. (2010). Teachers' mathematical knowledge, cognitive activation in the classroom, and student progress. *American Educational Research Journal, 47*(1), 133–180.

Clements, D. H., & Sarama, J. (2014). *Learning and teaching early math: The learning trajectories approach*. New York: Routledge.

Delaney S. F., Sleep L., Ball D. L., Bass H., Hill H. C. and Dean C. (2005, April). Validating "specialized mathematics knowledge for teaching": Evidence from mathematicians. Paper presented at the Annual Meeting of the American Educational Research Association, Montréal, Canada.

Döhrmann, M., Kaiser, G., & Blömeke, S. (2014). The conceptualisation of mathematics competencies in the international teacher education study TEDS-M. In S. Blömeke et al. (Eds.), *International Perspectives on Teacher Knowledge, Beliefs and Opportunities to Learn* (pp. 431–456). Springer Netherlands.

Eisenhart, M., Borko, H., Underhill, R., Brown, C., Jones, D., & Agard, P. (1993). Conceptual knowledge falls through the cracks: Complexities of learning to teach mathematics for understanding. *Journal for Research in Mathematics Education, 24*(1), 8–40.

Fuson, K. C. (1990). Issues in place-value and multidigit addition and subtraction learning and teaching. *Journal for Research in Mathematics Education, 21*(4), 273–280.

Hill, H. C., & Ball, D. L. (2004). Learning mathematics for teaching: Results from California's mathematics professional development institutes. *Journal for Research in Mathematics Education, 35*(5), 330–351.

Hill, H. C., Blunk, M. L., Charalambous, C. Y., Lewis, J. M., Phelps, G. C., Sleep, L., & Ball, D. L. (2008). Mathematical knowledge for teaching and the mathematical quality of instruction: An exploratory study. *Cognition and Instruction, 26*(4), 430–511.

Hill, H. C., Dean, C., & Goffney, I. M. (2007). Assessing elemental and structural validity: Data from teachers, non-teachers, and mathematicians. *Measurement, 5*(2–3), 81–92.

Hill, H. C., Rowan, B., & Ball, D. L. (2005). Effects of teachers' mathematical knowledge for teaching on student achievement. *American Educational Research Journal, 42*(2), 371–406.

Hill, H. C., Schilling, S. G., & Ball, D. L. (2004). Developing measures of teachers' mathematics knowledge for teaching. *The Elementary School Journal, 105*(1), 11–30.

Ma, L. (1999). *Knowing and teaching elementary mathematics: Teachers' understanding of fundamental mathematics in China and the United States*. Mahwah, NJ: Lawrence Erlbaum Associates.

McNamara, J., & Shaughnessy, M. M. (2010). *Beyond pizzas & pies: 10 essential strategies for supporting fraction sense, grades 3–5*. Math Solutions.

National Governors Association Center for Best Practices & Council of Chief State School Officers. (2010). *Common Core State Standards for Mathematics*. Washington, DC: Authors.

Richardson, K. (2012). *How children learn number concepts: A guide to the critical learning phases*. Bellingham: Math Perspectives Teacher Development Center.

Rowland, T., Huckstep, P., & Thwaites, A. (2005). Elementary teachers' mathematics subject knowledge: The knowledge quartet and the case of Naomi. *Journal of Mathematics Teacher Education, 8*(3), 255–281.

Shulman, L. S. (1986). Those who understand: Knowledge growth in teaching. *Educational Researcher, 15*(2) 4–14.

Shulman, L. (1987). Knowledge and teaching: Foundations of the new reform. *Harvard Educational Review, 57*(1), 1–23

Tatto, M. T., Ingvarson, L., Schwille, J., Peck, R., Senk, S. L., & Rowley, G. (2008). *Teacher education and development study in mathematics (TEDS-M): Policy, practice, and readiness to teach primary and secondary mathematics*. Amsterdam, The Netherlands: International Association for the Evaluation of Educational Achievement.

Wilson, S. M., Shulman, L. S., & Richert, A. E. (1987). "150 different ways" of knowing: Representations of knowledge in teaching. In J. Calderhead (Ed.), *Exploring teachers' thinking* (pp. 104–124). London: Cassell

# Assessing Mathematical Knowledge for Teaching: The Role of Teaching Context

Geoffrey Phelps
Educational Testing Service

Heather Howell
Educational Testing Service

**Abstract:** Assessments of mathematical knowledge for teaching (MKT), which are often designed to measure specialized types of mathematical knowledge, typically include a representation of teaching practice in the assessment task. This analysis makes use of an existing, validated set of 10 assessment tasks to both describe and explore the function of the teaching contexts represented. We found that teaching context serves a variety of functions, some more critical than others. These context features play an important role in both the design of assessments of MKT and the types of mathematical knowledge assessed.

*Keywords*: teacher content knowledge, mathematical knowledge for teaching, teacher assessment, mathematics, assessment

## Introduction

Mathematical Knowledge for Teaching (MKT) is the content knowledge used in recognizing, understanding, and responding to the mathematical problems and tasks encountered in teaching the subject (Ball & Bass, 2002; Ball, Thames & Phelps, 2008). Assessments of MKT are designed to measure the mathematical knowledge that teachers use in these teaching practices. A number of practice-based assessments of MKT have recently been developed for teachers of K-12 grades (Herbst & Kosko, 2014; Hill, Ball, & Schilling, 2008; Hill, Schilling, & Ball, 2004; Kersting, 2008; Krauss, Baumert, & Blum, 2008; McCrory, Floden, Ferrini-Mundy, Reckase, & Senk, 2012; Phelps, Weren, Croft, & Gitomer, 2014; Tatto et al., 2008).

We follow Ball, Thames, and Phelps (2008) in defining MKT to include the full range of mathematics content knowledge used in teaching. The most widely assessed component of MKT is the common content knowledge that is taught and learned as part of regular schooling and is familiar to most adults. There is a long history of assessing teachers' common mathematical knowledge (Hill, Sleep, Lewis, & Ball, 2007). Often these assessment tasks look identical to those on student assessments because the construct is essentially the content of the student curriculum, either at grade level or at a level above the assigned grade (Phelps, Howell, & Kirui, 2015).

MKT assessments have generally focused, however, on the specialized forms of content knowledge that only teachers need to use in the course of their day-to-day work (Ball et al., 2008). While definitions and focus vary in the literature, and the mapping of the MKT construct is likely somewhat dependent on curriculum and culture, most studies share a focus on MKT as a form of applied knowledge that goes beyond common content knowledge (Krauss et al., 2008; McCrory et al., 2012; Thompson, 2015; Turner & Rowland, 2008). MKT assessments typically present teachers with content tasks that are encountered in teaching, such as interpreting student thinking and work, selecting materials for instruction, explaining concepts and procedures, or evaluating whether to use a representation for a particular instructional purpose (Ball & Bass, 2002; Hill et al., 2004). And since these tasks often occur in complex instructional contexts, MKT assessments typically also provide key information about the teaching context, such as the learning goals that direct the teaching, details about a student's prior academic work, or how students are grouped and organized (Phelps et al., 2015). Assessments of MKT differ in how teaching practice is represented. Some provide written descriptions, while others incorporate video or animations depicting mathematics teaching (see, for example, Herbst & Kosko, 2014; Hill et al., 2004; Kersting, 2008). These features of context support test takers in recognizing the relevant aspects of the content task, understanding the content problem, or providing a response to the assessment question.

This contextualization of MKT assessment tasks is in part theoretically motivated. Ball and Bass (2002) argue that *how* teachers encounter mathematics in their teaching directly shapes the nature of the mathematical knowledge that is needed. The context used in many MKT assessment tasks defines both *what* kinds of content knowledge teachers need to use and *how* they use this knowledge. Largely missing, however, from the current literature on MKT assessment are well-articulated design arguments that make clear the links between the construct and assessment task design (Mislevy & Haertel, 2006). Given the central role of teaching in MKT, it seems likely that any endeavor to assess MKT would require consideration of how context functions in the design of MKT tasks (Phelps et al., 2015).

In this study, we take the first steps in this direction by presenting arguments and illustrations for how context functions in a set of elementary-level MKT assessment tasks, with a particular focus on how context enables tasks to measure MKT that goes beyond common content knowledge. We do not take up the question of whether other sub-components of MKT are distinctly measureable, as other studies have done (see, for example, Hill et al. (2008) and Krauss et al. (2008) for different approaches to the measurement of PCK as a distinct domain). Our argument is simply that context matters in the assessment of some components of MKT more than others; in particular it matters more for components that go beyond common content knowledge. Because these types of knowledge have been the objects of intense interest in teacher education it is worth attending closely to how context matters in their assessment.

The paper is organized as follows. First, we discuss the role of context in establishing the construct validity of MKT assessments using illustrative examples. We follow Messick's (1989) view of construct validity, which helps to determine how relevant and representative the tasks are in measuring MKT. We begin with an example that includes three tasks that assess similar content focused on exponential expressions but vary in how teaching context is represented in the task. This set of tasks provides a concrete illustration of major differences in context and its function. Next, we discuss two task examples in detail to illustrate the design and content focus of MKT assessment tasks and to make clear our arguments about the role that context plays in these assessment tasks. Finally we present a summary of how context functions across the 10 tasks and discuss the implications for assessing MKT.

**The Role of Teaching Context in Assessing MKT**

The appropriate use of teaching context in the assessment of MKT can help avoid threats to construct validity, namely construct-irrelevant variance and construct under-representation. Construct-irrelevant variance occurs when an assessment represents dimensions that are irrelevant to the correct interpretation of the construct, and construct under-representation occurs when an assessment does not adequately represent the dimensions of the construct that are the focus of the assessment (Messick, 1989). In respect to MKT, many assessments are designed to measure the MKT that is specialized to the work of teaching. In cases where teaching context is critical to assessing particular aspects of MKT, the absence of teaching context could lead to construct under-representation.

We begin with an illustration designed to highlight the various roles that context can play in the measurement of MKT. We present three related example tasks in Figure 1. The example in panel C was developed for the Measures of Effective Teaching project (Phelps et al., 2014) and is one of the 10 tasks analyzed in this study. Task selection and analysis is addressed in more detail in the methods section. The examples in panels A and B of Figure 1 are variants created by the authors for illustrative purposes to demonstrate both when teaching context does and does not support the assessment of MKT.

| A. Common Content Knowledge | B. Common Content Knowledge in a Teaching Context | C. Specialized Content Knowledge in a Teaching Context |
|---|---|---|
| Evaluate each of the following simple exponential expressions.<br><br>$3^3 =$<br><br>$2^3 =$<br><br>$2^2 =$ | Ms. Hupman is teaching an introductory lesson on exponents. She gives her students a set of problems to check their proficiency in evaluating simple exponential expressions. Ms. Hupman looks over the work from one of her students. For each of the answers, indicate if the student's evaluation is correct or incorrect.<br><br><table><tr><td></td><td>Correct</td><td>Incorrect</td></tr><tr><td>$3^3 = 9$</td><td>☐</td><td>☐</td></tr><tr><td>$2^3 = 6$</td><td>☐</td><td>☐</td></tr><tr><td>$2^2 = 4$</td><td>☐</td><td>☐</td></tr></table> | Ms. Hupman is teaching an introductory lesson on exponents. She wants to give her students a quick problem at the end of class to check their proficiency in evaluating simple exponential expressions. Of the following expressions, which would be <u>least</u> useful in assessing student proficiency in evaluating simple exponential expressions?<br><br>○ $3^3$<br><br>○ $2^3$<br><br>○ $2^2$ |
| Key: 27, 8, 4 | Key: incorrect, incorrect, correct | Key: $2^2$ |

*Figure 1.* Tasks to illustrate differences in types of content knowledge assessment.

Each of these three tasks involves the same underlying mathematical content, but they differ in whether and how each is situated in teaching. Task A does not include a context and simply requires the test taker to evaluate three exponential expressions. This task is not situated in teaching other than representing mathematics that is part of the grade school curriculum. However, the absence of context in this task is construct relevant because no context is required to assess whether teachers can do the work of the student curriculum.

Task B includes a context that shows a student's evaluation of three simple exponential expressions. The student has answered two problems incorrectly and one correctly. The test taker does not need to draw conclusions about *why* the student answered each correctly or incorrectly. He only needs to evaluate each problem and check the correct answer against the student's answer to determine whether the student's answer is correct or incorrect. While the look and feel of Tasks A and B are different, the mathematical work and knowledge required to answer is essentially the same. Both measure a test taker's ability to evaluate expressions. The context in Task B is arguably *construct-irrelevant* (Messick, 1989), meaning that its presence or absence does not relate directly to the skill of exponent arithmetic. However, the longer text included in Task B increases the reading burden on the test taker, raising the possibility that the task might unintentionally measure reading ability in addition to the skill of exponent arithmetic. Reading load is not necessarily problematic; the text is not excessive in length and the level of reading required may be well within the abilities of the tested population. But to the extent that such a task measures something unintended (in this case, reading ability), it can be a source of *construct-irrelevant variance* in the test scores (Messick, 1989).

Task C, like Task B, includes a written teaching scenario. But in this case, the context serves to direct the test taker to consider which expression would be a poor choice for teachers to use in understanding whether students know how to evaluate expressions. To respond to this task, the test taker needs to already know, or know how to figure out, what kinds of confusion students are likely to exhibit (e.g., confusion about which number is the base or exponent or confusion around what kind of operation is required to evaluate the expression). The test taker then needs to anticipate what the solution to each of the problems would be using the incorrect methods students might apply and from this figure out which problems reveal these confusions. The mathematical knowledge involved in responding to this task goes beyond the common content knowledge of how to evaluate exponents. The context that is included in this task is relatively minimal but clearly necessary; without the context the test taker lacks key information for comparing the problem choices. Unlike Task B, where the context is *irrelevant* to the content assessed, in Task C the context is *relevant* and arguably critical to the content knowledge that is being assessed.

The three tasks shown in Figure 1 are intended to illustrate a number of key points in the design of MKT assessment tasks. First, context is not always needed. Most notably, as illustrated in panel A, when teachers are simply doing the math that their students are learning, there is likely no need for context (Phelps, Howell, Schilling, & Liu, 2015). The context in Task B illustrates an authentic situation in teaching that requires the teacher to have common content knowledge. But from an assessment perspective, when measuring this type of MKT, it will often be more efficient to present the task without a context, as illustrated in Task A. A basic principle of assessment design is that irrelevant context should be avoided to the greatest extent possible so that only the intended construct is measured (Messick, 1989).

Second, as illustrated by task C, context can play a critical and relevant role in assessing the construct when the goal is to assess the components of MKT that go beyond the mathematics that students are expected to master (i.e., SCK or PCK in the Ball et al. (2008) model). In such cases, eliminating the context might shift the focus of the task in ways that leave the test taker unsure what is being asked or might fundamentally change the content assessed. Eliminating context entirely could reduce tests of MKT to assessing only the types of common content knowledge illustrated by task A, which would lead to tests that suffered from threats of construct under-representation (Messick, 1989).

Figure 1 also illustrates that it is not always simple to determine whether context is relevant. At first glance, Tasks B and C seem quite similar. It is only through analysis of the work that each task requires of the test taker and consideration of the measured construct that such a determination can be made. Consequently, from an assessment design perspective, it is critical to clarify how context that is included in an assessment task is relevant to the particular features of the construct being assessed (Mislevy & Haertel, 2006).

## Methods

Our goal in this study was to systematically investigate the ways in which teaching context can function in tasks designed to elicit the types of MKT that are particular to the work of teaching mathematics. While we follow general procedures for qualitative coding, our method differs from typical qualitative work in two key ways. First, our 'data' are the tasks themselves. We selected a set of tasks for which we have a large set of ancillary data showing that they perform well as measures of CKT and that context matters in how respondents reason through each task. We did not, however, examine teachers' actual response data in this particular study. Our claims therefore are built on arguments about task design and not on empirical data comprised of test takers' responses. Therefore, our results are the categories and associated characteristics of task design that emerged in the course of the close analysis of the MKT tasks. We think this type of close, rigorous analysis helps to call attention to aspects of task design that are otherwise largely invisible, even to test designers. We describe the process in some detail to help the reader follow our logic.

### Selection of MKT Tasks for Analysis

The analysis that follows focuses on a set of 10 mathematics tasks that were developed as part of the Measures of Effective Teaching project to measure elementary level MKT (Phelps et al., 2014). These tasks were chosen because we had strong evidence from a prior cognitive interview study that they situated test takers in teaching practice as designed (Gitomer, Phelps, Weren, Howell, & Croft, 2014; Howell, Phelps, Croft, Kirui, & Gitomer, 2013). As part of that study, we wrote rationales detailing the embedded assumptions about how context would function and about the construct each task measured. The study established that the alignment of participant reasoning to these rationales was strongly related to answering correctly or incorrectly. Across all mathematics task level interview responses (*n* = 640), 88% showed the desired pattern in which correct answers matched pre-specified correct reasoning and incorrect answers did not match that correct reasoning. For 97% of responses, participants reported that the task was an authentic representation of actual teaching practice. The study also found no evidence that reading load introduced construct-irrelevant variance by interfering with test takers' interaction with the assessment tasks (Gitomer et al., 2014; Howell et al., 2013). These response patterns led us to conclude that knowledgeable teachers were situated in context as specified by the task design.

It is worth clarifying that our goal was not to generalize to all MKT tasks or other such practice-based items. Instead we used strong tasks from a prior study with the goal of using this selection as a site for naming and defining important task design characteristics. Specifically, in order to understand how context can function, we required a set of tasks that measure more than common content knowledge, in which context is available to be analyzed, and for which we have some evidence that the context serves a function.

**Analytic Method**

As a first step in the analysis we expanded the written rationales used in the prior study to account more explicitly for context features and to understand better the role that context played in these tasks (Howell et al., 2013). We started by simply describing the context and its role in shaping how the test taker interacts with the content problem. These descriptions constituted the first step in our qualitative analysis and subsequently became objects of the second step of analysis. A summary of such a description is provided below for the task shown in figure 2.

To assess her students' prior knowledge about evaluating arithmetic expressions, Ms. Santiago assigned a worksheet of problems. She noticed that Alexis answered the first two incorrectly and the next two correctly.

1) $7 \times 2 - 6 + 3$   $= 5$

2) $9 - 5 + (16 \div 8)$   $= 2$

3) $9 + 24 \div 3 - 1$   $= 16$

4) $17 - (3 + 7 \times 2)$   $= 0$

Which of the remaining problems is Alexis likely to answer incorrectly?

O   $8 + 7 - 12 \div 3$

O   $13 - 3 \times 2 + 5$

O   $(27 \div 3 - 4) + 8$

O   $(16 - 12) \times 5 - 10$

*Figure 2.* The Santiago task.

To respond to this task, the test taker needs to analyze the four examples of Alexis' work, determine what she did to get the first two problems wrong, and then test any hypotheses about her confusion to see if they are consistent with answering the other problems correctly. The test taker needs to select an option that Alexis would answer incorrectly, assuming Alexis persists in the same error. However, the underlying, important task is to figure out what Alexis is misunderstanding. The assessment task is focused on the recurrent teaching practice of diagnosing student understandings or misunderstandings based on the written work they produce.

Analysis of the given problems reveals that in each of the incorrect problems, Alexis has added before subtracting. In the first problem she added 6 and 3 first and then subtracted the total of 9 rather than subtracting 6 and then adding 3. In the second problem she added 5 and 2 (where 2 is the result of 16 divided by 8) and then subtracted the total of 7 rather than subtracting 5 then adding 2. However, in the third and fourth problems this particular error does not lead to an incorrect answer. In the third problem, the ordering of the operations happens to be such that adding before subtracting is appropriate. In the fourth problem, the parentheses indicate that the expression inside should be added first before subtracting. There is not enough evidence to know *why* Alexis is making this error, although experienced teachers may recognize it as a possible overgeneralization of the use of the mnemonic PEMDAS[1] to dictate the order of operations. If we assume that Alexis will persist in the same error, the second answer option is the only option she would answer incorrectly because for each of the others, like the third and fourth given problems, adding before subtracting happens to be correct.

The scenario only specifies "arithmetic expressions" as the content topic under study, but the form in which the mathematics problems are written provides a great deal of subtle contextual information about the level of the students. Each expression is written out as a single line, using the division symbol ÷ and the multiplication symbol × rather than a fraction bar for division or a dot for multiplication. All four operations (addition, subtraction, multiplication, and division) are represented and parentheses are used, but there are no exponents. These details communicate to someone with knowledge about the teaching of this mathematics that the students are likely studying order of operations. Their use of the operations themselves is likely fluent at this point, but their ability to combine the operations correctly may not be. In the context of the assessment task, this is important because it makes some possible errors far less likely. For example, one could have assumed that Alexis misread the addition symbol or that she did not know how to perform the subtraction correctly, but this is an unlikely error for a student who is working with expressions of this type.

On the other hand, it is quite common for students at this level to make mistakes in the ordering of the operations. While the scenario does not state that this is an order of operations problem, the contextual clues embedded in the format of the content problems themselves make the work the test taker needs to do much easier by narrowing the field of all possible errors to a fairly small set of likely ones that need to be considered. This is a critical piece of information because it allows the test taker to rule out other competing, but unlikely theories. Again, one reason this set of assessment tasks was useful to study is that the prior interview work provides evidence to support such claims about the functioning of the context. And indeed in a prior study using this task, participants often referred explicitly to it being about order of operations, confirming this part of the design theory (Howell et al., 2013).

---

[1] PEMDAS is a mnemonic device commonly used in the U.S. to help students remember the order of operations. It stands for "parenthesis, exponents, multiplication, division, addition, subtraction," and is not strictly mathematically correct as written, although when used in instruction teachers generally qualify it by stating that the pairs "MD" and "AS" are performed in order, left to right, at the same time, not one before the other as the device implies.

The context also includes information about the student, Alexis, stating that she answered the first two problems incorrectly and the second two correctly. It is not strictly necessary to state which are correct and which incorrect, but providing the information up front may decrease the cognitive load on the test taker and encourage him to focus on the student's thinking rather than on whether the problems are correct. And pointing out that these are Alexis's answers also conveys a crucial piece of information about what the test taker needs to do by setting the condition to be met—the identified misconception must explain both Alexis's correct and incorrect work, and it must be a systematic error that the student makes consistently. A test taker who fails to attend to this aspect of the context may read through the problems assigning a unique diagnosis to each, or may cite difficulties students generally have with such problems without determining the specific difficulty Alexis is having. Both were patterns we observed in prior interview data and were associated with incorrect answers (Howell et al., 2013).

Finally, the assessment task presents an authentic scenario. Teachers frequently have to draw conclusions about student thinking from written work. The task of figuring out what Alexis is thinking seems not just plausible but worthwhile; teachers can't make informed decisions about next instructional steps without knowing first what their students understand and do not understand.

The summary above illustrates the type of descriptive account that was generated for each of the 10 tasks. These accounts provided rich descriptions of how the tasks functioned and more specifically the role that context played in these tasks. They also were used as the basis for generating provisional statements describing each context element. We then coded each identified context element inductively with short phrases describing the ways in which the context element functioned in the test taker's anticipated interaction with the task. We used a constant comparative method (Strauss & Corbin, 1990) to do this coding, which can be described in four steps: (1) independently analyzing a subset of tasks, (2) reconciling the coded elements and functions across tasks, (3) revising the list to reflect all elements and functions and testing the new categories by recoding the subset of tasks, and (4) expanding and iterating to a larger set of tasks until we had reached consensus on all codes for all context elements observed across all 10 tasks. Our goal in this work was not to achieve a particular level of coding reliability, but rather to generate a useful set of categories that captured the types of elements and functions we saw both in a given task and collectively across tasks. The short descriptors of the functions were then grouped together to form more general categories, and the entire set of tasks reviewed and recoded using these categories.

This process of task analysis generated three sets of categories that were relevant to describing the context and its function. Because we view these categories as an important outcome of this study, they are described in more detail below in the results section.

## Results

The results are organized in two main sections. The first section presents the categories that were derived inductively from the analysis of the 10 MKT tasks. The second section focuses on the use of these categories to describe the context features and their function across these 10 tasks. While we present counts across the set of tasks to illustrate the frequency, distribution, and co-occurrences we observed, we remind the reader that for a study of this type the main results are the identification and description of the categories themselves.

### Teaching Context and Function

**Context focus.** The various teaching contexts identified in the MKT tasks mapped onto three major components of instruction. These included features of *students* such as their history, learning needs, and actions; the *content* and how it is situated in the curriculum of school learning; and, the *setting*, which includes class size or grouping or mode of instruction such as lecture or discussion. Not only are these particular features central to instruction, but they have also recurred in many different heuristics and models used to characterize instruction (see, for example, Cohen, Raudenbush, & Ball, 2003; Hawkins, 1974; McDonald, 1992; Schwab, 1978). For each of the 10 MKT tasks, elements of the context could be identified as providing context for the *content*, *student*, or *setting* of instruction. These categories are useful for identifying the aspects of instruction that are the focus of the context features.

**Context Function.** The categories that were derived from the analysis describe the main function of the contexts identified in the 10 tasks. These categories are described in Table 1.

Table 1. *Context Functions.*

| | Context function | Description |
|---|---|---|
| **Critical functions** | Narrows a set of possibilities | Context that functions by narrowing the set of possibilities that the test taker must consider – e.g., narrowing the possible answer choices or eliminating one or more options. This can be quite subtle, as in cases where the specified level of a student or class sets an expectation for the level of sophistication one might expect in an answer, which in turn serves to eliminate some set of possibilities. Sometimes it might be some other list that is narrowed rather than the answer choices. For example, in the task shared in Figure 2 the test taker needs to figure out what error the student has made before even considering the options, and the content context serves to narrow the possible errors. |
| | Sets condition for the answer | Context that functions to specify, explicitly or implicitly, what condition the answer needs to meet to be correct. For example, in the task shared in Figure 3, the setting context sets a condition for the answer – i.e., that selected problem needs to be one for which the student's answer will reveal the suspected misconception to the teacher. |
| **Helpful functions** | Direct the test taker's focus | Context elements that encourage the test taker to focus (or not to focus) on a particular aspect of the task. For example, in the task shared in Figure 2, the statement that the student answered two problems correctly and two incorrectly is intended in part to cue the test taker to pay attention to the correct work and not just the incorrect work. |
| | Provides additional information | Context that provides additional information that is useful but not critical. This might include defining a term that some test takers may not know. Or it may include context that reduces cognitive demand by stating up front that a student's work is incorrect so that the test taker knows that figuring this out is not part of the work he needs to do. |
| | Reinforces critical information | Context that reinforces a key idea. This can help ensure that a test taker is directed to pay attention to critical information and thus raise the likelihood that the test taker engages in the assessment task as intended. |
| **Functions related to face validity** | Authenticity | Context that helps support an authentic representation of the work of teaching. Perceived authenticity can be a key motivating factor and enhance validity. |
| | Plausibility | Context that specifically helps to add plausibility to an element of the task that would not otherwise seem reasonable. For example, in the task shown in panel C of Figure 1, the specification that the problem is a quick check at the end of class makes it feel reasonable that the teacher has a need to diagnose understanding on the basis of a single answer alone. Without this information, the test taker might wonder why the teacher does not simply ask the students to explain their work. |
| | Motivation | Context that creates a situation in which the test taker can better recognize the importance the task. For example, tasks that give specifics about a student and their learning needs can motivate because there seems to be a real and pressing need to help the student. |

**Context Relevance.** Another pattern that emerged was one of relative levels of relevance or criticality. Some context functions, like those that set the condition the answer needs to meet to be correct, are essential for assessing the MKT content. Without that information the test taker would be unable to respond correctly and the particular type of MKT could not be assessed. Other functions were less essential in that it would still be possible to respond correctly absent that context. However, many of these context features were still quite helpful in directing the test taker and thus might serve to reduce cognitive load. For example, context that functions either to further define a key idea or direct the test taker to pay attention to something important falls into this second group. A third type of context functions to increase face validity, support the test taker's perception of authenticity, or to motivate in other ways that support completing an assessment task.

### Teaching Context in MKT Assessment Tasks

To make these three sets of categories more concrete, an example task (Figure 3) is used to illustrate the process and the types of decisions that the coding and classification entailed.

Mr. Chamberlain is concerned that his students' use of the calculator has led them to view the equal sign as a signal to carry out an operation rather than as a symbol indicating equality. Of the following missing-number problems, which would best assess whether students understand the mathematically correct meaning of the equal sign?

○ __ + __ = s 8

○ 7 + 5 = __ + 6

○ __ = s 7 + 9 + 5

○ 23 + 4 = __ = 4 + 23

*Figure 3.* The Chamberlain task.

The context for content in this task is given directly and indirectly. The scenario indicates that Mr. Chamberlain's concern is focused on the meaning of the equals sign. The format of the missing number equation problems communicates the level of the students as early elementary and signals that the use of the equal sign is likely new to them. This bolsters the authenticity and appropriateness of Mr. Chamberlain's concern as represented in the problem, as students often misunderstand the equals sign to be a command to perform an operation. It both makes sense that students working at this level would have this confusion and it conveys that the confusion is important for a teacher to attend to. Thus, in this case, the content context provides authenticity and contributes to the face validity of the task.

Unlike the task in Figure 2, in which information about the student was given directly, the student context in this problem is given indirectly in the form of the teacher's concern. What we know about the students is that they have used a calculator, and further that the teacher believes they may hold a particular misconception (that the equals sign is a command to perform an operation). Knowing that that the suspected misconception is connected to calculator use in this way provides key information to the test taker by defining, if indirectly, the operational view of the equals sign. Understanding the difference between the operational and equality views of the equal sign is key to answering correctly, and this piece of context reduces the cognitive load for a test taker unfamiliar with the misconception or with the terminology used to describe it. It also provides a plausible basis for the students to have that misconception, as calculator use is common and can lead to exactly this type of misunderstanding. The student context here serves dual functions. It supports the plausibility of the scenario, contributing additionally to face validity, and it also provides helpful but non-critical information to the test taker by defining a key idea.

We point out here an ambiguity in the coding and classification. One could argue that the teacher's concern is a part of the setting context, and not really information about the students. We acknowledge this, and use this example to draw attention to a necessary imprecision in the categories we have proposed. In many cases the distinctions are subtle and a piece of context might well fall into multiple categories. In fact, in this case we listed the teacher's concern about the operational view as setting context as well as coding the student's use of the calculator as student context. As a feature of the setting, the teacher's concern motivates the task by providing a plausible reason to care which problem is selected, further supporting face validity. More importantly, it sets the condition the answer needs to meet in order to be correct; the correct answer must be a problem that will reveal the given misconception to the teacher. This function of context (setting the condition the answer needs to meet to be correct) is at the highest level of relevance because it is critical that it be included in order for the task to function as designed. That the context is difficult to assign to the categories of setting or student is less important than the critical function it serves in orienting the test taker's thinking. We draw the reader's attention to the ambiguity here to illustrate clearly that our goal is not to create strict divisions between context types so much as to name categories that are useful for systematically analyzing or generating MKT tasks.

We also draw a distinction between the context that is situating the test taker and the actual knowledge or ability that the test taker must have in order to respond to the task correctly. This last piece of context sets the condition the answer needs to meet, and the test taker must distill this understanding from the context in order to answer correctly. But the test taker still needs to know which problem will meet that condition. While the context clues situate the test taker so that she is answering the right question, they do not answer the question for her. In this case, the test taker still needs to know or be able to anticipate that a student with the given misconception will likely write 12 in the blank on the second problem, having interpreted the equal sign as a command to add 5 and 7. For this option, 12 is incorrect because $5 + 7$ is not equal to $12 + 6$. While the student might think about the equal sign incorrectly in each of the other options, the answer the student gives would be the same as the correct answer and would not reveal the error to the teacher. This is the only problem that makes the misconception visible.

Table 2 gives an overview of the context features coded for each of the three MKT tasks that have been discussed in depth so far in the paper (Figures 1, 2, & 3). It is worth noting that while we made efforts to reach consensus in the coding, we do not propose that the context elements for which we coded are fixed or that there is always a clear classification. Rather, we find these elements useful in providing conceptual tools that help to identify and understand the function of context. Specifically, this makes these context elements more visible and provides a language that can be used to evaluate and critique the design of assessment tasks. The examples in Table 2 also illustrate that not every type of context element or function appears in every task. This is typical of what was represented across the set of analyzed tasks and suggests as a cautionary note that while the proposed categories are analytically useful, they are not strictly necessary. They do not form a template for assessment task construction.

Table 2. *Sample Coding Classifications.*

| Task Description | Content context and its function | Student context and its function | Setting context and its function |
|---|---|---|---|
| Ms. Hupman wants to select a brief assessment problem to ascertain whether her students understand how to evaluate exponential expressions. (Figure 1, panel C) | That the lesson is introductory **narrows the likely errors students would make**[1] to those above the level of arithmetic (students probably know how to multiply). | No student context is given. | That the problem is a quick proficiency check provides **plausibility**[3] for why the answer alone needs to convey information and also **sets the condition for the answer**[1] – that it must reveal to the teacher whether or not the student is proficient. The focus on the <u>least</u> useful problem **decreases authenticity**[3] as a teacher would generally look for the most useful, not the least. |
| One of Ms. Santiago's students has answered two order of operations problems correctly and two incorrectly, and the test taker must figure out what she has done wrong and predict which additional problems she will answer incorrectly. (Figure 2) | The types of mathematical symbols used ("x" for multiplication, for example) coupled with the specification that this is prior knowledge for the students **narrows the possible error types**[1] to exclude arithmetic errors and include errors related to ordering of steps. | The specification that the student answered two problems correctly and two incorrectly **encourages the test taker to attend to both**[2] the correct and the incorrect work, **suggests a systematic error**[2], and **sets the condition**[1] the selected error needs to meet – it needs to explain the given work. | That the student work shown was in response to a worksheet suggests that the teacher is looking at the work after the fact, with time to reflect, making the work needed to analyze the errors more **plausible**[3]. |

| | | | |
|---|---|---|---|
| Mr. Chamberlain is concerned that his students may have a specific misconception about the equal sign and must choose an assessment problem that would reveal to him whether or not they have that misconception. (Figure 3) | The format of the missing number equation problems communicates the level of the students as early elementary and hints that the use of the equal sign symbol is likely new to them, supporting the **authenticity**[3] of the teacher's concern that they might not understand it. | The teacher expresses concern that students use of the calculator may have caused them to have an operational view of the equals sign, providing a **plausible**[3] explanation for why they would misunderstand, and **defines**[2] for the test taker what is meant by the operational view. | The teacher's concern that the students may have an operational view of the equals sign provides **motivation**[3] for the task of teaching, as well as **setting the condition**[1] to evaluate the answers as those that reveal that incorrect operational view. |

*Note:* Bold text indicates the function of a context element: 1) critical context function, 2) useful context function, 3) face validity context function. A full version of this table and all tasks analyzed is available from the corresponding author upon request.

Table 3 provides an overview of how often each function type for each context category appeared across the 10 tasks analyzed. For many of these tasks, various context functions and features could appear multiple times. For example, the task presented in Figure 3 was coded as having content context that supports authenticity, student context that makes the situation more plausible and defines a key term, and setting context that motivates the situation as well as setting the condition the answer needs to meet. This particular task contributes one count to the content context category and two counts each to student and setting categories.

Table 3. *Context Type and Function for Ten MKT Tasks.*

| Type of function | Type of context | | | Total occurrences over 10 tasks |
|---|---|---|---|---|
| | Content | Student | Setting | Total |
| **Critical context functions:** | | | | |
| Narrows a set of possibilities | 5 | 2 | 0 | 7 |
| Sets the condition for the answer | 2 | 3 | 2 | 7 |
| | | | | |
| **Helpful context functions:** | | | | |
| Directs the test takers focus | 1 | 4 | 1 | 6 |
| Provides additional information | 4 | 4 | 0 | 8 |
| Reinforces critical information | 1 | 1 | 0 | 2 |
| | | | | |
| **Face Validity** | | | | |
| Authenticity | 1 | 2 | 2 | 5 |
| Plausibility | 0 | 1 | 8 | 9 |
| Motivation | 0 | 2 | 1 | 3 |

Our summary suggests that for the 10 tasks analyzed the context functions are relatively equally represented across all the major coding categories. While critical functions seem to occur slightly less often in the "setting" column than elsewhere, and face validity functions noticeably more, the overall distribution suggests that all three types of context elements can serve all functions and can also vary in their criticality. This suggests that teaching context, at least as it appears in these particular MKT tasks, can play a variety of functions across a number of major features of instruction.

## Discussion

Mathematical knowledge for teaching includes the full range of mathematics used in teaching the subject. This is a form of applied knowledge that teachers draw on and use as they engage in and carry out the many practices that make up the moment-to-moment and day-to-day work of mathematics teaching (Ball & Bass, 2002; Ball et al., 2008). In this study, we conducted an analysis of 10 tasks designed to assess MKT. These tasks assess types of MKT that go beyond the common content knowledge used in doing the work of the student curriculum (e.g., the first example in Figure 1), with the goal of measuring specialized types of MKT used in practices only encountered in mathematics teaching (Phelps et al., 2014). Because these tasks focus on types of MKT applied in teaching practice, they all include teaching context. We found that across these tasks the context served a number of different functions. In fact, for many tasks, the context served multiple functions. We coded almost 50 instances of context serving an identified function across just 10 tasks.

These context features focus on different aspects of instruction. We grouped these under the larger categories of content, student, and setting. These categories provide a useful set of lenses for considering which core aspects of instruction are represented in the context. It also seems likely that different types of MKT tasks might require context that focuses on aspects of instruction that did not come up in our analysis. For example, tasks like the Chamberlain task provide background information about the teacher's concern. This suggests that *teacher* might be an additional useful category that simply did not appear often in the set of tasks we examined. This category might include information such as teachers' pedagogical motivations, purposes, or constraints. Other MKT assessment tasks might call for a more fine-grained list of the major components or aspects of instruction, such as separate categories for curriculum materials and content.

We also identified across the context features a variety of functions (Table 1). Again, it is important to emphasize that these functions are almost certainly not exhaustive. Additional functions might be identified for a different set of MKT tasks. Although the list of context functions is likely incomplete, we think it nonetheless provides a useful start and important insights into the assessment of MKT. One insight that emerged from this analysis was that these functions could be placed into three larger groups describing the degree to which the context was critical to assessing the MKT construct. We discuss each group of functions briefly below.

We described one group as "functions related to face validity" (Table 1) because their only role was to support the test taker's perception of the situation as authentic, to make the work seem plausible, or to motivate. This group of context functions is arguably the least critical for supporting the test taker in providing an answer. In fact, in some situations, these context features may not be needed at all. If the test taker, for example, is familiar with the content and accepts that it is important and used in teaching, the context may do little more than add to reading load and may even introduce construct irrelevant variance. On the other hand, context that adds face validity can support the test taker in important ways. Michael Kane (2006), in his seminal chapter on validity, argues that tests that lack face validity can introduce construct-irrelevant variance since the test may in part measure a test taker's disengagement with the tasks rather than the construct of interest. Context associated with this group of context functions should be examined with special care to make sure that it plays a sufficiently important role to be included in the assessment task.

We described a second group of context functions as "helpful functions" (Table 1) because they served to support the test taker in providing an answer (e.g., directing the test taker's focus toward a particular aspect of the work or reinforcing critical information). While this type of context was not critical to answering the task, it played an important role, often reducing burden for the test taker. As was the case for the face validity functions, the context associated with this second group is not critical to answering. However, it is not obvious that the context is construct irrelevant, since it appears to support the test taker in productively and efficiently engaging the task.

We described the final group of functions as "critical functions" (Table 1) because the test taker needs to consider the associated context in order to provide an answer. This included cases where the context information narrows the answer possibilities or sets a condition the answer needs to meet. If the context were removed entirely for these instances then the MKT that was the focus of the task simply could not be assessed. In these cases the context is not only critical, but arguably an integral part of the construct itself (Phelps, Howell, & Kirui, 2015). Removing context from these tasks would fundamentally change the MKT assessed and would likely lead to tests that suffered from construct under-representation (Messick, 1989).

Our analysis also revealed that because context can simultaneously serve multiple functions of varying criticality, it cannot easily be labeled as strictly construct relevant or irrelevant. A passage that increases reading load may support the test taker's work in other ways. We also note that identifying context in an assessment task requires more than a surface analysis of its presence or absence. Tasks with a very limited instructional scenario may very well be rich in context, and others, like the task shown in Figure 1, Panel B, may have an instructional scenario that contributes little to the knowledge that the task assesses.

We recognize that both the specific ways that context functions and also their occurrences could vary for a different sets of MKT tasks. This analysis represents only a snapshot of possible context types, the ways in which they are hypothesized to function, and variation of each type. While we have no evidence that particular patterns or lack of patterns would generalize to other measurement situations, we do have evidence from a related study that the patterns are similar when looking at comparable measures in other subjects (Phelps, Howell, & Kirui, 2015).

In conclusion, we think that the approach to describing teaching context in this paper is likely to be useful in better understanding and evaluating MKT task design and as a basis for designing studies that systematically vary the use of context to further explore how those designs function. The analysis illuminates the relation between the types of knowledge a test taker uses in answering a task, the design of assessment tasks including relevant features of context, and the MKT domain assessed by the task. Explicit attention to the role that context plays in the design of MKT assessments offers the potential to better understand not only the content knowledge that is assessed in particular tasks, but also to begin to develop a theory of how teaching context itself may serve to define this knowledge.

# References

Ball, D. L., & Bass, H. (2002, May). Toward a practice-based theory of mathematical knowledge for teaching. In *Proceedings of the 2002 annual meeting of the Canadian Mathematics Education Study Group* (pp. 3–14).

Ball, D. L., Thames, M. H., & Phelps, G. (2008). Content knowledge for teaching: What makes it special? *Journal of Teacher Education, 59*(5), 389–407. doi: 10.1177/0022487108324554

Cohen, D. K., Raudenbush, S. W., & Ball, D. L. (2003). Resources, instruction, and research. *Educational Evaluation and Policy Analysis, 25*(2), 119–142. doi: 10.3102/01623737025002119

Gitomer, D., Phelps, G., Weren, B., Howell, H., & Croft, A. (2014). Evidence on the validity of content knowledge for teaching assessments. In T. Kane, K. Kerr, & R. Pianta (Eds.), *Designing teacher evaluation systems: New guidance from the Measures of Effective Teaching project* (pp. 493–528). San Francisco: Jossey-Bass.

Hawkins, D. (1974). I, thou, and it. In D. Hawkins (Ed.), *The informed vision: Essays on learning and human nature* (pp. 48–62). *New York: Agathon* Press.

Herbst, P., & Kosko, K. (2014). Mathematical knowledge for teaching and its specificity to high school geometry instruction. In J. Lo, K. Leatham, & L. Van Zoest (Eds.), *Research trends in mathematics teacher education* (pp. 23–46). New York, NY: Springer.

Hill, H. C., Ball, D. L., & Schilling, S. G. (2008). Unpacking pedagogical content knowledge: Conceptualizing and measuring teachers' topic-specific knowledge of students. *Journal for Research in Mathematics Education, 39*(4), 372–400. Retrieved from http://www.jstor.org/stable/40539304

Hill, H. C., Schilling, S. G., & Ball, D. L. (2004). Developing measures of teachers' mathematics knowledge for teaching. *The Elementary School Journal, 105*(1), 11–30. doi: 10.1086/428763

Hill, H. C., Sleep, L., Lewis, J., & Ball, D. L. (2007). Assessing teachers' mathematical knowledge: What knowledge matters. In F. Lester (Ed.), *Second handbook of research on mathematics teaching and learning* (pp. 111–156). Charlotte, NC: Information Age Publishing.

Howell, H., Phelps, G., Croft, A., Kirui, D., & Gitomer, D. (2013). *Cognitive interviews as a tool for investigating the validity of content knowledge for teaching assessments* (ETS Research Report 13–19). Princeton, NJ: Educational Testing Service.

Kane, M. T. (2006). Validation. In R. L. Brennan (Ed.), *Educational measurement* (4th ed., pp. 17–64). Westport, CT: Praeger.

Kersting, N. (2008). Using video clips of mathematics classroom instruction as item prompts to measure teachers' knowledge of teaching mathematics. *Educational and Psychological Measurement, 68*(5), 845–861. doi: 10.1177/0013164407313369

Krauss, S., Baumert, J., & Blum, W. (2008). Secondary mathematics teachers' pedagogical content knowledge and content knowledge: Validation of the COACTIV constructs. *ZDM Mathematics Education, 40*(5), 873–892. doi: 10.1007/s11858-008-0141-9

McCrory, R., Floden, R., Ferrini-Mundy, J., Reckase, M. D., & Senk, S. L. (2012). Knowledge of algebra for teaching: A framework of knowledge and practices. *Journal for Research in Mathematics Education, 43*(5), 584–615. doi: 10.5951/jresematheduc.43.5.0584

McDonald, J. (1992). *Teaching: Making sense of an uncertain craft.* New York: Teachers College Press.

Messick, S. (1989). Validity. In R. L. Linn (Ed.), *Educational measurement* (3rd ed., pp. 13–103). New York: American Council on Education/Macmillan.

Mislevy, R., & Haertel, G. (2006). Implications of Evidence-Centered Design for Educational Testing. *Educational Testing: Issues and Practice, 25*(4), 6–20. doi: 10.1111/j.1745-3992.2006.00075.x

Phelps, G., Howell, H., & Kirui, D. (2015). *Situating the assessment of content knowledge for teaching in the context of teaching practice* (ETS Research Report). Manuscript submitted for publication.

Phelps, G., Howell, H., Schilling, S. G., & Liu, S. (2015). *Exploring differences in mathematical knowledge for teaching for prospective and practicing teachers.* Manuscript in preparation. *Journal of Mathematics Teacher Education.*

Phelps, G., Weren, B., Croft, A., & Gitomer, D. (2014*). Developing content knowledge for teaching assessments for the Measures of Effective Teaching Study* (ETS Research Report 14-33). Princeton, NJ: Educational Testing Service. doi: 10.1002/ets2.12031

Schwab, J. J. (1978). The practical: A language for curriculum. In I. Westbury & N. Wilkof (Eds.), *Science, curriculum, and liberal education: Selected essays* (pp. 287–321). Chicago: University of Chicago Press.

Strauss, A., & Corbin, J. (1990). *Basics of qualitative research: Grounded theory procedures and techniques.* Newbury Park, CA: Sage Publications.

Tatto, M. T., Schwille, J., Senk, S., Ingvarson, L., Peck, R., & Rowley, G. (2008). *Teacher education and development study in mathematics (TEDS-M): Conceptual framework.* Lansing, MI: Teacher Education and Development International Study Center.

Thompson, P. W. (2015). Researching mathematical meanings for teaching. In English, L., & Kirshner, D. (Eds.), *Third Handbook of International Research in Mathematics Education* (pp. 435–461). London: Taylor and Francis.

Turner, F. and Rowland, T. (2008) *The knowledge quartet: a means of developing and deepening mathematical knowledge in teaching.* Mathematics Knowledge in Teaching Seminar Series: Developing and Deepening Mathematical Knowledge in Teaching (Seminar 5), Loughborough University, Loughborough.

# Interview Prompts to Uncover Mathematical Knowledge for Teaching:
## Focus on Providing Written Feedback

Yeon Kim (yeonkim10@silla.ac.kr)[1]
Silla University, South Korea

**Abstract:** One area of study that has been gathering enthusiastic attention and interest is mathematical knowledge for teaching (MKT). How to research MKT, however, is still unsettled despite the plethora of unexamined areas of practice. As one of ways to unearth and measure MKT, this study uses interview prompts designed to providing written feedback, as a target area of practice. This study specifies in what ways the interview prompts are used in order to provide a comprehensive method to researching MKT. From interviews across professional communities with different kinds of mathematical expertise, the author develops a conceptual model based on the tasks of teaching, elements of pedagogical context, and domains of MKT. This model provides the fluent character of proficient MKT decision making in teaching practice and explains key features of the design of prompts for investigating and measuring MKT. From the analysis, two claims emerged: bidirectional approaches to investigating MKT and continuous and spontaneous aspects of MKT.

*Keywords:* mathematical knowledge for teaching, providing written feedback, pedagogical context, measurement, conceptual study.

## Introduction

Mathematical knowledge that is specifically connected to the work of teaching has been investigated empirically and theoretically, leading to a significant progression of its conceptualization. In particular, Ball, Thames, and Phelps (2008) developed a practice-based theory of content knowledge for teaching and introduced *mathematical knowledge for teaching* (MKT) as the mathematical knowledge needed to carry out the work of teaching mathematics. Such knowledge has been studied by examining teaching practice, such as job analysis (Ball & Bass, 2003) and by developing its measurement (Hill, Schilling, & Ball, 2004). MKT has been identified as important in teaching mathematics (Lewis & Blunk, 2012) and in student achievement (Hill, Rowan, & Ball, 2005; Rockoff, Jacob, Kane, & Staiger, 2011).

Research on MKT has been conducted with practices of teaching that are prominent in mathematics classrooms, though not all practices. Ma (1999), for example, used four items with some exceptional tasks of teaching and mathematical demands, but did not include all topics and practices in and from teaching. Items developed by the University of Michigan include

---

[1] This research was supported in part by a grant from the National Science Foundation (Grants DRL-1008317). The author thanks members of the Mathematics Teaching and Learning to Teach Project for their help in developing aspects of this article. Errors are the responsibility of the author.

An earlier version of this research was presented with Mark Hoover at the 2015 KSME International Conference on Mathematics Education and at the 2015 annual meeting of the American Educational Research Association.

substantial tasks of teaching, but not a sufficiently organized approach to mathematical topics and teaching practice.[1] Ball et al. (2008) emphasized a practice-based approach to study content knowledge for teaching. While introducing their conceptualization about such special knowledge, they focused on several and seminal practices of teaching, such as presenting mathematical ideas, responding to students' "why" questions, choosing and developing useable definitions. Nonetheless, the conceptualization has not yet been broadened to explore all practices or all topics in terms of MKT. In other words, these studies do not cover the extensive terrain of mathematical demands in teaching across contexts. For sustainable and elaborated development of the study of MKT, the mathematical demands entailed in teaching now needs to be explored systemically using a clear and comprehensive map of the practice of teaching, mathematics, features of learners, and national or international curriculum or grade levels.

How MKT is studied has yet to receive sufficient attention. Ball and her colleagues studied MKT with a set of analytic tools they developed, using their wide range of experiences and disciplinary backgrounds, for coordinating mathematical and pedagogical perspectives (Ball et al., 2008; Thames, 2009). However, their experiences have not been shared in terms of researching MKT. A robust, reliable, and consistent study of MKT can be expected with an appropriate and good method. Substantial areas of mathematical demands in different practices of teaching still call for investigations from many researchers. Like the collaborative work on the Human Genome Project (International Human Genome Sequencing Consortium, 2001; Naidoo, Pawitan, Soong, Cooper, & Ku, 2011), research on MKT would need cooperative work for elaborate and systematic conceptualization. A major prerequisite for such collective work is the identification of a method to research MKT. If relevant methods of studying MKT are specified, research of MKT will be powerfully advanced. To systemically research MKT, a comprehensive method needs to be specified.

To address the problems, the current study focuses on both using interviews as a comprehensive way to study and measure MKT and researching mathematical demands in providing written feedback, which is an unexamined teaching practice. It does so by building on lessons from the interview prompts used in Ball (1988) and Ma (1999) and on items used to measure MKT (Ball, Bass, & Hill, 2004). Specifically, this paper explores the ways in which interview prompts are developed and used to provide the content and character of MKT, and what, through the use of such interview prompts, might be learned. Particularly, what is entailed mathematically and pedagogically in providing written feedback?

## Measuring MKT

Based on Shulman's (1986) notion of pedagogical content knowledge, Ball et al. (2008) specified several subdomains within pedagogical content knowledge (knowledge of content and students [KCS], knowledge of content and teaching [KCT], and knowledge of content and curriculum [KCC]). Further, they identified an important subdomain of "pure" content knowledge unique to the work of teaching, specialized content knowledge [SCK]. SCK is "distinct from the common content knowledge [CCK] needed by teachers and non-teachers alike" (p. 389). They argued that, "teachers' opportunities to learn mathematics for teaching could be better tuned if we could identify those types more clearly" (p. 399).

In earlier research, Ball (1988) created problems that were used in interviews with prospective teachers to explore their proficiency in meeting the mathematical demands of teaching. Ma (1999) extended the use of these interview prompts with practicing teachers in China. Their findings and arguments were critically informed by the carefully designed mathematical teaching problems represented in the interview prompts. Such mathematical teaching problems are, in short, tools for uncovering mathematics entailed in teaching. Based on analyses of teaching, Ball and Hill's *Learning Mathematics for Teaching* (LMT) project developed multiple-choice items to measure MKT (Ball, Hill, & Bass, 2005; Hill, Ball, & Schilling, 2008; Hill, Schilling, & Ball, 2004). Following the same approach, Herbst and Kosko (2014) developed MKT items for secondary geometry teaching.

To identify key design features of MKT items, my colleagues and I analyzed representations of teaching embedded in MKT assessment items.[2] We found that the reasoning of item doers was shaped by elements of pedagogical context: student background, teaching purposes, and classroom artifacts, as represented in the teaching situations described in the items. The pedagogical context given in items created situations that required doing mathematics while holding onto a specific pedagogical purpose; the context gave the tasks an MKT, rather than a disciplinary mathematics character. We also found that competent performance on the items depended on reasoning that used, in integral ways, features of the pedagogical context. This conceptualization of pedagogical context offers further ideas about how to design items in ways that measure mathematical knowledge that is fundamentally linked to teaching and not simply disciplinary knowledge that remains remote from effective teaching and learning.

The current study uses open-format prompts designed to investigate the work of teaching that is not readily evident from video of teaching (which was used to design the pedagogical contexts and mathematical tasks of teaching in LMT items). The pedagogical focus of the sub-study reported here is the task of providing written student feedback. This task of teaching requires consideration of the mathematics problem, students' responses to it, and decisions about how best to guide students toward the intended purpose of the problem. It does not mean to suggest that this task of teaching cannot be studied observationally or measured using multiple-choice items. Rather, it needs a means of uncovering more about what was involved in the task and have thus explored the use of designed interview prompts to expand the investigation of MKT. Hence, the current study investigates the following questions: What are the ways in which interview prompts can be developed and used to provide insight into the content and character of MKT? What is entailed, mathematically and pedagogically, in providing written feedback?

## Providing Written Feedback

The purpose of feedback is to reduce discrepancies between what is understood and what is intended to be understood (Hattie & Timperley, 2007). Feedback aims to modify a student's thinking to improve his or her learning (Shute, 2008). Apparently, major and complex tasks of teaching involved in the providing of feedback include the sizing up of students' current understandings, directing students to a desired goal, and deciding what information will be provided and in what ways. The complexity of providing feedback is evidenced by a large body of research that encompasses many conflicting findings and no consistent pattern of results. Nevertheless, Hattie and Timperley (2007) reviewed research on feedback and clarified that

> … feedback needs to be clear, purposeful, meaningful, and compatible with students' prior knowledge and to provide logical connections. It also needs to prompt active information processing on the part of learners, have low task complexity, relate to specific and clear goals, and provide little threat to the person at the self level. (p. 104)

Shute (2008) made a similar claim that feedback should be nonevaluative, supportive, timely, and specific. Based on a literature review of feedback that concentrates on specific information to a student about a particular response to a task, Shute (2008) suggested that feedback should address specific features of the student's work with a description of the what, how, and why of a given problem and suggest improvements that the student can manage. This clear feedback would reduce uncertainty in relation to how well students are performing on a problem and what needs to be accomplished to attain the goals.

Feedback, in fact, can promote learning if it is received mindfully (Bangert-Drowns, Kulik, Kulik, & Morgan, 1991). Yeager et al. (2014) emphasized that trust is crucial for successfully delivering written feedback, adding that "mistrust can lead people to view critical feedback as a sign of the evaluator's indifference, antipathy, or bias, leading them to dismiss rather than accept it" (p. 805). Harber et al. (2012) pointed out that many teachers tend to overpraise students for mediocre performance, particularly those subject to negative stereotypes, in order to enhance student self-esteem. However, unlike the teachers' intention, this overpraising in written feedback hinders the development of trust and reinforces minority students' perceptions that they are being viewed stereotypically (Croft & Schmader, 2012; Harber et al., 2012). Feedback should be based on a reflection of a teacher's high standards, not his or her bias, and offer students both an assurance about their potential to reach such high standards and the resources with which they might do so (Yeager et al., 2014). However, written feedback is more unbiased and objective than face-to-face feedback (Kluger & DeNisi, 1996; Shute, 2008).

Critical review about feedback, as specified previously, concentrates on features of feedback that function well to improve students' learning. However, it does not show what tasks of teaching are organically entailed in providing written feedback. The work of teaching includes the activities in which teachers engage and the responsibilities they have to teach content (Ball & Forzani, 2009). The work of teaching occurs in the dynamics initiated by a teacher before, during, and after instruction so as to help students learn the content (Sleep, 2009). Furthermore, in teaching mathematics, specific mathematical goals are critical throughout instruction (Schoenfeld, 2011). Recognizing the varied understandings of the mathematics, for example, probing so as to see what students do or do not know, and responding in ways that address students' errors and that build on student understanding help move students toward goals in instruction. Evidently, providing written feedback is purposeful work: control or modification of a student's ways of thinking or answers to learn something that a teacher desires. Providing written feedback assumes that a teacher gets and synthetically and analytically evaluates signals from students' responses or reasoning, tasks or problems given, and pedagogical situations enacted and created by the teachers and the students. Then, the teacher makes a decision to put forward certain comments to the individual student or to the whole group of students. This description offers a simple glimpse of a feedback loop because it is important to understand the circumstances that result in the differential outcomes and responses of students (Hattie & Timperley, 2007). This loop could be repeated in a lesson until the teacher is satisfied with the control or with the students' revised ideas or answers. Each teacher might have different perspectives and rationales on what feedback works well and what should be highlighted in which situations. However, providing written feedback can be specified with sub-tasks of teaching that are responsive and intrinsic to teaching and entail professional norms of teaching.

**Design and Analysis of the Prompts and Interviews**

**Design of the Interview Prompts**

Prompt development started with initial descriptions of the high-leverage practices given in the professionally vetted version on the *TeachingWorks* website.[3] The 19 high-leverage practices are tasks central to teaching, which are expected to increase the likelihood for students' learning across a broad range of subject areas, grade levels, and teaching contexts. These high-leverage practices are warranted by research evidence, wisdom of practice, and logic and were developed through many discussions with researchers, expert teachers, faculty members of teacher education, and education policy makers. Of the 19 practices, 16 are crucial in teaching mathematics, as shown in Figure 1. The figure illustrates both apparent interrelationships of the different practices of teaching and gives an overall picture of the practices of teaching mathematics. In other words, Figure 1 represents a comprehensive map of the practice of teaching, which the current study uses to study the mathematical demands entailed in teaching.

Figure 1 shows that providing oral and written feedback is based on assessing students' knowledge. Such an assessment includes selecting and using particular methods to check understanding and monitor student learning and composing, selecting, interpreting, and using information from methods of summative assessment. Providing feedback is influenced by appraising, choosing, and modifying tasks and texts for a specific learning goal. Furthermore, providing feedback may consist of leading a whole-class discussion and setting up and managing small group work. In other words, providing feedback is related to other practices in teaching

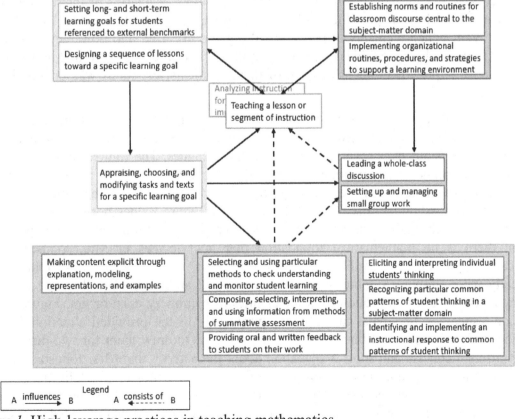

*Figure 1*. High-leverage practices in teaching mathematics

mathematics. Although providing feedback is typical work in teaching, it is crucial to approach this work from a holistic perspective about teaching practice.

*Teachingworks* explains the providing of oral and written feedback to students on their work as follows:

> Effective feedback helps focus students' attention on specific qualities of their work; it highlights areas needing improvement; and delineates ways to improve. Good feedback is specific, not overwhelming in scope, and focused on the academic task, and supports students' perceptions of their own capability. Giving skillful feedback requires the teacher to make strategic choices about the frequency, method, and content of feedback and to communicate in ways that are understandable by students.

This explanation highlights both feedback about a task and processing of a task, which Hattie and Timperley (2007) differentiated. The explanation also emphasizes the need to attend to the student's work and to give specific comments about that work to advance it in a way that they can truly accept and understand (Shute, 2008; Yeager et al., 2014).

To access potentially tacit knowledge and reasoning about what is involved in providing feedback, situations were generated using realistic pedagogical contexts focused on written feedback. On this point, Common Core State Standards were used as a second map to select mathematical topics and determine students' backgrounds. These standards outline a clear set of mathematical skills and knowledge that students will learn in a more organized way both during the school year and across grades. The standards' coherent composition and specific statements offer clear features of mathematical topics and possible information about students' backgrounds that were necessary in creating scenarios for interview prompts. Scenarios of instruction, which would require feedback, and possible student work were sketched. Different elements were considered as well: which elements of pedagogical context are used and how they flow in the scenarios and what mathematical ideas, reasoning, and practices are addressed and how they can be unveiled and interpreted. In deciding each of these elements, the focus was on keeping the situation realistic.

Figure 2 and Figure 3 offer an example that sheds light on the transformation of one initial interview prompt into one used in interviews. In fact, the transformation took place over the course of ten revisions over several meetings with colleagues. The figures illustrate some of the challenges that can occur in developing interview prompts. Both prompts are about providing written feedback as a major practice of teaching, lines of symmetry as a mathematical topic, and fourth-grade students as student background. Nonetheless, Figure 2 does not present a teaching purpose for the instruction; different triangles are used repeatedly without being developed in the instruction; there is no mathematical foundation that interviewees or Mrs. Johnson in the scenario can use to provide feedback; specifically, it is unclear what Mrs. Johnson did pedagogically when drawing symmetry lines on the board; and, at the fourth grade level, the congruency marks offer no indication of whether lines of symmetry exist or not. Figure 3—the final version—specifies first what the teaching purpose is in the provided situation (helping fourth-grade students understand lines of symmetry and how to draw them for two-dimensional figures); it offers a concise and short explanation about the activity that Mrs. Johnson had with her students (using squares and figuring out how to find lines of symmetry); a definition of a line of symmetry is given by the teacher in the scenario, which can be used by the interviewee to provide written feedback in the interview; and student work is provided that includes errors and a partial understanding about lines of symmetry. Phrases in the scenario have been revised repeatedly to create a succinct and unambiguous scenario that requires written feedback.

Three major questions were decided on: What does the interviewee first notice? What

In a lesson about lines of symmetry for triangles with her fourth grade students, Mrs. Johnson handed out the copies of an equilateral triangle, an isosceles triangle, and a scalene triangle, and students had time to try to fold each triangle into matching parts. Mrs. Johnson then explained lines of symmetry for triangles as lines across the triangles such that the triangles can be folded along the line into matching parts. She also drew symmetry lines on the board as shown in the following:

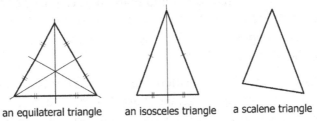

an equilateral triangle   an isosceles triangle   a scalene triangle

Then, to assess her students' understanding, Mrs. Johnson gave students time to individually practice drawing lines of symmetry for triangles. When she was monitoring, she noticed that several students drew lines on the textbook as following:

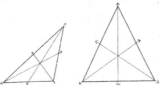

*Figure 2*. Example of initial interview prompt

Mrs. Johnson wanted to help her fourth-grade students understand lines of symmetry and how to draw them for two-dimensional figures. At the beginning of a lesson, she had the students cut out squares and look for ways to fold a square into two matching parts. The students identified two ways: folding it diagonally and folding it from the middle of one side to the middle of the opposite side. They used rulers to draw lines along their folds. Mrs. Johnson explained that these lines are called lines of symmetry and wrote the following definition on chart paper.

> A <u>line of symmetry</u> is a line across a figure such that the figure can be folded along the line into matching parts.

Then, the students did a similar activity with triangles, cutting out triangles and looking for ways to fold each triangle into two matching parts. Here, they noticed that some triangles do not have a line of symmetry.

At the end of this lesson, Mrs. Johnson gave her students a page of practice exercises, where she included three figures that are not triangles and squares. For these figures, one of her students handed in the following:

*Figure 3*. Example of interview prompt used in actual interviews

written feedback does he/she offer? What would he/she do for a fifteen-minute lesson? These three questions aimed to grasp an interviewee's knowledge and reasoning for providing written feedback. Along with the major questions, several minor questions were added to identify the interviewee's rationale behind his or her responses. Figure 4 includes a list of major and minor questions on the left and on the right one page of possible responses for each question, used to prepare interviews for the interviews. Minor questions helped interviewers prepare for the interviews as well, deciding which minor questions to ask depending on the interviewees' responses. The initial prompts were discussed with colleagues who had teaching experience in grades K-9 and who had studied mathematics and mathematics education. Such discussions were meant to help the researcher anticipate responses and revise the prompts based on whether the prompt would activate and support the task of providing feedback. In other words, collective work was initiated with the designing of interview prompts.

For this sub-study of providing written feedback, interviews were conducted for each interview prompt with two practicing mathematicians, two prospective teachers, and three expert teachers.[4] Interviewees were carefully selected and contacted. The mathematicians' responses in the interviews offered insight into how disciplinary thinking functions and what appropriately supporting learners means in instruction. The prospective teachers had no prior regular teaching experience and were attending the School of Education working toward their teaching certifications. They were considered a group that needed more professionality rather than what mathematicians and expert teachers performed in order to investigate mathematical reasoning and knowledge entailed in providing feedback and identify features of MKT in such a task of teaching. To recruit expert teachers, the project team first contacted researchers who were

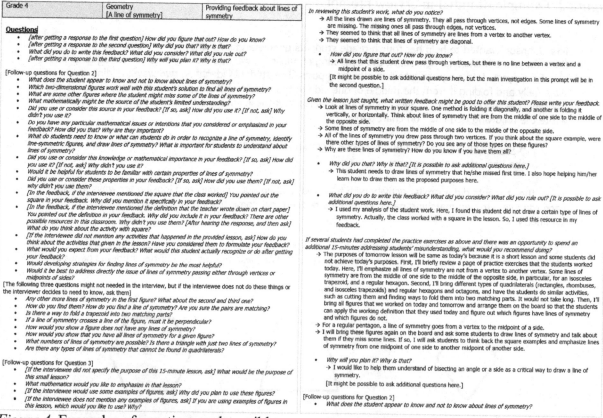

*Figure 4.* Examples of questions and possible responses

leading research about teaching, teachers, or professional development in the United States. Lists of teachers were gathered based on the researchers' recommendations of expert teachers. They were knowledgeable in mathematics as well as mathematical practice in instructional situations and each had more than ten years of teaching experience. The prospective and expert teachers' responses in the interviews provided insight into what professional reasoning runs in instructional contexts and how it plays out. Observing what these three groups of interviewees did was to help the research team perceive and specify clearly both mathematical and pedagogical reasoning that is engaged in the work of teaching mathematics. Furthermore, interviews were conducted by several interviewers. Each interview was conducted with one interviewee by one interviewer. As previously specified, the current study was collaborative group work, and the collectively performed interviews helped the project team to deepen their insights about mathematical demands entailed in providing written feedback.

### Vignette of One Interview and Its Analysis

Each interview was analyzed separately, and then all interviews and analyses were synthetically and analytically probed to investigate and characterize the mathematical demands of providing written feedback. The next section provides a more detailed explanation about the analysis, but this section shows a vignette of one interview and its analysis to illustrate a way of examining and researching MKT.

One interviewee had taught elementary students for approximately 25 years and, for the past 12 years, had worked as a math coach at the K-8th grade level. She also received an M.A. in teaching and learning mathematics. In the interview with the prompts shown in Figure 5, the interviewee immediately recognized that the student in the prompt used a system to make the list. The interviewee noted that the student changed the last two over and over again as well as reordered the first two letters. She was also curious about what made the student decide that the list was complete. The interviewee then wrote down as feedback: "I see you are using a system to find all of the orders. I noticed that your orders are coming in pairs: where` the last two groups switch order before the first two groups switch order. How would you know you have all the

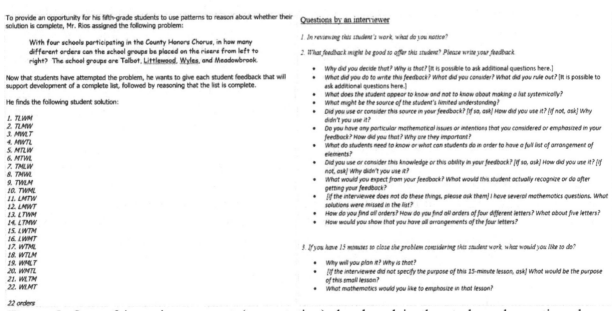

*Figure 5.* One of interview prompt (permutation) developed in the study and questions by an interviewer

choices for arranging the classes?" The interviewee emphasized the importance of the student's decision point about when he or she finished making a whole list. For an additional 15 minute-lesson, the interviewee planned to start the lesson by asking how many orders there were with W first and then L, for she expected the student to recognize there were six orders each with W and L, but only four orders with M. The interviewee would have asked, "Should there be six with M?" and "What makes you think there should be six?" And the interviewee would then let the student look at what he or she had gotten correct and what he or she had missed. Her reason for planning the lesson in this way was that there was a mistake in the middle of the list, but, at the bottom of the list, orders with W and L were listed very systemically. Because the student seemed to have a better sense of making a list with L and W, the interviewee expected that the student could figure out the regularity in orders with W and L and apply it to have a complete list.

The interviewee's responses were analyzed in terms of the mathematical work of teaching engaged in providing written feedback. First, there were shown several sub-tasks of teaching related to providing written feedback. (i) Identifying a mathematical feature that needs to be emphasized in the situation—the interviewee immediately recognized the main mathematical topic of this situation was having a complete list to show all the orders of the four schools. Throughout the interview, although the interviewee used different terms, such as "organization," "regularity," "system," and "efficient," she emphasized a systematic way to make a complete list with any number of cases from patterns involved in this situation. (ii) Playing with ways in which the student's work is produced from the given problem—the interviewee investigated the student's work and recognized that there is a systemic way to make the list. The interviewee also acknowledged that this student had some sense about permutation, but not enough to make a whole list. The interviewee also reflected on what this student's orders would be if there were two or three schools because the student might be able to make a generalizable list from the smaller example. (iii) Formulating critical feedback—rather than correcting the student's work or directly specifying what to do next, the interviewee wanted to give something that made the student think more broadly and deeply. She also clarified two elements in her feedback. One was what the teacher noticed in the student's work and the other was a question to help the student reflect on his or her work and reasoning and investigate and find a way to build a complete list.

The mathematical knowledge and reasoning entailed in providing written feedback were also analyzed. The interviewee was very confident that the most important thing in this situation was letting the student recognize whether or not he or she had all the orders and helping the student understand a generalizable way to make a full list from patterns involved in this situation. Her responses and comments were consistently geared toward the student. However, the interviewee did not ascertain what orders the student had missed until the interviewer asked about it. At the beginning of the interview, she briefly recognized that the orders starting with L and W were systematically listed, but those starting with T and M were not. Rather than just fixing the student's work, the interviewee put more value on mathematical generalization and working with the student's reasoning toward it. It was also interesting that the interviewee recognized completeness as a critical issue in this situation. However, the interviewee did not seem to know the total number of permutations using combinatorics (4!), but she could explain her reasoning to get a full list of orders by using a tree diagram.

The final thing examined in the analysis of this interview was distinctive mathematical knowledge and reasoning for teaching. The interviewee recognized that the last entries in the student's list were mathematically well organized, but she did not use it directly in her feedback. Rather, she focused on reasoning related to mathematical completeness and mathematical generalization to approach this situation, and she seemed to believe that mathematics instruction needed to concentrate on them rather than on merely getting correct answers. This might be the reason why she did not rush to solve the provided problem. She investigated the student's work and tried to specify possible reasoning that the student might have used in the work. Her feedback was specific enough to make the student think about what he or she had done. In general, her sense about providing written feedback is to give the student a chance to reflect on his or her work based on what the teacher noticed in the work rather than correcting the work.

## Analysis of the Interviews

Each interview was summarized and analyzed carefully by the interviewer using the following questions: How did the interviewee respond to each question of the prompt? What was the rationale for each response? What is the mathematical work of teaching as it relates to this teaching situation as suggested through this interview? What mathematical knowledge and reasoning are entailed in this work as suggested through this interview? What does this interview suggest about what might be distinctive about mathematical knowledge and reasoning for teaching? During this time, we continually revised the interview prompt so as to create with clear language realistic situations.

The summary notes were major resources for analyzing the data. During research meetings, the interviewers shared their experiences in the interviews and discussed major interests of the current research using the questions introduced in the previous paragraph. Figure 6 and Figure 7 show two summary notes, one from an interview conducted with an expert teacher and one from an interview conducted with a mathematician. One of the predominant features shown in the interview with the expert teacher was the teacher figuring out and using information from the pedagogical context. The mathematician, on the other hand, merely checked the correctness of the answer, neglecting to dig in and create feedback in the provided situation. In other words, the student work was mathematically analyzed by the expert teacher using pedagogical considerations of the teaching purpose; her analysis was greatly used to formulate feedback. In that situation, the bottom orders in the student's list constitute a critical clue that formulates feedback according to the given teaching purpose, "use patterns to reason about whether their solution is complete." Checking the correctness of the answer is insufficient to formulate feedback. This task of teaching requires complicated sub-tasks of teaching that are mathematically and pedagogically delicate and that also require comprehensive analyzing and decision making.

This research, again, does not aim to characterize the three different groups of interviewees in the context of providing written feedback. However, interviews with prospective teachers presented challenges to approaching the given situation and making a decision. Their reasoning, as revealed in the interviews, was inconsistent. Some prospective teachers only dug into the problem used in the given situation, just like learners would. Some prospective teachers solved the problem while others did not. Most of them recognized some information about the pedagogical context. Some of them used it to formulate feedback while others did not. It seems clear that one thing that is critical in providing feedback is seizing and analyzing mathematical and pedagogical information simultaneously, immediately and synthetically. It is a real challenge, however, to gather such information. Furthermore, interviewees in each group did not always respond to the prompts in the same way. In other words, mathematical, pedagogical, partial, analytical, and synthetic tendencies and professionality for figuring out the provided situation and making a decision were different across participants.

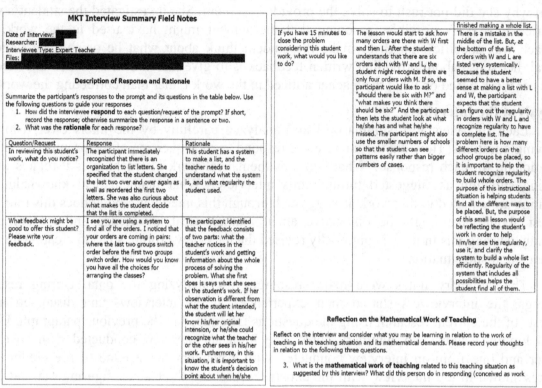

**MKT Interview Summary Field Notes**

Date of Interview:
Researcher:
Interviewee Type: Expert Teacher
Files:

**Description of Response and Rationale**

Summarize the participant's responses to the prompt and its questions in the table below. Use the following questions to guide your responses
1. How did the interviewee **respond** to each question/request of the prompt? If short, record the response; otherwise summarize the response in a sentence or two.
2. What was the **rationale** for each response?

| Question/Request | Response | Rationale |
|---|---|---|
| In reviewing this student's work, what do you notice? | The participant immediately recognized that there is an organization to list letters. She specified that the student changed the last two over and over again as well as reordered the first two letters. She was also curious about what makes the student decide that the list is completed. | This student has a system to make a list, and the teacher needs to understand what the system is, and what regularity the student used. |
| What feedback might be good to offer this student? Please write your feedback. | I see you are using a system to find all of the orders. I noticed that your orders are coming in pairs: where the last two groups switch order before the first two groups switch order. How would you know you have all the choices for arranging the classes? | The participant identified that the feedback consists of two parts: what the teacher notices in the student's work and getting information about the whole process of solving the problem. What she first does is says what she sees in the student's work. If her observation is different from what the student intended, the student will let her know his/her original intension, or he/she could recognize what the teacher or the other sees in his/her work. Furthermore, in this situation, it is important to know the student's decision point about when he/she |
| If you have 15 minutes to close the problem considering this student work, what would you like to do? | The lesson would start to ask how many orders are there with W first and then L. After the student understands that there are six orders each with W and L, the student might recognize there are only four orders with M. If so, the participant would like to ask "should there be six with M?" and "what makes you think there should be six?" And the participant then lets the student look at what he/she has and what he/she missed. The participant might also use the smaller numbers of schools so that the student can see patterns easily rather than bigger numbers of cases. | finished making a whole list. There is a mistake in the middle of the list. But, at the bottom of the list, orders with W and L are listed very systemically. Because the student seemed to have a better sense at making a list with L and W, the participant expects that the student can figure out the regularity in orders with W and L and recognize regularity to have a complete list. The problem here is how many different orders can the school groups be placed, so it is important to help the student recognize regularity to build whole orders. The purpose of this instructional situation is helping students find all the different ways to be placed. But, the purpose of this small lesson would be reflecting the student's work in order to help him/her see the regularity, use it, and clarify the system to build a whole list efficiently. Regularity of the system that includes all possibilities helps the student find all of them. |

**Reflection on the Mathematical Work of Teaching**

Reflect on the interview and consider what you may be learning in relation to the work of teaching in the teaching situation and its mathematical demands. Please record your thoughts in relation to the following three questions.

3. What is the **mathematical work of teaching** related to this teaching situation as suggested by this interview? What did this person do in responding (conceived as work of teaching)? What did the person's work suggest about the mathematical work of teaching? (one paragraph)

(i) Identifying a mathematical feature that needs to be emphasized in the situation: The participant immediately recognized the main mathematical topic of this situation is having a complete list to show all orders of four schools. Throughout the interview, although the participant used different terms, such as organization, regularity, system, and efficient, she emphasized a systematic way to make a complete list with any number of cases from patterns involved in this situation. (ii) Playing with ways in which the student's work is produced from the given problem: the participant investigated the student's work and recognized that there is a systemic way the student used to make a list. The participant also acknowledged that this student had some sense about permutation, but not enough to make a whole list. The participant also reflected that if there are two or three schools, what this student's orders would be because the student might be able to make a generalizable list. However, it does not seem that this participant critically use the possible sources of the student's misunderstanding. (iii) Formulating critical feedback: rather than correcting the student's work or directly specifying what to do next, the participant would like to give something that makes the student think more broadly and deeply. She also clarified the two elements as feedback. One is what the teacher noticed in the student's work and the other is a question to help the student reflect on his/her work and reasoning and investigate and find a way to build a complete list.

4. What **mathematical knowledge and reasoning** are entailed in this work as suggested by this interview? (one paragraph)

The participant was very confident that the most important thing in this situation is letting the student recognize whether or not he/she had all orders and helping the student understand a generalizable way to make a full list from patterns involved in this situation. Her responses and comments were consistently geared toward them. However, the participant did not investigate exactly what orders the student missed until before the interviewer asked it. At the beginning of the interview, she briefly recognized that the orders starting with L and W are systematically listed, but not those starting with T and M. Rather than just fixing the student's work, the participant put more value on mathematical generalization and working with the student's reasoning toward it. It was also interesting that the participant recognized completeness as a critical issue in this situation. However, the participant did not seem to know the total number of combinations based on combinatorics (4!) very well, but she could explain her reasoning to get a full list of orders by using a tree diagram.

5. What does this interview suggest about what might be **distinctive** about mathematical knowledge and reasoning for teaching? (This might be about something the interviewee evidenced, or about something noticeably absent or different from what would be expected.) (one paragraph)

The participant exactly recognized that the last entries in the student's list are mathematically well organized, and she did not use it directly in her feedback. Rather, she focused on reasoning related to mathematical completeness and mathematical generalization to approach this situation, and she seemed to believe that mathematics instruction needs to concentrate on them rather than just getting correct answers. It might be a reason why she did not rush to solve the provided problem. She investigated the student's work and tried to specify possible reasoning that the student might have been involved in the work. Her feedback was specific enough to initiate the student to think what he/she had done. In general, her sense about providing feedback is giving a chance to the student to reflect on his/her work based on what the teacher noticed in the work rather than correcting the work. However, she did not mention 4! as a way or an answer to figure out the total number of orders, and hesitated to specify a rule for it and she did not seemed to be confident in that moment.

*Figure 6.* Example of summary notes of one expert teacher

**MKT Interview Summary Field Notes**

Date of Interview: █████
Researcher: █████
Interviewee Type: Mathematician
Files: █████

**Description of Response and Rationale**

Summarize the participant's responses to the prompt and its questions in the table below. Use the following questions to guide your responses
1. How did the interviewee **respond** to each question/request of the prompt? If short, record the response; otherwise summarize the response in a sentence or two.
2. What was the **rationale** for each response?

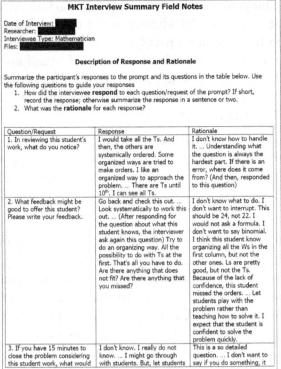

| Question/Request | Response | Rationale |
|---|---|---|
| 1. In reviewing this student's work, what do you notice? | I would take all the Ts. And then, the others are systemically ordered. Some organized ways are tried to make orders. I like an organized way to approach the problem. ... There are Ts until 10th. I can see all Ts. | I don't know how to handle it. ... Understanding what the question is always the hardest part. If there is an error, where does it come from? (And then, responded to this question) |
| 2. What feedback might be good to offer this student? Please write your feedback. | Go back and check this out. ... Look systematically to work this out. ... (After responding for the question about what this student knows, the interviewer ask again this question) Try to do an organizing way. All the possibility to do with Ts at the first. That's all you have to do. Are there anything that does not fit? Are there anything that you missed? | I don't know what to do. I don't want to interrupt. This should be 24, not 22. I would not ask a formula. I don't want to say binomial. I think this student know organizing all the Ws in the first column, but not the other ones. Ls are pretty good, but not the Ts. Because of the lack of confidence, this student missed the orders. ... Let students play with the problem rather than teaching how to solve it. I expect that the student is confident to solve the problem quickly. |
| 3. If you have 15 minutes to close the problem considering this student work, what would you like to do? | I don't know. I really do not know. ... I might go through with students. But, let students do this problem and argue each other. That would be wonderful. ... I might have some conversation with students, but I don't know. | This is a so detailed question. ... I don't want to say if you do something, it would be come up. ... I don't know because I have never been there. I don't have any strong opinion. ... I would do with similar problems. I mean examples are wonderful. (The interview asks "Do you have any particular example?") No. I don't have any now. (The interviewer asks "Would you use complex or easy one?") I don't know. |

**Reflection on the Mathematical Work of Teaching**

Reflect on the interview and consider what you may be learning in relation to the work of teaching in the teaching situation and its mathematical demands. Please record your thoughts in relation to the following three questions.

3. What is the **mathematical work of teaching** related to this teaching situation as suggested by this interview? What did this person do in responding (conceived as work of teaching)? What did the person's work suggest about the mathematical work of teaching? (one paragraph)

The interviewee seems to immediately recognize 22 orders is a wrong answer. However, analyzing student work or creating specific feedback were not shown. He strongly believed the power of students' argument rather than teaching particularly how to solve the problem, but did not provide any guidance to help the student move forward in his/her thinking. The teacher must recognize the error that the student is making. They must next be able to attribute the error to the student's thinking or reasoning. They must then be able to provide feedback that addresses the error in thinking or reasoning in a way that is actionable by the student. Several of these pieces were missing in this participant's response. For example, in his feedback, this participant asked the student to go back and check again, but did not provide any guidance to help the student move forward in their thinking.

4. What **mathematical knowledge and reasoning** are entailed in this work as suggested by this interview? (one paragraph)

The interviewee thought that 22 orders was incorrect and the orders were not well organized in the top ones. However, he did not try to figure out which orders the student was missing. Part of the mathematical work of teaching seems to be finding the correct solution. It seems like the mathematical knowledge and reasoning involved in this work includes being able to determine the understanding underlying a mistake and connect it to the student's broader thinking and reasoning.

5. What does this interview suggest about what might be **distinctive** about mathematical knowledge and reasoning for teaching? (This might be about something the interviewee evidenced, or about something noticeably absent or different from what would be expected.) (one paragraph)

Catching up the teaching purpose in the provided situation, analyzing the student work, noticing the student error and using student work to formulate feedback may be a distinctive feature of mathematical knowledge and reasoning for teaching. Furthermore, this knowledge may contain that the idea about a systemic approach is part of the content that needs to be taught to students. When asked about giving another problem, he agreed using another problems, but did not come up with appropriate problems (easy, difficult, complex, etc.). This seems to suggest that being aware of appropriate additional problems is distinctive about knowledge for teaching.

*Figure 7.* Example of summary notes of one mathematician

Based on investigating the distinctive features of mathematical knowledge and reasoning for teaching as it relates to providing written feedback using the interviews with the three different groups, my colleagues and I were able to build up our understanding of the main features of MKT entailed in providing written feedback. Throughout the analysis and discussion of summary notes, features related to confident and professional reasoning and performance in providing feedback were discovered. The first feature concerned whether or not interviewees had similar teaching experiences to those presented in the situation provided, showing that coherent and logical reasoning exists for approaching and examining provided instructional situations and making a decision based on the teaching purposes embedded in the situation. Confidence about such reasoning was critical as well. The teacher's confidence seemed to determine how well he or she explained the rationales of the analysis and decision making.

The second feature was the explicit recognition that providing feedback aimed to help extend students' understanding by offering specific advice rather than correcting their responses. The appropriateness of specific comments was also considered because comments that were too particular could reduce the chances that students would improve their response. Comments that were too vague offered no help to students.

The third feature was recognizing the instructional purpose in the provided situation and sticking to that purpose while investigating information, formulating feedback and specifying the rationales of such feedback. For example, the teaching purpose in the prompt shown in Figure 5 is to provide "an opportunity for his fifth-grade students to use patterns to reason about whether their solution is complete." This purpose includes pedagogical concerns (e.g., providing an opportunity for fifth-grade students) and mathematical topics (e.g., patterns and mathematical completeness). Immediate recognition of the purpose is prominent at the beginning of approaching the provided situation, and this recognition works critically throughout the providing of feedback.

The fourth feature is analytically and synthetically recognizing information from the pedagogical context. Figuring out and mathematically analyzing student work was critical. This task includes probing the reasoning behind the student work and having reasonable assumptions to explain what gives rise to students' wrong responses. Furthermore, other tasks related to recognizing information included looking for instructional resources, evaluating them to formulate feedback, and deciding which one should be used to create feedback. Identifying mathematical content areas related to the provided mathematical topic was important, too.

Based on the features discovered in the providing of feedback on a task of teaching, my colleagues and I have tried to develop a consistent and logical framework that illustrates proficiency in providing feedback in terms of sub-tasks of teaching. In the process of developing the framework, different domains of MKT used in the particular task emerged. Sub-domains of MKT are categorized by different features of "mathematical knowledge needed to perform the recurrent tasks of teaching mathematics to students" (Ball et al., 2008, p. 399). For example, CCK is the mathematical knowledge and skill used in settings other than teaching, such as "simply calculating an answer or, more generally, correctly solving mathematics problems" (Ball et al., 2008, p. 399). CCK seemed dominant in the mathematician's reasoning in the situation of formulating feedback while SCK, KCT and KCS seemed prevalent in the expert teachers' reasoning. More interestingly, while SCK seemed to be extensively implemented to investigate the mathematics embedded in the provided situation, KCT and KCS seemed to be at work when figuring out the pedagogical context and making a decision regarding the student in the provided situation. CCK seemed to function among SCK and KCS, but primarily garnered attention in identifying that the sub-domains of MKT overlap, through consecutive sub-tasks of teaching.

It was also discovered that different aspects of pedagogical context function across different tasks of teaching. For example, identifying the instructional purpose was closely related to teaching purpose and mathematical topic, and investigating student work was connected to mathematical topic. Student background, such as fifth-grade students, was significant to formulate feedback.

The mathematical work associated with the task of providing written feedback was iteratively analyzed using the interviews as empirical grounding, as explained in this section. The current study used an interactional conceptualization of teaching, as described in Cohen, Raudenbush, and Ball (2003) and a practice-based conceptualization of MKT by Ball and Bass (2003). In other words, rather than trying to describe the MKT held by interviewees or the particular ways in which they reasoned about the prompts, I was trying to use the ways in which they reasoned about the prompts to characterize professionally defensible knowledge and practice. The research questions for the larger study are as follows.

1. What is the mathematical work of teaching in unexamined areas of practice?
   a. areas perhaps not readily studied using video
   b. specific practices not previously studied
   c. key areas with distinctive MKT demands (such as impromptu talk)
2. What mathematical knowledge and reasoning are entailed in this work?
3. What is distinctive about mathematical knowledge and reasoning for teaching?
4. What are key features of the design of prompts and interview methods for uncovering MKT?

The methods of analysis for the study reported here are consistent with the job analysis and conceptualization described by Thames (2009), but are applied to the interview data rather than to video.

## Revision of the Prompts

Based on the analysis of the interview, the prompts were both revised and analyzed. The first major question—what does an interviewee notice first—was written in the document. It was decided, however, that the question should be removed and asked orally. The printed question seemed to force the interviewee to focus on noticing things within the provided situation rather than allowing the interviewee to get started in the given situation and find whatever the interviewee happened to pick up on. Moreover, conducting interviews helps the project team examine the prompts, specifically how providing written feedback is decomposed, what mathematical demands are entailed in each task of teaching, and which elements of pedagogical contexts function in each task of teaching. Although the prompts to investigate MKT were developed by the author and the project team, conducting and analyzing the interviews with the developed prompts helped extend our understanding of MKT as well. The characteristics found by the iterative process are specified in the following section.

## What is Entailed in Providing Written Feedback

From the iterative and synthetic analysis of the interviews and the interview prompts, it was found that skillfully providing students with written feedback requires teachers to draw on purpose and relevant information given in the pedagogical context and flexibly use knowledge resources across different domains of MKT. MKT seems to involve an ability to recognize important and adequate information about pedagogical context in teaching to make a decision about which actions to perform. To characterize this distinctive knowledge and reasoning with competent performance of the task, this study uses three features: tasks of teaching, pedagogical context, and domains of MKT.

## Sub-Tasks of Teaching

To develop a specification of the mathematical demands of providing feedback, the sub-tasks of teaching entailed in responding to the prompt were analyzed first. Specifying the work of teaching mathematics is critical to both characterize teaching in terms of the dual foci of mathematics and instruction that Ball (1993) specified as the nature of teaching mathematics. Providing written feedback is not just one simple task, but includes several combined tasks integral to developing feedback. Characterizing the work of teaching is a critical step in identifying MKT, as is specifying what this work is and what it requires. I found competently providing feedback, expressed in general terms, involves four sub-tasks:

1.  tracking on the instructional purpose of the problem and/or when in the students' learning trajectory the problem is being used, and what that implies for the mathematical territory of the problem;

2.  making sense of the student work in relation to the instructional goals, the mathematical structure and territory of the problem, and the multiple ways that the problem can be approached;

3.  identifying resources in the student work in relation to helping the student recognize the need for further work and have a way to make further progress;

4.  deciding on a clear aim for the feedback and using the resources to design feedback that supports students in being able to work toward the instructional goal.

These sub-tasks tend to unfold in a linear fashion, but each sub-task includes a non-linear fashion of tasks. Furthermore, the sub-tasks may unfold flexibly and may cycle. These can also be expressed in more particular terms for the given scenario in each specific prompt.[5] This general description of providing written feedback characterizes that the tasks of teaching include the providing of written feedback.

## Role of Pedagogical Context

As previously identified, one component that is salient to reasoning in teaching practice is pedagogical context. My analysis suggests four elements that can support MKT reasoning about providing feedback. These elements consist of the instructional purpose given in the scenario, the mathematical topic discussed in the provided instruction, instructional resources provided to support instruction, which can be used in instruction, and student background, which can offer information for feedback. These elements of pedagogical context provide support for reasoning about providing feedback. They might also be suited for other tasks of teaching.

Another issue is that these four elements of pedagogical context do not operate in all four sub-tasks simultaneously. In other words, there are targeted or untargeted elements in each aspect of the work. In the first sub-task, teaching purpose and mathematical topic are used to identify the instructional purpose and to consider the mathematical structure. Second, the mathematical topic and instructional resources are used to draw a map between the student work and the original problem and resources used in instruction. Teaching purpose is briefly used to interpret how well the student work matches the teaching purpose. Third, mathematical topic and instructional resources are reconsidered, but their uses are different. Here, they are used so as to enable the student to recognize the need for further work. All elements are used in different moments in the different sub-tasks, which are elaborated in the mathematical work of teaching, to formulate feedback. In particular, the instructional purpose provides direction for feedback.

**Overlapping Sub-Domains of MKT**

In parallel to the analysis of elements of pedagogical context, I examined each sub-task in terms of the domains of MKT. Each sub-task involves different demands in relation to MKT's sub-domains. First, to recognize the mathematical structure and identify resources, SCK and KCT play a role in identifying the instructional purpose, establishing core ideas of the provided problem, and determining which resources were mathematically used. The second sub-task entails the use of SCK, CCK, and KCS to identify relationships between the problem focused on and student work, solve the problem, identify which resources were used and how they were used in student responses, and generate possible reasons for the student work. The third sub-task requires SCK to find resources used to suggest ways to make further progress. The fourth sub-task involves SCK, KCS and KCT to encourage the student to review and develop her or his work. Providing written feedback entails SCK, CCK, KCS and KCT, though the use of these sub-domains shifts across its subtasks. SCK plays a major role, but CCK, KCS and KCT are also critical in providing feedback.

Figure 8 synthetically provides a view of the distinctive character of competent performance of the task conceptualized in relation to three basic features: the tasks of teaching, the pedagogical context, and the domains of MKT. This diagram, which is chronological from top to bottom, has a major axis with the sub-tasks of teaching for providing written feedback. Each task also entails different elements of the pedagogical context, which are represented by

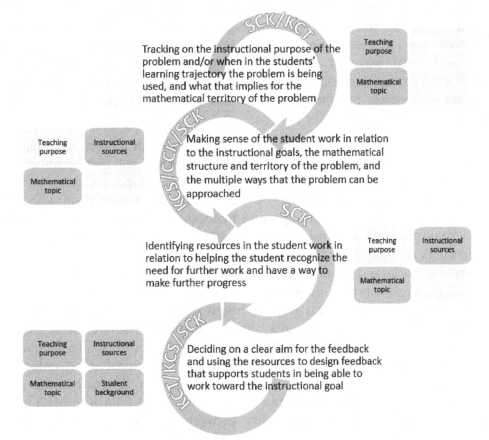

*Figure 8*. Proficiency in providing written feedback

dark or light gray rectangles depending on the targeted or untargeted elements. Furthermore, this diagram shows which different domains of MKT functions in each task of teaching, which are represented by abbreviations in the arcs.

## Discussion

This study aims to investigate the ways in which interview prompts are developed and used to provide the character of MKT. It also explores what is entailed mathematically and pedagogically in providing written feedback to demonstrate this type of method for studying MKT. From the analysis of the interview prompts and the interviews conducted with people who have different experiences related to teaching mathematics, the current study sets up the three interrelated aspects of mathematical demands: general description of sub-tasks of teaching entailed in providing written feedback, multiple and selective use of elements of pedagogical context in each sub-task, and continuous and simultaneous functions of different domains of MKT. Furthermore, this study specified how the interview prompts were used as a comprehensive way of studying MKT. Based on the analysis above, two issues have emerged related to investigating MKT and aspects of MKT.

In both generating and modifying prompts and probing interviews, prominence is given to the bidirectional approaches to investigating MKT. Developing prompts first aimed at creating situations that provided space for interviewees to organically provide written feedback. The development made certain assumptions about what providing written feedback involves, how interviewees might respond to prepared questions, and what reasoning would prevail. Because interview prompts touch on specific mathematical areas, the mathematical features embedded in the prompts were also carefully considered, including the mathematical facts, practices, and reasoning that would be related to teaching and learning in these particular instructional situations. Developing interview prompts entails analyzing the prompts themselves with such questions. Hypothesized ways of addressing the developed situations mathematically and pedagogically were set up and used to trim and elaborate the prompts. The prompts were designed for the interviews, but the process of creating the prompts also shed light on the instructional situations embedded in the prompt.

Conducting interviews aims to examine reasoning used in a given situation. Interviews with the three groups of people offered a sense of the pedagogical and mathematical reasoning at play in the given instructional situation. Although possible responses were prepared when developing the interview prompts, the interviews provided the project team with both a finer sense of how to revise the interview prompts and a better awareness of aspects of MKT entailed in providing written feedback than merely the step of developing the prompts. Conducting and analyzing interviews enabled the project team to scrutinize the interview prompts and trim them extensively. Going back and forth between creating and revising interview prompts and conducting and probing interviews ultimately unveiled the distinctive characteristics of MKT. This bidirectionality is also used as a way to design items and examine teaching (Jacobson, Remillard, Hoover, & Aaron, in press), and a common feature of MKT research. An interview is an efficient way of targeting and investigating a particular work of teaching in studying MKT. Creating the situation for an interview, however, requires intentional effort unlike the use of video clips, which show instruction clearly without requiring effort to create certain instructional situations. In this sense, the use of interviews in this study is not typical to MKT research and requires close attention to the bidirectionality between developing interview prompts and analyzing interviews.

The second major claim made by this study is that the features of MKT are continuous and simultaneous rather than separate and isolated. Ball (1993) described the nature of pedagogical deliberations and claimed to have understood more about the processes of pedagogical and mathematical deliberation in teaching mathematics. Shulman (1987) also claimed to have followed the process of pedagogical reasoning and actions that is used in teaching. He emphasized the continuity of the process of reasoning with the identification that "pedagogical reasoning and action involve a cycle through the activities of comprehension, transformation, instruction, evaluation, and reflection" (p. 14). These researchers pointed out the process in reasoning and performance engaged in teaching. The current study claims not only continuity in each of three different features but also simultaneousness among them.

To specify the complicated nature of teaching as an elaborated process in terms of tasks of teaching, domains of MKT, and pedagogical context, the current study suggests the three features as a way of describing the deliberation for professional and competent performance of providing written feedback, and to conceptually elaborate MKT. However, this does not mean that the three features operate independently. The three features are closely related to moment-to-moment teaching practice because the deliberations in teaching practice are continuous. Furthermore, elements of each feature functions continuously rather than separately or absently. Sub-tasks of teaching are continuously performed and elements of pedagogical contexts are always functioning. Different domains of MKT functions continuously and simultaneously. Different elements of the pedagogical context are continuously drawn on. This means that each feature includes continuity as well.

Teaching is intricate (Ball & Forzani, 2009). Therefore, investigating MKT requires a careful, analytic, and clear method to deepen and widen the study of nature of MKT by researchers interested in it. Also, an insightful lens is needed to scrutinize its intricacies and clarify the distinctive features of teaching mathematics in terms of MKT. Bidirectionality between the interview prompts and the interview as an approach to researching MKT offers a microscope by which we can discern the critical characters of MKT and teaching. As analysis using the bidirectional approach, the current study also claims that continuity and simultaneousness are major features among the three features used in this study to conceptualize the competent performance of providing written feedback. However, one of major limitations of the current study is that the method for studying MKT illustrated here requires use and validation by multiple groups of researchers to get universality as a robust method for studying MKT.

## Conclusions

This study has focused on providing written feedback and investigating key features in the design of prompts for uncovering MKT. This study argued that three dimensions—the decomposition of the task of providing feedback, elements of the pedagogical context, and sub-domains of MKT—are useful in characterizing the distinctive MKT reasoning involved in providing feedback and in designing pedagogical scenarios that can be used as part of a wide variety of tools for engaging, studying, and measuring MKT. This study has found the conceptual tools described here helpful across other datasets but there is a need to continue the analysis across other tasks of teaching and other data sources to both test and refine these initial ideas.

# References

Ball, D. L. (1988). *Knowledge and reasoning in mathematical pedagogy: Examining what prospective teachers bring to teacher education.* Unpublished doctoral dissertation, Michigan State University, Lansing.

Ball, D. L. (1993). Halves, pieces, and twoths: Constructing representational contexts in teaching fractions. In T. P. Carpenter, E. Fennema & T. A. Romberg (Eds.), *Rational numbers: An integration of research* (pp. 157–196). Hillsdale, NJ: Lawrenc Erlbaum.

Ball, D. L., & Bass, H. (2003). Toward a practice-based theory of mathematical knowledge for teaching. In B. Davis & E. Simmt (Eds.), *Proceedings of the 2002 annual meeting of the Canadian Mathematics Association Study Group* (pp. 3-14). Edmonton, AB: CMESG/GDEDM.

Ball, D. L., Bass, H., & Hill, H. C. (2004). Knowing and using mathematical knowledge in teaching: Learning what matters. In A. Buffler & R. Laugksch (Eds.), *Proceedings of the twelfth annual conference of the South African Association for Research in Mathematics, Science and Technology Education (SAARMSTE)* (pp. 51–65). Durban, South Africa: SAARMSTE.

Ball, D. L., & Forzani, F. (2009). The work of teaching and the challenge for teacher education. *Journal of Teacher Education, 60*(5), 497–511.

Ball, D. L., Hill, H. C., & Bass, H. (2005). Knowing mathematics for teaching: Who knows mathematics well enough to teach third grade, and how can we decide. *American Educator, 29*(3), 14–17, 20–22, 43–36.

Ball, D. L., Thames, M. H., & Phelps, G. (2008). Content knowledge for teaching: What makes it special? *Journal of Teacher Education, 59*(5), 389–407.

Bangert-Drowns, R. L., Kulik, C. C., Kulik, J. A., & Morgan, M. T. (1991). The instructional effect of feedback in test-like events. *Review of Educational Research, 61*(2), 213–238.

Cohen, D. K., Raudenbush, S. W., & Ball, D. L. (2003). Resources, instruction, and research. *Educational Evaluation and Policy Analysis, 25*(2), 119–142.

Croft, A., & Schmader, T. (2012). The feedback withholding bias: Minority students do not receive critical feedback from evaluators concerned about appearing racist. *Journal of Experimental Social Psychology, 48*(5), 1139–1144.

Harber, K. D., Gorman, J. L., Frank, P., Butisingh, S., Tsang, W., & Ouellette, R. (2012). Students' race and teachers' social support affect the positive feedback bias in public schools. *Journal of Educational Psychology, 104*(4), 1149–1161.

Hattie, J., & Timperley, H. (2007). The power of feedback. *Review of Educational Research, 77*(1), 81–112.

Herbst, P., & Kosko, K. (2014). Mathematical knowledge for teaching and its specificity to high school geometry instruction. In J. J. Lo, K. R. Leatham & L. R. Van Zoest (Eds.), *Research trends in mathematics teacher education* (pp. 23–45). Switzerland: Springer.

Hill, H. C., Ball, D. L., & Schilling, S. G. (2008). Unpacking pedagogical content knowledge: Conceptualizing and measuring teachers' topic-specific knowledge of students. *Journal for Research in Mathematics Education, 39*(4), 372–400.

Hill, H. C., Rowan, B., & Ball, D. L. (2005). Effects of teachers' mathematical knowledge for teaching on student achievement. *American Educational Research Journal, 42*(2), 371–406.

Hill, H. C., Schilling, S. G., & Ball, D. L. (2004). Developing measures of teachers' mathematics knowledge for teaching. *The Elementary School Journal, 105*(1), 11–30.

International Human Genome Sequencing Consortium. (2001). Initial sequencing and analysis of the human genome. *Nature, 409*(6822), 860–921.

Jacobson, E., Remillard, J., Hoover, M., & Aaron, W. (in press). The interaction between measure design and construct development: Building validity arguments. In A. Izsák, J. T. Remillard & J. Templin (Eds.), *Psychometric methods in mathematics education: Opportunities, challenges, and interdisciplinary collaborations. Journal for Research in Mathematics Education monograph series.* Reston, VA: National Council of Teachers of Mathematics.

Kluger, A. N., & DeNisi, A. (1996). The effects of feedback interventions on performance: A historical review, a meta-analysis, and a preliminary feedback intervention theory. *Psychological Bulletin, 119*(2), 254–284.

Lewis, J., & Blunk, M. (2012). Reading between the lines: Teaching linear algebra. *Journal of Curriculum Studies, 44*(4), 515–536.

Ma, L. (1999). *Knowing and teaching elementary mathematics: teachers' understanding of fundamental mathematics in China and the United States.* Mahawah, NJ: Lawrence Erlbaum Associates.

Naidoo, N., Pawitan, Y., Soong, R., Cooper, D., & Ku, C. (2011). Human genetics and genomics a decade after the release of the draft sequence of the human genome. *Human Genomics, 5*(6), 577–622.

Rockoff, J. E., Jacob, B. A., Kane, T. J., & Staiger, D. O. (2011). Can you recognize an effective teacher when you recruit one? *Education Finance and Policy, 6*(1), 43–74.

Schoenfeld, A. H. (2011). Toward professional development for teachers grounded in a theory of decision making. *ZDM, 43*(4), 457–469.

Shulman, L. S. (1986). Those who can understand: Knowledge growth in teaching. *Educational Research, 15*(2), 4–14.

Shulman, L. S. (1987). Knowledge and teaching: Foundations of the new reform. *Harvard Educational Review, 57*(1), 1–22.

Shute, V. J. (2008). Focus on formative feedback. *Review of Educational Research, 78*(1), 153–189.

Sleep, L. (2009). *Teaching to the mathematical point: Knowing and using mathematics in teaching.* Unpublished doctoral dissertation, University of Michigan, Ann Arbor.

Thames, M. H. (2009). *Coordinating mathematical and pedagogical perspectives in practice-based and discipline-grounded approaches to studying mathematical knowledge for teaching (K-8).* Unpublished doctoral dissertation, University of Michigan, Ann Arbor.

Yeager, D., Purdie-Vaughns, V., Garcia, J., Apfel, N., Brzustoski, P., Master, A., . . . Cohen, G. L. (2014). Breaking the cycle of mistrust: Wise interventions to provide critical feedback across the racial divide. *Journal of Experimental Psychology: General, 143*(2), 804–824.

Footnotes

[1] Their items mainly concentrated on the following practices of teaching in algebra and geometry for elementary levels: evaluating understanding, choosing examples – illustrating a concept, evaluating explanations, choosing representations, evaluating difficulty, choosing examples – selecting a problem for an exercise, and figuring out non-standard work.

[2] I would like to acknowledge the contribution of the MTLT project, particularly, Yvonne Lai, Erik Jacobson, and Mark Hoover for outlining the foundation of this work.

[3] Here is the link: http://teachingworks.soe.umich.edu/work-of-teaching/high-leverage-practices

[4] I would like to thank the MTLT project, particularly, Rachel Snider, Lindsey Mann, Joy Johnson, and Mark Hoover, for conducting interviews.

[5] Appendix shows particular description engaged in the interview prompts shown in Figure 5.

Appendix

Particular description of the work of providing written feedback in the interview prompt (permutation), shown in Figure 5

1. Tracking on the purpose of engaging students in using patterns to reason about the completeness of a solution to a permutation problem.

2. Working through the student solution, imagining how the student was likely thinking, and noticing that the student: (i) begins by swapping the two rightmost characters, then swapping the first two leftmost characters with the two rightmost and again swapping the two rightmost, then continuing to look for a new sequence on which to swap without repeating; (ii) shifts to the more systematic approach of fixing the first and recursively generating all swaps on the remaining three characters starting with the 11th ordering; and (iii) has missed two orderings beginning with "M" because the initial pattern did not have a systematic approach to determining the first two characters, but has been quite orderly and careful throughout.

3. Identifying that: (i) the solution is missing two orderings or that "M" is the start of only 4 orderings while the other letters have 6 and (ii) the second half of the list (for orders beginning with "L" and "W") competently uses a powerful standard system for finding all orderings. In addition, recognizing that the student's approach for the second half can be used to systematically list all solutions and reason that you have them all.

4. Deciding to have the student use the work in the second half of the list to reconsider or redo the work in the first half. Further, using the two missing orderings to get the student to realize the need for further work and drawing the student's attention to the pattern in the second half of the list and the idea of using that pattern to make a complete list and reason that it is complete.

# Why Defining the Construct Matters: An Examination of Teacher Knowledge Using Different Lenses on One Assessment

Chandra Hawley Orrill (corrill@umassd.edu)
University of Massachusetts Dartmouth

Allan S. Cohen (acohen@uga.edu)
University of Georgia

**Abstract:** What does it mean to align an assessment to the domain of interest? In this paper, we analyze teachers' performance on the Learning Mathematics for Teaching assessment of Proportional Reasoning. Using a mixture Rasch model, we analyze their performance on the entire assessment, then on two different subsets of items from the original assessment. We consider the affordances of different conceptualizations of the domain and consider the implications of the domain definition on the claims we can make about teacher performance. We use a single assessment to illustrate the differences in results that can arise based on the ways in which the domain of interest is conceptualized. Suggestions for test development are provided.

*Keywords:* assessment development, teacher knowledge assessment

## Introduction

While test performance is generally reported as if the score assigned a participant were the goal of the assessment, the actual interest is the inferences that can be made about a learner based on that score. It is critical to ensure an assessment is measuring what it is intended to measure if such inferences are to be accurate. Thus, assessments must be written in a way that allows accurate inferences to be drawn. If we are to make claims about quantities of or changes in participants' knowledge, alignment between the content and the underlying assumptions of the domain is critical. Further, if instruction is to be impacted in positive ways by assessment data, we need to ensure that scores accurately report knowledge of the intended construct. Thus, defining the construct one is interested in measuring is vital to the assessment process.

One particularly complex domain from a measurement perspective is that of teacher knowledge. This is complex because teacher knowledge is multidimensional. The specialized knowledge teachers need for teaching (SKT) necessarily includes content knowledge, pedagogical knowledge, and understandings of how students learn (e.g., Ball, Thames, & Phelps, 2008; Baumert et al., 2010; Manizade & Mason, 2011; Shulman, 1986; Silverman & Thompson, 2008). Measuring such knowledge requires adherence to a set of beliefs about the specific construct and how it is best tested, for example, using a paper-based assessment or using feedback in a video-based open-response system (e.g.,

*The Mathematics Enthusiast,* **ISSN 1551-3440, vol. 13, no. 1&2**, pp. 93–110
2016© The Author(s) & Dept. of Mathematical Sciences-The University of Montana

Kersting, Givvin, Sotelo, & Stigler, 2010). Despite the challenges of this complex knowledge domain, if we care about whether a teacher has the knowledge necessary to support student learning, assessments need to be written to address the domain.

In the case of SKT, we are faced with not only the complexity of the domain, but also the ambiguity of what it means in practical terms for a teacher to have or exhibit particular kinds of knowledge. In our work on proportional reasoning, we have chosen to focus on how teachers understand the mathematics of proportions rather than on their pedagogical understandings related to teaching such content. However, we also acknowledge that teachers need to be able to use the content in the process of teaching, thus our interest in assessment focuses on the knowledge teachers need as it is situated in tasks that ask teachers to make sense of student thinking, analyze multiple approaches to problems, or other authentic teaching activities. Fully defining the SKT in which we are interested is outside the domain of this article. However, we rely heavily on the work of Lamon (2007) and Lobato and Ellis (2010) in our definition. For example, we know that teachers need to have the ability to conceptually connect the two values in a ratio and to understand that a third, abstractable quantity results from that connection (e.g., Lobato & Ellis, 2010). Teachers also need to understand the multiplicative relationships inherent in proportional relationships (e.g., the constant of proportionality is the multiplicative relationship of one value in the ratio to another). And, they need to understand that this corresponds to the unit rate. Further, we assert that teachers should understand how proportional reasoning connects to other areas of mathematics such as fractions and geometric similarity (Lamon, 2007; Pitta-Pantazi & Christou, 2011). We rely on research on teacher knowledge that shows that teachers struggle to use unit rate when faced with values less than one (Harel & Behr, 1995; Post, Harel, Behr, & Lesh, 1988). And, we know from a series of small studies (e.g., Riley, 2010; Son, 2010) that teachers tend to rely on cross-multiplication, which seems to obscure the breadth of knowledge they may have about proportional relationships.

Considering only the domain of proportional reasoning for teachers, one could take a number of approaches to measuring different aspects of SKT. In this study, we considered one assessment's approach to measuring the construct of proportional reasoning knowledge for teaching through the lens of our emerging definition. By undertaking this effort, we were able to further define the domain of SKT for proportional reasoning and to consider valid measurement of that domain. We relied on an approach that searched for latent classes in the data and what those latent classes might reveal about the nature of teachers' knowledge. In this setting, latent classes are statistically determined groupings of participants who shared particular aspects of patterns in their responses to the assessment items. We suspected that the data available for this study might contain latent classes given results from previous research by Izsák, Orrill, Cohen, & Brown (2010) on teachers' understanding of rational numbers. That research indicated that in a similar assessment, a two-class model fit the data better than the one-class model.

## Mixture Rasch Model

For this analysis, we used the mixture Rasch model (described below) as a means of further examining the data provided through Item-Response Theory (IRT) by an assessment of teacher knowledge. The standard Rasch model is useful for tests that are designed to assess single categories of knowledge. With respect to measures of teachers' proportional reasoning, the standard model constrains the information about a teacher to a single estimate of proportional reasoning. As such, the standard model will not detect differences in the ways that examinees respond to individual items on the test. The Rasch model assumes that all examinees are drawn from a single population. When it is suspected that this may not be the case, such as when different groups of teachers have different patterns of responding to the items, then a mixture Rasch model (Rost, 1990, 1997) may be more useful.

The mixture Rasch examines patterns in responses to the assessment items that allow participants to be placed into latent classes. Latent classes are not determined a priori nor are they typically determined by more apparent commonalities such as ability, gender, or race. Members in a latent class are homogeneous on the characteristic that caused the latent class to form. Previous research with mixture IRT models in general, of which the mixture Rasch model is one example, has demonstrated that these models can address whether or not examinees exhibit the same response characteristics or whether there are groups of examinees that are latent and can only be identified by examining homogeneities in their response patterns. The groups are termed latent since they are not immediately visible simply by examining their responses. Observable characteristics like gender, height, and ethnicity are considered manifest. In contrast, characteristics such as differences in use of cognitive strategies for answering test items are considered latent until subsequent analysis can make them manifest. Previous research, for example, has found that latent classes of students that differ in their use of cognitive strategies for answering test questions can be detected (Bolt, Cohen, & Wollack, 2001; Embretson & Reise, 2000; Mislevy & Verhelst, 1990; Rost, 1990, 1997).

In a recent study of fraction knowledge that used the mixture Rasch analysis, we found that teachers in one latent class attended to the referent unit more than those in the other class (Izsák et al., 2010). These fine-grained understandings are important for mapping a domain of knowledge that teachers should have and for understanding what learning means within the domain. For example, we saw that teachers sometimes transitioned between latent classes following an intervention, albeit without exhibiting growth in their test scores. Similarly, some had significant changes in test scores without changing latent classes (e.g., Izsák, Jacobson, de Araujo, & Orrill, 2012). In our previous work, the latent classes were found to align with aspects of knowledge that are critical for meaningful understanding of fractions. We assert, therefore, that understanding latent class membership can provide important insights into the nature of teacher knowledge that ability scores alone cannot.

# Methods

## Sample

For the purposes of this analysis, we used the same dataset for all three analyses, but included different subsets of items for each version. While determining the latent class membership, item difficulty, and ability scores relied on the combined datasets (25 teachers from our sample plus 351 teachers from the LMT-PR sample), the analysis of the latent classes relied solely on an analysis of the 25 teachers participating in a larger study in which this work is situated. The 351 teachers in the national sample included teachers in grades 4-8 (Hill, 2008). The 25 teachers we analyzed in depth here were all teachers in grades 6-8. These 25 participants represented a convenience sample of practicing mathematics teachers drawn from across three states. They represented an array of schools including public, private, and charter schools situated in urban, suburban, and rural settings. Because the data are presented and analyzed in aggregated form throughout this study, we have not provided in-depth information about each teacher in the study.

## Instrument

The instrument of interest was the Learning Mathematics for Teaching form for Proportional Reasoning (LMT-PR: Learning Mathematics for Teaching, 2007), which included 73 unique items. Consistent with all LMT assessments, the LMT-PR seeks to measure a particular construct known as mathematical knowledge for teaching (MKT; Ball, et al., 2008). MKT emphasizes the use of knowledge in and for teaching rather than focusing on the broader body of knowledge teachers may have developed (Ball et al., 2008). Items on the LMT were designed to measure common content knowledge (CCK; math knowledge used outside of teaching) and specialized content knowledge (SCK; math knowledge specifically used in teaching) (Hill, 2008). The assessment was not created using a systematic conceptualization of the domain of proportional reasoning. Rather, the development team was opportunistic, using items that captured MKT across an array of topics with some breadth and an array of mathematical tasks of teaching (Thames, personal communication). The item development team aligned items to the *Principles and Standards for School Mathematics* (NCTM, 2000). The LMT-PR form included a wide range of questions including those asking teachers to solve proportions, to select harder or easier examples for students, to explain particular proportions ideas to students, to interpret a variety of graphs, and to determine whether given situations were directly/inversely proportional. As with other LMT forms, LMT-PR included questions that asked teachers to make sense of students' work, to interpret representations, and to select items appropriate for their students. It is the need for this array of understandings that makes the construct of teacher knowledge unique and challenging to measure.

We selected the LMT-PR both because it measures the content in which we are interested and because it situates many of the tasks in the work that teachers do, as described above. Our analysis was meant to highlight the importance of aligning constructs of interest to assessments and should not be interpreted as detracting from the important role the LMT has played in the assessment of teacher knowledge. The LMT has been one of the most widely used assessments of teacher knowledge and has been used as the basis for important research showing the alignment between teacher knowledge and student performance on standardized assessments (e.g., Hill, Rowan, &

Ball, 2005). Further, the LMT paved the way for the increasingly rich discussion of the construct of teacher knowledge in mathematics.

Table 1. *Items included in each version of the LMT-PR in this analysis.*

|  | LMT-PR 73 | LMT-PR 60 | LMT-PR 54 |
|---|---|---|---|
| Items Removed from full version of LMT-PR | All items included | 3a-c<br>10<br>15<br>24a-d<br>32a-d | 3a-c<br>5<br>8<br>10<br>12a, 12d<br>14b<br>15<br>18<br>24a-d<br>32a-d |

As shown in Table 1, each version of the LMT used in this study was created from the full assessment. Version 1 (LMT-PR 73) for this analysis considered all 73 items in the LMT-PR[1]. The second analysis (LMT-PR 60) removed 13 items that did not specifically measure proportional reasoning concepts. These items included several that asked teachers to interpret nonlinear graphs (see Figure 1 for an example). While these items featured covariational reasoning, they measured mathematics outside our construct of interest. If we wanted to carefully measure the domain to make inferences about teachers' proportional reasoning as we conceptualized it, then including items that did not specifically address proportional reasoning could create noise in the analysis.

---

[1] While the LMT is logically organized into sets of questions that appear to be testlets, in our past experience, those related questions have not had interdependent responses like testlets should. Participant responses on related questions appeared to be independent. Thus, it is appropriate to treat every item as a stand-alone item rather than considering them as testlets for the purposes of this analysis.

Ms. Reese and Mr. Ward celebrate student success by allowing students to eat small bags of popcorn at the end of the day. Describe what has happened over the course of the last six days in each of their classes.

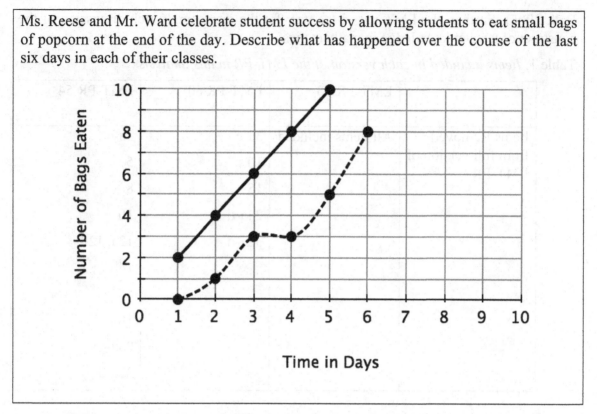

*Figure 1.* A task stem similar to those removed to create LMT-PR 60 for this study. (Note that actual LMT items are secure.)

For the third version, which included 54 items, we worked from LMT-PR 60 and removed those items reviewers suspected would lead to errors in measurement of the proportional reasoning domain based on our interpretation of the specialized knowledge teachers need for this domain. In particular, we were concerned with items that would lead to the right answers using incorrect reasoning or incorrect answers for reasons other than limitations in understanding.

One example of such an item was the single item from the LMT-PR for which we had item response interview data from a previous study. In that study, we asked 13 middle school teachers to think aloud on an item that asked them to explain why the cross-multiplication algorithm works. In our analysis of those interviews, we found that about half of the participants who answered the item incorrectly actually understood the intended correct answer as being the most mathematically precise. However, they chose a different response driven by their interpretation of the mathematics through the eyes of a teacher, even though the question did not necessarily ask them to. That is, they chose the less accurate answer because it more precisely reflected the way they would teach students about cross multiplication. This becomes particularly relevant to the discussion of our construct because the LMT team viewed this item as working appropriately because their definition of MKT is tightly tied to the ways in which teachers use their knowledge to teach (Thames, personal communication). Thus, the item would be seen as operating appropriately from the perspective that the teachers were not applying the most appropriate understandings in their teaching situations. However, from our perspective, which is grounded in knowledge in pieces (diSessa 1988; 2006), we assert that teachers may have a number of knowledge resources that are connected in ways that cause them to be invoked in some situations, but perhaps not in others. Working from this perspective, we interpreted items like the cross multiplication item as not working because we saw evidence that some teachers understood the mathematics of interest, but did not invoke that mathematics in expected ways in the situation presented in the item. By not using their understanding in expected ways, the participants appeared not to have the understanding, but based on their explanation, that is not an accurate assumption.

Based on our understanding and experiences working with teachers around proportional reasoning, we removed six potentially problematic items to create the 54-item version of the assessment.

**Data Analysis**

As mentioned above, data analysis was done using a mixture Rasch model (Rost, 1990, 1997). In the standard Rasch model (e.g., Hambleton & Swaminathan, 1985; Lord, 1980), item responses are typically scored dichotomously (e.g., 1 for a correct response and 0 for an incorrect response). The resulting data are then used to estimate parameters that describe ability for each examinee ($\theta_j$) and difficulty for each item ($b_i$). Ability ($\theta_j$) describes the amount of proportional reasoning knowledge possessed by person j. The scale for the model is centered at 0, and if $\theta_j = 0$, then person j is assumed to have the average amount of knowledge of proportional reasoning. The difficulty of the item, $b_i$, is expressed on the same scale as ability. The standard form of the Rasch model is given as

$$P_i(\theta_j) = \frac{\exp(\theta_j - b_i)}{1 + \exp(\theta_j - b_i)} \qquad (1)$$

The item difficulty, $b_i$, indicates the point on the scale at which persons with ability equal to the item difficulty have a 50-50 chance of answering that item correctly. Items that are more difficult have higher difficulty parameter estimates. Similarly, items that are easier have lower difficulty parameter estimates. As an example, for item 1a (Figure 1), the item difficulty score is -2.26 for Class 1; therefore, someone whose ability is -2.26 has a 50% likelihood of answering the question correctly. In contrast, item 8 shows that a member of Class 2 to have a 50% chance of answering the item correctly, the participant would have to have an ability score of 2.736.

In our study, we used the mixture Rasch model (Rost, 1990) as estimated using a Markov chain Monte Carlo algorithm as implemented in the computer software WinBUGS (Spiegelhalter, Thomas, Best and Lunn, 2003). The mixture Rasch model can be given as

$$P_i(\theta_j) = \frac{\exp(\theta_{jg} - b_{ig})}{1 + \exp(\theta_{jg} - b_{ig})}$$

where the subscript $g$ is used to indicate latent class. In this equation, it can be seen that ability for person $j$ and the difficulty for item $i$ differ depending on the latent class.

For each item, the mixture Rasch model estimated a separate item difficulty for each latent class, a separate probability of belonging in each latent class, and a separate ability estimate for each examinee. To determine the number of latent classes, we used the BIC (Bayesian Information Coefficient, Schwartz, 1973). This is a standard approach to determining fit for mixture IRT Models (Li, Cohen, Kim & Cho, 2009).

It is a well-known caution that that a statistical result does not necessarily indicate a meaningful result. In latent class analysis, for example, researchers have found that spurious latent classes may be found (Alexeev, Templin, & Cohen, 2011). In this study, the three different solutions detected in the different mixture Rasch model analyses were each composed of different test content and thus had different interpretations. Given that the mixture Rasch model was of the same family as the Rasch model used to originally calibrate the LMT items, that the latent classes detected in this study had a meaningful interpretation based on the membership of the different classes and that there were differences in performance on the content by members of each class on the different tests, it seems reasonable to infer that the latent classes were not spurious. Further, the classes we found had a clear meaning based on the membership of the classes and the differential responses of each latent class to the questions on the test. The membership of latent classes changed when the content of the test changed, which is consistent with the assumptions underlying latent class analyses. Finally, the responses offered by the members of the latent classes were helpful in determining how to characterize the different classes and the differences in performance on each of the test questions were useful in helping to characterize latent classes.

Once latent classes had been identified, we undertook analysis of the latent classes by considering those places where item difficulties varied significantly by class as well as those places where members of the lower-scoring class found items to be easier than those of the higher-scoring class. As shown in Figure 2, item difficulties were reported in standard deviation units, with easier items showing lower scores. Analyzing these items for trends in the knowledge used helps us better understand what makes the knowledge of the members of one class different from that of the other class. In our previous research (Izsák et al., 2010) we found that having item interview data substantially enhanced our ability to discriminate between classes. However, interview data were not available for the present study.

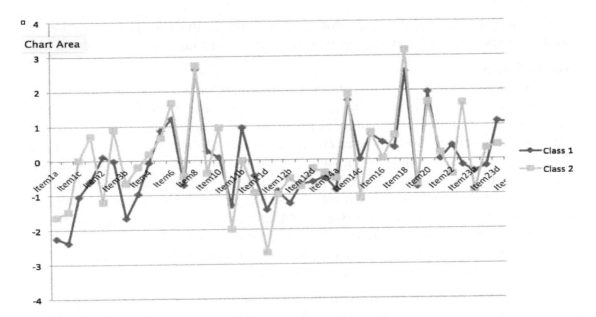

*Figure 2*. Plots of item difficulties for Class 1 and Class 2 in our analysis of LMT-PR. Overall, Class 2 scores are higher, but note there are items Class 1 members found easier.

In our initial analysis of the LMT-PR data from the combined dataset, we noted that the higher scoring of these classes seemed to find items focused on using cross-multiplication and other algorithms to be easier than their counterparts in the lower-scoring class. However, members of the lower-scoring class found items focused on understanding proportions and reasoning about them in a number of ways to be easier than those in the higher-scoring class. That is, the members of the higher-scoring class did better on items that allowed algorithmic thinking rather than solely focusing on reasoning about proportional relationships. These results suggested that the assessment itself may privilege teachers who are more facile with algorithms despite the field-wide emphasis on the importance of being able to reason about proportions (e.g., Lamon, 2007; Lobato & Ellis, 2010; NCTM, 2000). This finding raised interesting questions about the nature of the LMT-PR and what it might be measuring. This, in turn, led us to subsequent analyses focused primarily on 25 participants from whom we have collected completed LMT-PR forms, a prompted interview relying on a think-aloud protocol, and a face-to-face clinical interview. In the current study, we present our findings of the analysis of subsets of items on the LMT based on the performance of these 25 participants.

## Results

Our aim in conducting this analysis was to consider the implications of construct definition on inferences that can be made about teachers' knowledge. Specifically, we sought to use this case as an illustration of the importance of mapping an assessment to the construct of interest. In this results section, we present the findings of our analysis of each of the three LMT-PR versions, relying on our analysis of the sample of 25 teachers. We follow with a discussion of two key issues: the impact of the definition of the domain on the overall ability scores of participants and the need to constrain assessments to robust understandings of the mathematics of interest. In the Conclusions section, we highlight approaches that may support better assessment creation.

## Ability Scores

Consistent with our position outlined at the outset, determining the construct to be measured is important for making claims. As shown in Table 2, removing the items that did not fit our definition of proportional reasoning led to a significant increase—where significance is defined as 0.3 or greater—in the scores of three participants (12%). By removing the items that did not measure proportional reasoning and those that were determined to be problematic (LMT-PR 54), seven of the 25 participants (28%) had significant increases in their ability scores. This changed the nature of our interpretations of these teachers' knowledge. This was particularly important for Bridgette, Kelly, and Kathleen who were all within one standard deviation of the mean at the outset, but ended well over one standard deviation above the mean. Interestingly, all of the significant changes in scores were increases rather than decreases.

Table 2. *Ability scores of the 25 participants on the three versions of LMT-PR. Each ability column shows the participants' abilities along a single continuum. Scores are reported as logits. All names are pseudonyms.*

| | Name | Ability LMT-PR 73 | Ability LMT-PR 60 | Ability LMT-PR 54 |
|---|---|---|---|---|
| | | | | |
| 1 | Autumn | 1.03 | 1.17 | 1.30 |
| 2 | David | 0.72 | 0.58 | 0.76 |
| 3 | Alan | 1.88 | 2.20 | 2.74 |
| 4 | Ella | 2.56 | 2.55 | 3.01 |
| 5 | Mike | 2.12 | 1.90 | 1.97 |
| 6 | Bridgette | 0.50 | 0.68 | 0.86 |
| 7 | Allison | 1.03 | 0.97 | 1.07 |
| 8 | Larissa | 2.12 | 2.20 | 2.51 |
| 9 | Tori | 0.87 | 0.87 | 0.97 |
| 10 | Matt | 0.87 | 0.77 | 1.07 |
| 11 | Greg | 2.73 | 2.75 | 2.74 |
| 12 | Meagan | 0.65 | 0.68 | 0.86 |
| 13 | Brianna | 0.15 | 0.22 | 0.36 |
| 14 | Felicia | 1.88 | 1.76 | 1.97 |
| 15 | Eileen | 1.47 | 1.51 | 1.55 |
| 16 | Kelly | 0.50 | 0.97 | 1.19 |
| 17 | Todd | 1.57 | 1.51 | 1.68 |
| 18 | Kathleen | 0.80 | 0.97 | 1.30 |
| 19 | Patricia | 1.67 | 1.76 | 1.82 |
| 20 | Robyn | 1.67 | 1.63 | 1.82 |
| 21 | Peter | -0.84 | -0.83 | -0.58 |
| 22 | Nancy | 2.73 | 2.55 | 2.74 |
| 23 | Diana | 2.26 | 2.20 | 2.31 |
| 24 | Christine | -0.34 | -0.30 | -0.48 |

| 25 | Heather | 2.40 | 2.99 | 3.34 |

Table 3. *Participants' abilities for each version of the LMT-PR. The Ability scores reported here are for each latent class, but have been scaled to be comparable across latent classes.*

| | Name | Mixture Ability LMT-PR 73 | Class LMT-PR 73 | Mixture Ability LMT-PR 60 | Class LMT-PR 60 | Mixture Ability LMT-PR 54 | Class LMT-PR 54 |
|---|---|---|---|---|---|---|---|
| 1 | Autumn | 1.12 | 2.00 | 1.13 | 1.00 | 1.26 | 1.00 |
| 2 | David | 0.80 | 2.00 | 0.61 | 1.00 | 0.77 | 1.00 |
| 3 | Alan | 1.91 | 2.00 | 1.95 | 1.00 | 2.60 | 2.00 |
| 4 | Ella | 2.53 | 2.00 | 2.26 | 2.00 | 2.80 | 2.00 |
| 5 | Mike | 2.16 | 2.00 | 1.72 | 2.00 | 1.85 | 1.00 |
| 6 | Bridgette | 0.58 | 2.00 | 0.70 | 1.00 | 0.85 | 1.00 |
| 7 | Allison | 1.10 | 2.00 | 0.84 | 2.00 | 1.05 | 2.00 |
| 8 | Larissa | 2.15 | 2.00 | 1.96 | 1.00 | 2.40 | 2.00 |
| 9 | Tori | 0.98 | 2.00 | 0.85 | 1.00 | 0.96 | 1.00 |
| 10 | Matt | 0.96 | 2.00 | 0.77 | 1.00 | 1.05 | 1.00 |
| 11 | Greg | 2.66 | 2.00 | 2.38 | 1.00 | 2.57 | 2.00 |
| 12 | Meagan | 0.78 | 1.00 | 0.70 | 1.00 | 0.87 | 1.00 |
| 13 | Brianna | 0.34 | 1.00 | 0.30 | 1.00 | 0.43 | 1.00 |
| 14 | Felicia | 1.94 | 2.00 | 1.62 | 1.00 | 1.90 | 2.00 |
| 15 | Eileen | 1.54 | 2.00 | 1.41 | 1.00 | 1.49 | 1.00 |
| 16 | Kelly | 0.59 | 2.00 | 0.94 | 1.00 | 1.16 | 1.00 |
| 17 | Todd | 1.62 | 2.00 | 1.42 | 1.00 | 1.60 | 1.00 |
| 18 | Kathleen | 0.88 | 2.00 | 0.95 | 1.00 | 1.25 | 1.00 |
| 19 | Patricia | 1.74 | 2.00 | 1.61 | 1.00 | 1.75 | 2.00 |
| 20 | Robyn | 1.73 | 2.00 | 1.49 | 1.00 | 1.73 | 2.00 |
| 21 | Peter | -0.56 | 1.00 | -0.63 | 1.00 | -0.45 | 1.00 |
| 22 | Nancy | 2.65 | 2.00 | 2.28 | 2.00 | 2.56 | 2.00 |
| 23 | Diana | 2.27 | 2.00 | 1.95 | 1.00 | 2.19 | 2.00 |
| 24 | Christine | -0.13 | 1.00 | -0.16 | 1.00 | -0.34 | 1.00 |
| 25 | Heather | 2.39 | 1.00 | 2.60 | 2.00 | 3.02 | 2.00 |

## Latent Class Combined with Ability Scores

When we added latent class membership to the analysis, we were able to see that these three versions yielded different results (Table 3). In essence, they measured the construct differently. First, for each of the three versions, Class 2 was the higher scoring class (Table 4). This was true despite the fact that on LMT-PR 60 80% of the participants were in Class 1 whereas in LMT-PR 73 80% of the participants were in Class 2. While we initially suspected that the computer may have switched labels in the analysis (which does happen), our analysis of the responses from the members of these classes suggested that was not the case. In LMT-PR 73, our analysis showed that for the 25 participants, membership in Class 1 indicated more facility with ratio tables and combining ratios, which includes knowing that it is okay to add ratios to each other. Class 1 teachers also found easier those items that asked them to explain why particular relationships did or did not work as proportions (e.g., using scale factor, equivalence, division, or additive reasoning as rationales). Class 1 teachers found determining particular instances of inverse proportion to be easier than Class 2 teachers and they were better at setting up proportions for simple word problems (e.g., if two tickets cost $5, how much do 10 tickets cost). In contrast, Class 2 was better at items that involved algorithms, including those using unit rate, scale factors, and cross multiplication. They found complex equations to be easier to set-up and verify (e.g., situations in which two people cut grass at different rates and one needs to set up an equation to determine how long it will take to mow a particular number of lawns) as well as showing more flexibility in acceptable proportion set-ups (e.g., lbs/lbs, $/lbs, and $/$ are all acceptable). If the class labels had just been flipped, the groupings described for LMT-PR 73 could be reversed for LMT-PR 60, however, that was not the case. Instead, for LMT-PR 60, Class 1 seemed to be identified through items that involved explanations of why a situation is proportional, much like Class 1 for LMT-PR 73. In addition, Class 1 for LMT-PR 60 was better able to identify numbers that would make problems easier or harder for students, to identify situations in which additive reasoning was being improperly used, and to correctly interpret a variety of ways to solve proportions without a standard algorithm or equation. In contrast, Class 2 found easier those items that asked them to identify situations that were related in ways that were not proportional (e.g., linear relationships), those that asked them to model complex situations, and those that asked them to use scale factor reasoning for within measure space scale factors (e.g., given $3/17$ liters $= $5/x$ liters, one can divide $5 by $3 to see how many times larger 5 is than 3. Then multiply that result by 17 to determine $x$).

Class 1 and Class 2 meant different things in these analyses despite Class 2 including 80% of the participants in LMT-PR 73 and Class 1 including 80% in LMT-PR 60. In short, while it was clear that the class separations were unique to each version, the latent classes did not clearly organize the teachers in ways that might help us make inferences beyond whether the teacher was likely to be comfortable with algorithms.

Table 4. *Mean logit scale scores for each latent class for each version of LMT-PR*

|  | LMT-PR 73 | LMT-PR 60 | LMT-PR 54 |
|---|---|---|---|
| Class 1 mean | 0.565<br>*n*=20 | 1.096<br>*n*=5 | .909<br>*n*=14 |
| Class 2 mean | 1.595<br>*n*=5 | 1.939<br>*n*=20 | 2.233<br>*n*=11 |

More interesting was LMT-PR 54 because more teachers scored significantly higher on it and because it seemed to separate the teachers into latent classes in ways that might be more useful for measuring our construct given the more balanced distribution of teachers between the classes. Closer examination of LMT-PR 54 class membership highlighted more mathematical sensitivity in the separation of classes. Whereas LMT-PR73 and LMT-PR 60 both had clear separations, one of the clear distinguishing features between classes was that one class found using algorithms and algebraic approaches (such as modeling equations) to be easier than the other class. In LMT-PR 54, that distinction disappeared, which suggested that the privileging of algorithmic reasoning might have been mediated in this version of the assessment. Instead, we saw finer-grained and more conceptually grounded mathematical ideas separating the classes. For example, Class 1 found items that relied on scale factors determined by the within measure space relationship to be easier, whereas Class 2 found items related to scale factors determined between measure space to be easier (e.g., given $3/17$ liters $= \$5/x$ liters, one can divide 17 liters by $3 to determine the constant relationship, then multiply 5 by that value to answer for $x$). While both classes found particular items involving the combination of ratios to be easy, Class 2 seemed more able to both break down ratios, combine ratios, and to make sense of the ratio table representation. Class 2 found items that focused on why one cannot use addition to maintain equivalent ratios to be easier than Class 1. However, Class 1 was still better able to select explanations for why particular relationships did or did not work as proportions (e.g., using scale factor, equivalence, division, or additive reasoning as rationales). This was the one set of questions that Class1 consistently found easier across all versions. Class 1 in LMT-PR 54 also found the identification of easier or harder numbers (e.g., which set of numbers would make this problem harder for students to solve?) to be easier than Class 2 did. Interestingly, in LMT-PR 54, we see that Class 1 participants had an easier time with percentages as they related to proportions (for example, a sale price of 40% off is not the same as taking an additional 10% off of a 30% off price). This suggests that Class 1 might be more sensitive to questions grounded in the work of teaching (e.g., making pedagogical decisions) versus the work of solving problems.

In summary, LMT-PR 73 and LMT-PR 60 were more consistent with our initial analyses of LMT results. The class membership indicated that the higher scoring class was the one that found algorithms easier. It was not until we used the most narrowed form—the form that included only the 54 items related to proportional reasoning and removed items that seemed potentially problematic—that we were able to see differences in how participants reasoned about and with proportions and we were able to start seeing some separation tied to pedagogical concerns. This suggests that an assessment more focused on the construct of interest may yield more sensitive results.

## Discussion

In this paper, we used the case of the LMT-PR to highlight the importance of aligning the construct of interest to the assessment being used. Our analysis was aided by the use of psychometric models that allow us to vary from the unidimensional analysis to which traditional IRT scores are constrained. This allowed us to look at the same participants' performance on different versions of the same assessment.

Our findings indicated that the version of the assessment that was most tailored to our definition of the construct yielded clearer information about teachers' understandings of proportional reasoning and the higher overall scores for our participants than the less focused versions. Given the emergence of high stakes testing for teacher hiring and evaluation, being able to make clear, strong claims about teacher understanding is critical. For over one-quarter of the 25 participants in this study, scores varied significantly according to the particular items included in the assessment raising questions about what the versions of the assessment tell us about individual teacher knowledge. In contrast, at the group level (which is what the LMT is designed to measure) there was less variation in means. The group mean for the 25 participants on LMT-PR 73 was 1.39 and LMT-PR 54 was 1.49. We speculate that some of the limitation in variation is due to the overall skewing of our scores. The 25 participants we worked with were mathematically stronger than the national sample used to create the scale on which the scores were based.

In our analysis, we presented evidence that considering the assessment items' alignment to the construct of interest matters to the outcomes of that assessment and the inferences that can be made. The claims we can make about the participants changed based on the items considered, and the guidance provided by an analysis of a particular set of items could lead to potentially different implications for further learning opportunities for these participants.

## Importance of Aligning Assessments to the Defined the Construct

This study demonstrated that aligning an assessment as closely as possible to the defined construct can lead to significantly different interpretations of teachers' knowledge than those assessments that are less aligned. Thus, it is important for assessments to be as well aligned to the construct as possible. The LMT-PR was created with the NCTM standards at its heart (Hill, 2008), however, those standards allowed a broad interpretation of the domain of proportional reasoning, which led to item development that spanned a variety of topics. This led to both limits in the number of items focused on key proportional ideas and the number of items focused on the mathematics to which proportions are connected. For example, there are no questions linking proportions to slope and only two questions that link proportions to similarity. There are also no questions that ask teachers about the definition of a ratio and how that definition might be the same as or different from a fraction. This is a limitation in the LMT-PR's alignment to our construct of interest.

Of course, alignment also relies on robust definitions of the construct to be measured. This is problematic in domains like teacher knowledge where the constructs are not yet well defined. It is also difficult when using existing assessments that may not fit a construct as tightly as necessary.

Our approach of constraining a pre-existing assessment to key mathematical ideas related to the construct yielded more useful information for guiding subsequent instruction and/or making claims about participants' understandings. LMT-PR 54, the version most aligned with our conception of the specialized knowledge teachers should have of mathematics, yielded information at a finer grain size than the less focused versions. In general, it seems that Class 1 found pedagogically focused questions easier (such as selecting easier or harder numbers), whereas Class 2 had more facility with manipulating equivalent ratios. This information is at a fine-enough grain size to help a professional developer plan subsequent instruction.

Additional work would need to be done to translate the findings of this study into a format that would support instructional decision-making for professional development.

But, we assert that the analysis presented here suggests there is promise in using latent class measures, combined with using focused assessments, to provide results that can provide the basis for instructional decision making.

## Limitations

As with all studies, this one has a number of limitations. Here we present three major limitations. First, the national dataset included a number of 4th and 5th grade teachers for whom proportional reasoning is not in the content they teach. This may have led to very different results for the latent class analysis than if the teachers had all been from middle school. Further, our findings are limited by our lack of interview data with members of the latent classes. We could make stronger claims and, perhaps, find more similarities and differences between the latent classes were we able to hear why the teachers selected particular responses. Finally, we were limited in that the definition of the construct measured in the LMT is different from our own definition. This study sought to find the best fit between our construct definition and that LMT-PR, but that is still not as well aligned to our definition as if we had developed an assessment ourselves.

## Conclusion

Clearly, the development of assessments for measuring teacher knowledge is an area in need of much more consideration (Orrill & Cohen, in press). In the domain of proportional reasoning, for example, additional work is needed to define the specific construct of interest: the knowledge teachers should reasonably be expected to know.

Test developers and users could ensure alignment between their construct of interest and the assessment using one of many available techniques that rely on mapping the assessment to the domain of interest. For example, the construct of proportional reasoning as we defined it and the assessment used may have been achieved through reliance on systematic identification of the subconstructs of interest and intentional spread of items across the subconstructs (e.g., Izsák et al., 2010).

Another approach to mapping the domain would be through the use of a Q-matrix, which provides a confirmatory approach to measuring a domain. Q-matrices, critical for cognitive diagnostic models (e.g., Izsák & Templin, in press), require the research team to identify the subconstructs to be measured and indicate which of those subconstructs each item addresses. This allows tracking of all of the subconstructs intended to be measured, thus ensuring not only that the relevant subconstructs are being measured but also that they are being paired in multiple ways so that one idea does not obstruct the other. For example, if we are interested in the use of representations and the teachers' understandings of combining ratios, we would not want all of the combining ratios tasks to include ratio tables because that representation may be unfamiliar to teachers, thus masking the participants' actual knowledge of combining ratios.

In the end, careful consideration of the alignment of the assessment to the construct is important for making claims of validity. Given the widespread move toward using assessments of teacher knowledge for high stakes decision-making, this becomes even more important. The study presented here shows that the same teachers have the perception of more knowledge or less simply based on the items included in the analysis.

## Acknowledgements

Work on this paper was supported by the National Science Foundation through grant number DRL 1054170. The opinions expressed here are those of the authors and do not necessarily reflect the views of the NSF. The authors wish to thank Heather Hill for sharing the LMT-PR data with us as well as Mark Thames for his conversations about MKT. We also wish to thank Eric Gold, Dave Kamin, Tim Marum, Rob Nanna, Dennis Robinson, Ryan Robidoux, and Kaitlyn Walsh Rodrigues for their assistance in the initial analysis of the LMT-PR assessment.

## Works Cited

Alexeev, N., Templin, J., & Cohen, A. (2011). Spurious latent classes in the mixture Rasch model. *Journal of Educational Measurement, 48*, 313–332.

Ball, D. L., Thames, M. H., & Phelps, G. (2008). Content knowledge for teaching: What makes it special? *Journal of Teacher Education, 59,* 389–407.

Baumert, J., Kunter, M., Bum, W., Brunner, M., Voss, T., Jordan, A., Klusmann, U., …, & Tsai, Y. (2010). Teachers' mathematical knowledge, cognitive activation in the classroom, and student progress. *American Educational Research Journal, 47*(1), 133–180.

Bolt, D. M., Cohen, A. S., & Wollack, J. A. (2001). A mixture item response for multiple-choice data. *Journal of Educational and Behavioral Statistics, 26*, 381–409.

diSessa, A. A. (1988). Knowledge in pieces. In G. Forman & P. Pufall (Eds.), *Constructivism in the computer age* (pp. 49–70). Hillsdale, NJ: Lawrence Erlbaum Associates, Inc.

diSessa, A. A. (2006). A history of conceptual change research: Threads and fault lines. In R. K. Sawyer (Ed.). *The Cambridge handbook of the learning sciences* (pp. 265–282). New York: Cambridge University Press.

Embretson, S. E., & Reise, S. P. (2000). *Item response theory for psychologists*. Mawhah, NJ: Lawrence Erlbaum.

Hambleton, & Swaminathan, H. (1985). *Item response theory: Principles and application.* Boston, MA: Kluwer-Nijhoff Publishing.

Harel, G., & Behr, M. (1995) Teachers' solutions for multiplicative problems. *Hiroshima Journal of Mathematics Education, 3*, 31–51.

Hill, H. C. (2008). *Technical report on 2007 proportional reasoning pilot Mathematical Knowledge for Teaching (MKT) measures learning mathematics for teaching.* Ann Arbor, MI: University of Michigan.

Hill, H. C., Rowan, B., & Ball, D. L. (2005). Effects of teachers' mathematical knowledge for teaching on student achievement. *American Educational Research Journal, 42*(2), 371–406.

Izsák, A., Orrill, C. H., Cohen, A., & Brown, R. E. (2010). Measuring middle grades teachers' understanding of rational numbers with the mixture Rasch model. *Elementary School Journal, 110*(3), 279–300.

Izsák, A., Jacobson, E., de Araujo, Z., & Orrill, C. H. (2012). Measuring growth in mathematical knowledge for teaching fractions with drawn quantities. *Journal for Research in Mathematics Education, 43*(4), 391–427.

Izsák , A., & Templin, J. (in press). Coordinating descriptions of mathematical knowledge with psychometric models: Opportunities and challenges. In A. Izsák, J. T. Remillard, & J. Templin (Eds.), *Psychometric methods in mathematics education: Opportunities, challenges, and interdisciplinary collaborations* (pp. xxx–xxx). Journal for Research in Mathematics Education monograph series. Reston, VA: National Council of Teachers of Mathematics.

Kersting, N. B., Givvin, K. B., Sotelo, F. L., & Stigler, J. W. (2010). Using video to predict student learning of mathematics: Further explorations of a novel measure of teacher knowledge. *Journal of Teacher Education, 61*(1–2), 172–181.

Lamon, S. J. (2007). Rational numbers and proportional reasoning. In F. K. Lester (Ed.), *Second handbook of research on mathematics teaching and learning* (pp. 629–667). Charlotte, NC: Information Age Press.

Learning Mathematics for Teaching (2007). *Survey of teachers of mathematics: Form LMT PR-2007.* Ann Arbor, MI: University of Michigan.

Li, F., Cohen, A.S., Kim, S.-H., & Cho, S.-J. (2009). Model selection methods for dichotomous mixture IRT models. *Applied Psychological Measurement, 33*(5), 353–373.

Lobato, J., & Ellis, A. B. (2010). *Essential understandings: Ratios, proportions, and proportional reasoning.* In R. M. Zbieck (Series Ed.), *Essential understandings.* Reston, VA: National Council of Teachers of Mathematics.

Lord, F. N. (1980). *Applications of item response theory to practical testing problems.* Hillsdale, NJ: Erlbaum.

Manizade, A. G., & Mason, M. M. (2011). Using Delphi methodology to design assessments of teachers' pedagogical content knowledge. *Educational Studies in Mathematics, 76*, 1883–207. DOI: 10.1007/s10649-010-9276-z

Mislevy, R. J., & Verhelst, N. (1990). Modeling item responses when different subjects employ different solution strategies. *Psychometrika, 55*, 195–215.

National Council of Teachers of Mathematics (NCTM) (2000). *Principles and standards for school mathematics.* Reston, VA: Author.

National Governors Association & Council of Chief State School Officers (NGA & CCSSO) (2010). *Common core state standards* mathematics. Washington, DC: Author.

Orrill, C. H., & Cohen, A. (in press). Purpose and conceptualization: Examining assessment development questions through analysis of measures of teacher knowledge. To appear in *Journal for Research in Mathematics Education Monograph.*

Pitta-Pantazi, D., & Chritou, C. (2011). The structure of prospective kindergarten teachers' proportional reasoning. *Journal of Mathematics Teacher Education, 14*(2), 149–169.

Post, T., Harel, G., Behr, M., & Lesh, R. (1988). Intermediate teachers' knowledge of rational number concepts. In Fennema, et al. (Eds.), *Papers from First Wisconsin Symposium for Research on Teaching and Learning Mathematics* (pp. 194–219). Madison, WI: Wisconsin Center for Education Research.

Riley, K. R. (2010). Teachers' understanding of proportional reasoning. In P. Brosnan, D. B. Erchick, & L. Flevares (Eds.), *Proceedings of the 32$^{nd}$ annual meeting of the North American Chapter of the International Group for the Psychology of Mathematics Education* (pp. 1055–1061). Columbus, OH: The Ohio State University.

Rost, J. (1990). Rasch models in latent classes: An integration of two approaches to item analysis. *Applied Psychological Measurement, 14*, 271–282.

Rost, J. (1997). Logistic mixture models. In W.J. van der Linden & R.K. Hambleton (Eds.), *Handbook of modern item response theory* (pp. 449–463). New York: Springer.

Schwarz, G. (1978). Estimating the dimension of a model. *Annals of Statistics, 6*, 461–464.

Shulman, L. S. (1986). Those who understand: Knowledge growth in teaching. *Educational Researcher, 15*(2), 4–14.

Silverman, J., & Thompson, P. W. (2008). Toward a framework for the development of mathematical knowledge for teaching. *Journal of Mathematics Teacher Education, 11*, 499–511. DOI: 0857-008-9089-5

Son, J. (2010). Ratio and proportion: How prospective teachers respond to student errors in similar rectangles. In P. Brosnan, D. B. Erchick, & L. Flevares (Eds.), *Proceedings of the 32$^{nd}$ annual meeting of the North American Chapter of the International Group for the Psychology of Mathematics Education* (pp. 243–251). Columbus, OH: The Ohio State University.

Spielgelhalter, D., Thomas, A., Best, N., & Lunn, D. (2007). *WinBUGS with DoodleBUGS.* Medical Research Council, Imperial College and MRC, UK.

# Knowledge for Equitable Mathematics Teaching: The Case of Latino ELLs in U.S. Schools

Aaron T. Wilson (aaron.wilson@utrgv.edu)[1]
University of Texas Rio Grande Valley, USA

**Abstract:** This paper reports the exploration of an aspect of knowledge needed for equitable mathematics teaching. Pedagogical Content Knowledge for Teaching Mathematics to English Language Learners (PCK-MELL) was proposed as a theoretical knowledge construct, a subdomain of MKT, and the construct was investigated through a process of survey instrument development and administration. The survey contained items intended to measure teachers' knowledge of the obstacles encountered by ELLs in math classes, of the resources that ELLs draw upon, and of instructional strategies for teaching ELLs. Analysis of middle school mathematics teachers' responses ($N = 42$) offered insights into how to improve the reliability and measurement validity of this sort of instrument, as well as directions for further theory development.

*Key words:* English Language Learners; mathematical knowledge for teaching; deficits; affordances

## Introduction

Many mathematics teachers worldwide are finding new languages and new cultures in their classrooms. Since the 1970s the number of students who are English Language Learners (ELLs) in United States schools has grown and continues to grow dramatically (U.S. Department of Education, 2012; Payán & Nettles, 2008; Francis et al., 2006; Capps et al., 2005). Mathematics education researchers and others have observed, based upon the lower relative performance of these students on standardized exams of mathematics achievement in comparison with their mainstream (non-ELL) counterparts, that many of these ELLs have infrequently had equitable opportunity to learn mathematics in U.S. schools and have often been underserved by their teachers and schools (Center on Education Policy, 2010; Abedi & Herman, 2010). Furthermore, researchers have noted that using measurements of mathematical achievement to draw attention to disparities between student groups in fact offers little, if any, contribution toward promoting equity in education (Gutiérrez, 2008), and that focusing on achievement gaps may only perpetuate negative stereotypes. Yet, measurements of mathematical achievement have also shown that some teachers and school districts appear to serve their ELLs better than do others.

---

[1] A portion of this research was conducted in pursuit of a doctoral dissertation under the direction of M. Alejandra Sorto, Texas State University-San Marcos. The research was funded in part by the National Science Foundation Grant 1055067, M. Alejandra Sorto, principal investigator. Any opinions, findings, and conclusions or recommendations expressed in this material are those of the author and do not necessarily reflect the views of the National Science Foundation. The author would also like to express appreciation to Drs. Mark Hoover and Reidar Mosvold as well as to the other anonymous reviewers of The Mathematics Enthusiast for the thoughtful feedback given.

For example, consider the following phenomenon that occurred in a large southwestern state in which approximately 16% of all public school students are ELLs[2]. The Samsonville School District and the Wilkins School District (pseudonyms) both had significant percentages of ELL students, 25% and 33% of their approximately 80,000 and 50,000 respective student populations. Samsonville ELLs were approximately 85% Latino students, i.e., students from Spanish-speaking backgrounds, while Wilkins' ELLs were 99% Latinos. Sixty percent of Samsonville students were classified as economically disadvantaged compared to 95% of Wilkins students. Furthermore, less than 40% of Samsonville teachers were Latinos, while in the Wilkins district greater than 70% of teachers were Latinos. Although the characteristics of the students who were ELLs in these two school districts were largely equivalent in terms of language background and ethnicity, over several years Wilkins ELLs performed significantly higher than Samsonville ELLs on the mathematics portion of the state standardized test. To what could this differential performance be attributed?

In trying to explain the above phenomenon, one may ask the obvious question of whether the teachers in the Wilkins district do something differently than do the Samsonville teachers which results in the differential mathematics achievement among ELLs. One may also ask the question of whether there is something that the Wilkins teachers know about teaching mathematics to ELLs that capacitates them to more effectively instruct ELLs and that is less well known among Samsonville teachers. That is, one may seek to explain this phenomenon from the perspective of instructional practice or from the perspective of the teachers' knowledge that informs their practices.

There are also several reasons why it may be valuable to investigate a phenomenon such as this from the later perspective, that of teachers' knowledge. Researchers concerned with the assessment of ELLs have asserted that "with the rapid growth of ELL populations states should place a substantial focus on increasing teacher knowledge of current ELL issues…including pre-service teacher education and continuing teacher education" (Wolf, Herman, & Dietel, 2010, pp. 8–9). Hence, successfully characterizing knowledge that promotes achievement among ELLs would add to theories of mathematics teachers' knowledge. It would also fill a void of content for educator textbooks and professional development materials useful for equipping teachers for the work of teaching mathematics to ELLs (Watson et al., 2005). Ultimately and importantly, it would more fully complete the picture of essential elements that inform equitable mathematics teaching.

The research reported in this paper was done as part of a larger study that attempted to explain the difference in ELL achievement seen in the Samsonville and Wilkins school districts from the viewpoint of teacher qualities. That larger study looked at a number of important teacher and student variables in the hope of identifying the teacher qualities and instructional moves that resulted in higher mathematics achievement for ELLs. One among the many teacher variables to be measured in as a possible predictor of ELLs' achievement was teachers' mathematical knowledge for teaching, MKT (Ball, Thames, & Phelps, 2008). Yet, because of the

---

[2] This account regarding the Samsonville and Wilkins school districts arose through conversations between school district administrators and university researchers. Reporting of state standardized test results on the website of the state department of education website between 2002 and 2009 had revealed the higher performance of Wilkins ELLs on the math portion of the state standardized test. Administrators in some districts around the state and nation expressed interest in knowing what Wilkins teachers were "doing" with their ELLs. The study that resulted in part from those conversations was funded by the National Science Foundation (DRL-1055067).

special cultural and linguistic context of that study, the researchers sought an instrument that could capture teachers' MKT that informed their work with students who came to school with languages and prior mathematics learning that might vary from English and from traditional mathematical algorithms and notations used in U.S. schools. However, at the time of the study there were—and still are today—several limitations to doing an investigation of this sort, i.e., a study that can describe and measure the kind of mathematical knowledge needed by teachers of students who are still learning the language of instruction, in this case ELLs. The foremost of these is that, while there exist abundant strategies for teaching mathematics to ELLs, there is a shortage in theory about what effective math teachers of ELLs really need to know. Because of this lack of theory there is also a lack of research tools for investigating this knowledge. For example, most current instruments that measure mathematics teachers' knowledge – and the Learning Mathematics for Teaching (LMT) measures in particular (Hill & Ball, 2004) – do not address ELLs in particular, or aspects related to equity in general, and as a result, may fail to capture many aspects of knowledge for equitable teaching.

This paper narrates an initial attempt to create a research instrument capable of measuring knowledge for teaching mathematics as it is used by teachers in linguistically and culturally diverse classrooms containing large number of ELLs. The work reported here is fundamental in the sense that it was done in hopes of "laying the bearings" for observing MKT in linguistically and culturally diverse contexts and of building the capacity to study this aspect of knowledge for equitable mathematics teaching. Specifically, the goal of this study was to create a survey instrument that could be used to capture the particular kind of knowledge that mathematics teachers like those in the Wilkins district called upon when teaching ELLs. But how could that kind of mathematical knowledge be defined? To what extent could a survey be designed that measures that knowledge reliably? And further, to what extent does it even seem valid to define knowledge for teaching mathematics to ELLs as a special kind of knowledge? Is this a valid construct? Answers to these important questions would determine the usefulness of the survey to be created. Responses to questions like these are given here in the hope that the interested reader may appreciate the nuanced nature of the knowledge being addressed and in hopes that the foundational attempt described in this paper to measure that knowledge may only open the way for even more illuminating work along these lines.

To summarize, if we are to understand the differential performance that some mathematics teachers have with ELLs, like that described above, and to use this understanding to promote equitable access to mathematics instruction for all students, then there is the need to better understand the particular role that teachers' knowledge plays in the context of teaching math to ELLs. Thus, there is an initial need to first develop and test theory related to mathematics teachers' knowledge for teaching ELLs and to then develop research tools capable of observing and describing such knowledge. The ultimate goal of the study reported here was to produce a viable research instrument for measuring teachers' knowledge for teaching mathematics to ELLs and this process was guided by theory in survey and scale development (DeVellis, 2003; Dillman, Smyth, & Christian, 2009; Schuman & Presser, 1996). As a result, this work followed a somewhat linear fashion that commenced with a review of literature with the purpose of defining a theoretical construct, that is, defining what knowledge is needed by mathematics teachers of ELLs. Following this a test blueprint was outlined and original survey items were developed according to the knowledge domains defined in the blueprint. Finally responses were obtained from math teachers for the main purpose of evaluating the reliability and validity of the measure. The sections of this paper that follow elucidate the path that was taken in trying to capture this knowledge, showing in detail how knowledge for teaching ELLs mathematics was hypothesized based upon the research literature and classroom observations conducted, how the instrument was developed and administered, and finally what was learned about this knowledge as a result of administering the survey.

## Conceptualizing Mathematical Knowledge for Teaching ELLs and Developing a Measure
### Consulting the Literature: Connecting MKT to ELLs

As an entry into investigating the kind of knowledge needed for equitable mathematics teaching generally, by looking at mathematics teachers of ELLs specifically, this study began by examining the intersection between two existing strands of inquiry: research concerning mathematics teachers' knowledge and research concerning ELLs in the mathematics classroom. Highlights from some of the essential theory and findings that guided this study follow.

An important component of math teachers' knowledge is profound mathematical content knowledge (Ma, 1999, for example) and this kind of deep mathematical knowledge was assumed here to be important for math teachers of ELLs as well. Furthermore, it was assumed that, like all teachers, teachers of ELLs have *knowledge-in-practice* (Schon, 1983), i.e., knowledge that is gained by their practice of teaching ELLs and from their instructional experiences related to the particular linguistic and cultural background that their ELLs bring to the classroom. Shulman (1986) connected content knowledge to this kind of practice-based knowledge by explaining how teachers' knowledge of the content that they teach is shaped by the pedagogy that they practice. To Shulman (1986), *pedagogical content knowledge* (PCK) "includes an understanding of what makes the learning of specific topics easy or difficult" as well as "the most useful forms of representation of those ideas…—in a word, the ways of representing and formulating the subject that make it comprehensible to others" (p. 9). The above utterances concerning teachers' PCK, which have admittedly been reprinted in many a literature review because of their lucidity, are repeated here because of the role that the terms *easy*, *difficult*, and *representations* played in determining a connection between the two strands of research central to the present study and in informing survey development, as is explained in a later paragraph.

Recent research concerning mathematical knowledge for teaching, MKT, has given a robust explication of Shulman's (1986) PCK in the context of mathematics teaching (Ball, Thames, & Phelps, 2008; Hill, Ball, & Schilling, 2008; Hill, Blunk, Charalambous, Lewis, Phelps, Sleep, & Ball, 2008; Hill, Rowan, & Ball, 2005). These researchers proposed the existence of a system of related knowledge constructs, some more mathematical and some more pedagogical, that together constitute MKT. Furthermore, in their Learning Mathematics for Teaching (LMT) project, they have investigated this knowledge through the development of psychometric instruments, pen-and-paper (survey) tools, for measuring the knowledge in these domains (Hill, Schilling, & Ball, 2004).

Two subdomains of MKT, as theorized by the above researchers, were identified as particularly relevant to this study: knowledge of content and students (KCS) and knowledge of content and teaching (KCT). These researchers define KCS as "knowledge of how students think about, know, or learn this particular content" (Hill, Ball, & Schilling, 2008, p. 375). Furthermore, they define KCT as "mathematical knowledge of the design of instruction, [including] how to choose examples and representations, and how to guide student discussions toward accurate mathematical ideas" (Hill, Ball, Sleep, & Lewis, 2007). Within this model of PCK, the two domains of KCS and KCT seemed to be of special interest to the present study because they include awareness of students' background and awareness of instructional decisions appropriate to the teaching of specific mathematics topics. The particular linguistic and cultural qualities that ELLs bring to the learning of mathematics, which qualities often vary from those found in non-ELLs, seemed to have potentially the most practical impact on these two domains of teachers' PCK. Knowledge of mathematics and of ELLs and of appropriate instructional decisions that can be made on their behalf seemed central to the kind of knowledge under investigation by this study.

Moschkovich (2002) has investigated ELLs in the mathematics classroom extensively and was among the first mathematics education researchers to begin to make a shift in the perspective taken when studying ELLs. She took a sociocultural view of ELLs as mathematics learners and observed that, historically, much research concerning ELLs had taken a deficit perspective which focuses the research lens on the obstacles, sources of *difficulty*, that ELLs encounter in mathematics classrooms. Because of this lens, such studies, she noted, had failed to observe the resources that ELLs often draw upon to make *easier* the learning and expressing of mathematics. Factors that have been observed to cause difficulty for ELLs in mathematics classes are many, including, for example: language of instruction and limited English proficiency (Cuevas, 1984), word problems and linguistic complexity (Llabre & Cuevas, 1983; Martiniello, 2009), polysemy (Lager, 2006), and whole-class, teacher-centered instruction format (Chang, 2008), to name a few. Conversely, much research since the 1990s has taken note of factors that serve ELLs as resources upon which they can draw to do and express valid mathematics—even if at times using grammatically invalid English, or even no English at all. Such factors include: gesturing (Shein, 2012), first language and bilingualism (Gutiérrez, 2002; Khisty & Morales, 2004; Moschkovich, 2002; Sorto, Mejía Colindres, & Wilson, 2014), non-linguistic mathematical representations (Martiniello, 2009), and prior and cultural knowledge (Gutiérrez, 2002; Gutstein, Lipman, Hernandez, & Reyes, 1997; Henderson & Landesman, 1995). Studies mentioned in this paragraph represent ways in which researchers have investigated the obstacles that ELLs face in mathematics classrooms and have also recently begun to perceive the strengths that many ELLs possess for learning mathematics.

Here a connection can be made between mathematics teachers' knowledge and ELLs. Moschkovich (2002) saw a dichotomy in perspectives taken by mathematics education researchers concerned with ELL issues; they took either a deficit or an affordance perspective. And as has been mentioned above, Shulman (1986) also saw a dichotomy; he saw that teachers' knowledge includes an understanding of the things that make learning *easy* or *difficult*, as mentioned earlier. These two sets of extremes—deficits versus affordances and difficult versus easy—constitute the link found in the present study between the two bodies of research concerning mathematics teachers' knowledge and ELLs. More precisely, the two divergent perspectives concerning ELLs of deficits and affordances, with their different research perspectives, were taken as potentially illuminating the very domains of PCK—KCS and KCT—central to teaching mathematics to ELLs. It seemed that experienced mathematics teachers of ELLs should potentially have pedagogical content knowledge by which they perceive both the *obstacles* (deficits) that their ELLs face and also the *resources* (affordances) that their ELLs draw upon in mathematics classes. These two divergent research perspectives (and their respective results) informed the method used in this study significantly, as will now be seen.

## Developing the PCK-MELL Survey

In developing the survey in this study it was hoped that the understanding of knowledge needed by mathematics teachers of ELLs as well as the survey itself could be closely tied to current conceptions of mathematics teachers' knowledge, with a view to possibly investigating the relationship between this construct and related constructs in future studies. As a result, the theoretical basis of the survey was conceived of in connection with the theory behind MKT and the items themselves were written in a similar fashion as the multiple-choice LMT measures (Hill & Ball, 2004), although the measurement purpose and the mathematical topics and contexts were all novel. The process of survey development became a means of exploring this aspect of mathematics teachers' knowledge.

### Pedagogical Content Knowledge for Teaching Mathematics to ELLs

Drawing from a review of the literature, a sample of which is given above, a hypothetical framework of pedagogical content knowledge for teaching mathematics to ELLs (PCK-MELL) was first developed (Figure 1).

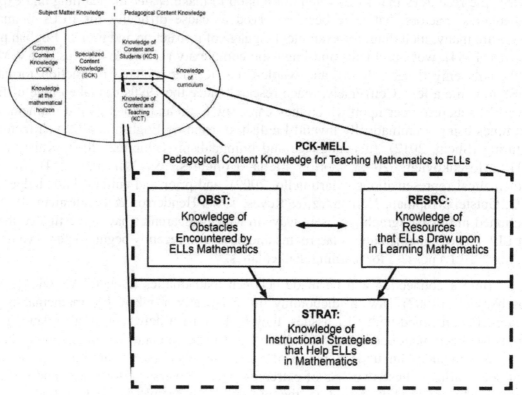

*Figure 1.* Hypothetical framework of PCK-MELL, proposed here as a subset of MKT (Hill, Ball, & Schilling, 2008).

The framework in Figure 1 makes several structural hypotheses regarding PCK-MELL. Firstly, it identifies the construct as a subset of pedagogical content knowledge and places it squarely within the larger framework of MKT (Hill, Ball, & Schilling, 2008). Secondly, it further embeds PCK-MELL within this framework by identifying it with two specific domains of MKT: knowledge of content and students (KCS) and knowledge of content and teaching (KCT). Thirdly, it posits three subdomains of PCK-MELL: knowledge of obstacles often encountered by

ELLs in math classes (OBST), knowledge of resources that ELLs draw upon when learning math (RESRC), and knowledge of instructional strategies for usage with ELLs in math classes (STRAT). Although not referenced in the brief summary of literature above, knowledge of instructional *strategies* was included because it seemed central to both Shulman's (1986) understanding of PCK and to Ball, Thames, & Phelps' (2008) theorization of MKT (specifically the KCT domain), and because a considerable number of strategies have been posited for teaching ELLs (see for example, Coggins, Kravin, Coates, & Carroll, 2007; Echevarría, Short, & Vogt, 2007). Finally, the arrows between the OBST and RESRC domains and between these and the STRAT domain above suggests a theoretical relationship between these three subdomains: knowledge of obstacles may be related to knowledge of resources, and knowledge in both of these subdomains may inform teachers' knowledge of, and especially selection of, strategies. (It may not be at once clear to the reader why the OBST and RESRC domains were suspected to be related; indeed, the affordance and deficit perspectives of researchers looking at ELL students seem to contradict each other significantly. This suspicion arose through reflection on the conceptual process of writing items in both of these domains and the grounds for this suspicion will be clearly explained in the presentation of the OBST survey item below.)

Furthermore, the following testing framework (Table 1) was crafted to serve as a guide to survey item development. The testing framework again delineates the three proposed knowledge domains of OBST, RESRC, and STRAT and then associates with these domains a number of specific aspects of knowledge within each domain. All of the aspects provided in this framework were derived specifically from research results. That is, in so far as it seemed possible to classify select findings from the literature as either identifying and explaining particular obstacles encountered by ELLs (such as linguistic complexity, Martiniello, 2009), or as describing resources that ELLs draw upon (such as their bilingualism, Moschkovich, 2002), or as enumerating instructional strategies for usage with ELLs (see for example, Coggins, Kravin, Coates, & Carroll, 2007), then those concepts from the literature concerning the mathematics education of ELLs were placed in the framework and hypothesized to be elements of PCK-MELL about which experienced teachers of ELLs should be familiar. At least three assumptions are being made here: first, that the research findings concerning ELLs accurately depict elements of the 'real' situation concerning many ELLs in mathematics classes, second, that the findings are generalizable in contexts beyond those in which the research occurred, and third, that teachers who have PCK-MELL will have gained, possibly through formal training but much more likely through actual teaching experience, an understanding of at least some of those very same elements described in the research literature and identified in the testing framework. In essence the reasoning behind the usage of the testing framework was as follows: if research such as Martiniello's (2009) finds that the linguistic complexity of word problems is an explanative factor in the differential (lower) performance of ELLs in mathematics, then not only is the difficulty caused by linguistic complexity something about which teachers of ELLs *should be aware*, but it even seems probable that experienced teachers of ELLs *will actually be aware* of ways in which linguistically complex word problems can cause difficulty for their students. The survey items were then written for the purpose of capturing this awareness (or lack of awareness). Assumptions like these may be dangerous. Yet, some assumptions had to be made since this effort was an initial attempt at hypothesizing about the very nature and contents of the knowledge domains in question. It was hoped that results from the survey administration would either validate, or indeed invalidate and lead to improvements of, the survey.

Two final notes regarding the testing framework are in order. First, the aspects given in the STRAT domain are merely an adaption of Chval & Chávez's (2011) synthesis of research-based strategies for instructing ELLs in mathematics. Many specific strategies for teaching ELLs have been posited and their list seemed to be general enough that specific tools such as "word walls", posters and manipulatives, for example, could be taken as mere instances of the broader aspects of strategic knowledge in the list. Finally and most importantly, it must be made clear that this testing framework is not now and was not believed at the time of the study to be exhaustive. There are, no doubt, many more aspects of the knowledge domains that are not represented here. The items in the testing framework are a best attempt given limited time and resources.

Table 1. *PCK-MELL Testing Framework*

| DOMAIN | | ASPECTS |
|---|---|---|
| **OBST** Knowledge of Obstacles encountered by ELLs in mathematics classes | | 1. Limited English proficiency in speaking, reading and writing |
| | | 2. Word problems<br>    a. Specific words (vocabulary), multiplicity of words (linguistic complexity), shifts of application, polysemy |
| | | 3. Classroom format<br>    a. High speech formats (direct teaching versus indirect or peer-based methods) |
| | | 4. Assessments<br>    a. Low performance because of:<br>        i. Word problems, time limitation, high stakes, cultural-irrelevance |
| **RESRC** Knowledge of Resources that ELLs draw upon in learning mathematics | | 1. Linguistic creativity, mathematics discursive and communicative ability<br>    a. Fluency in $L_1$ (i.e., first language)<br>    b. Bilingualism<br>    c. Usage of gestures, objects, and verbal *inventions* to convey meaning |
| | | 2. Linguistic and cultural identity<br>    a. Personal association with cultural icons, people, values, traditions, etc.<br>    b. Appreciation of first language |
| | | 3. Prior mathematical knowledge including knowledge of and fluency with alternative or "foreign" mathematical notations and algorithms |
| **STRAT** Knowledge of instructional strategies that help ELLs in mathematics | | 1. Usage of students' background knowledge—academic, linguistic and cultural—to promote understanding |
| | | 2. Maintenance of classroom environment rich in linguistic and mathematics content |
| | | 3. Emphasis on meanings of words and/or provisions for students' usage of multiple modes of communication to express mathematics |
| | | 4. Usage of visual supports to—gestures, objects, illustrations—to convey the meanings of classroom conversations |
| | | 5. Connection of mathematical language with multiple forms of mathematical representation |
| | | 6. Available visual display of classroom mathematics concepts, representations and words during instruction |
| | | 7. Rich usage of students' own mathematical writings and speech with opportunity for them to make revisions |

**Survey Development**

To inform and augment the development of the PCK-MELL theoretical and testing frameworks, more than thirty hours of original classroom observations were also conducted in middle school (and a few high school) mathematics classrooms composed of large numbers of ELLs with a view to 1) situating the development of survey items within authentic and mathematically specific teaching instances while 2) populating the testing framework with both research-based and classroom-observed aspects of the three knowledge domains. Then, based upon the testing framework and observations, a large number of survey items were developed, from which an approximately equivalent number of items across the three domains of OBST, RESRC, and STRAT were retained for the final 32-item instrument after a pilot version had been sent for review and comment to a number of mathematics educators and experts in ELL and teacher knowledge issues at different institutions in the United States. To give a sense of the final instrument which was the basis of the statistical analysis that follows, three exemplary survey items are given here.

The PCK-MELL survey item presented below was derived from a classroom observation in which a math teacher taught a geometry lesson to 9th and 10th grade students who were recent immigrants to the United States, more than 95% of which were Latinos. Both the mathematical situation and the answer options are authentic in the sense that the teaching situation and words used actually happened in a mathematics classroom. Furthermore, the mathematical topics and teaching tasks represented in this items are typical of the work of mathematics teaching. The "correct" answer to the item, as with all items in the survey, was based upon theory found in the research literature concerning effective (or ineffective) practice for teaching math to ELLs. The item below was designed to measure teachers' knowledge of the linguistic obstacles (OBST) that ELLs may encounter in mathematics classes.

---

Mrs. Tash's students, many of whom are Latino English Language Learners, have been working in groups to discover how surface area changes when the dimensions of an object are changed by a scale factor. On the previous class day, students used rulers and a rectangular prism to complete the table below.

| Picture of Solid | Name of Solid | Dimensions | Scale Factor | New Dimensions | New Surface Area, SA | Conclusions |
|---|---|---|---|---|---|---|
| ` | | | k = 2 | | | |

Which word in the table may cause the most trouble for the English Language Learners in the classroom?

    A.    Solid
    B.    Dimensions
    C.    Conclusion
    D.    Surface

---

*Figure 2*. Sample PCK-MELL survey item, OBST domain.

The item in Figure 2 presents several English mathematical words that could be unknown to English Language Learners. Among these four words, three of them (solid, dimensions, and conclusion) have direct Spanish-English cognates (*sólido*, *dimensiónes*, and *conclusión*) and may be more readily known to Latino ELLs as a result. Furthermore, in classrooms of Latino ELLs teachers' knowledge of the Spanish language may be helpful in answering the particular item above. To control for this effect knowledge of the Spanish language was assessed as a covariate on the survey as well. Yet it is also not difficult to imagine that even the observant, but non-Spanish-speaking teacher of ELLs may perceive *through teaching experience alone* both the English words that cause difficulty for ELLs and the words that ELLs can use to leverage their understanding. That is, non-Spanish-speaking teachers may also gain through experience an awareness of the English words that their Spanish-speaking ELLs stumble on or readily comprehend.

Furthermore, this item and the theory behind its hypothetically "correct" answer can be used to show why knowledge of *obstacles* encountered by ELLs in math was hypothesized to be related to knowledge of their *resources*: while language of instruction and linguistic complexity can cause difficulty for ELLs on one hand (Cuevas, 1984; Llabre & Cuevas, 1983; Martiniello, 2009), the ELLs' first language and their bilingualism can also give them access to the same mathematical ideas on the other (Gutiérrez, 2002; Khisty & Morales, 2004; Moschkovich, 2002; Sorto, Mejía Colindres, & Wilson, 2014). Teachers' knowledge of both of these functions of language in mathematics classrooms may be related. As a result, teachers' knowledge of the obstacles to and resources for ELLs' learning of mathematics may also be related.

The following figure represents a survey item in the RESRC domain.

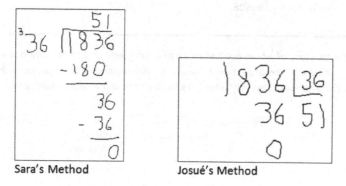

**To solve this problem, students had to divide 1836 by 36. Sara and Josué used these two different methods.**

Sara's Method          Josué's Method

**Who's division method is correct?**

A. Both methods are correct

B. Josué's

C. Sara's

*Figure 3.* Sample PCK-MELL survey item, RESRC domain.

The survey item above presents to two slightly different notations of the long division algorithm and then requires the respondent to select the correct method for performing the long division. Sara's method is equivalent to the standard algorithm traditionally used and taught in schools in the United States while Josue's method is an equivalent algorithm, but that uses notation commonly found in Central America. This item was inspired by conversations with

Latino immigrants to the United States for whom this notation had been the standard for long division. The item was included in the PCK-MELL survey as a potential indicator of teachers' practical knowledge of resources that some ELLs (especially Latino immigrant students) may bring to the math classroom, namely, their prior mathematical knowledge, including valid, yet perhaps alternative, mathematical algorithms and notation.

The final item below was intended to serve as a STRAT item, measuring teachers' knowledge of effective instructional strategies for teaching ELLs.

Mr. Garza asks his class a question about the contents of the grape-limeade drink below: "How do you find out what proportion the grape juice is of the drink?"

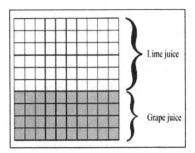

Ciarra, an English Language Learner, responds: "Cuenta los grises y lo divides entre cien."
Which is the BEST way for Mr. Garza to respond?

    A.   He should ask Ciarra to rephrase her answer using English

    B.   He should acknowledge that Ciarra tried to answer his question, but then ask another student to respond in English

    C.   He should acknowledge that Ciarra tried to answer his question, but then explain the correct answer in English so that all of the students understand

    D.   He should ask Ciarra to give her response again, but to use the pictured image and gestures to show what she means

*Figure 4.* Sample PCK-MELL survey item, STRAT domain.

The survey item above, taken from a situation that occurred during the observation of a 7th grade mathematics classroom having many ELLs, involves a strategic decision on the part of the teacher: how should the teacher respond to the student who has answered the English mathematics question using the Spanish language? (And does the fact that the student actually gave a concise and mathematically valid answer affect teachers' responses to the item? Again, responses to this item may also be informed by the teachers' knowledge or ignorance of the Spanish language and this effect was controlled for in asking respondents concerning their level of knowledge of that language.) Originally, this item was intended to serve as an indicator of teachers' knowledge of mathematics instructional strategies for teaching ELLs, the STRAT domain. Yet, as the theoretically "correct" answer to this question was actually option "D", then selecting the correct also implied knowledge that ELLs' first language was a resource in mathematics classes rather than a deficit. Not surprisingly then the item was later found to correlate more strongly with other items in the RESCRC domain. Furthermore, the intersection (overlapping) of knowledge domains required by items like these hints at a particular weakness of this initial exploratory effort at defining the construct: in hindsight, the domains were not altogether well-defined and items such as this, which may serve as indicators of knowledge in more than one domain, reflect the need for even more careful theory development and rigorous item-writing.

## Evaluating the Instrument

Following the theoretical development and writing of survey items, the intention in this study was to obtain enough responses in order to be able to evaluate the extent to which the instrument developed above measured the knowledge that it indeed was designed to measure. The survey instrument described above first underwent pilot study using pre-service mathematics teachers ($N$ = 142) at a major university in a southwestern U.S. state having large numbers of ELLs. Following this, responses to the final instrument were obtained via internet survey from forty-two ($N$ = 42) in-service middle school mathematics teachers around the state. Thirty-one of the teachers were female, eleven male, and they ranged widely in experience from novice teachers to teachers having over twenty years of classroom experience; 40.4% of the teachers had taught for ten or more years. Teachers that took the survey were from more than twenty-three different schools drawn from more than sixteen different school districts distributed across the state. Although an initial attempt at collecting a large random sample was made, and indeed the final sample bears similar demographic characteristics to that of the teacher population of the state, the final sample of 42 teachers was altogether one of convenience as all contacts were made with teachers through the ultimate mercy of cooperative school administrators. Brief results regarding the reliability and validity of the measures obtained follows.

### Reliability

As explained above, the PCK-MELL survey was built from a framework hypothesizing three underlying factors: OBST, RESRC and STRAT. Hence, the computation of Cronbach's (1951) α, a standard measure of internal consistency for surveys and which assumes the unidimensionality of the instrument, using all of the items as a single factor would in fact yield an inappropriate measure of reliability. Nevertheless, as this survey was exploratory in the sense that both the construct, the framework and the items were all novel, this alpha was computed at the outset as a starting point for further investigation of internal consistency. Cronbach's alpha, equivalent to Kuder-Richardson's formula 20 for dichotomously scored items (correct or incorrect), was .431 for the whole set of 32 items of the instrument. This alpha was taken as the first potential confirmation of multidimensionality.

To better understand the factor structure of the test, a combination of evidences from the theoretical orientation of the items, along with from Item Response Theory (IRT), factor analysis, and Cronbach's alpha were used to separate the items into different scales. First, the following two-parameter IRT model was used to compute difficulty and discrimination coefficients for the full set of 32 items: $p\left(x_j = 1 \mid \theta, a_j, \delta_j\right) = \dfrac{e^{a_j(\theta - \delta_j)}}{1 + e^{a_j(\theta - \delta_j)}}$. In comparison with classical test theoretic (CTT) measures, like Cronbach's alpha, which takes the observed score on the entire instrument as the unit of analysis, IRT models take the item as the unit of analysis (De Ayala, 2009). IRT is an important tool in psychometric test development. However, like Cronbach's alpha, IRT models assume unidimensionality. Hence, the initial IRT model was unstable, with about an equal number of items having positive discrimination coefficients and negative discrimination coefficients. The sign of these coefficients, positive or negative, was then used to relegate items to one of two scales, the first concerned more with the knowledge of both obstacles and resources (OBST/RESRC) and the second concerned more with knowledge of strategies (STRAT). The numerical assignment of items (using the positive or negative sign of the discrimination coefficient) held a high degree of agreement with the theoretical orientation of the items. That is, items with positive coefficients were more closely aligned theoretically with each other than with items having negative coefficients. Furthermore, the dual classification of items achieved through the IRT coefficients was also verified by confirmatory factor analysis (Principal Component Analysis with Promax rotation) in the case of all but one of the items. Finally, eleven (11) items were excluded from this analysis entirely because of weak correlation (Pearson $r$) with items in either of the two scales. The resulting two scales, OBST/RESRC and STRAT, had Cronbach's alpha reliabilities of .621 and .606, respectively. These two scales were used in the following assessment of the validity of the measures.

**Validity**

Measurement validity of the scales was investigated using the concept of a nomological net (Cronbach & Meehl, 1955) and its implied measures of convergent (and discriminant) validity (Campbell & Fiske, 1959). The survey included several questions related to teachers' education and licensure, to teachers' experience, and to teachers' linguistic knowledge, variables hypothesized in this study to be related with knowledge for teaching ELLs. The teachers' scores on the two scales (OBST/RESRC and STRAT), computed as a simple percentage of "correct" responses, were then regressed linearly on these variables. After entering all of the teacher variables, the percentages of variance in scores on the OBST/RESRC and STRAT scales explained (R-squared) by the models given in the table below were, respectively, 38.6% and 29.5%.

Several observations can be made from Table 2. At a glance it appeared that the OBST/RESRC scale was the more 'difficult' of the two, based upon the linear regression intercept; that is, controlling for all other factors, the average score on the OBST/RESRC scale was 28.7% correct compared to 64% on the STRAT scales. IRT item characteristic and test information curves confirmed this difference in difficulty. Furthermore, Table 2 shows that, for this group of teachers, some of the education and experience variables were significant predictors of differences in scores in the two knowledge scales of the PCK-MELL survey. Teachers that had experience in teaching math classes in which more than 40% of students were ELLs scored 19.4 percentage points higher ($p < .05$) on average on the OBST/RESRC scale than did teachers without this experience. Conversely, teachers that possessed a certification to teach English as a Second Language (ESL) scored 17.1% lower on average ($p < .10$) on this scale than did teachers not having this certification. Teachers that possessed additional certifications to work in other educational settings (e.g., administration) or content areas besides mathematics scored 12.7% higher on average ($p < .10$) on the STRAT scale than did teachers who held mathematics teaching certificates alone. Likewise, teachers that had been teaching mathematics for more than 5 years scored an average of 13.9 percentage points higher ($p < .10$) on the STRAT scale than did less experienced teachers. Finally, teachers that had completed any sort of professional development concerned specifically with ELLs scored 13.7% lower on average ($p < .10$) on the STRAT scale than did teachers not having this experience.

Table 2. *Linear Regression Model of OBST/RESRC and STRAT Scale Scores on Teacher Variables.*

| Measure | OBST/RESRC | | STRAT | |
|---|---|---|---|---|
| | Coefficient | SE | Coefficient | SE |
| Intercept | .287* | .106 | .640** | .106 |
| Teacher Education/Licensure | | | | |
| Have a Degree in Math | -.116 | .080 | -.054 | .080 |
| Certified to Teach Math | .053 | .088 | .015 | .089 |
| Certified to Teach ESL | -.171$^\dagger$ | .098 | -.008 | .099 |
| Had any ELL Professional Development | .013 | .067 | -.137$^\dagger$ | .067 |
| Possess Other Educational Certifications | .074 | .076 | .127$^\dagger$ | .076 |
| Teacher Experience | | | | |
| Taught Math for More than 5 Years | .066 | .072 | .139$^\dagger$ | .072 |
| Taught More than 3 Different Types of Courses | .122 | .095 | .071 | .096 |
| Taught a Class with More than 40% ELLs | .194* | .072 | -.004 | .072 |
| Teacher Linguistic Knowledge | | | | |
| Know Spanish "Well" | -.002 | .063 | -.004 | .063 |
| Speak a Language Other than English or Spanish | .084 | .079 | -.084 | .079 |

$\dagger p < .10$; $*p < .05$; $**p < .01$

## The PCK-MELL Survey as a Measurement Instrument

PCK-MELL, the underlying hypothetical construct at the heart this survey, was proposed to be a subset of MKT (Ball, Thames, & Phelps, 2008). Furthermore, the construct was initially theorized as composite of three subdomains: knowledge of obstacles that ELLs encounter in math classes (OBST), knowledge of resources that ELLs draw upon to help them learn mathematics (RESRC), and knowledge of instructional strategies for teaching ELLs mathematics (STRAT). Based upon the limited survey data obtained in this study and upon the theoretical relationship between items in the OBST and RESRC domains, the factor structure of the survey seemed to be binary, having not three scales, but two: OBST/RESRC and STRAT.

As was seen above, the reliabilities of these two scales were low by measurement standards. Although some have argued that, for dichotomously scored items, KR-20 greater than .5 is even acceptable (McGahee, T. W., & Ball, J., 2009), alpha of .70 is often considered a lower bound on acceptable internal consistency (Nunnally & Bernstein, 1994) and considerably higher consistency would be needed for commercial usage of such a measure. This means the measurements, and interpretations thereto, obtained in this study must be taken with a degree of skepticism. Nevertheless, many of the items designed to measure knowledge of content and students (KCS), one of the larger domains to which the OBST/RESRC and STRAT domains pertain, also showed very low reliability during testing, $\alpha < .70$ (Hill, Ball, & Schilling, 2008).

Hence, this study also echoes the complexity experienced by those researchers. Precise measurement of teachers' MKT is indeed a delicate, and at times elusive, accomplishment.

A goal of this paper has been to argue that it may be valid to define knowledge for teaching mathematics to ELLs as a construct subsumed with MKT, and as a part of knowledge for equitable mathematics teaching as a subset of MKT. Although the instrument in its current form was exploratory and is not ready for commercial usage, it may offer a valuable starting point for further theory and instrument development along these lines. Furthermore, some specific directions for improvement are evident. For example, as a relatively inexpensive improvement on reliability, the Spearman-Brown prophecy formula indicates that the mere addition of fewer than ten similar items to each of the scales would elevate reliability to greater than .70 (Brown, 1910; Spearman, 1910). Writing more items is an obvious improvement. But this study also points to the need for more focused research into the specific types of, dimensions of, and relationships between both the *obstacles* and *resources* encountered by ELLs in mathematics. That is, the theory behind this strand of knowledge needs development. Qualitative studies, including careful interviews and observations of teachers in classrooms of ELLs, would seem to offer promising methods to begin to better understand the relationship between ELLs' deficits and affordances, and teachers' understanding of these.

Additionally, the linear regression models presented in Table 2 may offer limited, initial insight into how knowledge in these domains operates. For the OBST/RESRC scale, of the variables potentially related to knowledge in this domain only experience teaching larger numbers of ELLs (greater than 40%) was a significant ($p < .05$) predictor of gain in the scale score for these teachers. Yet, this variable is probably the most important one if this type of PCK is related to actual teaching experience. Similarly for the STRAT scale, experience teaching math for more than five years was a moderate ($p < .10$) predictor of gain in scale score as was possession of multiple teaching certifications. These may, respectively, be proxies for length of teaching experience and breadth of teaching experience. In this case again knowledge in the STRAT domain may be related to actual teaching experience. Finally, it is interesting to note that two of the variables, one in each of the scales, were moderate ($p < .10$) predictors of a *decrease* in scales scores. For the OBST/RESRC and STRAT scales these were, respectively, certification to teach English as a second language (ESL) and whether or not the teacher had experienced any professional development for teaching ELLs. This particular finding is alarming; one would think that both of these variables would relate to increases in knowledge in the domains being tested, not decreases. Yet, if these two variables are seen as proxies for formal *preparation* to teach ELLs, then again the knowledge called upon by this exploratory instrument may be more a matter of actual experience than of formal education. At the least, it would seem that the topics presented in the instrument of this study were different than topics that the teachers in this study had seen in their ELL professional development experiences or ESL classes.

## Concluding Discussion

The PCK-MELL survey developed in this study gives an example of a way in which an aspect of knowledge for equitable mathematics teaching has been operationalized in the form of survey items yielding quantitative data and psychometric results that can lead to further theory development. Perhaps most importantly, this study has offered a way of thinking about knowledge needed by mathematics teachers of English Language Learners—and by extension, all second language learners in the mathematics classroom, including immigrant students in other countries—that builds upon and extends current research. Under the framework proposed herein, in addition to mathematics content knowledge, teachers of ELLs need: knowledge of *obstacles* to learning that are frequently encountered by ELLs in mathematics classes, knowledge of the *resources* that ELLs bring with them to the learning of mathematics, and knowledge of instructional *strategies* for teaching ELLs. This theoretical framework takes the bold, and admittedly controversial stance, of allowing both deficit and affordance perspectives of ELLs as mathematics learners, not advancing one perspective over the other or placing them in opposition, but rather positioning these two perspectives as two *sides of the same coin*: that is, both perspectives may together offer valuable insight for mathematics teachers of ELLs who should be aware both of obstacles to learning that their students are likely to face and also of the resources within and around those very students that can effectively capacitate them to do mathematics and to communicate mathematically. While this way of thinking about learners no doubt applies to all students—regardless of their language background—the viewpoint may be of special value to the particular linguistic experiences that ELLs and second language learners in all language contexts bring to learning mathematics.

This has been an initial attempt at advancing both the theory and the tools that can lead to better understanding teacher knowledge factors that impact achievement in mathematics among ELLs. The article offers an example of an exploratory process that was used to lay the bearings of a theoretical framework of knowledge needed for teaching mathematics to ELLs and also of some steps that were made toward developing a research tool to be used to study this knowledge. This study indicates the inadequacy of current theory and tools for studying aspects of mathematics teachers' knowledge related to equitable access for diverse student populations. It is hoped that this paper may make a small contribution toward bringing more clearly into focus a research agenda from which a viable theory of knowledge for equitable mathematics teaching in the context of linguistically and culturally rich and diverse populations of mathematics students will come forth.

# References

Ball, D. L., Thames, M. H., & Phelps, G. (2008). Content knowledge for teaching: What makes it special? *Journal of Teacher Education, 59*(5), 389–407.

Brown, W. (1910). Some experimental results in the correlation of mental abilities. *British Journal of Psychology, 3,* 296–322.

Campbell, D. T., & Fiske, D. W. (1959). Convergent and discriminant validation by the multitrait-multimethod matrix. *Psychological Bulletin, 56*(2), 81–105.

Capps, R., Fix, M., Murray, J., Ost, J., Passel, J., & Herwantoro, S. (2005). *The new demography of America's schools: Immigration and the No Child Left Behind Act.* Washington, DC: The Urban Institute.

Chang, M. (2008). Teacher instructional practices and language minority students: A longitudinal model. *Journal of Educational Research, 102*(2), 83–97.

Coggins, D., Kravin, D., Coates, G. D., & Carroll, M. D. (2007). *English language learners in the mathematics classroom.* Thousand Oaks, CA: Corwin Press.

Cronbach, L. J. (1951). Coefficient alpha and the internal structure of tests. *Psychometrika, 16,* 297–334.

Cronbach, L. J., & Meehl, P. E. (1955). Construct validity in psychological tests. *Psychological Bulletin, 52*(4), 281–302.

Cuevas, G. J. (1984). Mathematics learning in English as a second language. *Journal for Research in Mathematics Education, 15*(2, Minorities and Mathematics), 134–144.

De Ayala, R.J. (2009). *Theory and practice of item response theory.* New York, NY: The Guilford Press.

DeVellis, R. F. (2003). *Scale development: Theory and applications. Applied socialresearch methods series, vol. 26.* Thousand Oaks, CA: Sage.

Dillman, D. A., Smyth, J. D., & Christian, L. M. (2009). *Internet, mail, and mixed-mode surveys: The tailored design method.* Hoboken, NJ : John Wiley & Sons, Inc.

Echevarría, L., Short, D., & Vogt, M. (2007). *Making content comprehensible for English learners: The SIOP model* (3rd ed.). Boston: Pearson Educational Press.

Francis, D., Rivera, M., Lesaux, N., Kieffer, M., & Rivera, H. (2006). *Practical guidelines for the education of English language learners: research-based recommendations for instruction and academic interventions.* Portsmouth, NH: RMC Research Corporation, Center on Instruction.

Gutiérrez, R. (2002). Beyond essentialism: the complexity of language in teaching mathematics to latina/o students. *American Educational Research Journal, 39*(4), 1047–1088.

Gutiérrez, R. (2008). A "gap-gazing" fetish in mathematics education? Problematizing research on the achievement gap. *Journal for Research in Mathematics Education, 39*(4), 357–364.

Gutstein, E., Lipman, P., Hernandez, P., & Reyes, R. d. l. (1997). Culturally relevant mathematics teaching in a Mexican-American context. *Journal for Research in Mathematics Education, 28*(6), 709–737.

Henderson, R. W., & Landesman, E. M. (1995). Effects of thematically integrated mathematics instruction on students of Mexican descent. *The Journal of Educational Research, 88*(5), 290–300.

Hill, H.C., & Ball, D. L. (2004). Learning mathematics for teaching: Results from California'smathematics professional development institutes. *Journal for Research in Mathematics Education, 35*(5), 330–351.

Hill, H. C., Ball, D. L., & Schilling, S. G. (2008). Unpacking pedagogical content knowledge: Conceptualizing and measuring teachers' topic-specific knowledge of students. *Journal for Research in Mathematics Education, 39*(4), 372–400.

Hill, H.C., Ball, D.L., Sleep, L. & Lewis, J.M. (2007). Assessing teachers' mathematical knowledge: What knowledge matters and what evidence counts? In F. Lester (Ed.), *Handbook for Research on Mathematics Education (2nd ed)* (pp. 111–155). Charlotte, NC: Information Age Publishing.

Hill, H. C., Blunk, M. L., Charalambous, C. Y., Lewis, J. M., Phelps, G. C., Sleep, L., & Ball, D.L. (2008). Mathematical knowledge for teaching and the mathematical quality of instruction: An exploratory study. *Cognition and Instruction, 26*(4), 430–511.

Hill, H. C., Rowan, B., & Ball, D. L. (2005). Effects of teachers' mathematical knowledge for teaching on student achievement. *American Educational Research Journal, 42*(2), 371–406.

Hill, H. C., Schilling, S. G., & Ball, D. L. (2004). Developing measures of teachers' mathematics knowledge for teaching. *Elementary School Journal, 105*(1), 11–30.

Khisty, L. L., & Morales Jr., H. (2004). Discourse matters: Equity, access and Latinos learning mathematics. Retrieved September 1, 2011 from http://www.icme-organisers.dk/tsg25/subgroups/khisty.doc

Lager, C. A. (2006). Types of mathematics-language reading interactions that unnecessarily hinder algebra learning and assessment. *Reading Psychology, 27*(2–3), 165–204.

Llabre, M. M., & Cuevas, G. (1983). The effects of test language and mathematical skills assessed on the scores of bilingual Hispanic students. *Journal for Research in Mathematics Education, 14*(5), 318–324.

Ma, L. (1999). *Knowing and teaching elementary mathematics: Teachers' understanding of fundamental mathematics in China and the United States.* Mahway, NJ: Erlbaum.

Martiniello, M. (2009). Linguistic complexity, schematic representations, and differential item functioning for English language learners in math tests. *Educational Assessment, 14*(3–4), 160–179.

McGahee, T. W., & Ball, J. (2009). How to read and really use an item analysis. *Nurse Educator, 34*(4), 166–171.

Meskill, C. (2005). Infusing English language learner issues throughout professional educator curricula: The training all teachers project. *Teachers College Record, 107*(4), 739–756.

Moschkovich, J. N. (2002). A situated and sociocultural perspective on bilingual mathematics learners. *Mathematical Thinking and Learning,* 4(2 & 3), 189–212.

Nunnally, J. C., & Bernstein, I. H. (1994). Psychometric theory (3rd ed.). New York, NY:McGraw-Hill.

Payán, R. M., & Nettles, M. T. (2008, January). *Current state of English-language learners in the U.S. K–12 student population* (English-Language Learners Symposium Fact Sheet). Princeton, NJ: Educational Testing Service.

Schon, D. A. (1983). *The reflective practitioner: How professionals think in action.* New York: Basic Books.

Shulman, L. S. (1986). Those who understand: Knowledge growth in teaching. *Educational Researcher, 15*(2), 4–14.

Schuman, H., & Presser, S. (1996). *Questions and answers in attitude surveys: Experiments on question form, working, and context.* London: Sage.

Sorto, M. A., Mejía Colindres, C. A., & Wilson, A. T. (Sept., 2014). Uncovering and Eliciting Mathematical Perceptions in Linguistically Diverse Classrooms. *Mathematics Teaching in the Middle School.* 20(2), pp. 72-77.

Spearman, Charles C. (1910). Correlation calculated from faulty data. *British Journal of Psychology, 3,* 271–295.

U.S. Department of Education, Institute of Education Sciences, National Center for

Education Statistics, (2012). *The condition of education: Participation in education – elementary/secondary enrollment.* (Table-A-6-1. Number and percentage of children ages 5-17 who spoke only English at home, who spoke a language other than English at home and who spoke English with difficulty, and percent enrolled in school: Selected years, 1980 – 2009). Retrieved from http://nces.ed.gov/programs/coe/tables/table-lsm-1.asp.

Watson, S., Miller, T. L., Driver, J., Rutledge, V., & McAllister, D. (2005). English language learner representation in teacher education textbooks: A null curriculum? *Education, 126*(1), 148–157.

Wolf, M. K., Herman, J. L., Dietel, R. (2010). *Improving the validity of English language learner assessment systems. Policy brief 10, spring 2010.* National Center for Research on Evaluation, Standards, and Student Testing (CRESST).

# In-service Teachers' Reasoning about Scenarios of Teaching Mathematics to English Language Learners

Sultan Turkan
Educational Testing Service

**Abstract:** The student population in the U.S. and worldwide is becoming increasingly diverse, creating a need to support *all* learners, especially linguistically and culturally diverse subpopulations such as English language learners (ELLs). From a social equity standpoint, the need to support these learners is critical especially in mathematics classrooms. In the U.S, the demand for mathematics teachers who are adequately prepared to teach ELLs has in fact risen. Yet, little is known about what knowledge base is essential to teach mathematics to ELLs. Driven by the need to explore this knowledge base, in this paper I explore what is involved in reasoning about teaching mathematics to ELLs. To this end, a set of instructional scenarios illustrating the work of teaching mathematics to ELLs was utilized within an assessment environment. Interviews with 10 mathematics teachers reasoning about the scenarios showed that they drew on the information provided about ELLs' proficiency levels while reasoning through the scenarios. Also, teachers' reasoning seems to be qualified by the extent to which they could both use their content knowledge in mathematics and modify their instructional choices according to ELLs' language needs specified in the scenarios. This study motivates large-scale future studies examining what systematic teacher knowledge base might differentiate good teaching for ELLs from good teaching for *all* students.

*Key Terms:* Teaching mathematics, English language learners, teacher knowledge base, and instructional scenarios.

## Introduction

As the student population in U.S. classrooms becomes increasingly diverse, it is critical that all teachers develop the knowledge and skills to support English language learners (ELLs) in mainstream classrooms at every grade level. However, the majority of teachers are not adequately prepared to teach academic content to ELLs (Lucas & Villegas, 2011). Mathematics may be especially challenging for ELLs (Martiniello, 2008). The National Assessment of Education Progress (NAEP) mathematics test results in 2013 (United States Department of Education, 2015) showed that the achievement gap between non-ELL and ELL students both at the 4th and 8th grades did not significantly differ from the achievement gaps reported in previous years in 2011, 2009, 2000, or 1996. Challenges in learning mathematics might preclude access to mathematical and scientific fields, which raises a critical issue of equity (Moschkovich, 2002; Oakes, 1990; Secada, 1992). Teachers play a key role in leveling the playing field for ELLs in mathematics (Moschkovich, 1999; Secada, 1998). However, there is a clear gap between the demands of teaching ELLs and the supply of mainstream mathematics teachers who are adequately prepared to teach (Ballantyne, Sanderman, & Levy, 2008). As the ELL population has increased up to 10% of public school students (National Clearinghouse for English Language Acquisition, 2011), the demand for trained mathematics teachers has as well; many, if not most, teachers need to develop the knowledge and skills needed to teach in classrooms with increasingly diverse populations of students.

Despite the need, little is known about the knowledge base needed to teach mathematics to ELLs. Recent conceptualizations of the teacher knowledge base (Ball, Thames, & Phelps, 2008) have not addressed how manifestations of it change depending on who the students in the classroom are and what they bring to the learning experience. To inform research on the teacher knowledge base, most of the existing scholarship on educational linguistics (de Oliveira & Cheng, 2011; Fang, 2006; Lucas, Villegas, & Freedson-Gonzalez, 2008; Schleppegrell, 2001) suggests that language is central to teaching mathematics and that mathematics teachers, even at the secondary level, should know how to identify the linguistic demands in the mathematical content and model mathematics-specific language use. This line of work suggests that language serves as a medium to understanding and communicating the content in mathematics, which in turn is central to the work of teaching mathematics to ELLs.

However, our understandings as to what this knowledge base entails and how it gets or should get enacted in mathematics classrooms are limited. To better understand the knowledge base and reasoning involved in teaching mathematics to ELLs, eliciting teachers' reasoning around particular scenarios formatted in an assessment environment might help. As research (Ball & Hill, 2008; Jacobson, Remillard, Hoover, & Aaron, in press) suggests, there needs to be a reciprocal interplay between conceptualizations of teacher knowledge base and measurement in that the development of teacher knowledge assessments needs to be informed by conceptual work in the domain of effective and equitable teaching of mathematics. To facilitate this interplay, new lines of research also show that authentic classroom scenarios formatted into the assessment environment could serve as powerful tools for eliciting teacher reasoning as well as reflective teacher learning (Lai & Howell, 2014). Further, emerging research shows that instructional scenarios help to situate explorations about pre-service and in-service teachers' reasoning about teaching (Grossman et al., 2009; Masingila & Doerr, 2002). This study was therefore based on the premise that the instructional scenarios might be instrumental in explorations of the knowledge base needed to teach mathematics to ELLs.

Driven by the need to explore essential knowledge for teaching mathematics to ELLs, in this study, we utilized a set of instructional scenarios embedded in an assessment environment to explore what is involved in reasoning about teaching middle school mathematics to ELLs. First, I developed a theoretical lens explicating how understanding the language demands in any content area is central to teaching the content in diverse ways to diverse learners. Second, we developed scenarios illustrating the work of teaching mathematics to ELLs in an assessment environment and conducted interviews with mathematics teachers to explore their reasoning about the scenarios. The current paper is part of a larger study (Turkan, Croft, Bicknell, & Barnes, 2012) and sought to address the following research questions. In addressing these questions, teachers' strong and weak reasoning about the given instructional scenarios as evident in cognitive interviews was contrasted as a way of anchoring the exploration of knowledge essential for teaching mathematics to ELLs.

1. How do I conceptualize the knowledge for teaching mathematics to ELLs?
2. How do practicing mathematics teachers reason about the teaching of mathematics to ELLs?

Towards addressing the first research question, the next section provides an overview of the theoretical framework. The framework informed the development of the measure. Description of the measure and methods is followed by the discussion of results addressing the second question about how a sample of mathematics teachers reasoned through the teaching of mathematics to ELLs. Finally, directions for future research are shared.

## Theoretical Lens

Our assumption in identifying and proposing a knowledge base was that all teachers need to develop a specific knowledge base in order to make academic content accessible and meaningful to ELLs. This assumption has long been an argument among scholars that all teachers need to be provided with opportunities to develop an adequate knowledge base to make academic content accessible and meaningful to ELLs (de Jong & Harper, 2005, 2011; Lucas & Villegas, 2011).

Specifically, effective teachers of ELLs draw upon a specialized knowledge base in order to teach academic content in accessible ways called *disciplinary linguistic knowledge (DLK)* for teaching academic content to ELLs (Turkan, de Oliveira, Lee, & Phelps, 2014). These authors argued that DLK is needed to model for ELLs how language is used to communicate meaning and to engage them in disciplinary discourse. That is, DLK is a specialized knowledge that teachers need to have regarding how language is used to communicate the ideas and concepts within a particular disciplinary discourse. Discourse, in this context, refers to ways of knowing, constructing, and communicating knowledge. Involved within the discourse of an academic discipline is the academic register of the discipline that includes linguistic features specific to that discipline. DLK is an elaborated version of Reeves' (2009) use of the term, *Linguistic Knowledge for Teaching,* which is operationalized as the linguistic knowledge that English for Speakers of Other Languages (ESOL) teachers use to create opportunities for learners to communicate meaning. In this paper, however, I use DLK to refer to knowledge that *content* teachers need to facilitate ELLs' content understanding.

Additionally, this knowledge base is a facet of, but not equivalent to, content knowledge for teaching, CKT, (Ball et al., 2008). While the body of work on teacher knowledge measures and research on content knowledge for teaching (Ball, Hill, & Bass, 2005; Ball et al., 2008) is growing, it does not address the role of special student populations in how teachers attend to the tasks of teaching in the classroom. More specifically, CKT doesn't address what knowledge base content teachers draw on to teach special student populations such as ELLs. Outside general teacher education literature, several scholars (Bunch, 2013; Galguera, 2011) have conceptualized what knowledge base teachers need to teach content to ELLs. Bunch accounts for the pedagogical aspect of teaching content to ELLs while by proposing DLK, Turkan et al. (2014) complement Bunch's view that content teachers should also address ELLs' linguistic needs at the word, sentence, and discourse levels as these levels pertain to a particular discipline.

DLK encompasses teachers' knowledge of content-specific academic discourse as well as their ability to present conceptual language in multiple ways that are accessible to ELLs. Specific tasks of teaching subsumed therein include two subdomains: 1) identifying linguistic features of the disciplinary discourse and 2) modeling for ELLs how to communicate meaning in the discipline and engaging them in using the language of the discipline orally or in writing. These tasks are conceptualized further below. The measure of teaching quality for ELLs was developed (Turkan et al. 2012) according to this conceptualization of the targeted construct.

## Identifying Linguistic Features

The argument for the first subdomain is that DLK entails particular knowledge for identifying and unpacking the linguistic features and language demands of a disciplinary discourse to make the content accessible to ELLs. We argue that, as part of the unique knowledge base for teaching ELLs, teachers know which linguistic features associated with the particular content might cause or constitute ELLs' misunderstandings or misconceptions about the content. That is, to resolve misconceptions and facilitate ELLs' comprehension and interpretation of the content, linguistic features need to be unpacked. There are two premises in this argument: 1) there are linguistic features that convey meaning and content in a given discipline, 2) the linguistic choices are made in each discipline at the word, sentence, and discourse level to convey meaning.

The first premise is that teachers identify the linguistic features that are specific to a content area. In doing so, they decode or unpack the linguistic features of the discipline to build connections between content and meaning and particular linguistic features and structures that convey the particular meanings. By 'unpack', we mean that teachers make the linguistic *form-meaning* connections of the disciplinary discourse explicit for students. In other words, teachers make the content-related input comprehensible and accessible to ELLs (Cummins, 2000). *Form* here refers to a string of words grammatically put together to carry *meaning*. A string of forms (words, sentences) might carry various meanings depending on the context in which it is being used. For instance, McCarthy (1991) discusses how the following sentences might carry the meaning of question, statement, or command: "You don't love me: (a) question (b) statement" or "you eat it: (a) statement (b) command" (p. 9). Form-meaning connections (Schleppegrell, 2013) are made when teachers attend to the features of language (i.e., form) and they simultaneously model for ELLs the ways in which meaning and content is communicated in the particular discipline. For example, linguistic expression such as "three times as big as…" has a particular correspondence in mathematics, hence a certain mathematical form-meaning connection. As teachers identify the linguistic choices made to convey meaning in a particular discipline like mathematics, they also make the content comprehensible to ELLs by explicitly teaching specific language functions, forms, or meaning behind the text in order for ELLs to learn how the linguistic choices are being used to convey the particular meaning. Once ELLs can differentiate and be made aware of the linguistic choices through the comprehensible input received, they can more readily participate in producing or using the language orally or in writing to convey their understandings about the content.

The second premise is that the ways in which meaning is communicated in content areas are instantiated through linguistic choices, for example, at vocabulary, grammatical and syntactical structures, and discourse levels (Christie & Martin, 1997; Gebhard, Willett, Pablo, Caicedo, & Peidra, 2011; Schleppegrell, 2004). The linguistic choices operate at the word, sentence, and discourse levels: at the word (e.g., *square root* of 25), sentence (e.g., *taking the square root is the inverse operation of squaring)*, and discourse (e.g., *taking the square root involves finding the number that, when multiplied by itself, gives* 25) levels. The linguistic choices at these three levels exist to communicate meaning and perform language tasks and functions associated with the particular discipline such as writing laboratory reports in the science classroom (Schleppegrell, 2004), explaining solution processes or describing conjectures in the mathematics classroom (Moschkovich, 1999), writing personal recounts of an event in the English language arts classroom (Brisk & Zisselberger, 2011), and retelling events or presenting debates in the social studies classroom (Fang & Schleppegrell, 2008).

The above premises led us to argue that teachers should be able to identify the linguistic features specific to a content area so that they can decode or unpack the linguistic features of the discipline and build connections between content and meaning, on the one hand, and particular linguistic features and structures that convey the particular meanings, on the other; and that teachers should be able to explicitly teach what constitutes appropriate linguistic choices in their discipline.

## Modeling How to Communicate Meaning Orally or in Writing

The argument for the second subdomain is that DLK involves teachers' knowledge for modeling the ways in which the discourse of a discipline is constructed and for engaging ELLs in communicating meaning in the disciplinary discourse orally or in writing. When modeling the disciplinary discourse, teachers build on the mastery of the first domain to make the linguistic features of a content area explicit for ELLs. Further, teachers draw on DLK to engage ELLs in learning how the rules of the linguistic features function to convey meaning in the content area. In doing so, teachers encourage ELLs to explore and build *form-meaning* connections to read, write, listen, speak, and think in the language of the discipline. In this process, students produce work in writing or orally in which they demonstrate their knowledge of the discipline using the disciplinary language as it was modeled to them. Hence, students participate in using the language of the content for complex academic tasks such as "to generate new knowledge, create literature and art, and act on social realities" (Cummins, 2000). In the process of constructing *form-meaning* connections, students develop awareness of and knowledge about what linguistic features are used to represent the meanings and ideas associated with the content.

In this section, we have proposed that DLK encompasses teachers' knowledge of disciplinary discourse in order to represent the disciplinary content in accessible ways to ELLs. I have also argued that to make content accessible, every teacher needs to be able to identify and unpack the ways in which the linguistic features are connected to the meanings of a discipline. As teachers raise ELLs' awareness explicitly around how form-meaning connections are linked in the discourse of the discipline, they will be able to engage ELLs in reading, writing, listening, speaking, and thinking in the disciplinary discourse. Therefore, within the purview of DLK, teachers should know about the linguistic choices made in a given discipline and should be able to identify the language demands and unpack the meaning-form relationships for ELLs to model for them and engage them in using the discourse of the discipline.

## Methods

To explore and identify the knowledge base needed to teach mathematics to ELLs, I have grounded my study in practice-based reasoning about teaching. Specifically, teachers' reasoning about the given scenarios of teaching mathematics to ELLs was elicited through interviews. In doing so, I also explored how the high and low performing teachers, as identified by their number correct scores, reasoned about the scenarios. The scenarios were designed as part of the operationalization of the theoretical framework described above. Scenarios were formatted as part of the assessment development process, which served as a means to operationalizing and refining the domain. Next, the assessment development process is explained; later, the interview data collection procedures are presented along with the analysis.

The assessment development process started with the identification and verification of the domain of teacher knowledge and skills required to teach ELLs mathematics content effectively. I initially identified this domain through a review of the academic literature base on teaching mathematics to ELLs, as well as review of the state standards for teacher certification in the three content areas: mathematics, science, and English language arts. Based on the general and content-specific literature, we developed a set of 67 statements describing what content teachers should know and be able to do (for more information, see Turkan et al., 2012). Based on the description of the targeted domain, we categorized these statements into two subdomains: pedagogical knowledge and linguistic knowledge. Under each domain, there were statements generic to all content areas and specific to particular content areas. To validate these statements, we conducted a national survey of practitioners and teacher educators, receiving 269 responses. Subsequently, a panel of 14 teacher educators and teachers further validated the statements by reaching consensus that the statements supported the claims of the assessment under development (for more information about the panel, see Turkan et al., 2012). The principles of evidence-centered design (ECD; Mislevy, 1994; Mislevy, Almond, & Lukas, 2003) guided the panelists' review of the statements about linguistic and content knowledge necessary to teach ELLs. The panelists helped to unpack specifically what teachers would need to perform to manifest that they have the essential knowledge and skills. Based on the performance indicators the panel helped to further specify, the assessment was developed to measure the validated domain of teacher knowledge essential to teach content to ELLs.

During assessment development, we identified the language demands inherent to the content standards and topics selected from the 6th and 8th grade mathematics curricula. We selected middle grades based on the content and language demands—and informed by the panelists-- we identified a set of instructional practices representative of effective ELL teaching. With that, we developed authentic instructional scenarios embedded in an assessment environment (see appendix for two sample items). The scenarios included variant and nonvariant features. Nonvariant features were the effective mathematics teaching practices targeted for assessment. Variant features included description of the learning objective, the ELL student's task, as well as the characteristics of the focus ELL(s), which include their ELL proficiency level descriptors. Various indicators of ELL proficiency in at least one of the four skills (reading, writing, speaking, and listening) were specified. These indicators were provided to guide teachers' understanding about what ELL(s) of particular focus could do or could not do. The learning or teaching objectives were informed by the targeted 6th and 8th grade mathematics standards. This then determined the language demands that are embedded in the particular learning or teaching objectives. The language demands guided what tasks ELLs, characterized by specific linguistic skills in the instructional scenario, were confronted with in the classroom. For example, item 1, illustrated in the appendix, focused on the mathematical concept of 'distributive property' in a classroom scenario in which teachers are faced with a choice to either use the definition that they find or to provide an illustration of the concept to make it accessible for the ELLs described in the scenario. This focus then determined the language demands in explaining and representing the concept for understanding by the specified ELLs. Similarly, item 2, discussed in the results section, focused on a learning objective; namely, writing and solving equations based on word problems. In the particular classroom scenario, ELLs were characterized as having grade-level content knowledge but needing linguistic support in understanding and solving the given word problem. This item is representative of the other items developed to assess teachers' knowledge of content for identifying the language demands embedded in the content. That is, most of the items focused on teachers' knowledge and skills for identifying and connecting the content and language aspects of the content knowledge for teaching.

Approximately 60 items were developed to assess the DLK needed to teach mathematics to ELLs. Twenty of the 60 mathematics items were selected as a representative sample of the domains targeted for assessment. The sample was administered online to 60 middle school mathematics educators who teach in school districts that serve large numbers of students who are ELLs. Samples of mathematics teachers represented southern, western, and northeastern regions of the United States. These teachers had a minimum of five years teaching experience in predominantly ELL school districts. The initial analysis revealed a Cronbach's alpha ($\alpha = .50$) and only five items showed biserial correlation measures of item discrimination below 0.2.

## Interviews

To address how teachers reason about ELL teaching, we invited all 60 in-service mathematics teachers who took the assessment for a follow-up interview. Only 27 indicated interest to participate in the interview. We ranked the test performance of these 27 potential interview participants according to their number-correct scores. Two mathematics teachers got 14 while seven teachers got 11 of the 20 items correct, which was the mode on the test. Twelve teachers got fewer than seven items correct. Low and high performing teachers were identified by the number of items they got right out of the 20 multiple choice items. Those who got 10 and above items right were considered high performing while other teachers who got 10 and below items right were considered low performing on this particular test. This cut off decision was arbitrary based on the distribution of the number-correct scores. Based on this decision, we randomly selected five high performing and 5 low performing teachers, who agreed to participate in the interview. All 10 teachers had taught mathematics for a minimum of five years at predominantly ELL school districts where each teacher had at least five ELLs in their respective classrooms.

The goal of the semi-structured phone interviews was to allow teachers to explain their reasoning as each of them recalled their response to a sample of six items randomly selected from the 20 items. The aforementioned five items with low item discrimination characteristics were not included in the interviews. The remaining items were distributed across the interview participants in a way that each eligible item would be discussed by two teachers. The selected group of items was sent to the participants at least 24 hours ahead of the interviews so that they could have the chance to explain their reasoning about the items. The goal of the interviews was to elicit teachers' reasoning about the items through the use of retrospective questions (see appendix) more than to evaluate or elicit their ability to recall their exact reasoning at the time of taking the assessment.

**Analysis**

I coded the transcriptions of the interviews using NVivo 9 software. The a priori themes were 1) relevance of scenarios to teachers' work within the specified domains (i.e., identifying linguistic features, modeling how to communicate meaning orally and in writing), and 2) relevance of domains to the items. The subsequent analysis followed an "expanding frame" in which the analysis of the qualitative data began with a tight focus on one element or a few elements (Lindlof, 1995). Within this frame, as the researcher collects evidence, and sees new ways to consider, the frame of evidence is widened in analysis. Specifically, the analysis initially focused on understanding the extent to which teachers considered the ELL proficiency levels identified in the given scenario. As I tried to further address what is involved in this, other themes emerged such as teachers' views about ELL teaching and reasoning about the language demands of the mathematical content. I noted the themes as they came up across participating teachers and the items they discussed during the interviews. Comparison of reasoning between high and low-performing teachers helped to further explain the patterns noted across the participating teachers.

**Results**

In response to the research question, there were three main findings. One is that seven teachers agreed on considering what ELLs can or cannot do in their instructional decisions. Three teachers argued that all learners (i.e., not just ELLs) have difficulty with the language of mathematics. With this assumption, one teacher claimed that good ELL teaching is effective for all learners. The second finding is that the depth of teachers' content knowledge and sensitivity about ELLs' language needs played a noteworthy role in their reasoning. This may have been a consequence of the particular scenario in the assessment that was used for the interview. If the scenario elicited not only teachers' content knowledge but also their sensitivity about ELLs' language needs, the depth of teachers' content knowledge seemed to surface more noticeably. Third, all ten teachers believed that one effective way to teach mathematics to ELLs was to use pictures as a way of simplifying language and/or removing any language demands specific to mathematics.

**Role of ELL Proficiency Indicators**

Cognitive interviews with mathematics teachers who either performed well or poorly on the test questions revealed how teachers drew on the information about students' proficiency levels or other information given in the instructional scenarios to reason through the scenarios. Both high-and low-performing teachers drew on the information in their reasoning across all the six scenarios that they discussed during the interviews. I recorded the presence or absence of teachers' reference to the information provided in the scenario. When asked to rate the level of importance for identifying ELLs' proficiency levels and explain why, seven teachers agreed consistently across all the scenarios assigned to them that it was highly important to modify instruction accordingly. However, three of the low-performing teachers consistently explained across the six scenarios that language proficiency descriptors were of low relevance to mathematics instruction and they did not consider what ELLs can or cannot do in identifying the most appropriate instructional practices. One of these particular teachers' assumptions was that language of mathematics was challenging for all students and there is nothing special to the work of teaching mathematics to ELLs.

One teacher did not see the relevance of understanding what students could or could not do in their second language. The teacher reasoned that mathematics itself is a second language for everybody, not just for ELLs, and therefore ELLs should not receive special consideration during instruction above and beyond non-ELLs.

> I(nterviewer): Okay, on a scale of 1 to 5 with 5 being the most important, how important do you think it is for math teachers of ELLs to identify the language proficiency levels of their students?

> R(espondent): I think, well I think that; I would call it about a 2, meaning not as important, if I got your scale right. Because, you know, the way I think of math, and I tell them this, is that math is a language all by itself, and so everybody learning math is in fact kind of learning a second language. So in a math class, they may be using English to learn an additional language, that being math, so everybody's kind of at a disadvantage in a way.

Along similar lines, these three teachers thought that effective ELL teaching is good teaching for everyone. This reasoning might bear two related interpretations. One might be that teachers think that there is no uniqueness to the quality of teaching mathematics to ELLs. Another is that if a particular effective teaching benefits ELLs, it would benefit everyone. One teacher who represents this line of reasoning stated:

> R: No, but you know the thing is is that the way I conduct my classes, when I am doing whole group instruction, or we have a word problem, I basically teach those the same way whether I have ELLs in my class or not, in that we read them out loud, and making sure that even some kid that was raised in America on English may not understand some of the idiomatic ways of expressing ourselves that sometimes creep into these questions. So sometimes you know whether I have ELLs or not, I spend a second making sure that everybody understands what the question says. And the pictures and some of those things that help ELLs help everybody. So if you, to me, if you teach with ELLs in mind in general, you're helping everybody.

All in all, teachers found it helpful to consider what ELL can or cannot do in their instructional decisions. A few teachers, though, claimed that language of mathematics is difficult for all learners and that there is no distinctness to the work of teaching ELLs.

## Teachers' Reasoning about ELL Teaching

As for teachers' reasoning about the scenarios, there were several dynamics that played a role. One was the content and focus of the items. Since the items focused on teachers' content knowledge and guided teachers to pay attention to ELLs' proficiency levels, the strength of teachers' content knowledge determined the depth of their reasoning. This relationship was observed in varying degrees along a continuum of depth in all 10 teachers' responses. Here, the continuum of depth in teacher reasoning is presented by illustrating a few sample responses from high and low performing teachers.

To illustrate, item 1 (see appendix) contextualized the teacher's task as introducing the distributive property and providing examples to scaffold students' understanding of this property.

The teacher, in this scenario, has to have a good understanding of the concept 'distributive property' to make the connection that options A and B are two different but accurate representations of the same concept. One representation utilizes mathematical symbols (option A) while the other explains the concept with numerical and graphic illustrations. Item 1 assessed teachers' ability to represent the concept in relation to the particular characteristics of the students, tapping into knowledge of content and students. In terms of student characteristics, the scenario specified that the teacher has newly arrived ELLs whose placement test results indicate that they are weak in mathematics and low in English proficiency. Given the ELLs in the class, the teacher in the scenario decides to introduce distributive property with an illustration, instead of starting the lesson with a linguistically complex definition of distributive property. One of the teachers participating in the interview who had performed low on the overall assessment did not choose option B, intended key, because to her, *there was too much going on in that illustration* (see the original line from her below) and so it would be hard for ELLs to follow. According to this teacher, option A, on the other hand, conveyed the central conceptual understanding behind distributive property. However, this teacher seemed to have missed the nuance that option B conveyed the same conceptual understanding except with more illustration, which was more helpful for the particular ELLs than option A, as well as the more linguistically demanding illustrations given in options C and D. Option A was not helpful for the particular needs of the ELLs with weak mathematics and language skills as it just offered a symbolic representation of the definition for distributive property and so did not serve the instructional intentions of the teacher mentioned in the scenario. We observed that by choosing option A, the participating teacher did not combine her content knowledge and linguistic needs of the ELLs in the instructional scenario.

R: I'm almost positive A was my answer on the thing as well.

I: Okay. And why not B?

R: Too much going on. I just think there's way too much going on there. A is a clear example where you can see how the number outside the parentheses was moved to meet each of the numbers inside the parentheses. I just think showing them all those blocks, there's way too much going on there.

I: And you think A is the simplest way.

R: Um hmm.

I: But what other information in the item did you draw on when selecting A?

R: When selecting A?

I: Yeah.

R: Because they don't want to, they want to illustrate the distributive property and the distributive property says that you multiply the number outside the parentheses .... the number inside and add it together. I just think A clearly states that, you know, showing the mathematical symbols and everything like that and showing how one side of the equal sign equals the other.

Another teacher who scored higher on the test was swift in combining the characteristics of ELLs in the scenario with her content knowledge about mathematical representations of distributive property. This provided her the affordance to be able to reason that low English proficiency kids would still get lost with the teacher explanation that option A calls for, unless they were strong in mathematics and she eliminates this option.

I: Can you tell us in your own words what information from the question helped you to select an answer?

R: Basically the characteristics of the focus ELLs. In most questions, that was the thing that I cared the most about was the characteristics of the focus ELLs. These were newly-arrived again. Here, unlike the last one, you gave me more information about them – weak in math – which is not a characteristic of ELLs, that's a characteristic of most kids – and have low proficiency in English. So these are low proficiency English kids which meant to me that they needed pictures. You know the instructional scenario, there's the definition, the definitions, mathy definitions are horrifyingly difficult to follow even for people who speak English.

R:  And then when I got down to the choices, you know A is okay because, again, math itself is a language and I can, as a math teacher, clearly say that that's essentially the definition of the distributive property. But really I knew I didn't want to go straight into word problems down in C and D and using those pictures and B, which is my answer choice, that that seemed to me clearly the way to go.

Furthermore, seven teachers, regardless of how well they performed on the test, consistently conceived ELLs as the special student population that constantly needs linguistic support in the form of either linguistic simplification or removal of language demands specific to mathematics altogether. I interpreted this conception in a way that language of mathematics and mathematical content are separate entities. When one is present, the other one should be removed. This view, in turn, might have justified all ten teachers' reasoning in favor of using pictures or manipulatives, that is, anything not language-based, in relevant instructional scenarios. In one interview, for instance, the teacher was given a scenario in which a mathematics teacher gives the following word problem to her students (see item 2 in the appendix): "In 1996, the salary of the governor of New York was about $50,000 less than triple the salary of the governor of Arkansas. The total of the two salaries was $190,000. Find the 1996 salary of each state's governor." The teacher in the scenario wants to support the ELLs in being able to understand and solve the problem. ELLs in this class have grade level content knowledge, and their English reading proficiency has been identified at WIDA Reading Level 3[1](there are six WIDA levels). Options include a) providing ELLs with explanations of the difficult phrases in the problem, b) providing ELLs with visual illustration of the problem, c) giving ELLs a version of the problem in simplified English, and d) giving ELLs the equation that the problem is based on and have them solve the equation. The word problem included linguistically demanding phrases such as "$50,000 less than triple…" and "the salary of the governor of New York" that needed unpacking for the particular ELLs in the scenario to solve the equation. Given the scenario and ELL characteristics, option A was the best instructional strategy because providing a version of the problem in simplified English (option C) or visual illustration of the problem

---

[1] According to WIDA 6-8 grade cluster can-do descriptors, the ELLs with a WIDA reading level 3 can 1) Identify topic sentences, main ideas, and details in paragraphs; 2) Identify multiple meanings of words in context (e.g., "cell," "table"), 3) Use context clues; 4) Make predictions based on illustrated text; 5) Identify frequently used affixes and root words to make/extract meaning (e.g., "un-," "re-," "-ed"); 6) Differentiate between fact and opinion; 7) Answer questions about explicit information in texts; 8) Use English dictionaries and glossaries

(option B) would not significantly help the ELLs make the meaning-form connection between the mathematical concept and its linguistic representation in the word problem. Along the same lines, it would not be the best strategy to readily give the ELLs the equation (option D) without helping them understand the connection between the word problem and mathematical equation. The low performing teacher we interviewed explained that she would prefer the visuals over all the other options, as stated in the excerpt below. When prompted further, the teacher explains what she understands 'simplified English' as referring to "*taking difficult phrases and making them simplified or putting them in simplified forms*" and she says this strategy would be the same as providing explanations of the difficult phrases in the problem. In other words, the teacher discards the two options because she believes them to be identical.

I:      So just tell us what information this question gave you and what it asked you to do.

R:      Well it gave me a word problem and it first gave me some information about this particular ELL grade level class, that their content knowledge is at grade level and what their reading level is at. And it's asking which strategy I would use to support them in this word problem.

I:      Okay. So what was your answer?

R:      I don't remember. I remember doing it very late at night but I think what I would probably do is a visual and then the next explanation; and I wasn't sure on this if they want one answer or more than one. Because I might use a combination; now that I'm looking at it, I might use a visual and then explanation because of the difficult phrases.

I:      Yeah. If you had to go to choose one, which one would you say?

R:      I'd pick visual.

Overall, participating mathematics teachers seem to have taken into account the information about what ELLs can do at particular proficiency levels. Even though the majority believed in scaffolding ELLs' understanding of mathematics linguistically or otherwise, a few teachers also stated during the interviews that language of mathematics is hard for all learners, not just ELLs. Further, teachers' content knowledge might play a role in the depth of their reasoning about the language demands embedded in the mathematical content. Another pattern was that effective ELL teaching was equated to removing all the language demands or simplifying language to the extent possible through the use of pictures or visuals.

## Discussion

While presence of any systematic teacher knowledge base specific to teaching middle school mathematics to ELLs still requires further investigation, teachers drew on some sources of knowledge to reason through the scenarios. One belief that seemed to exist in teachers' reasoning is that effective ELL teaching is not unique above and beyond teaching all learners. While this view has conceptually and linguistically been debunked over the years (Harper & de Jong, 2004, 2009; Schleppegrell, 2001), only recently have we come to an empirical understanding that some teachers tend to be more effective with ELLs (Loeb, Soland, & Fox, 2014).

Loeb et al. (2014) examined the extent to which teachers' effectiveness is the same with ELLs as it is with non-ELLs and how much teacher effectiveness changes across classrooms with ELLs and non-ELLs. The premise here is that teacher effectiveness might vary depending on the specific group of students in the classroom. The findings showed that teachers who are effective with ELLs are also effective with non-ELLs and vice versa. However, when the authors regressed student test performance onto teacher characteristics such as Spanish fluency and attainment of a bilingual certification, they found that teachers with a command of Spanish proficiency and bilingual certification were more effective with ELLs, most of whom were Spanish speaking. This finding supports the emerging stance that good teaching for ELLs is characterized differently from good teaching for all (Harper & de Jong, 2004). Based on this understanding, scholars (Bunch, 2013; Galguera, 2011; Turkan et al., 2014) have attempted to identify pedagogical and linguistic aspects of the work of teaching content to ELLs.

From the current small-scale study, it is observed that the majority of the teachers view ELLs as a group of students who need specialized support in learning mathematics. Those who were observed to be skillfully integrating the information given about ELL characteristics with their content knowledge also performed high on the overall assessment. Teachers who had a better understanding about the content and language demands of the material were more flexibly able to reason through the scenarios. All in all, all the participating teachers integrated their existing knowledge or understanding about students, specifically, the given ELL characteristics into their reasoning about the scenarios. However, the extent to which they were able to integrate their knowledge of content and students was limited. One explanation for this might be that the depth of their specialized content knowledge determines the level of flexibility in which they could reason about the mathematical content in relation to its language demands. In other words, a unique contribution of this study might be that there is an interplay between teachers' content knowledge and DLK, which needs further investigation with more mathematics teachers on a larger scale.

These observations about the interviews might suggest that mainstream mathematics teachers need to draw on a systematic knowledge base to facilitate ELLs' linguistic and content-related challenges, as they are not routinely trained to differentiate their instructional practices according the needs of ELLs in the classrooms. Further research is needed to examine at a larger scale what mathematics teacher knowledge base differentiates good teaching for ELLs from good teaching for *all*.

## Conclusion and Future Directions for Research

There is both a growing population of ELLs in U.S. schools and a lack of preparation on behalf of the mathematics teachers who will be teaching them. This paper has barely scratched the surface of identifying and assessing the knowledge and skills needed to teach mathematics to ELLs. Identifying this knowledge base and testing its existence through assessment development is an essential first step towards understanding: What are the most effective pedagogically and linguistically responsive practices for teaching mathematics to ELLs? How can we meet the long term goal of preparing teachers for linguistically and culturally diverse learners? What will constitute quality professional development of mathematics teachers of ELLs?

# References

Ball, D. L., & Hill, H. C. (2008). Measuring teacher quality in practice. In D. H. Gitomer (Ed.), *Measurement issues and assessment for teaching quality* (pp. 80–98). Thousand Oaks, CA: Sage.

Ball, D. L., Hill, H. H., & Bass, H. (2005). Knowing mathematics for teaching: Who knows mathematics well enough to teach third grade, and how can we decide? *American Educators, 29*(1), 14–46.

Ball, D. L., Thames, M. H., & Phelps, G. (2008). Content knowledge for teaching: What makes it special? *Journal of Teacher Education, 59*(5), 389–407.

Ballantyne, K. G., Sanderman, A. R., & Levy, J. (2008). *Educating English language learners: Building teacher capacity* (Roundtable Report). Washington, DC: National Clearinghouse for English Language Acquisition & Language Instruction Educational Programs.

Brisk, M. E., & Zisselberger, M. (2011). We've let them in on the secret: Using SFL theory to improve the teaching of writing to bilingual learners. In T. Lucas (Ed.), *Teacher preparation for linguistically diverse classrooms: A resource for teacher educators* (pp. 55–73). New York, NY: Routledge.

Bunch, G. C. (2013). Pedagogical language knowledge: Preparing mainstream teachers for English learners in the new standards era. *Review of Research in Education, 37,* 298–341.

Christie, F., & Martin, J. R. (1997). *Genre and institutions: Social processes in the workplace and school.* London, UK: Cassell.

Cummins, J. (2000). *Language, power and pedagogy: Bilingual children in the crossfire.* Clevedon, UK: Multilingual Matters.

de Jong, E., & Harper, C. (2005). Preparing mainstream teachers for English language learners: Is being a good teacher good enough? *Teacher Education Quarterly, 32*(2), 101–124.

de Jong, E., & Harper, C. (2011). Accommodating diversity: Preservice teachers' views on effective practices for English language learners. In T. Lucas (Ed.), *Teacher preparation for linguistically diverse classrooms: A resource for teacher educators* (pp. 55–73). New York, NY: Routledge.

de Oliveira, L. C. & Cheng, D. (2011). Language and the multisemiotic nature of mathematics. *The Reading Matrix, 11*(3), 255–268.

Fang, Z. (2006). The language demands of science reading in middle school. *International Journal of Science Education, 28*(5), 491–520.

Fang, Z., & Schleppegrell, M. J. (2008). *Reading in secondary content areas: A language-based pedagogy.* Ann Arbor, MI: University of Michigan Press.

Galguera, T. (2011). Participant structures as professional learning tasks and the development of pedagogical language knowledge among preservice teachers. *Teacher Education Quarterly, 38,* 85–106.

Gebhard, M., Willett, J., Pablo, J., Caicedo, J., & Piedra, A. (2011). Systemic functional linguistics, teachers' professional development, and ELLs' academic literacy practices. In

T. Lucas (Ed.), *Teacher preparation for linguistically diverse classrooms: A resource for teacher educators* (pp. 55–73). New York, NY: Routledge.

Grossman, P., Compton, C., Igra, D., Ronfeldt, M., Shahan, E., & Williamson, P. W. (2009). Teaching practice: A cross-professional perspective. *Teachers College Record, 111*(9), 2055–2100.

Harper, C. A., & de Jong, E. (2004). Misconceptions about teaching English-language learners. *Journal of Adolescent & Adult Literacy, 48*(2), 152–162.

Harper, C. A., & de Jong, E. (2009). English language teacher expertise: the elephant in the room. *Language and Education, 23(*2), 137–151.

Jacobson, E., Remillard, J., Hoover, M., & Aaron, W. (in press). The interaction between measure design and construct development: Building validity arguments. In A. Izsák, J. T. Remillard, & J. Templin (Eds.), *Psychometric methods in mathematics education: Opportunities, challenges, and interdisciplinary collaborations* (pp. xxx–xxx). Journal for Research in Mathematics Education monograph series. Reston, VA: National Council of Teachers of Mathematics.

Lai, Y., & Howell, H. (2014). *Tasks assessing mathematical knowledge for teaching as representations of teaching practice.* Unpublished manuscript. University of Nebraska-Lincoln & Educational Testing Service.

Lindlof, T. R. (1995). *Qualitative communication methods.* Thousand Oaks, CA: Sage.

Loeb, S., Soland, J., & Fox, L. (2014). Is a good teacher a good teacher for all? Comparing value-added of teachers with their English learners and non-English learners. *Educational Evaluation and Policy Analysis, 36*(4), 457–475.

Lucas, T., & Villegas, A. M. (2011). A framework for preparing linguistically responsive teachers. In T. Lucas (Ed.), *Teacher preparation for linguistically diverse classrooms: A resource for teacher educators* (pp. 55–73). New York, NY: Routledge.

Lucas, T., Villegas, A. M., & Freedson-Gonzalez, M. (2008). Linguistically responsive teacher education preparing classroom teachers to teach English language learners. *Journal of Teacher Education, 59*(4), 361–373.

Martiniello, M. (2008). Language and the performance of English-language learners in math word problems. *Harvard Educational Review, 78*(2), 333–368.

Masingila, J. O., & Doerr, H. M. (2002). Understanding pre-service teachers' emerging practices through their analyses of a multimedia case study of practice. *Journal of Mathematics Teacher Education, 5,* 235–263.

McCarthy, M. (1991). *Discourse analysis for language teachers.* Cambridge, UK: Cambridge University Press.

Mislevy, R. J. (1994). Evidence and inference in educational assessment. *Psychometrika, 59,* 439–483.

Mislevy, R. J., Almond, R. G., & Lukas, J. F. (2003). *A brief introduction to evidence-centered design* (Research Report 03-16). Princeton, NJ: Educational Testing Service.

Moschkovich, J. N. (1999). Supporting the participation of English language learners in mathematical discussions. *For the Learning of Mathematics, 19*(1), 11–19.

Moschkovich, J. (2002). A situated and sociocultural perspective on bilingual mathematics learners. *Mathematical Thinking and Learning, 4*(2&3), 189–212.

National Clearinghouse for English Language Acquisition. (2011). *The growing numbers of English learner students*. Retrieved from http://www.ncela.gwu.edu/files/uploads/9/growingLEP_0809.pdf

Oakes, J. (1990). *Multiplying inequalities: The effects of race, social class and tracking on opportunities to learn mathematics and science*. Santa Monica, CA: Rand Corporation.

Reeves, J. R. (2009). A sociocultural perspective on ESOL teachers' linguistic knowledge for teaching. *Linguistics and Education, 20*, 109–125.

Schleppegrell, M. J. (2001). Linguistic features of the language of schooling. *Linguistics and Education, 12(*4), 431–459.

Schleppegrell, M. J. (2004). *The language of schooling: A functional linguistics perspective*. Mahwah, NJ: Erlbaum.

Schleppegrell, M. J. (2013). The role of metalanguage in supporting academic language development. *Language Learning, 63*(1), 153–170.

Secada, W. (1992). Race, ethnicity, social class, language and achievement in mathematics. In D. Grouws (Ed.), *Handbook for research on mathematics teaching and learning* (pp. 623–660). New York, NY: Macmillan.

Secada, W. G. (1998). School mathematics for language enriched pupils. In S. H. Fraud & O. Lee (Eds.), *Creating Florida's multilingual global work force: Educational policies and practices for students learning English as a new language*. Tallahassee, FL: Florida Department of Education.

Turkan, S., Croft, A., Bicknell, J., & Barnes, A. (2012). *Assessing quality in the teaching of content to English language learners*. (Research Report 12–10). Princeton, NJ: Educational Testing Service.

Turkan, S., de Oliveira, L., Lee, O., & Phelps, G. (2014). Proposing the knowledge base for teaching academic content to English language learners: Disciplinary linguistic knowledge. *Teachers College Record, 116*(3), 1–30.

U.S. Department of Education, Institute of Education Sciences, National Center for Education Statistics. (2015). Retrieved from http://www.nationsreportcard.gov/reading_math_g12_2013/#/changes-by-groups

# Appendix

## Interview questions

1. Can you tell in your own words what information the question gives you and what it asks you to do?

2. What was your answer?

3. What information in the item did you draw on when you selected this answer choice _____ (e.g., A)?

4. What information in the item helped you to eliminate the other choices _(e.g., B, C, D)? OR
- *You indicated that you chose __as the answer. Why do you think that's the best answer? Let's go through the other options. Why didn't you select__?*

5. Does the scenario present a challenge of teaching ELLs that you would encounter in your classroom? Which middle grade level is the scenario most appropriate?

6. Does the scenario present a learning challenge that you would encounter with your ELL(s) in your classroom?

7. On a scale of 1-5, how important do you think it is for (math or science) teachers of ELLs to identify the ELL language proficiency levels? On a scale of 1-5, how useful was it to be informed about the proficiency levels of the ELL(s) characterized in the item?

8. On a scale of 1 to 5, with 5 being the most important, how important do you think the knowledge being tested in this item is for math/science teachers who teach ELLs?

- How important is it for teachers of ELLs to know how to answer this question?
- How important is it for teachers of ELLs to be able to answer this question?

9. Was there anything about the question/scenario/problem that you found unclear or ambiguous?

**Item 1.**

**Read the learning objective, teacher task, characteristics of focus ELLs, and instructional scenario, and then answer the question that follows.**

**Learning objective:** Students will understand how to rewrite expressions using the distributive property.

**Teacher task:** The teacher is introducing students to the distributive property and providing examples to scaffold the students' understanding of the distributive property.

**Characteristics of focus ELLs:** ELLs in the class are newly arrived in the school. Their placement-test results indicate that they are weak in mathematics and have low proficiency in English.

**Instructional scenario:** While preparing for a lesson on the distributive property, a teacher looks up the definition given in the textbook. The definition is shown below.

*Distributive property—the product of a number and a sum is equal to the sum of the individual products of the addends and the number*

Rather than begin the lesson with the definition, the teacher wants to give an illustration of the distributive property.

**Which of the following examples should the teacher write on the board to best scaffold the ELLs' understanding?**

A.    $a\,(b + c) = ab + ac$

B.

$2 \times 3 + 2 \times 4$
$\phantom{2\times}6\phantom{\times} + \phantom{2}8\phantom{\times} = 14$

$2 \times 7 = 14$
$2\,(3 + 4) = 2\,(3) + 2\,(4)$

C.    At a store, each book costs \$4, and each video game costs \$15. Ellen is buying a book and a video game for each of her 3 children. How much does Ellen spend altogether?

D.    The ingredients in a cookie recipe include 4 cups of flour and 2 eggs. Bob is doubling the recipe. How many cups of flour and how many eggs will Bob need to make the cookies?

**Item 2.**

Read the learning objective, student task, characteristics of focus ELLs, and instructional scenario, and then answer the question that follows.

**Learning objective:** All students will understand how to write and solve equations based on word problems.

**Student task:** Students will write and then solve an equation based on a word problem.

**Characteristic of focus ELLs:** ELLs in the class have grade-level content knowledge, and their English reading proficiency has been identified as being at WIDA Reading Level 3.

**Instructional scenario:** During a unit on solving word problems that involve multi-step equations, a teacher gives the problem below to the students.

> *In 1996, the salary of the governor of New York was about $50,000 less than triple the salary of the governor of Arkansas. The total of the two salaries was $190,000. Find the 1996 salary of each state's governor.*

**Which of the following is the best strategy to support the ELLs in being able to understand and solve the problem?**

A. Provide ELLs with explanations of the difficult phrases in the problem.
B. Provide ELLs with a visual illustration of the problem.
C. Give ELLs a version of the problem in simplified English.
D. Give ELLs the equation that the problem is based on and have them solve the equation.

# Teachers and their Educators – Views on Contents and their Development Needs in Mathematics Teacher Education

Mika Koponen, Mervi A. Asikainen, Antti Viholainen and Pekka E. Hirvonen
University of Eastern Finland

**Abstract:** Finland has scored well in international assessments (e.g. PISA, TIMSS), and the pressure to attain excellent scores has activated a drive toward even more effective mathematics teacher education. This article presents the results of a qualitative assessment of the mathematics teacher education provided by the University of Eastern Finland. In this study, the views held by practicing teachers (N=101) and teacher educators (N=19) are compared so that the outstanding development needs of mathematics teacher education in terms of their contents can be revealed. The data was gathered via an electronic survey and was mainly analyzed using data-driven methods. In addition, framework provided by Mathematical Knowledge for Teaching (MKT) was used to categorize the respondents' views regarding the contents of mathematics teacher education and to develop general guidelines for the reform of mathematics teacher education. The results indicate that mathematics teacher education should include pure mathematical content (Common Content Knowledge, CCK) and mathematical content that will have been designed only for future teachers (Specialized Content Knowledge, SCK). Teacher educators and practicing teachers both held the view that the relevance of CCK studies depend on the connections between university and school mathematics. Pedagogical studies should also be reformed because practicing teachers have realized that effective teaching (Knowledge of Content and Teaching, KCT) requires knowledge about learning mathematics (Knowledge of Content and Students, KCS) that is not offered in the current educational system on a sufficiently broad basis. In this study, suggestions for developing mathematics teacher education were mostly connected to four domains of MKT: (CCK, SCK, KCT and KCS). Interestingly those domains are the same domains which has been empirically tested and better conceptualized.

Keywords: Mathematical Knowledge for Teaching, MKT, mathematics teacher education, evaluating teacher education, contents of mathematics teacher education.

## Introduction

Finland has scored well in international assessments (e.g., PISA, TIMSS), and the Finnish school system has been rated as being of top quality. Finnish teacher education has also been evaluated as high in quality from an international perspective (Kivirauma & Ruoho, 2007; Tryggvason, 2009). An important reason for this success is that Finnish teachers are educated both systematically and extensively, and every qualified teacher must have a Master's degree (Tryggvason, 2009). It is claimed that Finnish teacher education is the result of a long-term, research-based development (Tryggvason, 2009). However, the voices of practicing mathematics teachers and teacher educators have not received attention enough in the research field. Are these two groups satisfied with the current contents of mathematics teacher education and what kind of needs of development they see at the moment?

In the present study we focus on practicing mathematics teachers' and teacher educators' views on mathematics teacher education. The practicing teachers participating in this study graduated in the period of 2002–2012 and they nowadays teach at school level, which enables them to evaluate the contents of teacher education from a perspective of the teacher's profession. In addition, when the survey was implemented the teacher educators were actively working as teacher educators. We were interested in discovering how these two subject groups saw the present contents of the Mathematics Teacher Education Program (MTEP) at the University of Eastern Finland and also in how they would develop the teacher education program. We sought answers to the following research questions:

1. How do teacher educators and practicing mathematics teachers regard the course contents of mathematics teacher education?
2. What kind of recommendations would teacher educators and practicing mathematics teachers make for improving mathematics teacher education program?

The views held by practicing teachers and teacher educators play an import role in developing teacher education. There may be a possibility that the contents are not regarded as being as useful as teacher educators assume. It is also possible that practicing teachers and teacher educators hold conflicting views about the contents. Hence, the views of both groups are important in order to be able to form a coherent picture of the current status of teacher education and to construct an extensive basis for the development work.

Our methodical aim has been to test a theoretical framework called *Mathematical Knowledge for Teaching (MKT)* (Ball, Thames & Phelps, 2008) through the process of categorizing practicing teachers' and teacher educators' views. This framework appeared to be promising for categorizing these views, since it has previously worked relatively well in classifying teacher knowledge (see Markworth, Goodwin, & Glisson, 2009; Fauskanger, Jakobsen, Mosvold, & Bjuland, 2012).

## Conceptualizing the Teaching of Mathematics

### Mathematical knowledge for teaching

There was an increasing interest in the 1980s in teacher qualifications and methods of effective teaching that would influence student learning. Lee Shulman proposed that a teacher also needs to possess other types of knowledge than pure subject matter knowledge in order to teach so that students would understand. In 1986 Lee Shulman introduced a new term, *pedagogical content knowledge (PCK)*. According to Shulman (1986), teachers must have an integrated knowledge of subject and pedagogy, some kind of amalgam knowledge. Initially, Shulman considered PCK to be a topic-specific subcategory of content knowledge, which included two further subcategories: knowledge of representations and knowledge of learning difficulties and strategies for overcoming them. Shulman's later model consisted of seven categories, of which PCK was one, with no subcategories (Shulman, 1987). By proposing PCK as one out of seven categories of conceptualization, Shulman neglected the potential for integration among these categories and the hierarchies that might exist between them, and left the task of further development of the concept to other researchers (Hashweh, 2005).

Shulman's conceptualization has been criticized for its restricted and ambiguous definitions of categories (Ball et al., 2008, Hashweh, 2005). Ball et al. (2008) claim that the terms PCK and *content knowledge* are frequently confused with common pedagogical skills. Meredith (1995) argues that PCK as defined by Shulman simply implies one type of pedagogy rooted in particular representations of prior knowledge. Meredith suggests that learners have a built-in competence for constructing their own understanding of subject matter, but Shulman's PCK seems not to encompass alternative views of teaching. Meredith argues that Shulman's definition of PCK leads to teaching methods where the teacher will explain and illustrate procedures while learners practice the procedures by using examples. Thus, the teacher's role can be seen as transmitting mathematical knowledge and helping learners to acquire understanding.

Shulman's conceptualization has also been claimed to ignore the interaction between the different categories, assuming that knowledge is static rather than possessing a dynamic nature (Hashweh, 2005; Fennema & Franke, 1992). Fennema and Franke (1992) argue that teacher knowledge frequently changes in light of classroom interaction experiences, and hence teachers' beliefs should form an important part of the conceptualization. According to Fennema and Franke, teacher knowledge can be divided into four parts: *knowledge of content, knowledge of pedagogy, knowledge of students' cognitions*, and *teachers' beliefs*. At the center of this model is *context specific knowledge*, which can be seen as dynamic knowledge, since it occurs in the context of the classroom. In this model, PCK consists of teachers' knowledge of teaching procedures, such as effective strategies for planning, classroom routines, behavior management techniques, classroom organization procedures, and motivational techniques. Fennema and Franke (1992) see teacher knowledge as interactive and dynamic in nature and they suggest that no single domain of teacher knowledge plays a particular role in the effective teaching of mathematics.

Rowland, Turner, Thwaites & Huckstep (2009) developed *The Knowledge Quartet* conceptualization, which was based on Shulman's conceptualization (1986) with respect to Fenneman and Franke conceptualization (1992). The Knowledge Quartet was generated by categorizing elementary teachers' classroom actions. The main aim of the research work was to investigate the relation between the teacher's subject matter and PCK knowledge. Detailed analysis of the elementary mathematics lessons taught by pre-service teachers resulted in the identification of teacher knowledge framework. Rowland et al. (2009) suggest that the framework can be used to classify teachers' actions in the context of a classroom.

One of the most promising recent efforts in discovering the kind of knowledge and skills that are needed for high-quality mathematics teaching has been the theoretical framework known as *Mathematical knowledge for teaching (MKT)*, as posited by Ball and her associates[1]. In this model, subject matter knowledge is categorized into three domains: *common content knowledge (CCK), horizon content knowledge (HCK), and specialized content knowledge (SCK)* (see Figure 1). In addition, PCK consists of three parts: *knowledge of content and student (KCS), knowledge of content and teaching (KCT)*, and *knowledge of content and curriculum (KCC)*. The domains CCK, HCK, and SCK are subject matter knowledge that requires no knowledge concerning either the students or pedagogy. In addition, the domains of KCS, KCT, and KCC are the kind of knowledge that requires an integrated knowledge made up of subject matter knowledge and pedagogical knowledge (Sleep, 2009), as in Shulman's (1986) conceptualization. According to Sleep (2009), four of the domains (CCK, SCK, KCS and KCT) have been empirically tested and better conceptualized, while two of the domains (HCK and KCC) are still in the earlier stages of conceptualization.

---

[1] The University of Michigan projects *Mathematics Teaching and Learning to Teach project (MTLT)* and *Learning Mathematics for Teaching project (LMT)* produced plenty of details to form MKT, e.g., Hill & Ball, 2004; Hill, Schilling & Ball, 2004; Hill, Rowan & Ball, 2005; Hill & Lubienski, 2007; Hill, Ball, Sleep et al., 2007; Hill, 2007; Schilling, 2007; Schilling, Blunk & Hill, 2007; Schilling & Hill, 2007; Hill, Ball, Blunk, et al., 2007; Hill, Dean & Goffney, 2007; Hill, Ball & Schilling, 2008; Delaney, Ball, Hill et al., 2008; Stylianides & Ball, 2008; Hill, Blunk, Charalambous et al., 2008; Ball, Thames & Phelps, 2008; Ball & Forzani, 2009; Thames & Ball, 2010.

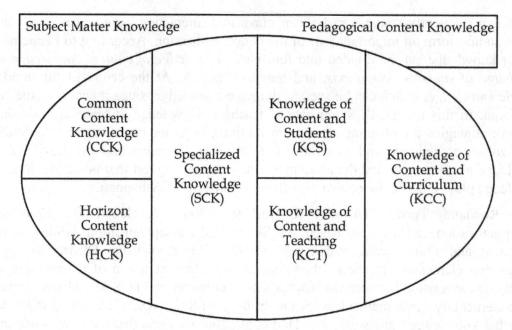

*Figure 1.* Domains of Mathematical Knowledge for Teaching (MKT) by Ball et al. (2008)

CCK consists of mathematical knowledge and skills used in any settings, including in settings other than teaching, and it includes calculating, solving problems, and other common mathematical knowledge that is not unique to teaching (Ball et al., 2008). SCK is mathematical knowledge and skills that are peculiar to teaching, and is typically not intended for other settings than teaching (Ball et al., 2008). In other words, SCK consists of the mathematical knowledge and skills that a mathematician does not need, while at the same time they are needed by a teacher in order to practice effective teaching. HCK consists of mathematical knowledge of the mathematical structures and also awareness of how mathematical topics are related to each other in a curriculum (Ball et al., 2008). This means that a teacher needs to know how topics are related to each other at different school levels and how mathematics is actually constructed.

KCS consists of amalgam knowledge of students, learning, and mathematics (Ball et al., 2008). A teacher must be able to anticipate students' difficulties, hear and respond to students' thinking, and choose suitable examples and presentations while teaching. A teacher's action in planning and teaching requires awareness of students' conceptions and misconceptions of different mathematical topics. KCT is also amalgam knowledge of teaching and mathematics (Ball et al. 2008). Teachers need KCT knowledge in choosing proper activities, exercises and representations for different topics. Teachers need KCT knowledge for both planning and teaching. One important part of this knowledge for teachers is to recognize situations where teachers should diverge from their original planning, for example, if a student makes a mathematical discovery.

KCC represents amalgam knowledge of mathematics and curriculum. According to Sleep (2009), a teacher needs to know the contents of the curriculum, but Ball et al. (2008) offer only a restricted definition of KCC and hence the kind of knowledge and skills that KCC includes remains unclear. Our preliminary analysis of the data in the present study showed that if MKT is used to organize practicing teachers' and educators' views, the KCC domain has to be modified. The practicing teachers and teacher educators mentioned skills and knowledge related to teaching equipment. Hence, our conceptualization states that KCC also includes knowledge and skills related to teaching materials (including textbooks, other materials, etc.), teaching instruments (blackboard, overhead projector, etc.), and technology (computer, smart board, calculators, software, etc.).

## The Evolution of MKT

The development of MKT started with the study of classroom actions with a view to identifying the knowledge needed for teaching mathematics (Ball & Bass, 2003). This work continued with the formation of hypothetical characterizations of MKT (e.g. Ball, Hill & Bass, 2005; Ball et al., 2008). Thereafter, Hill, Schilling, and Ball (2004) developed specific measurements of MKT that could be used to test this hypothetical characterization. In the case of validating measurements, the Michigan group tested measurements against practice (Hill, Blunk, Charalambos, et al., 2008) and also against students' achievements (Hill, Rowan & Ball, 2005). Thereafter, MKT has been used to develop the contents of teacher education in ways that should help teachers to acquire the knowledge required for teaching mathematics (Ball, Sleep, Boerst & Bass, 2009).

Markworth, Goodwin and Glisson (2009) have used MKT to evaluate what student teachers have learned during a teaching practicum course. They coded interview responses and conversational topics on the basis of the domains of MKT. By using MKT in their analysis, Markworth, Goodwin and Glisson (2009) were able to capture more detailed information about the subject matter knowledge and pedagogical content knowledge that student teachers had gained during the teaching practicum course.

In the course of this study, practicing teachers and teacher educators suggested various recommendations for improving the mathematics teacher education program. To identify these suggestions systematically, we used MKT in a similar way to that of Markworth, Goodwin and Glisson (2009). This meant that suggested recommendations for improving mathematics teacher education could be classified in terms of six domains of MKT.

## Method

### Context

This study was implemented at the University of Eastern Finland, which offers two programs for students of mathematics: one for mathematicians and another for teachers. The programs are almost identical in their respective amounts of mathematics courses, but they differ in minor subjects. In the present study, we concentrate on the program for teachers, Mathematics Teacher Education Program, MTEP.

MTEP includes a Bachelor's degree (180 cp[2]) and a Master's degree (120 cp). Both degrees are required for a student to qualify as a mathematics teacher in Finland. MTEP includes mathematical studies (130 cp), pedagogical studies (60 cp), and studies in one or two minor subjects (60 cp each). Most mathematical studies are traditional mathematics courses, which are compulsory for both future teachers and mathematicians (e.g. *calculus, analysis, algebra, differential equations,* etc.).

The pedagogical studies include theoretical studies focusing on teaching and learning (30 cp), the didactics of mathematics (10 cp), and teaching practice (20 cp). Teaching and learning courses are intended to all subject teachers and courses are concerning teaching and learning in general. However, the following courses which are intended only to forthcoming subject teachers of mathematics enables taking into account the special aspects of mathematics. Teaching practice is undertaken at the university teacher training school. Student teachers plan their own teaching sequences or lessons under the guidance of a subject teacher. Student teachers' lessons are evaluated and feedback is also provided. The amount of student teaching is approximately 50

---

[2] One *credit point* (*cp*) is the equivalent of 25 hours of study. The recommendation is to complete 60 cp of studies per year.

lessons. The training school teachers' task is to guide student teachers in addition to performing their own ordinary teaching work.

Student teachers can choose to study any school subject as a minor subject, but the most typical choices are physics or chemistry, or both. In its entirety, MTEP provides students with the competence to teach mathematics and minor subjects at lower or upper secondary schools and vocational schools.

## Sample

The data was collected in the course of two separate electronic surveys conducted in 2012–2013. The first survey was aimed at mathematics teachers who had graduated from the UEF during the period 2002–2012. Our sample (N=101) includes 54% of all teachers who have graduated from UEF during period 2002–2012. In the sample, the majors taken by our respondents were 72% (73) mathematics, 20% (20) physics and 8% (8) chemistry, which makes the sample similar to the distribution of graduated teachers according to their major subject. All of the respondents, with the exception of one, had had previous experience of teaching mathematics at school or they were working as teachers when the survey was implemented.

The second survey targeted the teacher educators in mathematics at the UEF, who taught either mathematical studies or pedagogical studies or were guiding teaching practice. Our sample (N=19) includes 79% of all of the teacher educators in mathematics at the UEF. In the sample, 74% (14) of the teacher educators taught mathematics and 26% (5) worked in pedagogical studies or teaching practice. To conceal the respondents' identities, the teacher educators in the fields of both pedagogical studies and teaching practice were placed in the same category.

## Instrument and data analysis

The study was implemented with the aid of a survey that included statements about the knowledge and skills learned in MTEP and open questions about the present state and the future of MTEP.

The survey conducted with practicing teachers included three open questions about MTEP.

1. Evaluate the contents of mathematical studies in MTEP, especially with regard to the work of a mathematics teacher.
2. Evaluate the contents of the pedagogical studies and teaching practice in MTEP, especially with regard to the work of a mathematics teacher.
3. Suggestions regarding the development of mathematics teacher education would also be appreciated.

The survey involving the UEF teacher educators included two open questions.

4. Evaluate the contents of the studies you teach, especially with regard to the work of a mathematics teacher.
5. Please make a suggestion regarding the development of mathematics teacher education would also be appreciated.

The data was analyzed using qualitative content analysis (Tesch, 1990; Hickey & Kipping, 1996; Mayring, 2000; Hsieh & Shannon, 2005). Hsieh and Shannon (2005) have identified three different approaches to qualitative content analysis that can be used to interpret meaning from the content of text data: conventional, directed or summative (Hsieh & Shannon, 2005).

Our analysis started with reading the data several times to achieve immersion and obtain a sense of the whole (Tesch, 1990). Then, practicing teachers' and teacher educators' perceptions about the contents of MTEP (Questions 1, 2 and 4) were analyzed with *Conventional Content Analysis* (Hsieh & Shannon, 2005). In the conventional content analysis, coding categories are derived directly from the text data. In our data, respondents' personal experience or more like attitudes towards contents emerged clearly from data. Each respondent was placed in one of these categories (Figure 2).

A majority of the practicing teachers' mentioned only issues that should be developed in the contents of MTEP (Questions 1 and 2), and therefore their responses were placed in the category of *In need of development*. The contents of this category were analyzed with directed content analysis, which is a more structured process than the conventional approach (Hickey & Kipping, 1996; Hsieh & Shannon, 2005). Direct content analysis starts with a theory, which is used for coding text data (Hsieh & Shannon, 2005). Generally, a goal of the directed approach is to validate or conceptually extend a theoretical framework or theory (Hsieh & Shannon, 2005). *Mathematical Knowledge for Teaching (MKT)* framework was a starting point for designing the survey, and each statement was designed to be interconnected to the domain of MKT. In the planning, we noticed a possibility for using MKT for directed content analysis. A pre-analysis of the data indicated that all the issues in question 1 and many of the issues in question 2 can be categorized with MKT. Issues beyond MKT were categorized with the conventional content analysis in case of question 2.

Both the surveys also included blank spaces for other suggestions for the development of the teacher education program (Questions 3 and 5). Many of the respondents did, however, mention the same issues which they already mentioned in the previous question related to the contents. Therefore we used directed content analysis similarly as in the categorization of the suggestions related to the six domains of MKT. Suggestions beyond MKT were categorized with the conventional content analysis. Previous questions in the survey covered the majority of respondents' ideas, and so there were only a few new ideas among these suggestions.

---

### CONVENTIONAL CONTENT ANALYSIS

Respondents were categorized in six categories based on their answers. Categories were derived from the data. Each respondent was categorized in one category only.

| Positive | Neutral | Analytic | Negative | In need of development | No answer |
|---|---|---|---|---|---|
| "Only positive issues mentioned" | "Neutral issues mentioned, but without taking a stand on any of them" | "At least one positive and one negative issue mentioned" | "Only negative issues mentioned" | "Only issues that need development mentioned" | "Blank or irrelevant response" |

---

### DIRECTED CONTENT ANALYSIS

Directed content analysis focused on issues requiring development. Categories are derived from Mathematical Knowledge for Teaching framework. Each issue was categorized in one category only.

| Common Content Knowledge | Horizon Content Knowledge | Specialized Content Knowledge | Knowledge of Content and Students | Knowledge of Content and Teaching | Knowledge of Content and Curriculum |
|---|---|---|---|---|---|

*Figure 2*. Text data analysis was performed with conventional and direct content analysis (Hsieh & Shannon, 2005).

## Results

The results of the study are presented in two parts. First, we discuss how teacher educators and practicing mathematics teachers view the contents of mathematics teacher education. Second, our discussion focuses on teacher educators' and practicing mathematics teachers' ideas for developing teacher education. Suggestions for developing mathematics teacher education will be represented in tables where the categories have been provided mainly by MKT.

### Views on the contents of mathematics teacher education

**Practicing teachers' views concerning the contents of mathematics studies.** The categorization of practicing mathematics teachers' views concerning the contents of mathematical showed that one fifth of the respondents (21%) viewed the contents neutrally. Half of them considered the number of mathematics courses appropriate for teachers, while the other half gave no reasons for their responses. A small minority of the respondents (7%) did not consider the contents of the mathematics courses useful for teachers. In most cases, the reason for this was that the courses were considered to provide too complex a discussion of mathematics in comparison with the mathematics needed in a teacher's work. No fully positive views appeared in the categorized responses.

A majority of the practicing teachers (59%) provided only suggestions related to developing the present contents of mathematics courses. These suggestions were analyzed again by using MKT. Most of these suggestions (79%) were related to improving student teachers' subject matter knowledge, while one fifth of them were concerned with developing student teachers' pedagogical knowledge and skills (see Table 1).

Table 1. *Categorization of practicing teachers' (N=60) suggestions for developing the content of mathematical studies. Each respondent was permitted to mention more than one issue.*

| Category | Domain of MKT | f |
|---|---|---|
| **Subject matter knowledge and skills** | *Common content knowledge (CCK)* | |
| | • Present course contents are not linked with school mathematics | 12 |
| | • Present course contents are not the same as in schools | 11 |
| | • More geometry | 4 |
| | • More financial and statistical mathematics | 2 |
| | • Wider knowledge of mathematical concepts | 2 |
| | *Specialized content knowledge (SCK)* | |
| | • Present mathematical studies should be separate for student teachers and mathematicians | 20 |
| | • More school mathematics needed | 12 |
| | *Horizon content knowledge (HCK)* | |
| | • Present course contents are not linked with each other | 4 |
| | • More skills concerned with teaching students at different levels | 1 |
| **Pedagogical content knowledge and skills** | *Knowledge of content and students (KCS)* | |
| | • More studies concerning learning difficulties in mathematics | 4 |
| | • More skills concerned with teaching students at different levels | 3 |
| | *Knowledge of content and teaching (KCT)* | |
| | • More courses about didactic mathematics | 6 |
| | • More courses about how to differentiate teaching | 2 |
| | • More skills to motivate students in mathematics | 1 |
| | • More studies about teaching problem solving | 1 |
| | *Knowledge of content and curriculum (KCC)* | |
| | • More courses about using technology in teaching mathematics | 1 |

*Common content knowledge (CCK)*. The practicing teachers mentioned that the present contents of mathematical studies are not the same as the contents of school mathematics. They were disappointed that they had studied so much mathematics that they had never used in their school teaching. The practicing teachers also mentioned that the present contents did not link properly with school mathematics. Some teachers claimed that presentations at university were either symbolic and theoretical or too complex in comparison with school mathematics, and hence it was hard to see how the course contents were linked with school mathematics. The practicing teachers mentioned that they lacked the competence to teach geometry and financial or statistical mathematics, and so they suggested that MTEP should include more courses in those domains. It is evident that practicing teachers need mathematical content knowledge (Ball et al., 2008), but the opinions of the practicing teachers indicated that to be useful for future teachers, the contents should be linked to school mathematics.

*Specialized content knowledge (SCK)*. The practicing teachers suggested that mathematical studies should be arranged differently for future teachers and for mathematicians. They argued that the current integrated mathematical courses do not support future teachers properly. Some practicing teachers said that at university the focus of mathematics was proving and presenting results, whereas school mathematics consisted of rather more than that. Almost all the respondents suggested that the mathematical representations of the course contents should be modified with respect to teachers' actual work. According to them, in the current situation MTEP includes too much pure mathematics and not enough school mathematics. Most of them recalled that in MTEP there was a course called *School mathematics*, which they found important and useful. The contents of the course were the same as in actual school mathematics, and the implementation of the course resembled the mathematics teaching conducted in schools. In consequence, they argued that they learned the contents of such courses well and that they had been able to use the course contents in their teaching work. All of them argued that there should be more courses of this kind in MTEP. All of these practicing teachers' views are linked to the definition of SCK (Ball et al., 2008): practicing teachers need mathematical knowledge that is particular to the needs of teachers.

*Horizontal content knowledge (HCK).* The practicing teachers argued that the contents of university mathematics courses were not interconnected or that the links could not be detected during the courses. In their view, courses that were in fact extensions of each other (e.g., calculus 1, calculus 2) were separate courses; alternatively, they were unable to detect the ways in which new mathematical concepts could be constructed on the basis of previously learned concepts. In the view of the respondents, the mathematical knowledge base ought to resemble a network, while, for them, the contents of MTEP did not support the construction of that kind of concept. Mathematical knowledge lacking a proper understanding of the structure of mathematics can be identified as the major challenge in the domain of HCK (see Ball & Bass, 2009).

*Knowledge of content and students (KCS).* The practicing teachers claimed that pedagogical issues can also be discussed during a mathematics course. They mentioned that they did not develop any clear idea of how students were actually learning mathematics during the mathematics courses. Some practicing teachers mentioned that they had too little competence in the issues concerned with mathematical learning difficulties. Some teachers also argued that they need more skills related to teaching mathematics to both weak and talented students at the same time.

*Knowledge of content and teaching (KCT).* The practicing teachers argued that issues concerned with teaching mathematics can also be handled in mathematical studies. They mentioned that didactic mathematics and studies about how to differentiate teaching should be included in mathematical studies.

*Knowledge of content and curriculum (KCC).* One practicing teacher argued that future mathematics teachers needed more enhanced skills concerned with the use of technology in teaching mathematics since teachers were increasingly using technology in schools.

**Practicing teachers' views about the contents of pedagogical studies and teaching practice.** The categorization of practicing mathematics teachers' views about the contents of pedagogical studies and teaching practice demonstrated that the respondents' views were diverse. Small minorities of the respondents viewed these studies positively (2%) as useful for teachers; or neutrally (9%), often without providing reasons; or negatively (7%), considering the courses useless for teachers; but most of them (67%) consider that there was a need for development in these studies. 3 The categorization of their suggestions is presented in Table 2. These suggestions mainly concerned pedagogical content knowledge and skills (51%), and development of the structure of mathematics teacher education (40%).

---

[3] 12% of respondents did not answer this question, and the responses of 3% were irrelevant.

Table 2. *Categorization of practicing teachers' (N=68) views on how to develop the content of pedagogical studies and teaching practice. Each respondent was permitted to mention more than one issue.*

| Category | Domain of MKT | f |
|---|---|---|
| **Pedagogical content knowledge and skills** | *Knowledge of content and students (KCS)* | |
| | • More studies of learning difficulties in mathematics | 13 |
| | • More skills concerned with how to handle students at different levels | 10 |
| | • More skills concerned with evaluating students' knowledge and skills | 6 |
| | • More studies concerned with other learning theories | 2 |
| | *Knowledge of content and teaching (KCT)* | |
| | • More studies of teaching mathematics; didactic mathematics | 10 |
| | • More training in the planning and teaching of complete courses | 6 |
| | • More studies of how to differentiate teaching | 5 |
| | • More skills concerned with motivating students of mathematics | 4 |
| | • More courses about functional teaching methods, teaching problem-solving, or visualizing mathematics | 3 |
| | • More studies of how to link learning theories to practice | 1 |
| | *Knowledge of content and curriculum (KCC)* | |
| | • More skills and knowledge to produce teaching materials of their own | 4 |
| | • More studies of using technology in teaching mathematics | 3 |
| **Structure of mathematics teacher education** | *Amount of studies* | |
| | • More teaching practice | 21 |
| | • More pedagogical studies | 5 |
| | • Compulsory update education after some years of teaching | 1 |
| | • More studies of how to teach minor subjects | 1 |
| | *Quality of studies* | |
| | • Linking theory to practice | 25 |
| | • Educators of practice teachers should give more advice about didactic issues | 1 |
| | *Developing curriculum of MTEP* | |
| | • Contents of pedagogical studies and practice should be better integrated | 1 |
| **General issues** | *The other knowledge and skills* | |
| | • More studies of teachers' extramural duties | 10 |
| | *Common issues* | |
| | • Departments' cooperation should be improved | 1 |

*Knowledge of content and students (KCS).* The practicing teachers suggested that the pedagogical studies and teaching practice should include more courses about the learning difficulties encountered in mathematics. Some teachers said that they were struggling with students who probably had learning difficulties and hence they needed more skills in order to be able to recognize and handle such students. Some of the practicing teachers also mentioned that the skills concerned with teaching students at different levels would be useful for them because student groups were often very heterogeneous. The practicing teachers also mentioned that they needed more knowledge and skills for evaluating student learning and more knowledge concerned with various learning theories, since students learn in different ways.

*Knowledge of content and teaching (KCT).* The practicing teachers demanded more skills for teaching mathematics. Some of them mentioned that the courses in didactic mathematics were useful for them, but that they needed more knowledge of this kind. The practicing teachers argued that their studies did not include enough courses on planning and teaching complete courses. They stressed that planning was the first thing that new teachers needed to undertake after graduation. The practicing teachers also mentioned that more skills for differentiating teaching and increasing student motivation would be of assistance.

*Knowledge of content and curriculum (KCC).* The practicing teachers said that textbooks or other printed material did not always fit their ideas about teaching, and so they would need more skills to design their own teaching materials. The practicing teachers also mentioned that they needed more knowledge and skills concerned with using technology in teaching mathematics in a pedagogically reasonable way.

*The number of courses.* The practicing teachers said that both the teaching practice and the pedagogical studies were very useful and suggested that their number should be increased in MTEP. They felt that the teaching practice was a good place for trying out new teaching methods or for trying to transform pedagogical knowledge into practice. They also told about the use of useful and functional teaching methods learnt during their teaching practice in their actual work. Some of them mentioned encountering similar situations in the classrooms to those that had been discussed in the pedagogical studies, which had helped them to better understand the relevance of the pedagogical studies.

*The quality of courses.* The practicing teachers argued that the courses in the pedagogical studies and even courses about teaching and learning were too theoretical, which made linking theory with practice difficult. They described a feeling of learning a lot during these studies, but without having the necessary skills to apply this knowledge in classroom situations. Some teachers even felt that the pedagogical studies were useless because they had too few links with real-life teaching situations.

*Other knowledge and skills.* Some practicing teachers said that they were surprised by the duties that teachers had outside the classroom. They suggested that these issues should be discussed in mathematics teacher education.

**Teacher educators' views on the contents of their own courses.** One fourth of the teacher educators (26%) viewed their own courses positively and considered that the courses were useful for future teachers (see Table 3). Many of them (42%) viewed their own courses neutrally and regarded the courses as having been only partly useful. Some teacher educators (16%) viewed their own courses negatively and indicated problems in the contents of courses that made them not very useful for teachers.

Table 3. *Teacher educators' (N=19) views about their courses and their suitability for future mathematics teachers. ME = teacher educator in mathematical studies, PTE = teacher educator in pedagogical studies and teaching practice.*

| Class and justification | ME (N=14) | PTE (N=5) |
|---|---|---|
| *Positive 26% (5)* | | |
| • Contents increase pure mathematical knowledge and the teaching methods used teach how to teach mathematics | 2 | 0 |
| • Contents are the same as in school mathematics | 1 | 0 |
| • No justification | 0 | 2 |
| *Neutral 42% (8)* | | |
| • Contents are not the same as in school mathematics, but studies develop mathematical thinking | 2 | 1 |
| • Only some parts of contents link with school mathematics | 2 | 0 |
| • Some contents go beyond school mathematics or general knowledge for teachers | 2 | 0 |
| • Courses are non-compulsory for student teachers and therefore their contents are not useful for teachers | 1 | 0 |
| *Negative 16% (3)* | | |
| • Some courses are simply all-around education for teachers and in some courses there is not enough time to teach important issues | 1 | 0 |
| • Students' knowledge is poor at the beginning of courses and therefore they cannot learn the contents | 0 | 1 |
| • Teachers learn contents but they have insufficient skills for using this knowledge in school teaching | 1 | 0 |
| *Empty 16%(3)* | 2 | 1 |

*Positive (5).* There were three mathematics educators (MEs) and two pedagogical studies and teaching practice educators (PTEs) who considered that the contents of their own course were useful for teachers. The PTEs did not justify their views, but the MEs argued that the contents of their courses increased student teachers' mathematical knowledge and the teaching methods modeled the way to teach mathematics. The MEs underlined the significance of presenting things; they argued that it was important for student learning that the educator demonstrated how things worked. One ME argued that the contents of his courses were the same as in school mathematics and hence the contents were useful for future teachers.

*Neutral (8).* Seven MEs and one PTE considered the contents of their courses only partly useful for future teachers. One PT and two MEs claimed that, despite the contents not being the same as for school mathematics, the courses nevertheless developed students' mathematical thinking, which was also important for future teachers. Two MEs claimed that only some parts of the contents were linked with school mathematics and therefore these parts were useful for future teachers. Another ME thought that the contents are "good to know", but were unnecessary for teachers. Two MEs argued that the contents of their courses offered teachers only general knowledge since the contents were not specialized for use by teachers or the contents went beyond school mathematics. Both of them justified their views with the argument that future teachers needed a wide knowledge base in mathematics.

*Negative (3).* Two MEs and one PTE argued that the contents of their courses did not fully support future mathematics teachers. One ME claimed that in some courses s/he had too little time to teach issues that were important for teachers, while in other courses the contents were simply general knowledge for teachers. Another mathematics ME claimed that teachers usually learned the contents of his/her courses, but the course nevertheless did not provide them with the competence to apply this knowledge in their own teaching. One PE claimed that students had acquired insufficient earlier knowledge to learn the contents of his/her courses.

**Teacher educators' and practicing teachers' suggestions for developing mathematics teacher education**

The second research question was concerned with how practicing mathematics teachers and mathematics teacher educators would develop mathematics teacher education.

The practicing mathematics teachers made numerous suggestions for developing mathematics teacher education. The categorization of the suggestions in Table 4 shows that it would be valuable to develop teacher education both at the general level and also in terms of supporting future mathematics teachers' subject matter knowledge and pedagogical knowledge and skills. More than half of the suggestions (60%) concerned the contents of teacher education that could be categorized with MKT. One third of the suggestions (28%) focused on the quality of the teaching or the quantity of the studies that were categorized as ideas for developing teacher education program. A minority of the suggestions (12%) concerned a number of general issues related to teacher education.

Suggestions for improving the contents of teacher education mostly concerned pedagogical knowledge and skills. Practicing teachers suggested that they would add courses about learning difficulties in mathematics, the evaluation of students' mathematical know-how, and how to teach students with different levels of mathematical knowledge and skills. The practicing teachers also hoped that differentiating mathematics teaching would be discussed during teacher education, since classroom situations required that kind of competence from a teacher. They also suggested that the learning theories courses should be modified so that they would become easily applicable to one's own teaching. The practicing teachers would also add future teachers' knowledge and skills related to using technology in teaching mathematics because technology was assuming a more important role both in the classrooms and in society. Almost all of the suggestions concerning subject matter knowledge dealt with separate mathematics studies programs for future mathematicians and teachers. One common argument was that future teachers needed a different kind of mathematical knowledge from that used by mathematicians.

The ideas that were presented regarding development of the teacher education program concerned both the quality of teaching and the quantity of studies. The practicing teachers thought that the quality of teacher education could be increased by improving students' learning. This could be achieved by modifying present teaching methods as well applying new interactive teaching methods that would include discussions. The practicing teachers also argued that the studies should be modified to be less theoretical because they felt that the present studies were too theoretical, causing the students problems in understanding the course contents to any depth. The practicing teachers also thought that the teaching practice supported their teacher growth. However, they considered that the length of the teaching practice could be increased.

Table 4. *Categorization of practicing mathematics teachers' (N=101) suggestions for developing mathematics teacher education. Each respondent was permitted to mention more than one issue.*

| Category | Suggestions for improving education | f |
|---|---|---|
| **Pedagogical knowledge and skills** | *Knowledge of contents and students (KCS)* | |
| | • Courses about learning difficulties in mathematics | 7 |
| | • How to test students' knowledge and skills | 3 |
| | • How to teach students at different stages of learning | 2 |
| | *Knowledge of contents and teaching (KCT)* | |
| | • How to differentiate teaching | 5 |
| | • How to bridge the gap between learning theories and practice | 4 |
| | • How to produce and use one's own teaching materials | 2 |
| | • Functional learning methods | 1 |
| | • Learner-centered teaching methods | 1 |
| | • Special education in mathematics | 1 |
| | • How to plan and teach complete courses | 1 |
| | *Knowledge of contents and curriculum (KCC)* | |
| | • How to use technology in teaching mathematics | 6 |
| | • Curricular knowledge | 1 |
| | • Teaching methods based on technology | 1 |
| **Subject matter knowledge and skills** | *Specialized content knowledge (SCK)* | |
| | • Separate mathematics courses for teachers and mathematicians | 16 |
| | • Problem-solving | 1 |
| | • Mathematics in different professions | 1 |
| | *Horizon content knowledge (HCK)* | |
| | • Mathematical concepts at different school levels | 1 |
| | • The structure of mathematics | 1 |
| **Ideas for developing the teacher education program** | *Quality of teaching* | |
| | • New teaching methods (e.g., more discussion) | 10 |
| | • Less theory – more practice – linking theory and practice | 6 |
| | • Integrating lectures and doing exercises in mathematical courses | 1 |
| | *Quantity of studies* | |
| | • More teaching practice | 7 |
| | • More pedagogical courses | 1 |
| | • More teaching practice and less pedagogical studies | 1 |
| **General issues** | *Improving cooperation* | |
| | • Cooperation between different departments | 4 |
| | • Cooperation between university and schools | 1 |
| | • Cooperation between students and educators; paying attention to students' suggestions regarding development | 1 |
| | *Special suggestions* | |
| | • Teachers should be specialized in teaching at different school levels | 1 |
| | • Subject teachers' major should be in education | 1 |
| | • Compulsory updating of education after some years of work experience | 1 |
| | *Beyond MKT knowledge* | |
| | • More knowledge about teachers' duties out of class | 2 |
| | *Uncategorized responses* | |
| | • Irrelevant | 4 |
| | • Blank | 27 |

The teacher educators saw less reason for development than did the practicing teachers. Most of the teacher educators suggested developing mathematics teacher education by improving student teachers' subject matter studies (see Table 5).

Table 5. *Categorization of teacher educators' (N=19) suggestions for developing mathematics teacher education.*

| Category | Suggestion for improving education | f |
|---|---|---|
| **Subject matter knowledge and skills** | *Specialized content knowledge (SCK)* | |
| | • Contents of mathematics courses should be revised to be useful for future teachers | 4 |
| | • Mathematics courses should be separately designed for teachers and mathematicians | 3 |
| | *Common content knowledge (CCK)* | |
| | • More courses in mathematics | 2 |
| | *Horizon content knowledge (HCK)* | |
| | • New course on the structures of mathematics | 1 |
| **Developing the teacher education program** | *Updating structure of studies* | |
| | • Combining mathematics and pedagogics courses as an integrated unit | 1 |
| | • Re-scheduling courses in mathematics, pedagogics, and teaching practice | 1 |
| | *Uncategorized responses* | |
| | • Irrelevant | 2 |
| | • Blank | 5 |

Most of the suggestions concerned modifying future teachers' mathematical studies. Some educators suggested that the contents of present courses should be revised from the viewpoint of teacher's work and current school curricula. Many of the respondents would develop courses to increase future teachers' specialized content knowledge (SCK). Some educators also suggested that mathematics studies should be separately designed for future teachers and mathematicians, an idea that was also put forward by the practicing teachers. However, a few teacher educators argued that pure mathematics was the basis of good teaching and therefore the quantity of pure mathematics studies should be increased. One educator suggested that there was a need for developing a mathematics course whose rationale would be to link together the various domains of mathematics. We categorized this as an example of improving students' horizon content knowledge.

Two teacher educators considered that the structure of the mathematics teacher education should be updated. Another educator suggested that mathematical and pedagogical studies should not be organized separately, since their separation prevented the possibility of linking theory and practice. Another teacher educator argued that studies should be better scheduled to help student teachers to acquire an integrated knowledge of subject matter and pedagogy. It should also be noted that a third of the respondents provided no suggestions for developing mathematics teacher education, and hence it remains unknown whether these respondents were satisfied with the current teacher education or not.

## Discussion

This study has investigated teacher educators' and practicing mathematics teachers' views of the contents and the development needs of mathematics teacher education as provided by the University of Eastern Finland. Practicing teachers and teacher educators made various recommendations for improving mathematics teacher education program. We consider that we have been able to identify systematically and in a detailed way the kind of subject matter knowledge and pedagogical content knowledge that these recommendations concern by classifying them in terms of the domains of MKT. Challenges concerning the content of mathematics teacher education seem to become more explicit when subject matter knowledge and pedagogical content knowledge are divided into more detailed components. Markworth, Goodwin and Glisson (2009) found similar benefits when they used MKT to evaluate a single course in mathematics teacher education. The combined results show that a majority of the recommendations concerning the issues that will need to be examined in mathematics teacher education are closely related to four domains of Mathematical Knowledge for Teaching (CCK, SCK, KCS, and KCT). Interestingly, these four domains have been more empirically tested and better conceptualized than the other two domains (Sleep, 2009).

Our results indicate that the majority of practicing mathematics teachers do not regard the present contents of mathematics studies to be fully functional for future mathematics teachers. The practicing teachers suggested, for instance, separate courses for future mathematics teachers and mathematicians, and the possibility of taking school mathematics into account in the teaching of mathematics courses. These ideas were also proposed by some of the teacher educators. These findings are broadly in line with the well-known recommendations by other mathematics educators (e.g., Ball et al., 2008).

Both the practicing teachers and the teacher educators argued that course contents are purely general knowledge for future teachers if there were no explicit links with school mathematics, and hence these contents were not regarded as useful for future teachers. The practicing teachers argued that the links between university and school mathematics were difficult to perceive if the university course contents were set at too high a level compared with the usual school contents. The findings show that the pure mathematical contents, i.e., the common content knowledge (CCK) of the mathematics teacher education, should be carefully examined so that the most relevant mathematical contents for future mathematics teachers could be discovered. It is well known that a weak knowledge of mathematics on the part of teachers has a negative influence on teaching (McDiarmid, Ball & Anderson, 1989), but, on the other hand, a competence solely in mathematics is insufficient enough for good teaching (Hodgen, 2011). It seems that the relevance of mathematics courses depends on how explicitly the link between university and school mathematics is stressed in mathematics courses.

The results suggest that, in addition to pure mathematical contents, mathematics teacher education should include mathematical contents designed specifically for teachers. The practicing teachers argued that they needed mathematical knowledge and skills that were different from the skills and knowledge useful for mathematicians. This reflects the well-known ideas embodied in *pedagogical content knowledge* (Shulman, 1987), or *Specialized content knowledge, SCK* (Ball et al., 2008), i.e., an area of knowledge for teachers that also separates researchers from teachers. The practicing teachers and teacher educators suggested that the mathematical courses should be at least partly separate for future teachers and mathematicians. In practice, this would mean that more resources would be needed for mathematics teacher education, which might be a challenge.

Many of the teacher educators who participated in this study espoused the traditional view of development that emphasizes improving future teachers' subject matter knowledge (SMK) (Ball, 2003). Some educators viewed that good teaching requires knowledge of pure mathematics (CCK) and therefore they suggested that pure mathematical contents should be increased. On the other hand, many educators viewed that future mathematicians and future teachers need different kind of mathematical knowledge (SCK) and therefore they suggested that some of the present contents should be modified to be more suitable for teachers or new courses should be developed for teachers. Some educators viewed that forming integrated knowledge of pedagogy and mathematics (SCK) is one challenge for the mathematics teacher education, and therefore they suggested that the present courses should be re-scheduled or integrated.

On the other hand, a majority of the practicing teachers observed that there was a wider need for development than simply reforming the mathematical contents. The majority of practicing teachers demanded more courses concerned with teaching mathematics, students' learning difficulties in mathematics, and how to differentiate mathematics teaching. These knowledge domains can be identified as *Knowledge of content and teaching* (KCT) and *Knowledge of content and students* (KCS). The practicing teachers pointed out that they needed to alternate the knowledge and skills of teaching and learning in many classroom situations, and they seemed to consider that the KCS and KCT knowledge types were interconnected especially in classroom *actions* (see Fernández, Figueiras, Deulofeu, et al., 2011; Ball et al., 2008). Many practicing teachers considered that they had learned pedagogical and mathematical issues in the course of their teacher education and that they had found teaching practice a very useful experience, but still they had difficulty in forming an integrated understanding of pedagogy and mathematics (see also Korthagen & Kessels, 1999; Sharp, 2004).

This linkage of theory and practice (Carlson, 1999; Tryggvason, 2009) seems to be a major challenge in mathematics teacher education, since it concerns not only the pedagogical and mathematical studies but also the teaching practice. Earlier research work has shown that solving the problem will not be simple. According to Verloop, Driel, and Meijer (2001), it is still difficult to foresee how teacher knowledge can be clarified clear for future teachers in their teaching practice. One of the problems appears to arise from the teacher educators' knowledge: not even experienced educators in the field of teaching practice have a clear grasp of the types of knowledge that teaching procedures involve, which makes it difficult to make the connection between theory and practice visible to student teachers (Verloop et al., 2001; Asikainen, Pehkonen & Hirvonen, 2013).

Filling the gap between theory and practice is a demanding task because teaching practice comprises only a small proportion of the teacher education studies as a whole. Hence, it is almost unrealistic to suggest that the gap could be fulfilled during the teaching practice. Our results suggest that the links between theory and practice should be made visible in all of the components of the teacher education so as to support future teacher development. Numerous suggestions have been made for the solution of this problem, e.g., by approaching it from practice to theory (Carlson, 1999), by developing the pedagogy of teacher education (Korthagen & Kessells, 1999), or by taking problem-solving into account in mathematics teacher education (Leikin & Levav-Waynberg, 2007). There is a possibility that contents, teaching methods, and the learning process may all be involved in the solution.

We have come to the realization that one of the key factors in reforming mathematical studies is the performance of a detailed analysis and comparison of curricula in university and school mathematics. In fact, it would seem obvious that the pure mathematical contents (CCK) should be the same as the topics in school mathematics or, at the very least, explicit links should exist between university and school mathematics. Another challenge is to design and develop special content knowledge (SCK) courses for future teachers. As yet, there is no general consensus about the knowledge and skills included in SCK (see e.g., Carrillo, Climent, Contreras & Muñoz-Catalán, 2013; Flores, Escudero & Carillo, 2013) but there should be no problem in designing new courses for future teachers, since there would be no harm caused if the subject matter and pedagogical contents are mixed. But as far as conceptualizing MKT is concerned, there is still work to be done to reach a consensus about this type of knowledge.

Although teacher education and teachers' knowledge are related (Darling-Hammond, Chung, & Frelow, 2002), more research into the challenges revealed by individual teacher education programs will be needed in order to construct a broader picture of this multifaceted phenomenon. Individual reports may act as an important part in this process by evaluating and improving mathematics teacher education before all of the universal challenges facing mathematics teacher education have been fully recognized. Although the present study has concerned only a single mathematics teacher education program, we would suggest that the following issues may prove to be more general challenges facing all mathematics teacher education programs:

☐ The connections between university mathematics and school mathematics are not self-evident for student teachers. Teachers need pure mathematical knowledge, e.g. Common Content Knowledge (Ball et al., 2008) and Subject Matter Knowledge (Shulman, 1986;

1987). However, student teachers may find that the mathematics studied at university level is too advanced and has no clearly visible connections to the mathematics taught in school.

- Specific mathematical knowledge is missed from teacher education, while the contents of mathematical courses focus too largely on pure mathematics. In addition to mathematical content knowledge, teachers also need specific mathematical knowledge, e.g. Specialized Content Knowledge (Ball et al., 2008) or School Mathematics (O'Meara, 2010), because they need to carry out a variety of different activities (e.g., producing teaching materials, formulating and marking exams) for which pure mathematical knowledge is insufficient.

- Teachers may have too few tools to be able to teach "good and poor" students at the same time. In the classroom teachers are simultaneously attempting to evaluate their students' starting levels, to recognize their individual learning habits, and also to implement different teaching strategies that will match up to the pertaining situation. The knowledge required in these situations can be referred to as Pedagogical Content Knowledge (Shulman, 1986; 1987) or as both Knowledge of Content and Students and Knowledge of Content and Teaching (Ball et al., 2008).

- Courses in teacher education may be too theoretical (e.g., Carlson, 1999; Korthagen & Kessells, 1999). Student teachers may feel that mathematical and pedagogical courses and also teaching practice are too far removed from teachers' actual work.

The results of this study encourage us in the development work of mathematics teacher education although the circumstances are still difficult at the starting point. The most demanding part has been and will be to evaluate what the personnel in mathematics teacher education teach and what kind of methods they use. We believe that assessment, feedback, and the teacher education personnel themselves and their cooperation are important components in the process of improving teacher education. It is common sense that there is always a possibility of improvement, and therefore the development must begin from critical thinking: what can we do better? With this article, we should like to encourage other researchers to evaluate and develop teacher education, and hence we would close with words that are too frequently dead and buried:

*- Without criticism, development dies –*

## Acknowledgement

We would like to thank *The Finnish Cultural Foundation, North Karelia Regional Fund* for funding our research project *Evaluating and Improving Mathematics Teacher Education*.

# References

Asikainen, M. A., Pehkonen, E., & Hirvonen, P. E., (2013). Finnish Mentor Mathematics Teachers' Views of Teacher Knowledge Required for Teaching Mathematics. *Higher Education Studies, 3*(1), 79–91.

Ball, D. L., & Forzani, F. (2009). The work of teaching and the challenge for teacher education. *Journal of Teacher Education, 60*(5), 497–511.

Ball, D. L., Hill, H.C, & Bass, H. (2005). Knowing mathematics for teaching: Who knows mathematics well enough to teach third grade, and how can we decide? *American Educator, 29*(1), 14–17, 20–22, 43–46.

Ball, D. L., Sleep, L., Boerst, T. A., & Bass, H. (2009). Combining the development of practice and the practice of development in teacher education. *The Elementary School Journal, 109*(5), 458–474.

Ball, D. L., Thames, M.A. & Phelps, G. (2008). Content knowledge for teaching: What makes it special? *Journal of Teacher Education, 59*(5), 389–407.

Ball, D. L., & Bass, H. (2003). Toward a practice-based theory of mathematical knowledge for teaching. In E. Simmt & B. Davis (Eds.), *Proceedings of the 2002 Annual Meeting of the Canadian Mathematics Education Study Group,* (pp. 3–14). Edmonton, AB: CMESG/GCEDM.

Carlson, H. L. (1999). From practice to theory: A social constructivist approach to teacher education. *Teachers and Teaching: theory and practice, 5*(2), 203–218.

Carrillo, J., Climent, N., Contreras, L. C., & Muñoz-Catalán, M. C. (2013). Determining Specialised Knowledge For Mathematics Teaching. In B. Ubuz, C. Haser & M. A. Mariotti (Eds.), *Proceedings of the Eighteenth Congress of the European Society for Research in Mathematics Education* (pp. 2985–2994). Ankara, Turkey: European Society for Research in Mathematics Education.

Darling-Hammond, L., Chung, R., & Frelow, F. (2002). Variation in Teacher Preparation How Well Do Different Pathways Prepare Teachers to Teach? *Journal of Teacher Education, 53*(4), 286–302.

Delaney, S., Ball, D. L, Hill, H. C., Schilling, S. G., & Zopf, D. (2008). "Mathematical knowledge for teaching": Adapting U.S. measures for use in Ireland. *Journal of Mathematics Teacher Education, 11*(3), 171–197.

Fauskanger, J., Jakobsen, A., Mosvold, R., & Bjuland, R. (2012). Analysis of psychometric properties as part of an iterative adaptation process of MKT items for use in other countries. *ZDM, 44*(3), 387–399.

Fennema, E. & Franke, L. M. (1992). Teachers' knowledge and its impact. In D. A. Grouws (Ed.), *Handbook of research on mathematics teaching and learning* (pp. 147–164). New York, NY: Macmillan.

Fernández, S., Figueiras, L., Deulofeu, J., & Martínez, M. (2011). Re-defining HCK to approach transition. In M. Pytlak, T. Rowland, & E. Swoboda (Eds.), *Proceedings of the Seventh Congress of the European Society for Research in Mathematics Education* (pp. 2640–2649). University of Rzeszów, Poland.

Flores, E., Escudero, D., & Carrillo, J. (2013). A theoretical review of specialised content Knowledge. In B. Ubuz, C. Haser, & M.A. Mariotti (Eds.), *Proceedings of the Eighteenth*

*Congress of the European Society for Research in Mathematics Education* (pp. 3055–3064). Ankara, Turkey: European Society for Research in Mathematics Education.

Hashweh, M. (2005) Teacher pedagogical constructions: a reconfiguration of pedagogical content knowledge. *Teachers and Teaching: Theory and Practice, 11*(3), 273–292

Hill, H. C. & Ball, D. L. (2004). Learning mathematics for teaching: Results from California's Mathematics Professional Development Institutes. *Journal for Research in Mathematics Education, 35*(5), 330–351.

Hill, H. C., Ball, D. L., & Schilling, S. G. (2008). Unpacking pedagogical content knowledge: Conceptualizing and measuring teachers' topic-specific knowledge of students. *Journal for Research in Mathematics Education, 39*(4), 372–400.

Hill, H. C., Blunk, M. L., Charalambous, C. Y., Lewis, J. M., Phelps, G. C., Sleep, L., & Ball, D. L. (2008). Mathematical knowledge for teaching and the mathematical quality of instruction: An exploratory study. *Cognition and Instruction, 26*(4), 430–511.

Hill, H. C., & Lubienski, S.T. (2007). Teachers' mathematics knowledge for teaching and school context: A study of California teachers. *Educational Policy, 21*(5), 747–768.

Hill, H. C. (2007). Mathematical knowledge of middle school teachers: Implications for the No Child Left Behind Policy initiative. *Educational Evaluation and Policy Analysis, 29*(2), 95–114.

Hill, H. C., Ball, D. L., Blunk, M. Goffney, I. M., & Rowan, B. (2007). Validating the ecological assumption: The relationship of measure scores to classroom teaching and student learning. *Measurement: Interdisciplinary Research and Perspectives, 5*(2–3), 107–117.

Hill, H. C., Ball, D. L., Sleep, L., & Lewis, J. M. (2007). Assessing Teachers' Mathematical Knowledge: What Knowledge Matters and What Evidence Counts? In F. Lester (Ed.), *Handbook for Research on Mathematics Education* (2nd ed.) (pp. 111–155). Charlotte, NC: Information Age Publishing.

Hill, H. C., Dean, C., & Goffney, I. M. (2007). Assessing Elemental and Structural Validity: Data from Teachers, Non-teachers, and Mathematicians. *Measurement: Interdisciplinary Research and Perspectives, 5*(2–3), 81–92.

Hill, H. C., Rowan, B., & Ball, D. L. (2005). Effects of teachers' mathematical knowledge for teaching on student achievement. *American Educational Research Journal, 42*(2), 371–406.

Hill, H. C., Schilling, S. G., & Ball, D. L. (2004). Developing measures of teachers' mathematics knowledge for teaching. *Elementary School Journal*, 105, 11–30.

Hodgen, J. (2011). Knowing and identity: a situated theory of mathematics knowledge in teaching. In T. Rowland & K. Ruthven (Eds.), *Mathematical knowledge in teaching* (pp. 27–42). Dordrecht: Springer.

Korthagen, F., & Kessels, J. (1999). Linking theory to practice: Changing the pedagogy of teacher education. *Educational Researcher, 28*(4), 4–17.

Leikin, R., & Levav-Waynberg, A. (2007). Exploring mathematics teacher knowledge to explain the gap between theory-based recommendations and school practice in the use of connecting tasks. *Educational Studies in Mathematics, 66*(3), 349–371.

Markworth, K., Goodwin, T., & Glisson, K. (2009). The development of mathematical knowledge for teaching in the student teaching practicum. In D. S. Mewborn & H. S. Lee (Eds.), *Scholarly Practices and Inquiry in the Preparation of Mathematics Teachers* (pp. 67–83). San Diego, CA: Association of Mathematics Teacher Educators.

McDiarmid, G. W., Ball, D. L., & Andersen, C. W. (1989). Why staying one chapter ahead doesn't really work: subject-specific pedagogy. *The national center for research on teacher education*. Issue paper 88-6.

Meredith, A. (1995). Terry's learning: some limitations of Shulman's pedagogical content knowledge. *Cambridge Journal of Education, 25*(2), 175–187.

O'Meara, N. (2010). *Improving mathematics teaching at second level through the design of a model of teacher knowledge and an intervention aimed at developing teachers' knowledge.* University of Limerick. Dissertation.

Rowland, T., Turner, F., Thwaites, A., & Huckstep, P. (2009). *Developing Primary Mathematics Teaching: reflecting on practice with the Knowledge Quartet.* London: Sage.

Schilling, S. G. (2007). The role of psychometric modeling in test validation: An application of multidimensional item response theory. *Measurement: Interdisciplinary Research and Perspectives, 5*(2–3), 93–106.

Schilling, S. G., & Hill, H. C. (2007). Assessing measures of mathematical knowledge for teaching: A validity argument approach. *Measurement: Interdisciplinary Research and Perspectives, 5*(2–3), 70–80.

Schilling, S. G., Blunk, M., & Hill, H. C. (2007). Test validation and the MKT measures: Generalizations and conclusions. *Measurement: Interdisciplinary Research and Perspectives, 5*(2–3), 118–128.

Sharp, J. (2004). Spherical Geometry as a Professional Development Context for K-12 Mathematics Teachers. In T. Watanabe & D. R. Thompson (Eds.), *The Work of Mathematics Teacher Educators: Exchanging Ideas for Effective Practice* (pp. 103–118). Association of Mathematics Teacher Educators.

Shulman, L. S. (1986). Those who understand: Knowledge growth in teaching. *Educational Researcher, 15*(2), 4–14.

Shulman, L. S. (1987). Knowledge and teaching: Foundations of the new reform. *Harvard Educational Review, 57*, 1–22.

Sleep, L. (2009). *Teaching to the mathematical point: Knowing and using mathematics in teaching.* University of Michigan. Dissertation.

Stylianides, A. J., & Ball, D. L. (2008). Understanding and describing mathematical knowledge for teaching: Knowledge about proof for engaging students in the activity of proving. *Journal of Mathematics Teacher Education, 11*(4), 307–332.

Thames, M. H., & Ball, D. L. (2010). What mathematical knowledge does teaching require? Knowing mathematics in and for teaching. *Teaching Children Mathematics, 17*(4), 220–225.

Tryggvason, M-T. (2009). Why is Finnish teacher education successful? Some goals Finnish teacher educators have for their teaching. *European Journal of Teacher Education, 32*(4), 369–382.

Verloop, N., Van Driel, J., & Meijer, P. (2001). Teacher knowledge and the knowledge base of teaching. *International Journal of Educational Research, 35*(5), 441–461.

# Use of Mathematical Tasks of Teaching and the Corresponding LMT Measures in the Malawi Context

Mercy Kazima
University of Malawi, Malawi

Arne Jakobsen
University of Stavanger, Norway

Dun Nkhoma Kasoka
University of Malawi, Malawi

**Abstract:** We discuss the adaptation and piloting of the previously developed U.S.-specific measures of mathematical knowledge for teaching to the Malawi context. The purpose is to produce measures that can be used to evaluate changes in mathematical knowledge for teaching gained through primary teacher education, thus informing teacher educators on the most effective evidence-based practices. By interviewing 14 teachers, we first examine whether the 16 recurrent mathematical tasks of teaching tasks identified in the U.S. are applicable to the Malawi context. This is followed by the discussion of the adaptability of the U.S. developed number concept and operations LMT measures. Next, we report on the item psychometric properties estimated from a pilot study in which 351 preservice primary school teachers participated at the end of their coursework. Our findings suggest that all the 16 tasks of teaching mathematics are applicable to the Malawi context, albeit to varying degrees, and should be complemented by additional tasks suggested by the Malawi teachers. For the LMT measures, we found that the majority of the LMT items psychometrically function well in the Malawi context and that item difficulty estimated in Malawi was strongly correlated with that reported in the U.S. We thus argue that there is some generality to the mathematics teaching tasks across the two contexts, as well as some specificity to Malawi, and that the adapted LMT measures can be used in a Malawi context.

***Keywords:*** Teacher knowledge, Mathematical knowledge for teaching, Primary teacher education, Malawi.

## Introduction

Teacher knowledge is important for both teaching and learning. Since Shulman (1986) introduced the concept of pedagogical content knowledge, his ideas have triggered widespread interest among researchers and practitioners alike. In addition, different conceptualizations of teacher knowledge have emerged, such as Knowledge Quartet (Rowland, Huckstep, & Thwaites, 2005), Knowledge for Teaching (Davis & Simmt, 2006), and Mathematical Knowledge for Teaching (Ball, Thames, & Phelps, 2008). Despite different views on categorizations, researchers seem to agree that teacher knowledge plays a key role in student learning (e.g., An, Kulm, & Wu, 2004; Rowan, Correnti, & Miller, 2002; Wright, Horn, & Sanders, 1997), and much progress has been made in understanding the professional mathematical knowledge that teachers need in order to perform the recurrent tasks of teaching mathematics. There is also an extensive body of knowledge on how this knowledge is acquired, how it can be measured, and how it relates to teaching and student learning (e.g.,

*The Mathematics Enthusiast,* **ISSN 1551-3440,** vol. 13, no. 1&2, pp. 171–186
2016© The Author(s) & Dept. of Mathematical Sciences-The University of Montana

Ball et al., 2008; Hill, Rowan, & Ball, 2005; Rowland et al., 2005). Nonetheless, a greater understanding of how this knowledge might differ as context changes is necessary.

Different approaches to measuring teacher knowledge are presently in use, one of which is based on analyzing school mathematics curriculum and developing measures that would test knowledge and the teaching of that curriculum. The main drawback of this approach is that it requires many, often inaccurate, assumptions about outcomes of teaching that is aligned with the curriculum. Ball and colleagues (2008) developed an alternative practice-based approach in the United States. According to the authors, their aim was to "unearth the ways in which mathematics is involved in contending with the regular day-to-day, moment-to-moment demands of teaching" (p. 395). Using this approach enabled these researchers to define the theory of mathematical knowledge for teaching (MKT) (Ball & Bass, 2003). In particular, they identified a list of 16 mathematical teaching tasks that are part of the work teachers routinely do (Ball et al., 2008). In connection to this work, measures of mathematical knowledge for teaching were developed as a part of the Learning Mathematics for Teaching (LMT) project at the University of Michigan. The LMT project developed items in three content areas, namely (i) number concepts and operations (NCOP); (ii) geometry; and (iii) patterns, functions, and algebra (a sample of released items can be found in Ball and Hill (2008)).

The LMT measures[1] have been adapted for use in different contexts outside the U.S., for example, in Ireland (Delaney, 2012), Norway (Fauskanger, Jakobsen, Mosvold, & Bjuland, 2012), Indonesia (Ng, 2012), South Korea (Kwon, Thames, & Pang, 2012), and Ghana (Cole, 2012). In South African context, Adler and Patahuddin (2012) researched mathematical knowledge for teaching, reporting that the LMT items "have much potential in provoking teachers' talk and their mathematical reasoning in relation to practice-based scenarios; and exploring with teachers a range of connected knowledge related to the teaching of a particular concept or topic is most important resource for teachers" (p. 17).

It is important to note that the mathematical knowledge for teaching theory and the associated LMT measures were developed from classroom observation in the U.S. and were not intended for use in other cultures. Thus, we were interested in exploring whether the 16 tasks of teaching identified by Ball et al. (2008) also are applicable in a Malawi context. By applicable we mean if teachers in Malawi would do the same or similar tasks as part of their work of teaching mathematics in schools. Our interest in mathematical knowledge for teaching, with associated tasks of teaching and measures, developed from our work in teacher education in Malawi. With the overarching project goal of improving quality and capacity of mathematics teacher education in Malawi, we are interested to learn more about the development of preservice teachers' mathematical knowledge for teaching through their teacher education. Knowing what works and what does not in teacher education in Malawi is the first step in the process of improving the quality of teacher education. In our view, measures of mathematical knowledge for teaching can help answer these questions. In this larger study, we plan to measure preservice teacher mathematical knowledge for teaching before and after coursework, using a pre- and post-test. This can help us determine knowledge growth in preservice teachers, and consequently inform our practice. Since the LMT measures were developed from studying work and identifying tasks of teaching in the U.S. context, our first objective was to ascertain whether these tasks are applicable to the Malawi context.

---

[1] The measures (or instruments) developed as part of the Learning Mathematics for Teaching Project at the University of Michigan are also sometimes called LMT items, MKT items, or MKT measures. In this paper, will use the terms LMT measures/instruments.

These findings elucidated the suitability of adapting the LMT measures for use in Malawi. Our next goal was to outline the method of adaptation that the measures underwent. This paper is therefore guided by the following research questions:

1. Are the tasks of teaching mathematics identified in the U.S. applicable in Malawi?

2. What can we learn about the adaptability of LMT measures to the Malawi context from psychometric properties estimated in a pilot study?

Other researchers have raised similar questions when adapting the LMT measures to their contexts (for example, Cole, 2012; Delaney, 2012; Fauskanger et al., 2012; Ng, 2012). We focus specifically on the Malawi context, acknowledging the cultural differences between Malawi and other cultures. Furthermore, in their work, Hoover, Mosvold, and Fauskanger (2014) called for "increased efforts to identify professionally defensible mathematical tasks of teaching that can serve as a common foundation for conceptualizing and measuring mathematical knowledge for teaching internationally" (p. 7). Our study is an attempt to respond to the call. By adapting the measures and piloting a set of 88 items, our study will also make an argument for use of the LMT measures in a Malawi context.

## Adapting LMT Measures to Different Cultural Contexts

Many researchers agree that teacher knowledge of mathematics is influential in shaping teaching practices that in turn affect student achievement (e.g., Hill, Ball, & Schilling, 2008). One of the reasons the work of Ball and colleagues has received extensive attention is that the authors based their findings on a thorough study of actual classroom teaching. Because their research was conducted in the U.S. context, it is debatable whether the findings are generalizable. For example, Andrew (2011) argued that "teacher's mathematical knowledge, as manifested in their observable behavior, is a cultural construction" (p. 100). While we raise similar questions about generalizing the theory of mathematical knowledge for teaching to the Malawi context, we also draw upon findings of similar studies in this filed. We agree with the view that the practice of teaching is a cultural activity, making teaching of mathematics culturally specific, even though the content is general (Delaney, 2012).

According to the results reported by Cole (2012), the LMT measures can be used in the Ghanaian context after careful adaptation. However, so far Ghana is the only African country where the LMT measures have been used. Moreover, Cole and other researchers have demonstrated that adaptation of LMT measures is not easy and requires a process that considers all differences between the contexts, including culture, language, and teaching practices (Delaney, 2012; Fauskanger et al., 2012). In our study, we observed that the differences were manageable and hence proceeded with the adaptation.

Malda et al. (2008) emphasized on fairness of adapted measures, noting that "it is unfair to assess intelligence of children from Africa with a test that has been validated in a Western culture . . . , with a population of children exposed to very different educational and material environments at home and school" (p. 452). Similarly, it could be argued that it is not fair to assess Malawian teachers' knowledge using measures developed for the vastly different U.S. context. As Hoover et al. (2014) pointed out, such an argument focuses on differences in the practice of teaching in different cultures. The authors further argued that, to evaluate such arguments, we need to consider "the underlying concept of *work of teaching* and *tasks of teaching* used in MKT assessment items and to ask whether, or to what extent, such concepts, as defined therein, are meaningful across cultural contexts" (p. 8) [emphasis in original]. In this study, we consider whether the items we adapted are meaningful in the Malawi context.

## Malawi Context

Malawi, formerly known as Nyasaland, was a British protectorate that gained its independence from Britain in 1964. English remains the official language and is also the school language from fifth year of primary school onwards. Malawi's school system comprises of eight years of primary and four years of secondary school. Primary school education is free and easily accessible, neither of which is the case for secondary school education (Kazima & Mussa, 2011). Currently recommended age for enrollment into the first grade of primary school is six years. Thus, primary school students are aged 6 to 13 years old, whereas secondary school serves 14 to 17-year-olds. However, in reality, many classrooms are attended by children of various ages because some children commence education when they are older than six and some repeat classes. At the end of secondary school, students take Malawi Schools Certificate of Education (MSCE) national examinations, equivalent to the Ordinary level (General Certificate of Education). Tertiary education, including teacher education, requires passing the MSCE examinations. Teacher education for primary schools is offered by teacher education colleges, most of which are government owned, while some are private institutions. All teacher colleges follow one curriculum, referred to as the Initial Primary Teacher Education program. It is a two-year full time program, comprising of college coursework in the first year and teaching practice, which is the focus of the second year (Malawi Institute of Education, 2010). Students that successfully complete the program are awarded Primary Teacher's Certificate. There is no subject specialization for primary teachers; all preservice teachers learn all subject areas, including mathematics, and they are expected to teach all subjects in primary schools (Malawi Institute of Education, 2010). As a part of this study, we piloted the adaptability of LMT measures on preservice teachers at the end of their coursework, as a part of the first year of the teacher education program.

## Methods

### Mathematical Tasks of Teaching

In order to answer our first research question about the applicability of U.S.-specific tasks of teaching in the Malawi context, we asked 14 experienced teachers to respond to a questionnaire and followed this with a group discussion. All study participants were practicing teachers from various primary schools in Malawi and were drawn from an in-service upgrading course at the University of Malawi. The questionnaire consisted of two parts, whereby Part A listed all the 16 core tasks of teaching mathematics (Ball et al., 2008). The respondents were asked to indicate which ones were applicable to them as mathematics teachers and were required to elaborate on their responses. Verbal instructions were given to the teachers prior to completing the questionnaire. In particular, they were informed that they should draw from their experience and consider whether each of the tasks is something they do as a part of their work of teaching mathematics. In Part B of the questionnaire, the teachers were instructed to note tasks of teaching mathematics that they perform that are not included on the list. All study participants completed the questionnaire at the same time, in one room. Subsequently, eight teachers took part in a group discussion.

### LMT Measures

We answered the second research question in two phases—Phase I, which was the actual adaptation of instruments, and Phase II, pilot testing of the adapted instrument.

**Phase I: Adapting measures**. Phase I was performed in three stages, the first of which consisted of selecting the most appropriate instruments from those available. We limited our study to measures of number concepts and operations (NCOP) content area. The

LMT project provided us with ten forms,[2] with items from the content area number concept and operations, and corresponding questions related to content knowledge (CK) and knowledge of content and students (KCS) (see, for example, Ball et al., 2008). Upon closer inspection, we noted some repetition and modification of stems and items from the 2001 to the 2004 version. After aligning items in each form to Malawi's mathematics curriculum for Initial Primary Teacher Education, we observed that Form A from the 2001 instrument (NCOP-CK_2001A) had the closest and the most comprehensive range of items covering the curriculum. Therefore, we selected this form and its corresponding Form B (NCOP-CK_2001B) as a starting point for adaption. There were six common items between the two forms and because we wanted to pilot as many items as possible, we replaced the common items in Form B with similar items from the remaining forms. We also added new stems to both forms to cover the concepts of division, multiples, and factors, which were not sufficiently covered by the original forms. These stems were also taken from some of the remaining forms. In total, our new Form A had 25 stems and 46 items, while Form B had 24 stems with 42 items, making a set of 88 items. The original 2001 Form A has 13 stems with 26 items, while Form B has 15 stems with 24 items. Having two forms at this stage allowed us to pilot as many items as possible. We considered the alternative of having all items in one form, but decided that it was not appropriate, since it would make the form too long. Another argument for having two forms at this stage was that the project would eventually need to group all items into two comparable forms that could be used in pre- and post-tests.

The second stage of adaptation involved contextualizing the U.S.-based instruments, both stems and items, to the Malawi context. This was done by changing some words, phrases, and names of places, people, and objects to what we think would be familiar in Malawi context, as shown in Table 1.

Table 1. *Examples of Changes Made from U.S. to Malawi Context.*

| Category of change | U.S. | Malawi |
|---|---|---|
| General context:<br>☐ Names of people (42 instances) | Ms. Jamison | Mrs. Banda |
| | Chad | Chisomo |
| ☐ Names of objects (21 instances) | Pizza | Bread |
| | Field | Farm |
| School context:<br>(11 instances) | Scoring/Reviewing | Marking |
| | Rules of thumb | Simple rules |
| | Student papers | Student notebooks |
| | State assessment | MANEB examinations |
| | Quiz | Test |
| | Mini-lessons for students focused on particular difficulties | Revision lessons |
| Item Numbering | Numbering not continuous | Continuous numbering |

[2] These are three 2001 NCOP-CK forms (A, B, and C), three 2001 NCOP-KCS forms (A, B, and C), two 2002 NCOP-CK forms, and two 2004 NCOP-CK forms.

The third and final stage required checking and modifying the mathematical content of each stem and the corresponding items in order to ensure that they reflect the Malawi curriculum. The modifications made addressed "changes related to school cultural context" and "changes related to mathematical substance" (Delaney et al., 2008, p. 182). After initial changes were made to both forms, we sought input from four experienced primary school teachers via semi-structured individual interviews. Their feedback helped us modify the items further, thus ensuring relevance of the content, wording, representations, and notations (Kasoka, Kazima, & Jakobsen, 2016).

**Phase II: Pilot testing**. The two adapted (and extended) forms (Form A and B) were subjected to a pilot test by 351 preservice primary school teachers. The two forms were administrated to all preservice teachers at one Teacher Education College at the end of their first-year coursework. They were all informed of the objectives of the study and the instructions were carefully explained to them. The two forms were randomly distributed to the students, whereby 212 students answered Form A while 139 answered Form B. Ideally, we wanted the two groups to be equal in size, but due to incorrectly estimating the total number of students at the teacher college, this was not achieved. We had no control over the distribution of the two forms within the group because the forms were mixed in advance for randomness. While the preservice teachers were allowed as much time as they needed to complete the forms, the majority took 1–2 hours. After the test, fifteen randomly selected students were asked to comment on the test and the specific items. Their input allowed us to assess the suitability of the items for testing teacher knowledge in Malawi.

### Results and Findings

**Mathematical Tasks of Teaching**

Table 2 shows the results of Part A of the questionnaire answered by the 14 teachers. The results reported pertain to the frequencies, the number of teachers that indicated that the task was applicable (yes), or not applicable (no), or did not respond to the question (non-response).

As can be seen from Table 2, there was no task that none of the teachers identified with, and all 16 tasks were considered applicable to Malawi school context by at least five of the teachers. In other words, the teachers viewed these tasks as something they do as a part of their work of teaching mathematics. This finding seems to suggest that U.S. and Malawi contexts share some similarities. However, there were variations in the frequencies, as some tasks were considered applicable by a greater number of teachers than others were. Thus, in subsequent analysis, we consider these tasks as most applicable to the Malawi context. We also identified three tasks that all teachers felt were applicable to Malawi, namely "presenting mathematical ideas," "finding an example to make a specific mathematical point," and "appraising and adapting the mathematical content of textbooks." All these tasks involve typical traditional classroom practices of mathematics teachers—explaining mathematics content or procedure, illustrating to students using examples and often from textbooks, and asking students to practice exercises from textbooks. Hence, it is not surprising that all the teachers identified these as aspects of the work they do when teaching mathematics, thus making these tasks most applicable to the Malawi school context.

Table 2. *Frequencies of Responses to Applicability of Tasks to the Malawi School Context.*

| Task Description (from Ball et al., 2008) | Is task applicable to Malawi school context? | | |
| --- | --- | --- | --- |
| | Yes (%) | No (%) | Non-response (%) |
| Presenting mathematical ideas | 14 (100) | 0 (0) | 0 (0) |
| Finding an example to make a specific mathematical point | 14 (100) | 0 (0) | 0 (0) |
| Appraising and adapting the mathematical content of textbooks | 14 (100) | 0 (0) | 0 (0) |
| Connecting a topic being taught to topics from prior or future years | 13 (93) | 0 (0) | 1 (7) |
| Recognizing what is involved in using a particular representation | 12 (86) | 1 (7) | 1 (7) |
| Modifying tasks to be either easier or harder | 12 (86) | 1 (7) | 1 (7) |
| Responding to students' "why" questions | 11 (79) | 3 (21) | 0 (0) |
| Linking representations to underlying ideas and to other representations | 11 (79) | 2 (14) | 1 (7) |
| Giving or evaluating mathematical explanations | 10 (71) | 2 (14) | 2 (14) |
| Choosing and developing useable definitions | 10 (71) | 2 (14) | 2 (14) |
| Asking productive mathematical questions | 10 (71) | 2 (14) | 2 (14) |
| Selecting representations for particular purposes | 10 (71) | 2 (14) | 2 (14) |
| Evaluating the plausibility of students' claims (often quickly) | 9 (64) | 2 (14) | 3 (21) |
| Using mathematical notation and language and critiquing its use | 8 (57) | 3 (21) | 3 (21) |
| Inspecting equivalences | 6 (43) | 5 (36) | 3 (21) |
| Explaining mathematical goals and purposes to parents | 5 (36) | 8 (57) | 1 (7) |

In sum, 12 of the 16 tasks were identified as applicable by more than 70% of the teachers. Of the remaining four tasks with lower frequencies, only two—"inspecting equivalences" and "explaining mathematical goals and purposes to parents"—were identified as applicable to the Malawi context by less than half of the teachers. Those that deemed the task applicable explained that they discuss teaching mathematics (and other subjects) with parents during gatherings, such as Parents and Teachers Association meetings. During the subsequent group discussion, it was clear that some schools do indeed inform parents about the teaching of mathematics.

It is also important to note that not all teachers responded to all questions and some failed to elaborate on their responses. This issue was raised in the group discussion, where the participants revealed that, if they were unsure about some of the tasks, they felt that it was appropriate not to respond to the question.

**Tasks Teachers Did Not Find Applicable**

We also examined the tasks that some teachers did not view as something they do. Looking at Table 2, it is evident that, for 12 of the 16 tasks, at least one teacher responded that he/she does not see it as applicable to the Malawi context. We have presented some of these tasks in Table 3 in descending order of frequencies. We have also included examples of the reasons that teachers gave. It is important to note that, following the group discussion with eight teachers, no new information arose. Nevertheless, the discussion helped explain some of the written responses.

Table 3. *Examples of Teachers' Reasons for Tasks Not Being Applicable to Malawi Contexts.*

| Mathematical task of teaching | Freq. for not applicable | Reasons |
|---|---|---|
| Explaining mathematical goals and purposes to parents | 8 (57%) | Parents rarely know what their children learn in school. Majority of teachers do not have time to deal with parents. Parents are not involved in class work. Learners' guardians do not ask about mathematics. Teachers do not discuss mathematics with parents. |
| Inspecting equivalences | 5 (36%) | Teachers' guide matches content to find links Associating concepts is not done by teachers. We do not do this in primary school. |
| Responding to students' "why" questions | 3 (21%) | Teachers just give explanations. Learners ask how to solve problems. Teachers do not do this. |
| Using mathematical notation and language and critiquing its use | 3 (21%) | Teachers do not critique teachers' guide. Notation given in textbooks. Guided by teachers' guide. |
| Choosing and developing useable definitions | 2 (14%) | Definitions are given in textbooks. Teachers guide has definitions. |
| Selecting representations for particular purposes | 2 (14%) | Teachers' guide has representations. Use text books. |
| Modifying tasks to be either easier or harder | 1 (7%) | Textbook includes both easy and hard problems. |

More than half of the teachers indicated that the task of "explaining mathematical goals and purposes to parents" is not applicable to the Malawi school context, as this is not something they do as a part of their work. The reasons given suggest that the teachers do not consider this as a mathematical task of teaching because, according to the teachers, parents are not involved in what goes on in the mathematics classroom; thus, teachers do not talk to parents about mathematics. During the group discussion, some teachers said that they have not seen this practice in any school they had worked in. They explained that they have meetings with parents where they discuss students' welfare, but do not address any academic issues. From our experience of the Malawi context, some schools would not involve parents into their students' academic progress due to the low literacy levels of parents and guardians, especially in rural areas. Another reason might be teachers' lack of subject matter knowledge or confidence in their own knowledge of mathematics. This was not mentioned by the teachers that took part in this study, but seems plausible that, if teachers are not confident of their knowledge or are aware of their limitations, they might not be willing to discuss their work with parents. Interestingly, one teacher stated, "learners' guardians do not ask about mathematics," which seems to suggest that, if the parents asked about goals and purposes of mathematics, the teacher would explain them. Another teacher noted, "majority of teachers do not have time to deal with parents," suggesting that lack of time is the only reason for this omission. Nevertheless, it is evident that some teachers do not perceive parent involvement as a task related to teaching mathematics in Malawi primary schools.

The mathematical task of "inspecting equivalences" was found not applicable by five teachers, two of whom did not give any reasons. The remaining three explained that this is done for the teachers but not by the teachers themselves. One teacher clearly stated that this is done by the teachers' guides, which are books for teachers that accompany textbooks for each grade.

Most of the reasons given pertain to teachers' guides and textbooks, which are perceived as sufficient, thus absolving the teachers from the responsibility for checking the equivalencies, as this is done by the authors of these curriculum materials. For example, the tasks of "using mathematical notation and language and critiquing its use," "choosing and developing useable definitions," and "selecting representations for particular purposes" were not perceived by the teachers as something they should do, as they are guided by what is written in textbooks and in teachers' guides. It appears that the teachers follow the textbook and teachers' guide diligently without questioning their applicability. This is common among Malawi teachers, who tend to take teachers' guide and textbooks as prescriptions of what and how to teach and not as suggestions. It might be important to note that the Malawi Institute of Education—a government institution—is the sole provider of textbooks and accompanying teachers' guides for primary schools in Malawi and these are made available to the teachers. Hence, for most of the teachers, these are the only resources they have and use. There are other possible reasons for not identifying with the tasks, although these are not mentioned by the teachers. For example, limited knowledge of the mathematical content by the teachers would compromise their ability to see equivalences, develop useable definitions, select representations, or critique use of mathematical notations. All these require deep understanding of the mathematics involved and might explain the heavy reliance on teachers' guides.

Similar to reasons for finding tasks applicable to the Malawi context, some teachers did not give the reason for selecting "no" as their response. It is likely that these teachers did not clearly understand the task or their own perception of it, and thus felt that it would not be appropriate to elaborate on their response.

## Other Mathematical Teaching Tasks

Table 4 shows results pertaining to Part B of questionnaire, where teachers were asked to indicate tasks of teaching mathematics in Malawi schools that are not included in the original list of 16.

Table 4. *Frequencies of Responses to Other Mathematical Teaching Tasks in the Malawi Context.*

| Other mathematical tasks of teaching mentioned | Frequency (%) |
|---|---|
| Making teaching and learning aids (TALULAR) | 14 (100) |
| Assessing learners (preparing tests, marking scripts) | 10 (71) |
| Using different teaching methods | 8 (57) |
| Ordering content/sequencing topics | 6 (43) |
| Relating mathematics to context and other subjects | 4 (29) |

As can be seen from the results presented in Table 4, five additional tasks of teaching were mentioned by at least four of the teachers. All participating teachers mentioned "making teaching and learning aids" and they referred to this as TALULAR, which stands for Teaching And Learning Using Locally Available Resources. TALULAR is a common concept in Malawi schools and is emphasized in teacher education. Thus, teachers take it as part of their job to make their own resources for teaching (Malawi Institute of Education, 2010). We acknowledge that use of teaching aids per se is not part of Ball et al.'s (2008) conceptualization of mathematical tasks of teaching. However, we contend that, in the Malawi context, where the teachers are expected to make the teaching aids from their local resources, this involves some mathematical reasoning and is hence integral part of their mathematical tasks of teaching.

The next common additional task was "assessing learners," most likely because assessment of learners is very important in Malawi schools, where learners take school tests at the end of each term, as well as during the term. Furthermore, learners' progression from one grade to the next is dependent on passing such tests. It is therefore not surprising that Malawi teachers consider "assessing learners" as one of distinct mathematical tasks of teaching. Indeed, the reasons teachers gave for explicitly noting this task was that it is their job to prepare tests, evaluate learners' work, and give feedback to their students. However, this is not a distinct task because the work involved in assessing students is reflected in some of the tasks in the original list of 16, such as "evaluating the plausibility of students' claims" and "asking productive mathematical questions."

More than half of the teachers mentioned "using different teaching methods" as another distinct mathematical task of teaching, as all respondents felt that it was linked to using learner-centered approaches in teaching. This reflects the current curriculum and initiatives in Malawi schools, which emphasize learner-centered teaching, as opposed to the traditional teacher-centered teaching methods (Malawi Institute of Education, 2014). We also acknowledge here that teaching methods do not fit with the conceptualization of mathematical tasks for teaching by Ball et al. (2008).

The last two suggested tasks "relating mathematics to context and other subjects" and "ordering content/sequencing topics" are partly covered by the task of "connecting a topic being taught to topics from prior or future years." While during the group discussion, teachers acknowledged this, they nonetheless felt that their suggestions extended beyond what is covered in the 16 tasks. For example, comparing "connecting a topic being taught to topics from prior or future years" with "relating mathematics to context and other subjects," one teacher said:

. . . that task is about scope and sequence, what we teach before and after . . . it is talking about scope and sequence . . . this we are saying linking to everyday life and other learning areas, science, agriculture . . . it is not same thing

The fact that the teachers were suggesting other tasks that are partly covered in the 16 tasks emphasizes our finding from Part A that, in general, the teachers find the tasks originally developed in the U.S. applicable to the Malawi context. The parts that are not covered seem to be additional tasks for Malawi and therefore worth paying attention to. We acknowledge that Ball et al. (2008) do not claim that the list of 16 is exhaustive, and our goal is not to show that the Malawi context has more tasks. Rather, our aim is to illustrate that there seems to be notable similarities between the two contexts, while acknowledging some specificities of the Malawi context.

## Discussion on Mathematical Tasks of Teaching

These findings are drawn from teachers' responses to the questionnaire and the information gained during the group discussion, rather than from observations of actual teaching of mathematics in Malawi schools. Some might dispute this approach because it is based on teachers' views and not their practice. While we acknowledge the limitations of our approach, we argue that experienced teachers would be able to consider each of the tasks presented to them and draw from their experience to determine whether the task is something they do as part of their work of teaching mathematics. We emphasize that the goal of this work was to gain a better understanding of what teachers perceive as tasks of teaching in the Malawi context. On one hand, the fact that all 16 tasks were recognized by at least a third of the teachers informs us that the recurrent tasks of teaching are generally applicable to both the U.S. and the Malawi context. On the other hand, the fact that some tasks were not recognized by a relatively large number of teachers suggests that there is some specificity to the Malawi context. Thus, our findings help elucidate some differences between the two contexts, which support the claim that teaching is culturally specific (Delaney, 2012). The additional tasks of teaching described by teachers, especially the making of teaching aids (which was mentioned by all teachers), emphasize further the cultural specificity of teaching mathematics in the Malawi context. Thus, while we acknowledge that Ball et al. (2008) do not claim that the list of 16 is exhaustive, we also highlight the fact that some of the additional tasks the Malawi teachers suggested are covered in the original list of 16.

## Piloting Adapted Measures

As explained earlier, the adapted forms were piloted on preservice teachers, whose responses were analyzed via the BILOG-MG software version 3.0 (Zimowski, Muraki, Islevy, & Bock, 2003) using the 2-parameter logistic (2PL) IRT model. As suggested by Hambleton, Swaminathan, and Rogers (1991), a model-data fit was investigated using multiple methodologies, starting from the assumptions of the IRT model, including unidimensionality. Since the two forms were answered by two equivalent groups (drawn from the same sample), we used equivalent groups equating with no overlapping items, placing the items on the two forms on the same scale. The 2PL IRT model produces two parameters for each item— parameter $a$ that defines item discrimination, and parameter $b$, often referred to as item difficulty, as it shows the location where a respondent with ability of $b$ will have a 50% chance of answering the item correctly (Edwards, 2009). Item characteristics on this scale typically range from -3 to +3, with items associated with higher values being more difficult than those with lower values. We used the values of these parameters to investigate problematic items in the Malawian context, and to identify items that are candidates for inclusion in the two forms that can be used for measuring teacher knowledge. We also studied point biserial correlation to identify problematic items (Fauskanger et al., 2012; Ng, 2012).

All the preservice teachers were also given an IRT score, placing them along the ability interval with 0 as the mean ability level and 1.0 as the standard deviation. It is, however, important to note that the preservice teachers' scores are not the focus of this paper, even though the psychometric properties of the items are.

The results yielded by the IRT analysis revealed an IRT reliability of 0.874, whereby maximum information was obtained at 1.25 standard deviation above the mean score. This indicated that this set of items is optimal for assessing more knowledgeable preservice teachers. BILOG flagged two items that had problematic point biserial (less than -0.15) and these were omitted from further calculation (Item 11b and 24d, both from Form B). In the next phase of data analysis, we examined the relative item difficulty distribution reported in the U.S. and estimated in the Malawi context. Even though the two scales are not directly comparable, one can expect—if the items function similarly in the Malawi as in the U.S. context—that there are some similarities among the relative ordering of item difficulties when looking at the full set of items (Fauskanger et al., 2012). More specifically, we posit that, if items are ordered from easy (low value of the $b$ parameter/difficulty) to hard (more difficult, high value of the $b$ parameter), one should expect—if the psychometrical properties of these items are similar across the two populations—that the order should be maintained (Fauskanger et al., 2012). The strong correlation pertaining to the relative item difficulty in the U.S. and Malawi context indicates that the relative order of the item difficulties is strongly maintained. This finding suggests that the items seem to function in a similar manner in the two populations.

For this set of items (Form A and Form B), the item difficulty found in the U.S. ranged from -4.281 to 3.961, with the average item difficulty of -0.535. In Malawi, the relative item difficulty ranged from -5.56 to 4.98, with the average difficulty of 1.119. A scatterplot of relative item difficulty found in the U.S. and in Malawi is depicted in Figure 1.

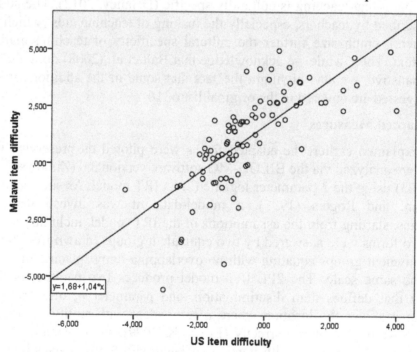

*Figure 1.* A scatterplot of the relative item difficulty found in the Malawi context, plotted against item difficulty found in the U.S.

We found a strong correlation between the item difficulty reported in the U.S. and that estimated in the Malawi context (Pearson correlation $r = .738$, $p < .0005$). The average item difficulty for this set of items was -0.535 for the U.S., while 1.119 was estimated in the Malawi context.

## Discussion of the Psychometric Results

The primary purpose of the psychometric analysis was to ensure that the adapted items function well psychometrical in the Malawi context, and to finalize two forms with items—including anchoring items—that can be used for further studies in Malawi. For the set of items used in this pilot, it seems clear from the results that some of the more difficult items should be taken out of the set. Firstly, the point of maximum information (0.874) and the average item difficulty (1.119) indicated that the set of items was skewed in favor of the more knowledgeable respondents, implying that, overall, there were too many difficult items in the set. We were ideally seeking to compose two forms, both with approximately 30 items (including five repeating items as anchoring items) in a pre- and post-test study. Because of this aim, we decided to remove some items that appeared too difficult for Malawi preservice teachers. The main objective was to attain item distribution on each form with majority of items around mean ability of zero. As a result of these considerations, Item 2b, 3, 5b, 6, 7, 13, 18, 21e, and 24 from Form A, and Item 3, 8, 10, 11b, 14b, 16a, 17, 18, 20, 21, 22, and 24d from Form B were removed. We then added five items from our Form B to Form A, and one item from Form A to Form B. After this Form A had 38 items and Form B 35 items including six anchoring items on both forms. The resulting revised set of items had an average item difficulty of 0.854 in Malawi, versus -0.770 obtained in the U.S. The final item distribution of the new forms is given in Table 5 below, which also provides the average difficulty level when the items are distributed to the two forms.

Table 5. *Distribution of Item Difficulty for the Malawi Context.*

| Difficulty level | Number of items | | |
|---|---|---|---|
| | **Form A** | **Form B** | **Common items** |
| $b < -3$ | 2 | 1 | |
| $-3 \leq b < -2$ | - | 1 | |
| $-2 \leq b < -1$ | 4 | 4 | 1 |
| $-1 \leq b < 0$ | 7 | 7 | 1 |
| $0 < b \leq 1$ | 9 | 9 | |
| $1 < b \leq 2$ | 10 | 8 | 3 |
| $2 < b \leq 3$ | 5 | 4 | 1 |
| $b > 3$ | 1 | 1 | |
| **Total items** | **38** | **35** | **6** |
| Total stems | 20 | 20 | 6 |
| **Average Malawi** | **0.823** | **0.425** | |
| **Average U.S.** | **-0.83** | **-0.929** | |

## Conclusion

In conclusion, we revisit our research questions and briefly explain how we have answered them in this work based on the findings we obtained. With respect to the first question—*Are the tasks of teaching mathematics identified in the U.S. applicable in*

*Malawi?*—our findings show that the 16 tasks are applicable in the Malawi school context. However, some of the tasks are more commonly recognized by teachers as applicable to the Malawi context, while other tasks are found less relevant. The Malawi school context is very different from the U.S. context and it is interesting that the work of teaching seems generally similar. It is also intriguing to observe that, while there seems to be generality between the two contexts, there is also some specificity to the Malawi setting. In Malawi, making teaching aids from local resources is seen as integral part of teaching mathematics. It is important to note that this study does not draw any inferences about the way the tasks are carried out. It also does not extant to the assessment of their implementation by teachers in the Malawi classroom. Thus, while answering the call for efforts to identify mathematical tasks of teaching as a basis for conceptualizing MKT (Hoover et al., 2014), the study does not provide empirical evidence. However, it does offer a valuable foundation for further studies in this field, where empirical evidence can be gathered in order to contribute towards building a shared understanding of "common tasks of teaching mathematics" and consequently development of "internationally shared measures of MKT" (Hoover et al., 2014, p. 9).

For the second research question—*What can we learn about the adaptability of LMT measures to the Malawi context from psychometric properties estimated in a pilot study?*—we found, after piloting the set of 88 adapted items, that majority of items psychometrically function well in the Malawi context. The item difficulty reported in the U.S. was strongly correlated with that pertaining to the Malawi context. We also found that the set of items was slightly skewed towards the "difficult" side of the ability scale. Thus, by removing some of the more difficult items, we were able to group the set of items into two adapted forms—Form A and Form B—that can be used for further studies in the Malawi school context.

## References

Adler, J., & Patahuddin, S. (2012). Recontexualising items that measure mathematical knowledge for teaching into scenario based interviews: an investigation. *Journal of Education, 56,* 17–43.

An, S., Kulm, G., & Wu, Z. (2004). The pedagogical content knowledge of middle school, mathematics teachers in China and the U.S. *Journal of Mathematics Teacher Education, 7,* 145–172.

Andrews, P. (2011). The cultural location of teachers' mathematical knowledge: Another hidden variable in mathematics education research. In T. Rowland & K. Ruthven (Eds.), *Mathematical Knowledge in Teaching* (pp. 99–118). New York, NY: Springer Publications.

Ball, D. L., & Bass, H. (2003). Toward a practice-based theory of mathematical knowledge for teaching. In B. Davis & E. Simmt (Eds.), *Proceedings of the 2002 annual meeting of the Canadian Mathematics Education Study Group* (pp. 3–14). Edmonton, Alberta, Canada: Canadian Mathematics Education Study Group (Groupe Canadien d'étude en didactique des mathématiques).

Ball, D. L., & Hill, H. C. (2008). *Mathematical knowledge for teaching (MKT) measures. Mathematics released items 2008.* Ann Arbor, MI: University of Michigan.

Ball, D. L., Thames, M. H., & Phelps, G. (2008). Content knowledge for teaching: What makes it special? *Journal of Teacher Education, 59*(5), 389–407.

Cole, Y. (2012). Assessing elemental validity: the transfer and use of mathematical knowledge for teaching measures in Ghana. *ZDM—The International Journal on Mathematics Education, 44*, 415–426.

Davis, B., & Simmt, E. (2006). Mathematics-for-teaching: An ongoing investigation of the mathematics that teachers (need to) know. *Educational Studies in Mathematics, 61*(3), 293–319.

Delaney, S. (2012). A validation study of the use of mathematical knowledge for teaching measures in Ireland. *ZDM—The International Journal on Mathematics Education, 44*, 427–441.

Delaney, S., Ball, D., Hill, H., Schilling, S., & Zopf, D. (2008). "Mathematical knowledge for teaching": Adapting U.S. measures for use in Ireland. *Journal of Mathematics Teacher Education, 11*(3), 171–197.

Edwards, M. C. (2009). An introduction to Item Response Theory using the need for cognition scale. *Social and Personality Psychology Compass, 3*(4), 507–529.

Fauskanger, J., Jakobsen, A., Mosvold, R., & Bjuland, R. (2012). Analysis of psychometric properties as part of an iterative adaptation process of MKT items for use in other countries. *ZDM—The International Journal on Mathematics Education, 44*, 387–399.

Hambleton, R. K., Swaminathan, H., & Rogers, H. J. (1991). *Fundamentals of item response theory*. Newbury Park, CA: Sage Publications.

Hill, H. C., Rowan, B., & Ball, D. L. (2005). Effects of teachers' mathematical knowledge for teaching on student achievement. *American Educational Research Journal, 42*(2), 371–406.

Hill, H. C., Ball, D. L., & Schilling, S. G. (2008). Unpacking pedagogical content knowledge: Conceptualizing and measuring teachers' topic-specific knowledge of students. *Journal for Research in Mathematics Education, 39*, 372–400.

Hoover, M., Mosvold, R., & Fauskanger, J. (2014). Common tasks of teaching as a resource for measuring professional content knowledge internationally. *Nordic Studies in Mathematics Education, 19*(3–4), 7–20.

Kasoka, D., Kazima, M., & Jakobsen, A. (forthcoming 2016). Adaptation of mathematical knowledge for teaching number concepts and operations measures for use in Malawi. To appear in *Proceedings of the 24st Annual meeting of the Southern African Association for Research in Mathematics, Science and Technology, 12$^{th}$–15$^{th}$ January, 2016)*. Pretoria, South Africa: SAARMSTE.

Kazima, M., & Mussa, C. (2011). Equity and quality issues in mathematics education in Malawi schools. In B. Atweh, M. Graven & P. Valero (Eds.), *Managing Equity and Quality in Mathematics Education* (pp. 163–176). New York, NY: Springer Publications.

Kwon, M., Thames, M. H., & Pang, J. (2012). To change or not to change: Adapting mathematical knowledge for teaching (MKT) measures for use in Korea. *ZDM—Mathematics Education, 44*, 371–385.

Malawi Institute of Education (2010). *Initial Primary Teacher Education Programme Handbook*. Domasi, Malawi: Malawi Institute of Education.

Malawi Institute of Education. (2014). *Improved teaching and learning methods handbook*. Domasi, Malawi: Malawi Institute of Education.

Malda, M., Van de Vijver, F. J. R., Srinivasan, K, Transler, C, Sukumar, P., & Rao, K. (2008). Adapting a cognitive test for a different culture: An illustration of qualitative procedures. *Psychology Science Quarterly, 50*(4), 451–468.

Ng, D. (2012). Using the MKT measures to reveal Indonesian teachers' mathematical knowledge: Challenges and potentials. *ZDM—The International Journal on Mathematics Education, 44*, 401–413.

Rowan, B., Correnti, R., & Miller, R. J. (2002). What large-scale, survey research tells us about teacher effects on student achievement: Insights from the Prospects study of elementary schools. *Teachers College Record, 104*(8), 1525–1567.

Rowland, T., Huckstep, P., & Thwaites, A. (2005). Elementary teachers' mathematics subject knowledge: The knowledge quartet and the case of Naomi. *Journal of Mathematics Teacher Education, 8*(3), 255–281.

Shulman, L. (1986). Those who understand: Knowledge growth in teaching. *Educational Researcher, 15*(2), 4–14.

Wright, S. P., Horn S. P., & Sanders, W. L. (1997). Teacher and classroom context effects on student achievement: Implications for teacher evaluation. *Journal of Personnel Evaluation in Education, 11*, 57–67.

Zimowski, M. F., Muraki, E., Islevy, R. J., & Bock, R. D. (2003). *BILOG-MG 3 for Windows: Multiple-group IRT analysis and test maintenance for binary items [Computer software]*. Lincolnwood, IL: Scientific Software International, Inc.

## Acknowledgements

This work was kindly funded by Norwegian Programme for Capacity Building in Higher Education and Research for Development (NORHED) through the project: *Improving quality and capacity of mathematics teacher education in Malawi—a* collaboration between University of Malawi, Malawi and University of Stavanger, Norway.